Opening Speech

Dear participants and other attendants, ladies and gentlemen.

On behalf of the organizing committee, and in representing the Board of the Delft University, I may cordially welcome you at this 5th International Workshop on Traffic and Granular Flow, here in Delft.

We are proud to have you here in Delft, coming from nearly all continents, to exchange and share your new scientific findings in the areas of modeling and analysis of traffic flow and granular flow. Certainly, these are two heavily interrelated fields with a lot of interesting commonalities, not only because both deal with the movement of particles in a stream. There are lots of instances where both fields offer ample opportunities for mutual scientific stimulation and enrichment such as among other matters the analytical treatment of problems and the computational approaches in solutions.

We are proud to offer in Delft a venue for this outstanding gathering of top scholars and scientists in traffic and granular flow study. The TU Delft did'nt had a second of hesitation as we were asked by the organizing committee to host this prestigious Workshop, now for the 5th time in a growing series.

Let me briefly **recall**, especially for the newcomers, the four earlier predecessors of this conference of today: It started in 1995 with the first Workshop held in Jülich in Germany; where for the first time it was tried to work out the similarities and differences of traffic and granular flows. A strong impetus was at that time the similarities in computational approaches in these fields. The *second* of this series was held in Duisburg, also Germany, in 1997 where special attention was devoted to the connections with collective motion in biological systems. The *third* Workshop took place in Stuttgart in 1999, trying to stress the connections with socio-economic systems, a subject of special concern in the Physics Department at Stuttgart University. After three occasions in Germany, the organizers felt strong enough to organize the fourth Conference in this successful series outside of Europe, namely in Nagoya, Japan, in 2001. All four conferences not only showed lively debates and exchange of ideas, but the presented papers also were reviewed and finally published in well-edited proceedings, published in book form by highly esteemed publishers such as World Scientific and Springer. This way of publishing guarantees a broad distribution among the interested scientists. Your numerous attendance at this fifth Traffic and Granular Flow Workshop proves that this subject did not loose a grain of its attraction, I would say, to the contrary. I discovered that many of you already were participants in the earlier workshops, even several of you already were there at the first one in 1995. Also some subjects already become traditional, such as demonstrated by the extensive appearance of contributions on the cellular automata as a typical example of a granular approach.

At the same time I observe lots of contributions adopting the continuity approximation to granular flows. This continuity in subjects and attendance is a good omen for the coming three days.

I think the **multidisciplinary nature** of this workshop also highly contributes to its special value. We have attendants from the fields of Transportation Engineering, various branches of Physics, from Computer Science and Mathematics, but also from Mechanical and Hydraulic Engineering. This variety certainly will offer rich opportunities for mutual analogies and cross-overs with respect to theoretical, methodological and computational findings.

Looking back at those earlier Workshops, I was struck by the high quality contributions of outstanding scientists again and again. In browsing through these earlier proceedings, I found to my surprise a note I appeared to have written during my winter holidays 1996 in the Bavarian Alps, in which I synthesized for my own use the various contributions on cellular automata at the very first workshop in Jülich. It was in fact my first encounter with this important modeling tool that appeared to gain widespread application in later years until today, also at this conference. This first proceeding also is highly recommended to you because of a very interesting Prologue in it, written by two famous scholars in this field, Ilya Prigogine and the late Bob Herman, both from University of Texas, and both well-known for their successful approach to apply principles from physics to vehicular traffic. A nice statement from this Prologue still holds today: "It came to us as a surprise that many complex systems can be successfully modeled by very simple models still keeping close to reality". Cellular automata models among other cases certainly belong to these simple but astonishingly realistic models.

This quote reminds me to a famous statement attributed to Einstein: "A model (or a theory) should be as simple as possible, …. but not simpler". This should be a guide for all of us. In fact, what Einstein expressed is the well-known methodological principle of parsimony. At such occasions as today, it is always tempting to try to summarize the commonalities of traffic and granular flows, but also to highlight clear differences, all for the sake of a better understanding of each others work, and to emphasise exchange opportunities. Traffic flows, because of being controlled by human beings, only partly obey the classical physical laws. The laws of gravity and inertia, and also conservation of mass do apply, but conservation of momentum and energy don't, as is true for Newton's third law of reaction on action. Traffic flows clearly also are anisotrope: the particles show unequal directional properties with respect to distances, speeds, and speed variance. In addition, participants in the traffic flow have an own will. They deliberately choose their own speeds, their own accelerations, and their own destinations. Their interaction behavior is non-local: they react at other particles some distance apart in the flow, and they may anticipate at future conditions.

Let me on the other hand try in my opening talk to speculate on potential **exchange opportunities** between the traffic flow and granular flow fields. The es-

TRAFFIC AND GRANULAR FLOW '03

S. P. Hoogendoorn   S. Luding
P. H. L. Bovy   M. Schreckenberg   D. E. Wolf
Editors

TRAFFIC AND GRANULAR FLOW '03

With 344 Figures, 135 in Colour

 Springer

Editors

Serge P. Hoogendoorn
Faculty of Civil Engineering and Geosciences
Delft University of Technology
Stevinweg 1
2628 CN Delft, The Netherlands
e-mail: s.hoogendoorn@citg.tudelft.nl

Stefan Luding
Particle Technology
DelftChemTech, TU Delft
Julianalaan 136
2628 BL Delft, The Netherlands
e-mail: s.luding@tnw.tudelft.nl

Piet H. L. Bovy
Faculty of Civil Engineering and Geosciences
Delft University of Technology
P.O. Box 5048
2600 GA Delft, The Netherlands
e-mail: P.H.L.Bovy@citg.tudelft.nl

Michael Schreckenberg
Physik von Transport und Verkehr
Universität Duisburg-Essen
Lotharstr. 1
47048 Duisburg, Germany
e-mail: schreckenberg@uni-duisburg.de

Dietrich E. Wolf
Theoretische Physik
Universität Duisburg-Essen
Lotharstr. 1
47048 Duisburg, Germany
e-mail: wolf@comphys.uni-duisburg.de

Library of Congress Control Number: 2005926660

Mathematics Subject Classification (2000): 37Mxx, 90Bxx, 65Cxx, 60H35, 60K30, 68U20, 76T25

ISBN-10 3-540-25814-0   Springer Berlin Heidelberg New York
ISBN-13 978-3-540-25814-8   Springer Berlin Heidelberg New York

Springer is a part of Springer Science+Business Media
springeronline.com
© Springer-Verlag Berlin Heidelberg 2005
Printed in The Netherlands

Cover design: *design & production* GmbH, Heidelberg
Typeset by the authors using a Springer TeX macro package
Production: LE-TeX Jelonek, Schmidt & Vöckler GbR, Leipzig

Printed on acid-free paper   46/3142/YL - 5 4 3 2 1 0

sential difference being that the traffic particles, apart from following physical laws, also are governed by human rules, by human decision making. It can be observed that traffic scientists more and more tend to follow true granular (that is discrete) descriptions of their flows of interest, away from the still widely adopted continuity approximation for these flows. That means that they are dealing with the movements of individual subjects or vehicles, by using microscopic approaches such as microscopic simulations. They look at a flow as a multi-body system. An important reason for that continual shift from a continuous to a discrete flow description is not only in the improvement of our mathematical and computational skills or computer power. It certainly is also an expression of increased insight, and of a conviction, that each particle in a traffic flow in fact is a unique entity, especially with respect to its decision making behaviour. More and more, traffic science is shifting from considering particles as physical entities, obeying physical laws, towards particles as actors, as agents, that also obey behavioral rules. These agents may be characterized as players having idiosyncratic properties and having there own strategies of making choices in order to achieve the best outcome for themselves.

Mathematical game theory proves to be a highly valuable tool in analyzing such a social multi-body system, also in case of a traffic stream. Well-known in this respect is the description of traffic flow through a network in time and space, as a so-called "N-person non-co-operative game", giving rise to a socalled Nash equilibrium. At a finer level of detail, the encounters of traffic participants, such as passing of pedestrians or merging of cars in a flow, the negotiation among these particles, e.g. for priority, can nicely be modeled with differential game theory. Traffic science in the past benefited a lot from physics such as by adopting various approaches from statistical mechanics. I wonder now whether the opposite may be valuable as well, that is, whether the prevailing agent paradigm of particles negotiating for space or for priority, might be a fruitful approach also for the non-human granular flow field.

Now let us **start** with this fifth conference.

Delft, October 1st, 2003   Piet H.L. Bovy

Preface

These proceedings are the fifth in the series *Traffic and Granular Flow*, and we hope they will be as useful a reference as their predecessors. Both the realistic modelling of granular media and traffic flow present important challenges at the borderline between physics and engineering, and enormous progress has been made since 1995, when this series started. Still the research on these topics is thriving, so that this book again contains many new results.

Some highlights addressed at this conference were the influence of long range electric and magnetic forces and ambient fluids on granular media, new precise traffic measurements, and experiments on the complex decision making of drivers. No doubt the "hot topics" addressed in granular matter research have diverged from those in traffic since the days when the obvious analogies between traffic jams on highways and dissipative clustering in granular flow intrigued both communities alike. However, now just this diversity became a stimulating feature of the conference. Many of us feel that our joint interest in complex systems, where many simple agents, be it vehicles or particles, give rise to surprising and fascinating phenomena, is ample justification for bringing these communities together: *Traffic and Granular Flow* has fostered cooperation and friendship across the scientific disciplines.

We are very grateful for the generous support of several sponsors: Delft University of Technology, Department Transport & Planning, FOM, and Burgerscentrum. Moreover, this conference has benefited from a satellite workshop at the Lorentz Center in Leiden "Cooperative Grains: From Granular Matter to Nano Materials" coorganized by the Collaborative Research Center SFB 445 (Nano-Particles from the Gasphase: Formation, Structure, Properties) at the University Duisburg-Essen.

Special thanks go to Serge Hoogendoorn, the conference secretary Nicole Fontein, for her competent and thoughtful help, and to Birgit Dahm, who compiled the manuscripts into this book.

The next conference *Traffic and Granular Flow '05* will take place in Berlin, and we hope to see many of you again there.

Delft / Duisburg, March 2005

Serge P. Hoogendoorn
Stefan Luding
Piet H.L. Bovy
Michael Schreckenberg
Dietrich E. Wolf

Contents

Traffic Networks and Driver's Behaviour

Pedestrian Dynamics

Granular

Empirical Traffic Data

Lane-Change Maneuvers Consuming Freeway Capacity

B. Coifman, S. Krishnamurthy, and X. Wang

Department of Civil and Environmental Engineering and Geodetic Science,
Department of Electrical Engineering – The Ohio State University
Columbus, OH, 43210 – USA
Coifman.1@OSU.edu, krishnamurthy.26@osu.edu, wang.833@osu.edu

Summary. Conventional traffic flow theory dictates that flow on a freeway is usually constrained only by a small number of critical locations or bottlenecks. When active, these bottlenecks cause queues that can stretch for several miles and reduce flow on other parts of the network. Bottlenecks are often thought to arise over short distances and are usually modeled as if they occur at discrete points since the resulting queues are thought to be much longer then the bottleneck region. This paper presents evidence that the delay causing phenomena may actually occur over extended distances. Some of which may occur downstream of the apparent bottleneck where drivers are accelerating away from the queue, while related phenomena are observed in the queue, over a mile upstream of the apparent bottleneck. It is shown that lane change maneuvers are responsible for some of the losses, reducing travel speed and consuming capacity when vehicles enter a given lane. These losses in one lane are not fully balanced by gains in other lanes.
Keywords: freeway traffic, traffic flow theory, bottlenecks, congestion

1 Introduction

Conventional traffic flow theory dictates that flow on a freeway is usually constrained only by a small number of critical locations or bottlenecks. When active, these bottlenecks cause queues that can stretch for several miles and reduce flow on other parts of the network. Bottlenecks are often thought to arise over short distances and are usually modeled as if they occur at discrete points since the resulting queues are thought to be much longer then the bottleneck region.

This conventional model of bottleneck operation proved to be insufficient when attempting to localize precisely the source of delay through a major freeway interchange on Interstate 71 (I71) in Columbus, Ohio, USA. Using probe vehicle data, the apparent bottleneck location drifted from day to day, falling in a range of almost a half mile without any clear geometric explanation such as an on-ramp or lane drop. Further investigation of vehicle detector data upstream of the bottleneck revealed that velocity increased as traffic approached the bottleneck, which is consistent with earlier findings. Using loop detector data, both Hall et al (1992) and Cassidy and Mauch (2001) observed average velocity dropping, flow dropping, and occupancy increasing in the queue as one moved further upstream of a bottleneck. The worsening conditions were attributed to vehicles entering on the ramps and consuming some of the capacity of the bottleneck that would otherwise be available for the mainline traffic upstream of the ramp.

Unlike the earlier research, however, the dropping velocity as one moves upstream of the bottleneck on I71 cannot be explained by ramp flows alone, e.g., the phenomena was observed between two consecutive stations that have conservation of flow. As will be shown in this paper, by measuring vehicle inflow, i.e., the difference between the number of vehicles that enter a lane and those that leave the lane between two locations (Coifman, 2003), closer investigation reveals that lane change maneuvers reduced the effective capacity in adjacent lanes. Of course the total number of lane change maneuvers may be greater than the inflow, since one exiting vehicle could be replaced by another entering, but the results do not depend on the exact number of lane change maneuvers. In any event, the phenomena was observed both within the half mile first identified from the probe vehicle data as containing the bottleneck location, and a mile upstream of the apparent bottleneck. This latter point indicates that the bottleneck mechanisms result in complicated events that can trigger a drop in capacity far upstream of the initial source of delay.

The remainder of this paper begins by examining the probe vehicle data and providing a detailed overview of the freeway corridor to localize the bottleneck. It then uses loop detector data to study the evolution of the queue and examine features that may impact vehicle throughput. Finally, it closes with a brief summary and conclusions.

2 Analysis

We examined northbound I71 in Columbus, Ohio first by collecting Global Positioning Satellite (GPS) data from over 70 probe vehicle runs passing through the corridor. The GPS data used in this study were collected on preselected Tuesdays, Wednesdays, and Thursdays over a period of approximately one year. From these GPS data, we were able to localize the apparent location of a major bottleneck. However, as noted earlier, the precise location appeared to drift from one run to the next, falling in a range of almost a half mile. Then to gain a more complete temporal picture, we examined the corresponding loop detector data.

2.1   Bottleneck Localization

The probe vehicle runs were sorted based on traffic conditions. A few of the runs indicated the presence of queued conditions at the downstream end of the corridor and they were omitted from further study since the subject bottleneck was not the source of the queue. Of the 66 remaining runs, 29 did not exhibit any queuing and 37 showed some degree of queuing arising from a bottleneck within the study corridor. Taking the average within each of these sets, Fig. 1A shows the spatial evolution of the two averages. Note that to the eye the active bottleneck average ("congested") becomes indistinguishable from the inactive bottleneck average ("free flow") at approximately mile 5.3. When the bottleneck is active, the average velocity decreases as one moves upstream of this location, with a significant drop at mile 4.5.

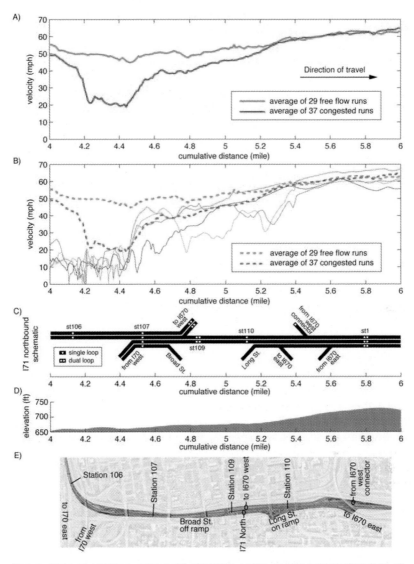

Figure 1, (A) Average velocity versus time for 66 probe vehicle runs through a bottleneck on northbound I71, (B) superimposing four individual runs on part A, (C) the corresponding schematic and (D) elevation diagram of the freeway segment. Finally, (E), an aerial photo showing the geometry of the freeway segment.

For this study we examined five days in detail that were selected because the queue was particularly long, extending upstream of mile 4, namely December 5, 2001; February 28, 2002; April 23, 2002; May 7, 2002; and July 23, 2002. For four of these days the long queue is evident in the probe vehicle data, as shown in Fig. 1B. All four of these runs were also included in the congested average curve. From these individual runs the reader can see that the exact location of the bottleneck changes from run to run, with the given congested run becoming indistinguishable from the free flow average at different locations between mile 5 and 5.5.

To place the data in context, Fig. 1C shows a straight line schematic of the corridor at the same scale as the previous plots. The central business district

(CBD) ends around mile 5.2 and I670 eastbound leads to the suburbs. Following the Ohio Department of Transportation (ODOT) convention, lanes are numbered consecutively starting with one in the median lane and increasing to the outside shoulder. The probe vehicle drivers were instructed to drive in lane 2, i.e., second from top in this diagram. Several loop detector stations are in the corridor, as indicated on the schematic. The averages in Fig. 1A suggest that the bottleneck is somewhere between mile 4.5 where a significant increase in average congested run velocity corresponds to the addition of two lanes from I70 westbound and mile 5.3 where the two curves become indistinguishable from one another just downstream of a diverge to I670 eastbound. Using the detector on the ramp from the I670 westbound connector, no velocity drops were evident on the subject days, verifying that queues did not back up on to the ramp and providing further evidence that this ramp was downstream of the bottleneck(s). For completeness, Figs. 1D-E show the elevation and geometry, respectively, for this corridor. Note that the latter is shown at a different scale than the other plots in this figure.

2.2   Queue Evolution

Detector stations 109 and 110 are within the region identified as containing the bottleneck in section 2.1. Fig. 2 shows plots of time series velocity and flow sampled every five minutes for the entire day in lane 2 at these stations on May 7. The general trends are similar on the four other days in the study. In Fig. 2B we see low flows before 5:00 and then a sharp increase in demand during the morning peak, rising to 2500 vehicles per hour (veh/hr), which many practitioners believe is the maximum capacity of a freeway lane. As will be discussed shortly, it is believed that this peak represents capacity conditions and the true bottleneck is very close to station 110 in the morning. Fig. 2A shows that velocity starts dropping to 40 mph just before this peak flow is reached. Although not explicitly shown in this paper, these morning peak data fall in the upper portion of the velocity versus flow curve. Thus, the data from this period do not correspond to the congested regime, rather, the drop in velocity is likely due to conditions approaching theoretical capacity within the uncongested regime. Velocity was measured directly from the dual loops at station 109 while it was estimated from the single loops at station 110 (as well as the other stations) using the methodology presented in Coifman et al (2003). As was shown in the earlier paper, this estimation technique yields good results except during low flow, and is evident by the noise in the velocity time series from station 110 during the low flow period prior to 5:00.

In theory the Long St. on-ramp, which enters just downstream of station 110, should have sufficient capacity for all entering vehicles since the additional lane continues for over 10 miles. However, drivers from I71 wishing to reach I670 eastbound must cross this lane in under 0.2 miles. In the morning peak both the Long St. on-ramp and I670 eastbound off-ramp should see lower demand than the evening peak since the flow is away from the CBD. We do not have detector data from either of the two ramps, but except for a brief disturbance during the morning peak on this day, the high velocity data at stations 109 and 110 support this supposition. As already noted, these data fall within the uncongested regime of the velocity versus flow plane. It is further reinforced by the fact that changes in the time series flow and occupancy are positively correlated during this period, further indicating non-queued conditions.[1] In the evening peak, flow begins to climb and then flow and velocity usually drop into a queued state, around 15:00 and shortly after 16:00, respectively, in the sample time series. A return to free flow velocities around 17:00 indicates that the queue has receded past these stations. The flow typically increases concurrent with this velocity recovery, suggesting that a downstream restriction is alleviated rather than upstream demand dropping. We believe that either the I670 eastbound off-ramp is backing up, or more likely, a high demand from Long St. prevents I71 drivers from finding gaps to merge into the outside lane and reach I670 during the evening peak. The exact source of the evening bottleneck is the subject of on-going research and for this paper the analysis in this link is restricted to the morning peak.

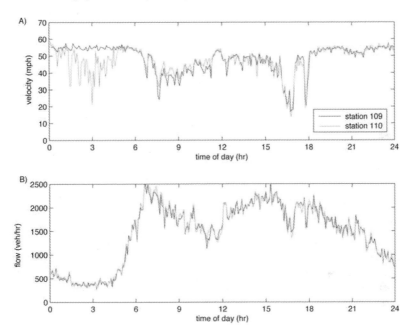

Figure 2,   Five minute average (A) velocity and (B) flow for lane 2 at stations 109 and 110, all day, on May 7, 2002.

   Extending our analysis beyond lane 2 and examining another morning peak, absent the brief disturbance seen at 8:00 in Fig. 2A, Fig. 3A and B show the time series velocity during the morning peak in lane 1 and 2, respectively. Each plot includes data from the two stations, 109 and 110, for the given lane to facilitate comparison. One can observe the velocity dropping slightly in both lanes at both stations shortly before 6:30. These plots also show that in each lane the drop is on the order of 4mph greater at station 109 than at 110. Figs. 3C-D show that in each lane the drop in velocity occurs just before flow reaches the maximum rate in lane 2, and this rate is sustained for

---

[1] During non-queued conditions, all commonly accepted flow-occupancy relationships indicate that increasing occupancy corresponds to increasing flow, while during queued conditions increasing occupancy corresponds to decreasing flow.

over 30 minutes. At both stations the peak flow in lane 2 is roughly 500 veh/hr greater than that of lane 1.

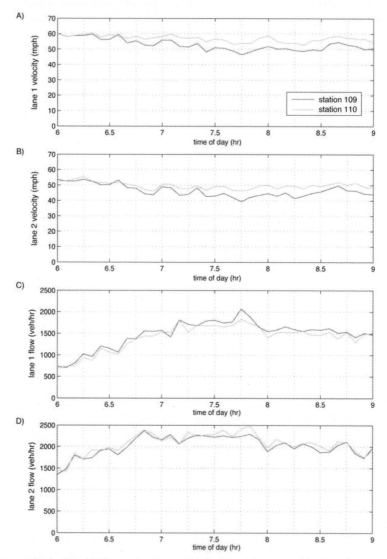

Figure 3,   Station 109 and 110 five minute average velocity in (A) lane 1, (B) lane 2, and flow in (C) lane 1, (D) lane 2, during the morning peak on February 28, 2002.

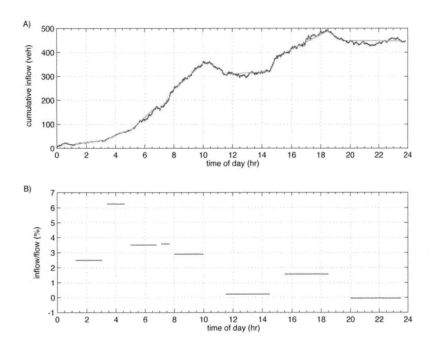

Figure 4,   (A) cumulative inflow to lane 2 between stations 109 and 110 and (B) the corresponding inflow as a percentage of flow, all day on February 28, 2002.

Conventional traffic flow theory would tell us that the increase in velocity from station 109 to 110 indicates that these two stations are downstream of the bottleneck and that the plots show traffic accelerating away from the queue.[2] Yet the flow in lane 2 at station 109 is near the theoretical capacity of a freeway lane and as shown in Fig. 4A, measuring vehicle inflow to this lane (Coifman, 2003), between 6:00 and 9:00 approximately 190 additional vehicles pass station 110 than 109 in lane 2 (although not shown, an equivalent reduction is seen in lane 1). Using the straight-line approximations superimposed on Fig. 4A, the percentage of flow that is attributable to the inflow was calculated and is shown in Fig. 4B. The merging vehicles represent just over three percent of the flow in lane 2 at station 110 between 6:00 and 9:00. The rising flow in lane 2 as one travels downstream of the apparent bottleneck and the proximity to lane capacity represents another mechanism that can limit flow and potentially give rise to the disturbances that evolve into stop and go traffic. Either there is a sequence of short bottlenecks that trade dominance in restricting flow over a short distance, or the actual bottleneck spans a non-negligible distance. If the former, lane 1 drivers take advantage of an upstream restriction to merge into lane 2, if the latter, these lane changing drivers are causing delay to lane 2 drivers and stretch the bottleneck over an extended distance.

In either case, this long bottleneck theory implies that all traffic states below pure free flow may limit upstream throughput. In the subject link, vehi-

---

[2] Again, although not shown in a figure, these morning peak data fall in the upper portion of the velocity versus flow curve.

cles entering lane 2 would delay drivers upstream because each entering driver consumes a minimum of one unit headway of time at station 110, analogous to the impact of on-ramp inflow degrading upstream conditions in a queue, as discussed in Hall et al (1992) and Cassidy and Mauch (2001). There are no ramps in this link, rather, the entering traffic comes from the adjacent lane. It is hypothesized that drivers in lane 1 assume lower velocity than dictated by conditions in the lane, presumably because they are matching velocity with lane 2 either to merge or because the drivers do not like to pass traffic in adjacent lanes with a significant difference in relative velocity. This coupling hypothesis will be the subject of future research, for this paper, it is sufficient to note that the delays to lane 2 drivers are not balanced by improvements to lane 1 drivers.

Returning the focus to lane 2, the accelerating drivers experience a delay compared to traversing the entire link at the velocity observed at station 110. Assuming they accelerate at a constant rate, each driver would experience the following delay,

$$\Delta t = \frac{2x}{v_{109} + v_{110}} - \frac{x}{v_{110}}$$

where,

$\quad x$ = the distance between the two stations, and

$\quad v_i$ = velocity at station i.

During very high flow and queued conditions, each entering vehicle would delay the following driver and these delays would propagate upstream. If each entering vehicle resulted in exactly one unit headway delay, the location of lane changes were uniformly distributed between the two stations, and the signals travel at empirically established velocities of well formed signals (on the order of 14 mph, Mauch and Cassidy, 2002), one would expect the delay arising from a single vehicle entrance to impact drivers in the link for just over 40 sec. The exact location of the lane change maneuvers are unknown and it is conceivable that more occur closer to station 110, thus potentially increasing the expected dwell time of the disturbance in the link. Meanwhile, the average time between lane change maneuvers is on the order of 55 sec on this day. So the expected number of such unit headway delays a driver must pass through is just below one. If the unit headway was simply the inverse of the maximum observed flow, 1.5 sec, the expected delay is on the order of $\Delta t \approx 1.3 \, \text{sec}$ in this case. These findings do not refute the long bottleneck theory that vehicles entering lane 2 between stations 109 and 110 during uncongested, high flow periods restrict flow upstream of 109. If conditions downstream of the apparent bottleneck impact the flow of traffic through the constriction, then that could help explain the formation and behavior of bottlenecks, but further research is needed to eliminate alternative explanations.

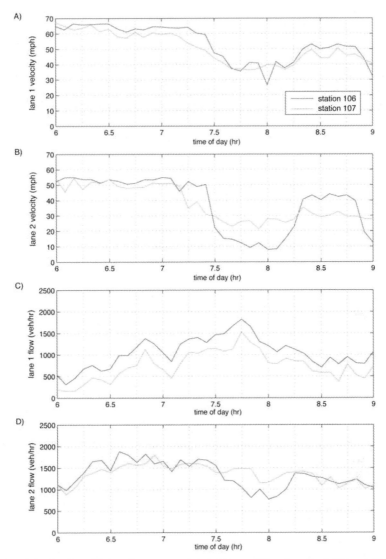

Figure 5, Station 106 and 107 five minute average velocity in (A) lane 1, (B) lane 2, and flow in (C) lane 1, (D) lane 2, during the morning peak on May 7, 2002.

Moving further upstream and returning to May 7, 2002, Fig. 5 shows the time series velocity and flow in lanes 1 and 2 at stations 106 and 107 (lanes 3 and 4 at station 107 are omitted for clarity). On this morning, the two lanes at station 107 exhibit a slow drop in velocity starting shortly after 7:00. At 7:30 velocity drops much more suddenly at station 106 as the queue passes over. Unlike station 107 in which velocity dropped by roughly the same amount in both lanes, at station 106 the velocity drop in lane 2 is roughly twice that of lane 1. Also at station 106, the flow drops in lane 2 shortly after the velocity does. This drop is consistent with what one would expect since a queue grows only as long as demand exceeds capacity. Thus, the demand flow upstream of the queue should be larger than the flow within the queue after it grows past the station. But the two following features are unexpected. First, Fig. 5D compares the lane 2 flow at both stations, while the station 107 flow remains relatively constant, station 106 flow actually drops below the downstream flow, which is also in the queue. Secondly, as shown in Fig. 5C, the flow in lane 1 at station 106 actually increases after the velocity drop.

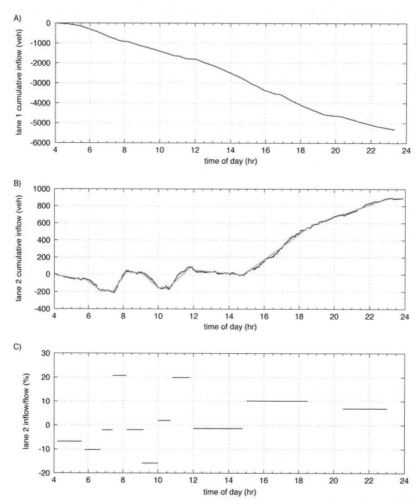

Figure 6, (A) cumulative inflow to lane 1 between stations 106 and 107, (B) cumulative inflow to lane 2 between stations 106 and 107 and (B) the corresponding inflow as a percentage of flow for lane 2, all day on May 7, 2002.

Two lanes enter the link just upstream of station 107 from eastbound I70, precluding conservation of flow for lane-to-lane comparisons. However, one can still measure the net lane inflow, as shown in Fig. 6A-B. Between 7:30 and 8:00, lanes 1 and 2 exhibit an inflow of -130 and 200 vehicles, respectively. Fig. 6C shows that the inflow to lane 2 accounts for approximately 20 percent of the lane flow at station 107. Meanwhile, Fig. 5A shows that velocity in lane 1 at station 106 drops down to that of station 107 after the queue passes over the upstream station, while Fig. 5B shows that at the same time the velocity in lane 2 at station 106 drops more than 10 mph below that of station 107. In other words, vehicles departing lane 1 do not appear to benefit upstream drivers while vehicles entering lane 2 appear to delay upstream drivers.

3 Conclusions

This paper used probe vehicle and loop detector data to examine more closely some of the mechanisms that occur in bottlenecks and consume freeway capacity. Section 2 used one case study to examine the small delay occurring between stations 109 and 110, though the results were typical of the morning peak on all five days that were studied. Evidence was presented to suggest that although velocity was near free flow and conditions may indicate vehicle discharge downstream of a queue, lane change maneuvers within this link may actually be restricting flow through the apparent bottleneck upstream, i.e., the long bottleneck theory. If the theory proves robust, it may help explain the formation backwards propagating *stop and go* waves and other bottleneck phenomena. Future research will test further this hypothesis.

A second case study found that vehicle maneuvers within a queue can also impact the throughput, as observed between stations 106 and 107. Each of the five days examined was selected because a queue reached station 106 and the case study presented above was representative of the behavior whenever a queue reached station 106 on the subject days. This point indicates that the bottleneck mechanisms result in complicated events that can trigger a drop in capacity far upstream of the initial source of delay. Understanding these capacity drops upstream of the primary bottleneck will help explain how disturbances arise in traffic queues. Eliminating these capacity drops might help drivers who exit the queue at a ramp or diverge upstream of the bottleneck. But this solution alone is not expected to increase the number of vehicles that can pass the primary bottleneck since it continues to restrict capacity even though the upstream capacity drops reduce the demand on the primary bottleneck. Furthermore, after concentrating the capacity drop to a single location, the queue will still spread upstream and could continue to impact those drivers that exit before the primary bottleneck.

In the freeway interchange we studied, we believe the queue arises due to vehicles changing lanes to reach an off-ramp. By channelizing the queue, i.e., restricting lane change maneuvers that enter the lane adjacent to the off-ramp, it may be possible to reduce queue spillover to the other lanes and thus, increase the effective capacity of those remaining lanes. However, drivers are unlikely to accept such a complicated control scheme and the speed differential between adjacent lanes may cause new problems. A more feasible solution is to control the percentage of flow that reaches the bottleneck location from each origin and increase the net percentage of vehicles that are not destined for the off-ramp.

As one should expect, the case studies found that drivers entering a lane consume capacity and delay upstream drivers. But both case studies revealed the unexpected result that the degraded conditions due to vehicles entering a lane are not balanced by improvements to the lane from which the vehicles departed.

Acknowledgements

The authors would like to acknowledge the support of the Ohio Department of Transportation in this research. This material is based upon work supported in part by the National Science Foundation under Grant No. 0133278.

References

1. Cassidy, M.J., Mauch, M. (2001). An observed traffic pattern in long freeway queues, *Transportation Research: Part A*, 35(2), pp 143-156.
2. Coifman, B. (2003). Estimating Density and Lane Inflow on a Freeway Segment, *Transportation Research: Part A*, 37(8), pp 689-701.
3. Coifman, B., Dhoorjaty, S., and Lee, Z. (2003). Estimating Median Velocity Instead of Mean Velocity at Single Loop Detectors, *Transportation Research: Part C*, 11(3-4), pp 211-222.
4. Hall, F.L., Hurdle, V., Banks, J. (1992). Synthesis of recent work on the nature of speed-flow and flow-occupancy (or density) relationships on freeways, *Transportation Research Record 1365*, pp 12-17.
5. Mauch, M., Cassidy, M.J. (2002). Freeway traffic oscillations: observations and predictions, *International Symposium of Traffic and Transportation Theory*, (M.A.P. Taylor, Ed.) Elsevier, Amsterdam, pp 653-674.

Empirical Description of Car-Following

P. Wagner

Institut for Transport Research – German Aerospace Centre
Rutherfordstr. 2, 12489 Berlin – Germany
peter.wagner@dlr.de

Summary. This contribution reports a recently recorded data-set that helps to understand the car-following process better. Since the empirical basis of most traffic flow models can be called weak, these findings may help to design better models. They demonstrate, that the process of car-following seems to have a very interestic stochastic dynamics. Especially, and different from most of the existing models the data show clearly that car following cannot be described by a noisy fixed point dynamics. This is because the acceleration of the cars is smooth and roughly constant for a certain time, with fast changes to a new value at so called action points.

Furthermore, the data can be used for calibration and validation purposes of the various models. This has been done with very interesting results indicating that all the models tested behave very similar and cannot describe the data better than with 15 % difference between models and data.

Keywords: calibration/validation, microscopic traffic flow models, phase transitions in traffic flow

1 The Empirics of Car-Following

Car-following data have been around for a very long time, however they are rarely analyzed thoroughly. Usually, they are difficult to obtain, because people who collected them do not like to share them. However, this changes. In [1] an interesting approach has been taken: a platoon of ten cars, each equipped with differential GPS, drove along a test track and records some hours of car-following data with 0.1 s time resolution, and an accuracy of 0.1 m. Therefore, the data for each car is its 3d position (x_n, y_n, z_n), the speed v_n and the angle of direction ϕ_n of movement; from that the usual description (x_n, v_n) has been computed. Note, that the data contain only the gross headway $\Delta x_n = x_{n-1} - x_n$, i.e. the distance from GPS-receiver to GPS-receiver, but not the car lengthes.

In the following, these data together with other data from single car recordings will be analysed. By considering other data as well, the danger of finding very special results related to this particular data-set is minimized. Nevertheless, a few odd things may be present in this data, if apparent, they will be cited below.

In Fig. 1, the so called driving relation, i.e. the dependence of distance Δx upon speed of the lead car V [1] is plotted. What is interesting is that the

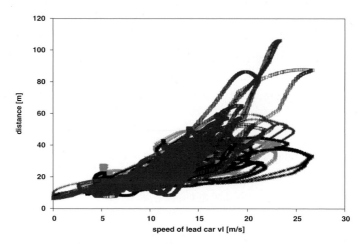

Fig. 1. The driving relation $\Delta x(V)$. Different colours are data from different drivers.

difference between different drivers seems to be very small, they all behave very similar on this level of description.

Next, the speed differences between the lead and the following car $\Delta v = V - v$ will be analyzed. Their distribution is displayed in Fig. 2. The distribution can be approximated for small distances $\Delta x < 30$ m between the cars as:

$$p(\Delta v) \propto \exp(-|\Delta v|). \tag{1}$$

For larger distances, i.e. distances bigger than $30 \ldots 40$ m it is more like:

$$p(\Delta v) \propto \exp(-(\Delta v)^2), \tag{2}$$

which may be a consequence of the rapidly decreasing accuracy of human perception. Another interpretation might be a certain discreteness of human actions: a distribution of this type can be generated if the acceleration applied to the car consists of several discrete levels, rather than a continuous distribution.

The same type of distribution of Δv can be found in other data-sets as well, for instance in radar data recorded by us[2], in American car following data and in single car data from counting loop devices [2].

---

[1] To simplify notation, sometimes the index n is dropped. In this case, capital letters refer to the lead car.

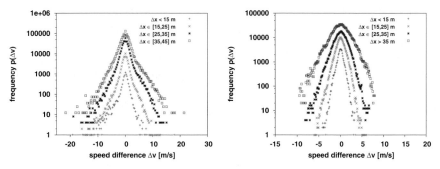

Fig. 2. Frequency distribution $p(\Delta v)$ of velocity differences for different distances Δx between the cars. Left: freeway data, right: GPS-data as described in this contribution.

This distribution has to be discussed together with the prominent oscillations in car-following (they are also fairly robust and can be found in any data-set about car-following), as can be seen in Fig. 3.

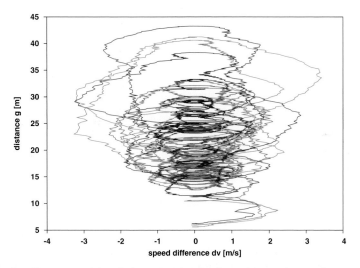

Fig. 3. Oscillations in $(\Delta v, \Delta x)$–space. At the first glance, the oscillations are the most apparent structure, but looking twice it could be seen that the majority of the data are distributed along the line $\Delta v = 0$. The three colours code for different drivers.

The distribution of the speed-differences indicate that the oscillations are not real oscillations in the sense of a limit cycle, furthermore, the non-analytic behaviour at $\Delta v = 0$ indicate something non-standard. In case of a noisy limit-cycle, at least $p(\Delta v)$ should not be peaked that sharply at $\Delta v = 0$. On the other hand, it is not possible to have a fixed point $\Delta v = 0$ plus some white noise disturbing the system: in this case a normal distribution around $\Delta v = 0$ results [3]. So far, the only mechanism found so far that is compatible with these observations is a so called dynamical trap [3, 4], a poster will be presented at this conference, too.

Despite the fact that drivers like $\Delta v = 0$, they do not very well in achieving that goal. Even if the lead car moves practically with constant speed, the fluctuations of each of the cars that follow gets larger. However, they are probably bounded, i.e. human driving might not be completely stable but it is platoon stable. (However, the ten cars are not enough to be sure; additionally, they are driving in a relatively relaxed style with headways bigger than those observed on freeways.) By analyzing parts of the data-set where the first car

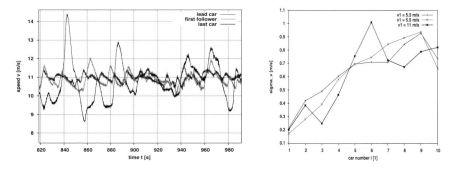

Fig. 4. Increasing fluctuations as function of car number behind a lead car that moves with roughly constant speed. Left: time series of the speeds of the lead car, the first and the last following car. Clearly visible are the bigger fluctuations in the speed of the following cars. The right plot shows the standard deviation of speed as function of the position i in the platoon $\sigma_v(i)$.

drives with constant speed one obtains the standard deviation of the following cars as a function of car number, see Fig. 4. The fluctuations seem to settle to an asymptotic value, however this data set is definitely not suitable to draw more solid conclusions about platoon stability.

The headway distributions, where headway means in the following gross time headway $T \approx \Delta x / V$ are well-known in traffic science: except for very small or for very large densities, they are assumed to be Pearson type III (traffic engineering language) or gamma distributed (physics language)[2]. Physicists use the scaled version (by introducing $s := T/\langle T \rangle$) of the gamma distribution, however, for traffic it is wise to additionally introduce a cut-off $d = T_{\min}/\langle T \rangle$ at small headways:

---

[2] All kinds of functions have been hypothesized so far for this distribution, without reaching common agreement.

$$p(s) = \frac{(s-d)^{\alpha-1}}{\Gamma(\alpha)\langle T \rangle} \exp(-(s-d)) \tag{3}$$

The single car data from certain highways, as well as the values often reported in the traffic engineering literature are compatible with small values of α, we found $\alpha \approx 1 \dots 3$. (But note [5] for slightly different results.) The car-following data reported here show a slightly different picture, see Fig. 5: the distributions are much more symmetric, which happens with larger values of $\alpha \approx 6 \dots 8$. However, for values that large, a (log-)normal distribution is certainly an

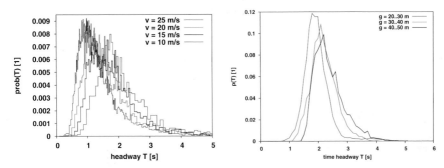

Fig. 5. The headway distributions for freeway traffic (left) and for the GPS car-following data described in this text (right).

alternative. If that result turns out to be robust, then the empirically observed distributions on roads can be understood as a kind of superposition of the more Gaussian-like distribution found in the car-following data. An empirical result that points into this direction and is therefore compatible with this idea is described in [6].

Finally, the acceleration will be discussed. Before doing so, a short account is given how the acceleration is extracted from the speeds of the cars. The basic instrument to do so is a Savitzky-Golay filter [7]. This filter is a numerically very efficient method to fit a low-order polynomial (of order p) to the data in a certain window around any given data-point, interpreting the fitted value as a data-point in the smoothed time-series:

$$v_i^{\text{smooth}} = \sum_{j=-n1}^{n2} c_j v_{i+j} \tag{4}$$

A different set of filter coefficients computes the first (or higher) derivative of the speed directly. For the plot in Fig. 6, $p = 2$ and a symmetric window of size $w = 11$ (1 second data) have been applied, larger windows obviously give smoother curves. From the time-series, it could be seen that there are different episodes, with certain "decision-points" where the acceleration changes. Those

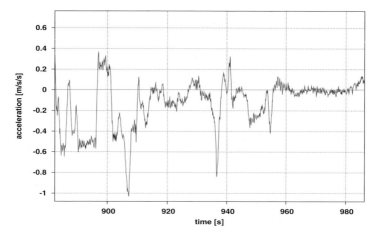

Fig. 6. The acceleration as function of time. The jumps in the acceleration could be seen clearly. The fluctuations at the acceleration plateaus are due to numerical noise.

decision points called action points have been proposed already in [8], so this name will be used in the following. Three different strategies could be identified: there are periods with fairly constant acceleration (the noise is pure measurement and numeric noise from the differentiation of the speed), short peaks in acceleration (in both directions), and periods where acceleration is changing linearly in time. The latter ones can be related to periods where the driver seems to do nothing, causing a small deceleration due to air-resistance, so it is in a certain sense a constant acceleration, too.

Finally, the average behaviour of the acceleration as function of distance and speed difference have been analyzed. This is shown in Fig. 7. It is important to point out that this is an average behaviour, because all the data from the nine drivers have been averaged to compile this plot. Note, that the standard deviation of the acceleration is constant (as function of Δv, Δx and fairly big (about $\sigma_a \approx 0.5 m/s^2$).

Apparently, the acceleration can be described by a simple function

$$a(g, \Delta v) \propto g_0 + a_g g + a_{\Delta v} \Delta v, \tag{5}$$

except for that hole at small distances and very negative Δv which might constitute a dangerous situation. Of course, the pre-factors can and will depend on speed and other circumstances. Nevertheless, it seems not unlikely, that human drivers follow some fairly simple rules like the linear Eq. (5).

More can be learned when analysing the distribution of the action points. To do this, the action points have been identified by a fully automatic method. It computes first the second derivation $\ddot{a}$ of the speeds (the same Savitzky-Golay technique had been used, with larger windows sizes $n = 21$ and poly-

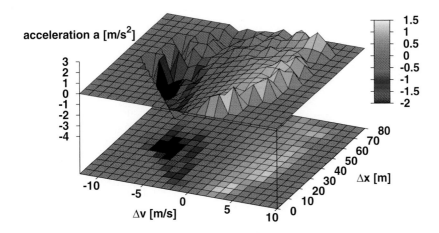

Fig. 7. The acceleration as function of distance and velocity difference. Note, that $a(\Delta v, \Delta x)$ is an average over all nine drivers, and all data to have sufficient statistics.

nomial order $p = 4$), looks for minima/maxima of $\dot{a}$ that are above the noise levels (indicating a fast change of acceleration) and finally identifying the action-point as all those pieces of the speed time-series, which are above noise level and contain at least one maximum/minimum value of $\dot{a}$. (Without the last step, a lot of spurious action-points results, because minima/maxima of $\dot{a}$ are numerous due to noise.)

First of all, the time between subsequent action points can be shown to follow a distribution that is similar to the distribution of the time headways. This can be interpreted as generated by a random process where a new action point is drawn randomly, with one exception: the likelihood of a new action point is suppressed directly after the driver has changed his acceleration. More interesting is the distribution of the action-points in the $\Delta v, \Delta x$ phase space. The models that use action points [9, 10] assume that there are certain thresholds for perceptions, which are the trigger-points for the action points. The most important threshold is the one for speed difference. Obviously, humans cannot recognize arbitrarily small speed differences, because the corresponding signal is the change of size of the object on the retina. Therefore, to at least recognize such a speed difference, it must be larger than a certain threshold [11]:

$$\theta_{\Delta v} \propto \frac{\Delta v}{(\Delta x)^\gamma} \qquad (6)$$

It is easy to construct a simple toy model with those thresholds and compare its outcome to the empirical data. This is done in Fig. 8, it could be seen that the data cannot be described by a driver model assuming these thresholds. Definitely, this creates a certain dilemma. The results from psycho-physiology demonstrate that those thresholds are there, but drivers seem to drive without using them.

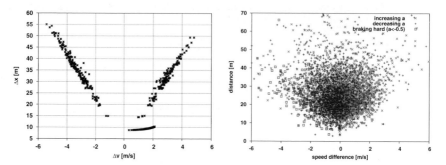

Fig. 8. The distribution of action points for a simple model with perception thresholds (left), in comparison to the empirical data (right).

2 Calibration & Validation

Originally, the goal of this work was to use these data to calibrate and validate microscopic traffic flow models. Recent work [12, 13] for the car-following process (considering only two cars at a time) show that the models tested so far cannot reproduce the distances better than with 10 % error. (Mostly, the error was 16 % for calibration and 25 % for validation, respectively, for sufficiently long time series.) Error means root-mean-squared error (RMSE), defined as the following distance between data and simulation

$$e = \sqrt{\frac{1}{N} \sum_{i=1}^{N} \left(q_i^{(\text{sim})} - q_i^{(\text{data})} \right)^2}. \qquad (7)$$

Here, q can be anything, most models perform well with the speeds, but worse with the distances. However, when going to more macroscopic measures like travel times, the error figures do not change much.

At the related poster [14] more details on the calibration issue are given. It seems, that all models that have been tested so far are equivalent, see Fig. 9. The differences between the models are smaller than the differences between different drivers.

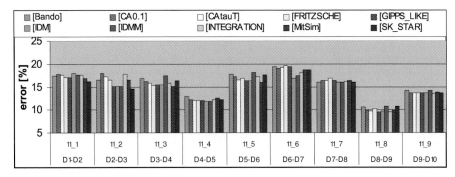

Fig. 9. Result of the calibration of ten different microscopic car following models. The normalized RMSE for a data-set consisting of nine driver pairs are displayed. Note, that the error figures range from 10 to 20 % between the drivers, while the errors of the models for one particular driver are much more similar.

Complementary to this work, now the whole platoon of ten cars will be analyzed. Here, only the following models will be considered:

Newell This model assumes, that the trajectory $x_f(t + \tau)$ of the following car just "mirrors" the trajectory $x_l(t)$ of the lead car[15]:

$$x_f(t + \tau) = x_l(t) + d \qquad (8)$$

It can be shown, that this is equivalent to

$$\dot{v} = \frac{1}{T}\left(\frac{\Delta x - d}{\tau} - v\right) \qquad (9)$$

which is a simple optimal velocity (OVM) model. It is the microscopic analogue of the LWR theory of traffic flow. When discretizing in time and space, one ends up with a model that is similar (but not exactly the same) to the deterministic version of the well-known NaSch–CA [16].

IDM This is the model described in [17]. It reads:

$$\dot{v} = a\left(1 - \left(\frac{v}{v_{max}}\right)^4 - \left(\frac{g^\star}{g}\right)^2\right) \qquad (10)$$

$$g^\star = g_0 + vT + \frac{v\Delta v}{2\sqrt{ab}} \qquad (11)$$

and is usually the model that performs best in our tests.

SK★ This is a model that has been invented to better understand what is wrong with the car-following models. A convenient starting point is [18, 19]. There it is assumed, that cars maintain a so called safe velocity, that can be computed by setting up an equation for the braking distances $d(v) = v^2/(2b)$:

$$\frac{v^2}{2b} + g^\star \le \frac{V^2}{2b} + g \tag{12}$$

Here, $g^\star$ is a desired distance, originally [19] it was choosen to be $v\tau$. Here it is assumed, that $g^\star$ is driven by a stochastic process. In its simplest form, it may be specified by a certain probability p_c to compute a new desired gap $g^\star$:

$$g^\star = V\left(\tau_0 + \delta\tau\xi\right) \tag{13}$$

where ξ is a random number in $[0, 1]$. Inserting this into the model's equation yields a stochastic, time-continuous traffic flow model:

$$\dot{v} = \frac{1}{T}\left(v_{\text{safe}}(g, V) - v\right). \tag{14}$$

Here, $\dot{v}$ is bounded by a positive constant $\dot{v} \le a(v)$. The safe speed is given by $v_{\text{safe}} = -b\tau + \sqrt{b^2\tau^2 + V^2 + 2gb}$, where b is the maximum deceleration, τ the reaction time, and T the relaxation time. Note, that mechanisms like the ones described here can be inserted into other models as well.

When doing the same calibration naively, i.e. with using RMSE as error metric, the models' calibration results are around 25%. However, if car-following is a stochastic process, it is not the best idea to use RMSE as an error measure. What happens, can be seen in Fig. 10a. The RMSE-minimum

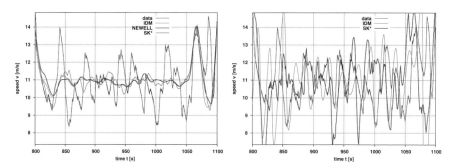

Fig. 10. Optimizing the parameters of various models using an RMSE error function as defined above (a, left), and, for comparison, with an error function based on the Kolmogorov Smirnov (KS) distance (b, right). As could be expected, the KS-distance preserves the fluctuations much better.

gives the correct mean-values. If the dynamical system behind would have been deterministic and well-behaved[3], this ansatz should give a really good fit. However, the fluctuations seem to be important. They are captured by

---

[3] A counter-example, where such an approach goes awfully wrong is the logistic map [20] in the chaotic regime. (Chaos is an example for not well-behaved).

introducing an error measure based on the Kolmogorov Smirnov statistics. By using a window of a certain size, this measure computes the maximum deviation between the two frequency distributions of the gaps within the windows, the one from the data and the one from the simulation, respectively. This is then averaged over all the windows. Right now, those windows have been choosen to be partially overlapping. Other settings probably have to be checked. Although the fit is worse in the RMSE-sense, the corresponding simulation results look better, it seems that the behaviour of the drivers has been captured better.

3 Conclusion

To conclude, it may be stated that the empirical data presented here rule out most of the existing microscopic traffic flow models. Of course, this does not mean that they are completely nonsense, however, some extensions are needed if they can be regarded as a serious approximation to microscopic reality. (Their abilities to describe macroscopic features are a different and highly interesting story.) The models which claim that there is a fixed point in the car-following dynamic are wrong because it is clearly demonstrated, that such a fixed point is not there. Additionally, almost all models assume a continuous control by the driver, described by an ODE, which is not existent either: the control of the car by the driver happens at definite points in time, the so called action points. Other results indicate that the prediction of the action point models that the action points are distributed along certain thresholds in $(\Delta v, \Delta x)$–space is wrong, too.

With a different argument, optimal velocity models can be ruled out altogether. The main argument is that the acceleration unambigously depends on the velocity difference, this is much more clear than the dependence on distance. A nice example for this dependence is shown in Fig. 11. OV-models only depend on the distance, so they are wrong microscopically.

However, it is really challenging to construct enhanced models. The simple ansatz

$$\dot{v} = \Omega(\Delta v)\left(f_v(V - v) + f_g(g^\star - g)\right) + D\xi(t) \tag{15}$$

leads to trajectories $v(t)$ that are not continuously differentiable which is microscopically not acceptable. A number of remedies have to be explored into more detail: one is to add a colored noise process instead of a white noise process to the acceleration equation. A different, albeit similar idea is to introduce time-varying parameters, especially $g^\star$ is a good candidate. Another idea might be to follow the reasoning introduced in [21]: building a model that is formulated as an equation for $\dot{a}$. To find out which of those ideas will finally lead to the correct microscopic formulation of car following is an exciting research task.

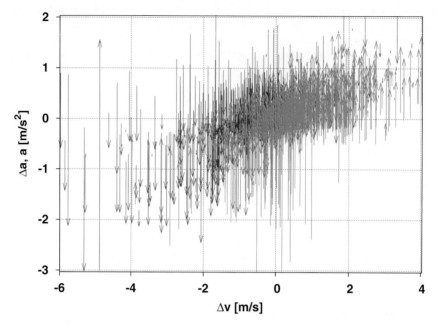

Fig. 11. The jump of acceleration at the action points, as function of speed difference Δv. Red arrows show action points where the acceleration is reduced, while green arrows show increasing acceleration. A hysteresis can be seen, and also the roughly linear dependence $a \propto \Delta v$.

Acknowledgement

This work would not have been possible without the great support from my collaborators Elmar Brockfeld, Daniel Krajzewicz, Stefan Lorkowski, and Nils Eissfeldt in Berlin and Cologne.

The data used in this work have been provided by Takashi Nakatsuji, Michael Schreckenberg, and the DLR-team (Manfred Hofmeister, Ibro Pandzo, and Ronny Terzer). Many thanks to them all.

References

1. G. S. Gurusinghe, T. Nakatsuji, Y. Azuta, P. Ranjitkar, and Y. Tanaboriboon. Multiple car following data using real time kinematic global positioning system. *Transportation Research Records*, 2003.
2. Institute for Transport Research. Clearinghouse for traffic data. http://www.clearingstelle-verkehr.de, accessed Aug. 2004.
3. I. Lubashevsky, M. Hajimahmoodzadeh, Albert Katsnelson, and P. Wagner. Noise-induced phase transitions in systems of elements with motivated behavior, this volume.

4. S. Kalenkov, I. A. Lubashevsky, R. Mahnke, and P. Wagner. Long-lived states in synchronized flow: empirical prompt and dynamical trap model. *Physical Review E*, 66:016117, 2002.

5. M. Krbalek and D. Helbing. Determination of interaction potentials in freeway traffic from steady-state statistics. *Physica A*, 333:370–378, 2004. cond-mat/0301484.

6. M. J. Cassidy. Driver memory: motorist selection and retention of individualized headways in highway traffic. *Transportation Research A*, 32:129–137, 1998.

7. W. H. Press, S. A. Teukolsky, W. A. Vetterling, and B. P. Flannery. *Numerical Recipes in C*, volume 2nd edition. Cambridge University Press, 2002.

8. E. P. Todosiev and L. C. Barbosa. *Traffic Engineering*, 34:17–20, 1963/64.

9. Rainer Wiedemann. Simulation des Straßenverkehrsflußes. Technical report, Institut für Verkehrswesen, Universität Karlsruhe, 1974. Heft 8 der Schriftenreihe des IfV, in German.

10. H.-T. Fritzsche. A model for traffic simulation. *Transportation Engineering And Control*, (5):317–321, 1994.

11. M. W. Szeto and D. C. Gazis. Application of kalman filtering to the surveillance and control of traffic systems. *Transportation Science*, 6:419 – 439, 1972.

12. E. Brockfeld, R. D. Kühne, A. Skabardonis, and P. Wagner. Towards a benchmarking of microscopic traffic flow models. *Transportation Research Records*, 1852:124 – 129, 2003.

13. E. Brockfeld, R. D. Kühne, and P. Wagner. Calibration and validation of microscopic traffic flow models. *Transportation Research Records*, 1876, 2004.

14. E. Brockfeld and P. Wagner. Testing microscopic traffic flow models, this volume.

15. G. F. Newell. A simplified car-following theory: A lower order model. *Transportation Research B*, 36:195 – 205, 2002.

16. K. Nagel and M. Schreckenberg. A cellular automaton model for freeway traffic. *J. Physique I*, 2:2221, 1992.

17. M. Treiber, A. Hennecke, and D. Helbing. Derivation, properties, and simulation of a gas-kinetic-based, non-local traffic model. *Physical Review E*, 59:239–253, 1999.

18. P. G. Gipps. A behavioural car following model for computer simulation. *Transportation Research B*, 15:105–111, 1981.

19. S. Krauß, P. Wagner, and C. Gawron. Metastable states in a microscopic model of traffic flow. *Physical Review E*, 55:5597–5605, 1997.

20. Werner Horbelt. *Maximum likelihood estimation in dynamical systems*. PhD thesis, Albert-Ludwigs-Universität Freiburg im Breisgau, Germany, 2001.

21. I. Lubashevsky, P. Wagner, and R. Mahnke. A bounded rational driver model. *Euro. Phys. J. B*, 32:243 – 247, 2003.

Optimization Potential of a Highway Network: An Empirical Study

W. Knospe[1], L. Santen[2], A. Schadschneider[3], and M. Schreckenberg[1]

[1] Physik von Transport und Verkehr – Universität Duisburg-Essen
47048 Duisburg – Germany
`schreckenberg@uni-duisburg.de`,`wolfgang.knospe@vodafone.com`

[2] Fachrichtung Theoretische Physik – Universität des Saarlandes
Postfach 151150, 66041 Saarbrücken – Germany
`santen@lusi.uni-sb.de`

[3] Institut für Theoretische Physik – Universität zu Köln,
Zülpicher Str. 77, 50937 Köln – Germany
`as@thp.uni-koeln.de`

Summary. We present an analysis of traffic data of the highway network of North-Rhine-Westphalia in order to identify and characterize the sections of the network which limit the performance, i.e., the bottlenecks. It is clarified whether the bottlenecks are of topological nature or if they are constituted by on-ramps. This allows to judge possible optimization mechanisms and reveals in which areas of the network they have to be applied. Our results support previous empirical observations and theoretical studies indicating that the overall travel-time of vehicles in a traffic network can be optimized by means of ramp metering control systems.
Keywords: highway network, capacity analysis, bottlenecks, ramp metering

1 Introduction

The increasing amount of vehicular traffic led to the (over-)saturation of many freeway networks. Therefore, in order to use the existing network more efficiently, several control strategies like signal control, variable message signs, route guidance, motorway-to-motorway control and ramp metering [1] have been proposed for the optimization of the highway's performance and for the amelioration of the system's traffic state.

Empirical studies of parts of a highway network [2] have revealed the benefits of ramp metering strategies for the total mean speed and the total travel-time for both, the vehicles on the highway *and* the queuing vehicles on the on-ramp. These empirical findings are underscored by several theoretical investigations of simplified [3, 4] as well as well as of more realistic [5–7] traffic models, which all generate stable stationary states with high throughput. In case of open boundary conditions, which are appropriate if one compares to empirical data, generally the steady states are controlled by the in- and output currents [3–5, 7, 8].

The improvement of the system's throughput can be explained by considering fluctuations of the traffic flow. Obviously, free flow is the most desirable traffic state since large flows can be obtained. However, at increasing traffic demand, synchronized traffic forms in the vicinity of on-ramps, which enables a large amount of vehicles to drive with a velocity considerably larger than in a jam but smaller than in free flow, and leads to a flow comparable to free flow [9]. Synchronized states are not very stable, i.e., already small perturbations can cause wide jams [9]. Due to its role as a precursor of traffic jams, the onset of synchronized traffic can be used in order to activate ramp-metering: If synchronized traffic is observed at a bottleneck control mechanisms will help to stabilize a high flow state and, thus, ensure large throughputs of the highway at large densities. In particular, by means of ramp metering, it is possible to tune the injection rate of the on-ramp in order to affect the vehicle-vehicle interactions. Increasing the interactions will improve the vehicle coupling and help to synchronize them [10]. However, strong interactions lead to perturbations of the flow and, thus, to wide jams. The variation of the on-ramp flow therefore helps to adjust the interactions and to stabilize the synchronized state. As a result, the density fluctuations are reduced, and the performance of the highway is improved.

In this study we present an extensive analysis of a highway network which gives insight about the spreading of jams in the network [11]. In contrast to former analyses which are restricted to only small parts of a highway system, here, the coverage of the network with inductive loops allows to analyze the overall traffic state on a global scale. This enables us to identify and characterize its bottlenecks. In particular, the question is raised whether bottlenecks are of topological nature or if they are constituted by on-ramps with a large inflow. Topological bottlenecks are static (e.g. lane reductions) whereas dynamical bottlenecks (like ramps) in principle can be influenced externally, e.g., by controlling their strength. Moreover, the properties of the bottlenecks may give evidence if the network's state can be optimized by local control devices only or whether global traffic management strategies are necessary [12].

2 Network Characteristics

We examine the highway network of North Rhine-Westphalia (Fig. 1) which has a total length of about 2250 km and includes 67 highway intersections and 830 on- and off-ramps. The data set is provided by about 400 inductive loops which send minute aggregated data of the flow, the occupancy and the velocity online (that is every minute) via permanent lines from the traffic control centers in Recklinghausen and Leverkusen. For the sake of simple data handling, the driving directions are classified into south/west and north/east. As one can see in Fig. 2, most of the detectors are concentrated around the inner network. This is justified by the distribution of the traffic volume, as we will show later. The recording of the data started on 10-10-2000. For the analysis of the network, a period of 265 days from 10-10-2000 to 07-01-2001 was considered.

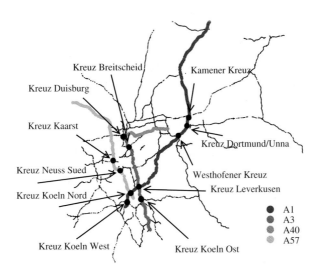

Fig. 1. Schematic map of the highway network of North Rhine-Westphalia. The positions of the the main highways, the A1, the A3, the A57 and the A40, and important crossings are indicated.

The highway network of North Rhine-Westphalia has an average traffic load of about 30 000 veh/24h per measurement section for one driving direction. Note that the traffic volume is calculated per measurement section and therefore sums the flow of two or three lanes. However, there are large differences between the highways (Fig. 3). Obviously, only a few sections with very large traffic volumes exist which concentrate around the main urban areas. In detail, the Kölner Ring, especially the section between the highway intersections *Kreuz Leverkusen* and *Kreuz Köln Ost*, large parts of the A3 and the A40, the A57 near Neuss and the A2 near the highway intersection *Kreuz Duisburg* have an average load of 40 000 − 80 000 veh/24h. The traffic volume decreases with increasing distance from these conurbations. In addition, there are only small differences between the traffic volumes for the driving directions south/west and north/east.

Fig. 2. Position of the inductive loops (black dots). The main highways indicated in the previous figure are densely equipped with detection devices.

About 15% of the vehicles are trucks. Fig. 3 shows the distribution of the relative number of trucks per measurement section. Obviously, the share of trucks on the traffic volume increases with the distance to the conurbations, while the absolute number of trucks does not change significantly on the main highways. Thus, the long-distance traffic is mainly dominated by trucks, while at short distances commuter traffic leads to large traffic loads of the network.

3 Bottlenecks

Since the traffic volumes are simple integrated quantities, it is possible that a large traffic volume can indicate that a two-lane road is often jammed, while the capacity of a three-lane road is still not reached. Therefore, in order to allow an evaluation of the traffic load, the probability to find a jam is calculated. This has the advantage that traffic volumes which are larger than the highway capacity can be related directly to a large jam probability, and thus bottlenecks of the highway network can clearly be identified. Due to the size of the analyzed data set it is necessary to introduce a simple criterion to identify jams. Our criterion is the following: A sector is considered to be jammed (during the observation period) if the density is larger than 50%

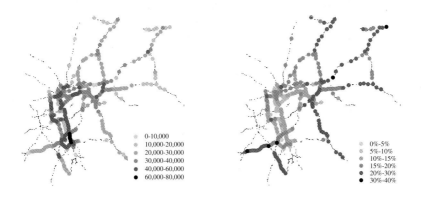

0-10,000	0%-5%
10,000-20,000	5%-10%
20,000-30,000	10%-15%
30,000-40,000	15%-20%
40,000-60,000	20%-30%
60,000-80,000	30%-40%

Fig. 3. Left: Number of vehicles per measurement location per 24 hours and driving direction. Right: Relative number of trucks per measurement section.

at least for a single one-minute interval. [4] This criterion was motivated by former empirical studies [13] that indicated that densities higher than 50% are typically neither observed in free-flow nor in synchronized traffic. The threshold value as well as the chosen time interval are, of course, somehow arbitrary. Our results are, however, rather insensitive upon an increase of the density threshold. By contrast, time intervals longer than one minute may lead to problems, because jams are often not compact but include regions of free flow traffic. This implies that the density threshold has to be lowered if longer time intervals are used, but lower density threshold lead to problems in discriminating between jams and synchronized traffic.

The usage of the density as an indicator for a jam has some advantages compared to the mean velocity and the flow. First, the flow as well as the velocity depend on speed limits which vary in the network. Therefore, it is not possible to distinguish whether a small velocity is due to a large density or a large velocity flucutation in a synchronized state. Second, a small flow can indicate both a jam or just free flow with only a few cars. The jam probability is therefore calculated as the relation of the number of days a jam was found to the total number of days of the observation period (265 days).

Fig. 4 represents the jam probability of the network for the driving direction south/west. Obviously, large jam probabilities can be found in a region of the inner network, where the conurbations of the Ruhrgebiet (region 1), the areas of Dortmund (region 2), Düsseldorf and Krefeld (region 3) and Köln

---

[4] Here the density is given as occupancy, i.e. the fraction of time a detector is covered by a vehicle, averaged over 1 minute intervals.

(Cologne) (region 4) are located. As one can see in Fig. 2, most of the inductive loops are concentrated in this regions, which are indeed the most crowded sectors of the network. In contrast, in regions which are less equipped with counting devices, only few jams can be found.

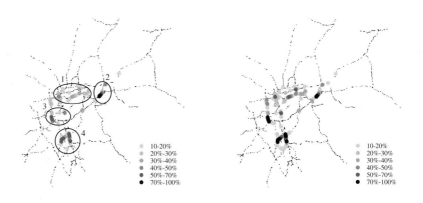

Fig. 4. Left: Jam probability for the driving direction south/west. The encircled regions are the Ruhrgebiet (region 1), the area of Dortmund (region 2), the area of Krefeld and Düsseldorf (region 3) and the area of Köln (region 4). Right: Jam probability on wednesday for the driving direction south/west.

Large differences of the jam probability between the areas can be observed. Only $10\% - 30\%$ of the observation days were jammed in the area of the Ruhrgebiet (region 1), that is every third day a jam occurred. In contrast, in the areas of Dortmund (region 2), Krefeld and Düsseldorf (region 3) and Köln (region 4), more than 50% of the days showed jams. Especially in the area of Dortmund, between the *Westhofener Kreuz* and the *Kreuz Dortmund/Unna*, nearly 5 days a weak jams can be measured.

Next, we analyze *when* the jams typically occur. First, we expect significant differences between the weekend and working days. Indeed, the exclusion of weekends leads to an increased jam probability on weekdays compared to the probability averaged over the whole observation period. Second, we investigated the daily variations of the jam probabilities. During the week from monday to friday only minor differences in the jam probability exist, but one has to distinguish the period from monday to thursday and the friday as we will show below. The jam probabilities we obtained are qualitatively the same as in case of an average over the whole data set (see Fig. 4, right), the same four regions show a large jam probability that is the Ruhrgebiet and the areas of Dortmund, Krefeld and Düsseldorf and Köln. The actual value of the jam probability, however, is increased and the difference to the remainder of the network is now more pronounced. During the weekend the jam probability is considerably reduced. Nevertheless, there are a few locations in the network where high jam probabilities can be observed on sundays, i.e., between the *Westhofener Kreuz* and the *Kreuz Dortmund/Unna*. Remarkably the same classification scheme was found to be appropriate for city data as discussed in [14, 15].

Next, we analyse the intraday variations of the data set. We further divided the time-series into five time intervals, in order to capture the main rush-hours. The first interval covers the time from 0 to 6 a.m., the second interval from 6 to 11 a.m., the third interval from 11 a.m. to 1 p.m., the fourth interval from 1 to 8 p.m. and the fifth interval from 8 to 12 p.m. Indeed, the division of a day is somehow arbitrary, but a finer discretization of a day will not change the results significantly. Moreover, analyses of city data revealed the existence of basically two rush-hour peaks in the time-series of the traffic volume at about 8 a.m. and 4 p.m. [14, 15].

Figure 5 shows the jam probability of the whole data set in the driving direction south/west for two time intervals. During the first time interval, that is 0 to 6 a.m., no jams could be measured. The main traffic volume occurs in the second and in the fourth interval, which include the rush-hours. From 11 a.m. to 1 p.m. only a few jams are visible, while from 8 p.m. to 12 p.m. just one region shows jams, namely the region between the *Westhofener Kreuz* and the *Kreuz Dortmund/Unna*. The jams therefore occur mainly in the morning and afternoon rush-hours. There are, however, parts of the network where the load is so large, that the probability to find a jam is high throughout the whole day.

The same picture can be drawn from the analysis of the data set classified in single days which are divided into the five time intervals. Due to the exclusion of the weekends, the values of the jam probability increase significantly. Congestion is mainly observed during the rush-hours, while the regions 2 and 4 show continuously a large traffic load. Again, there are only small differences between the weekdays. On fridays on the one hand, the load of the network in the time between 6 and 11 a.m. is smaller, but, on the other hand, is increased significantly between 1 and 8 p.m. compared to monday to thursday. Moreover, on sundays, the large jam probability in the area of Dortmund can be traced back to congestions between 1 to 8 p.m.

An analogous analysis of the data for the driving direction north/east (Fig. 6) shows the same four regions with large jam probabilities. Now the area of the Ruhrgebiet (region 1) and the area Krefeld and Düsseldorf (region 3) have a larger jam probability in the direction north/east, while in the direction south/west the area of Köln (region 4) shows a larger traffic volume. In the area of Dortmund (region 2), a large jam probability can be found in the south of the *Westhofener Kreuz*. Because north of the Westhofener Kreuz

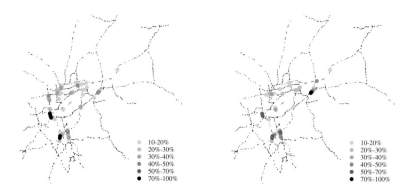

Fig. 5. Left: Jam probability for the driving direction south/west averaged over the whole observation period for the time interval 6-11 a.m. Right: Jam probability for the driving direction south/west averaged over the whole observation period for the time interval 1-8 p.m.

in the driving direction south a large jam probability was measured, too, one can clearly identify the *Westhofener Kreuz* as a dominant bottleneck.

Again, the distinction between the days shows no difference to the analysis of the whole observation period with the exception of a larger jam probability in the single regions. In addition, the main traffic volume can be measured in the morning and afternoon rush-hours.

4 Spatial Extension of Jams

Calculating the jam probability in the four regions, one can observe that consecutive measurement sections are often jammed even at the same time. This may be a consequence of jams that have a large spatial extension. However, as one can see further, a sequence of jammed detectors is always restricted by highway intersections, not by on- and off-ramps. But more importantly a jam which branches from one highway to another highway via an intersection has been observed only in a single case (see below).

In order to determine the spatial extension of a jam, it is necessary to consider temporal information: If for a given day a jam was detected at a detector, the detector is considered to be active (1) if not as being passive (0). In this way we generated a binary time-series which allows the calculation of the spatial correlation between different measurement locations. The explicit calculation of the correlation of the density time-series, in contrast, has the

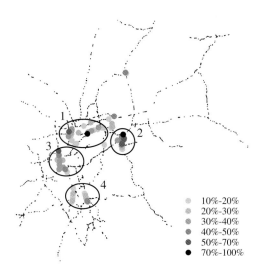

10%-20%
20%-30%
30%-40%
40%-50%
50%-70%
70%-100%

Fig. 6. Jam probability for the driving direction north/east averaged over the whole data set. The encircled regions are the Ruhrgebiet (region 1), the area of Dortmund (region 2), the area of Krefeld and Düsseldorf (region 3) and the area of Köln (region 4).

disadvantage that the duration of jams decreases in upstream direction, which leads to vanishing correlations.

Although the activity gives only a rough estimate of the jamming state at an inductive loop, the typical spatial extension of jams can clearly be seen. The large jam probability in region 2 in the driving direction south/west can be related to a jamming of large parts of the A1 between the *Westhofener Kreuz* and the *Kreuz Dortmund/Unna* (Fig. 7) since strong correlations are observed. The same picture can be drawn in the area of Köln (region 4) where a jam may cover a distance from the *Kreuz Köln West* and from the *Kreuz Köln Ost* to the *Kreuz Leverkusen*. The largest extension of a jam can be found on the A40 (region 1) where strong correlations between the *Kreuz Duisburg* and the *Kreuz Bochum* are measured. Obviously, there are also correlations with highways that are very far away from the reference detector, but which are not connected by a sequence of succeeding inductive loops with a large jam probability. The correlations are simply a consequence of a large traffic load of the network which leads to many jams on various highway sections. Since the number of jams counted at the reference detector is large, correlations of a free flow signal can be excluded.

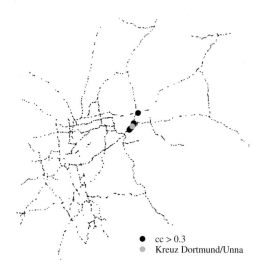

Fig. 7. Spatial correlations between the binary time-series, which encode the occurrence of a jam. The (equal time) correlations are taken between the measurement devices at highways in south/west direction and the detector at the intersection *Kreuz Dortmund/Unna*. Correlation values larger than 0.3 are shown.

The correlation analysis of the driving direction north/east reveals strong values on the A57 near the *Kreuz Kaarst* (Fig. 8, left) and again on the A40 between the *Kreuz Bochum* and the *Kreuz Duisburg* (Fig. 8, right). A jam measured at the *Westhofener Kreuz* (Fig. 9, left) can branch due to the highways intersection on both, the highway A45 and the highway A1.

The analysis of the activity for one day periods suppresses temporal differences, which arise due to the morning or afternoon rush-hours. Therefore, the activity is classified into the five periods used for the jam probability calculation. This additional consideration of the temporal occurrence of the activity allows a better determination of the spatial correlation of the measurement locations.

The picture of large congested areas on the highways can be confirmed by the classified activity correlation analysis. Especially the A40 between the

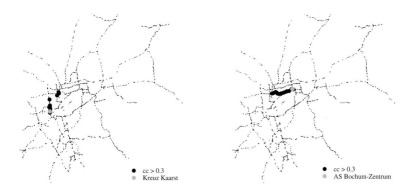

Fig. 8. Left: Spatial correlation in the driving direction north/east near the highway intersection *Kreuz Kaarst.* Correlation values larger than 0.3 are shown. Right: Spatial correlation of the activity in the driving direction north/east near AS Bochum-Zentrum. Correlation values larger than 0.3 are shown.

Kreuz Bochum and the *Kreuz Duisburg* in the directions east and west, the A57 in the vicinity of the *Kreuz Kaarst* in the direction north and the area of Köln in the direction south/west (Fig. 9, right) show strong correlations of consecutive measurement locations. Since the finer discretization of the activity time-series does not change the results significantly, the classification of the activity into five intervals is sufficient for the proper recording of the temporal occurrence of jams. Thus, although the classified activity is only a rough estimate for the density time-series of a detector, the correlation analysis nevertheless allows the determination of the spatial extension of jams.

The correlation analysis supports the results drawn from the jam probability calculation. Although jams can have a large spatial extension, branching of a jam via a highway intersection is rarely observed. As a consequence one can conclude that the bottlenecks are, apart from the *Westhofener Kreuz*, a result of perturbations due to on- and off-ramps rather then local capacity restrictions. As one can see in Fig. 10 most of the sources of the network with a large injection rate are concentrated in the four regions. In addition, most of the sinks can be found there [5]. Therefore, traffic in these four regions is determined by strong fluctuations of the traffic demand and, thus, the flow, resulting in jams with a large spatial extension. However, these jams

---

[5] The source and sink rate is calculated by summing up the flow of all lanes at a measurement section. The difference of the total flow between two succeeding sections gives the loss or gain of the number of vehicles.

Fig. 9. Spatial correlation of the activity in the driving direction north/east near the highway intersection *Westhofener Kreuz*. Right: Spatial correlation of the classified activity in the driving direction south/west near the highway intersection *Kreuz Köln-Nord*. Correlation values larger than 0.3 are shown.

are restricted to single highways and do not affect traffic in other parts of the network.

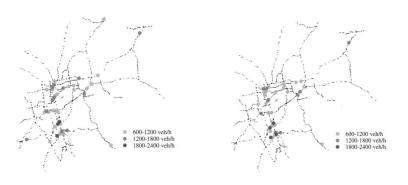

Fig. 10. Source (left) and sink rate (right) in the driving direction south/west.

We found one example for a topological bottleneck (Fig. 11, left). It is located north of the junction Oberhausen-Lirich on the A3 and is generated by a reduction from three to two lanes in each driving direction. In September 2001 this bottleneck was eliminated by the extension to three lanes. Due to the road works, jams regularly emerged in the south of the bottleneck. These jams could have an extension up to the *Kreuz Breitscheid*, passing the *Kreuz Kaiserberg* undisturbed. For comparison, Fig. 11 (right) shows the same part of the network after the road constructions have been finished. Obviously, the probability to find a jam on the A3 south of the junction Oberhausen-Lirich is reduced drastically. A systematical impact on other parts of the highway network, however, cannot be observed.

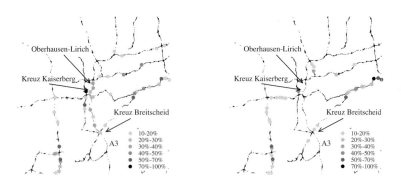

Fig. 11. Left: Jam probability in the driving direction north/east in October 2000 during the road constructions north of the junction Oberhausen-Lirich. Right: Jam probability in the driving direction north/east in October 2001 after finishing the road construction at Oberhausen-Lirich.

5 Conclusion

The analysis of the highway network of North Rhine-Westphalia leads to the following results: The main traffic load concentrates on the inner network where the conurbations of the Ruhrgebiet (region 1), the areas of Dortmund (region 2), Düsseldorf and Krefeld (region 3) and Köln (region 4) are located. In these regions, traffic is predominantly determined by commuter or local traffic. In contrast, the outer regions of the network are dominated by long-distance traffic, indicated by a large truck share on the total traffic volume. Jams can be measured very often in the central part of the network i.e., jam probabilities of more than 50% have been measured. This picture can be confirmed by a more detailed analysis of the traffic data which takes the daily occurrence of jams into account. Like in [14, 15] we find three distinct classes, namely traffic on Monday to Thursday, Friday, and Saturday and Sunday, respectively. However, the time a jam emerges is considered by the subdivision of one day into five intervals. As a result, most of the jams can be traced back on the morning and afternoon rush-hours, but for the areas of Köln (region 4) and Dortmund (region 2), the probability to find a jam is large during the whole day. In addition, on fridays, the main traffic load is shifted from the morning to the afternoon rush-hour.

The correlation analysis of the activity takes the temporal occurrence of a jam at consecutive measurement locations into account. Large correlations between succeeding detectors in the four regions can be measured. Nevertheless, the jams are mainly restricted to single highways and do not branch on other highways via intersections. The only bottleneck which can be identified is the highway intersection *Westhofener Kreuz*. A more detailed calculation of the correlation by means of a classified activity confirms these results.

The bottlenecks of the network are therefore predominantly on- and off-ramps rather than topological peculiarities of the highway system. It is the large in- or outflow at ramps that perturbs the stream of vehicles on the main highway. As proposed in [2, 6], it has to be expected that ramp metering systems are able to counteract these destabilization of the flow and reduce the formation of jams. In this way a restricted flow on the ramps may lead to a significant increase of the capacity of the main highways. Of special interest for a capacity optimization is the existence of some "hot spots" in the network: The highest jam probabilities are spatially and temporally well localized. These are the parts of the network, where the flow has to be optimized. Presently these sections self-organize in a congested state, leading fluxes that are far below their capacity. The situation can be improved a lot by controling the number of entering cars and by optimizing the traffic stream at the on- and off-ramps at this section. The small number of bottlenecks that are present in the network shows that it is possible to improve the capacity of the network with a reasonable technical effort. Of particular interest are sections where the main input is from other, less crowded, highways. In these cases a restricted input does not lead to a collapse of the urban traffic.

There is of course another way to avoid congestion in conurbations. Our analysis shows that the jams are caused mainly by commuters. This means that a better usage of public transport in the urban areas would indeed reduce significantly the congestion at the conurbations.

Summarizing our analysis has shown that the bottlenecks of a highway network can be found by means of a simple statistical analysis. The distribution of the bottlenecks in the network indicates that optimization strategies can be successfully applied in order to increase the capacity of the network.

Acknowledgments

We are grateful to the Landesbetrieb Straßenbau NRW for data support and to the Ministry of Transport, Energy, and Spatial Planning as well as to the Federal Ministry of Education and Research of Germany for financial support (the latter within the BMBF project "DAISY"). L.S. acknowledges support from the DFG under Grant No. SA864/2-1.

References

1. M. Papageorgiou. An integrated control approach for traffic corridors. *Transp. Res. C*, 3:19–30, 1995.
2. H. H. Salem and M. Papageorgiou. Ramp metring impact on urban corridor traffic: Field results. *Transp. Res. A*, 29:303–319, 1995.
3. A.B. Kolomeisky, G. M. Schütz, E. B. Kolomeisky, and J. P. Straley. Phase diagram of one-dimensional driven lattice gases with open boundaries. *J. Phys. A*, 31:6911–6919, 1998.
4. C. Appert and L. Santen. Boundary induced phase transitions in driven lattice gases with metastable states. *Phys. Rev. Lett.*, 86:2498–2501, 2001.
5. V. Popkov, L. Santen, A. Schadschneider, and G. M. Schütz. Empirical evidence for a boundary-induced nonequilibrium phase transition. *J. Phys. A*, 34:L45–L52, 2001.
6. B.A. Huberman and D. Helbing. Economics-based optimization of unstable flows. *Europhys. Lett.*, 47:196–202, 1999.
7. R. Barlovic, T. Huisinga, A. Schadschneider, and M. Schreckenberg. Open boundaries in a cellular automaton model for traffic flow with metastable states. *Phys. Rev. E*, 66:046113, 2002.
8. G. M. Schütz D. Helbing, D. Mukamel. Global phase diagram of a one-dimensional driven lattice gas. *Phys. Rev. Lett.*, 82:10, 1999.
9. B.S. Kerner. Complexity of synchronized flow and related problems for basic assumptions of traffic flow theories. *Netw. Spat. Econom.*, 1:35–76, 2001.
10. J. T. Ariaratnam and S. H. Strogatz. Phase diagram for the winfree model of coupled nonlinear oscillators. *Phys. Rev. Lett.*, 86:4278–4281, 2001.
11. A. Schadschneider, W. Knospe, L. Santen, and M. Schreckenberg. Optimization of highway networks and traffic forecasting. *Physica A*, 346:165–173, 2005.
12. E. Brockfeld, R. Barlovic, A. Schadschneider, and M. Schreckenberg. Optimizing traffic lights in a cellular automaton model for city traffic. *Phys. Rev. E*, 64:056132, 2001.
13. W. Knospe, L. Santen, A. Schadschneider, and M. Schreckenberg. Single-vehicle data of highway traffic: Microscopic description of traffic phases. *Phys. Rev. E*, 65:056133, 2002.

14. R. Chrobok, O. Kaumann, J. Wahle, and M. Schreckenberg. Three categories of traffic data: Historical, current, and predictive. In E. Schnieder and U. Becker, editors, *9th IFAC Symposium Control in Transportation Systems*. IFAC (Braunschweig), 2000.

15. R. Chrobok, J. Wahle, and M. Schreckenberg. Traffic forecast using simulations of large scale networks. In B. Stone, P. Conroy, and A. Broggi, editors, *4th International IEEE Conference on Intelligent Transportation Systems*. IEEE (Oakland), 2001.

Observation, Theory and Experiment for Freeway Traffic as Physics of Many-Body System

Y. Sugiyama[1], A. Nakayama[2], M. Fukui[3], K. Hasebe[4], M. Kikuchi[5], K. Nishinari[6], S.-i. Tadaki[7], and S. Yukawa[8]

[1] Graduate School of Information Science – Nagoya University
Nagoya 464-8601 – Japan
genbey@eken.phys.nagoya-u.ac.jp

[2] Gifu Keizai University
Ohgaki, Gifu 503-8550 – Japan
g44153g@cc.nagoya-u.ac.jp

[3] Nakanihon Automotive College
Motosu-gun, Gifu 503-8550 – Japan
g44153g@cc.nagoya-u.ac.jp

[4] Faculty of Business Administration – Aichi University
Miyoshi, Aichi 3 – Japan
hasebe@vega.aichi-u.ac.jp

[5] Cybermedia Center – Osaka University
Toyonaka 560-8531 – Japan
kikuchi@cmc.osaka-u.ac.jp

[6] Department of Applied Mathematics and Informatics – Ryukoku University
Shiga 520-2194 – Japan
knishi@rins.ryukoku.ac.jp

[7] Computer and Network Center – Saga University
Saga 840-8502 – Japan
tadaki@cc.saga-u.ac.jp

[8] Department of Applied Physics – University of Tokyo
Bunkyo 113-8656 – Japan
yukawa@ap.t.u-tokyo.ac.jp

Summary. The importance of physical viewpoint for understanding the phenomena of freeway traffic is emphasized using the simulations of a mathematical model and the experiment. The traffic flow is understood as a many-body system of moving particles with the asymmetric interaction and the emergence of jam is the pattern formation as the phase transition of non-equilibrium system by the effect of collective motions.

Keywords: traffic flow, phase transition, many-body system, simulation, experiment

1 Introduction

1.1 Observational Data

The phenomena of traffic flow provide several interesting problems in physics. One of the most attractive problems is the behavior of freeway traffic, which includes the emergence of "phantom jam".

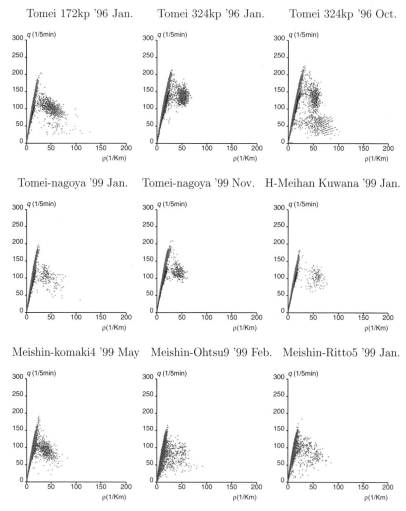

Fig. 1. Q-k diagrams by one-month data at several points in several times. The original data is measured by Japanese Highway Cooperation.

The well known property of the relation of vehicle-density and flow-rate, which is so called the q-k relation (fundamental diagram), is a very universal phenomenon in freeway traffic. The traffic flow is generally divided into two parts in terms of the vehicle-density, so called the free flow and the jam flow. As in Fig. 1, the shape of q-k curve is similar in any highway, and especially the critical vehicle-density which separates the free flow from the jam flow, has almost the same value at any point in any highway, wherever jamming phenomenon usually appears. Moreover, the velocity of jam cluster is almost the same in several cases, as in Fig. 2. The universality of such properties indicates the traffic phenomena can be studied from the physical point of view.

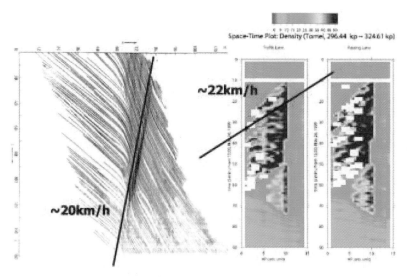

Fig. 2. The jam measured using aerial photograph in 1965 by J. Treiterer et al. Interin Report EEs 278-3, Columbus; Ohio State University, 1970 (Left). The jam measured by loop-coil detectors in Tomei Highway by Japanese Highway Coopera- tion in 1999 (Right).

1.2 Basic Concepts for Physics of Traffic

Physically, traffic flow is nothing but the moving particles with the asymmetric interaction for exclusive effect. The realistic interpretation of the property means simply that a vehicle slows down at a short headway distance. This is enough for us to describe the basic properties for freeway traffic flow. The physical aspect of forming jam can be considered as the phase transition or pattern formation of non-equilibrium system of such interacting particles. The phenomena are caused by the effect of collective motion in many-body system. Then, we introduce the minimal model for traffic flow.

2 Mathematical Model of Traffic Flow (OV Model)

On the basis of the guiding principle of constructing a model for freeway traffic, we express the minimal model as the following mathematical formulation. [1–3]. The model is so called Optimal Velocity Model, which is very simple but equipped with the essential feature to explain the behavior of traffic flow.

The OV model is 1-dimensional chain of interacting particles (vehicles) described by the set of following equations,

$$\frac{d^2 x_n}{dt^2} = a\left\{ V(\Delta x_n) - \frac{dx_n}{dt} \right\}, \tag{1}$$

where x_n and $\Delta x_n \equiv x_{n+1} - x_n$ are the position and the headway of nth vehicle respectively. $V(\Delta x)$ is called the OV function, which represents the optimal velocity of the vehicle with headway Δx. The parameter a is called sensitivity. A driver controls a vehicle to reduce the difference between the real velocity and the optimal velocity determined by OV function.

The OV function should satisfy several conditions. The function is monotonically increasing, which means that if a headway is small a velocity should be small, and that if a headway is large a velocity is allowed to become larger. But a vehicle should have the maximum velocity, so the function has the maximum value V^{max} with $\Delta x_n \to \infty$. The function makes a vehicle gradually accelerating in order to prevent from the crash to the vehicle ahead. So, the OV function is a sigmoidal function, which is typically taken as a hyperbolic tangent function,

$$V(\Delta x) = v_0\{\tanh(\Delta x - c) + \tanh c\}. \tag{2}$$

3 Analysis of OV Model in Periodic Boundary

In the periodic boundary condition with no bottleneck, we have two kinds of solutions. One is the homogeneous flow solution, which can be identified as the free flow in real traffic, and the other is the jam flow solution.

3.1 The Instability of Homogeneous Flow and the Enhancement of Fluctuation

The homogeneous flow solution is expressed as

$$x_n = bn + V(b)t + \text{const.}, \tag{3}$$

where all the vehicles run with the same headway b at the same velocity $V(b)$. The linear analysis derives the condition that the homogeneous flow solution is unstable. The condition is described using a mode of (wave-length)$^{-1}$, $k = 2\pi n/N (n = 1, 2, \cdots, N)$ for the frequency of the vehicle-density wave as

$$a < 2V'(b)\cos^2\left(\frac{k}{2}\right),\tag{4}$$

where V' is the derivative of V [1]. The condition means that the modes of the long wave-length of fluctuation, which can be hardly recognized by drivers, make the homogeneous flow unstable. Actually, such a long wave-length mode plays the essential role for the enhancement of fluctuation as in Fig. 3, which breaks the stability of a state in many-body system and makes another state stable. The break-down phenomenon of the free flow is explained by the effect of such collective motions, which is mathematically derived in the successful models of traffic flow.

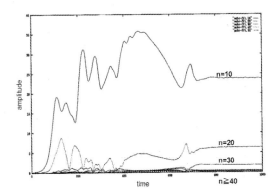

Fig. 3. The evolution of Fourier mode amplitudes for $k = 2\pi n/N$ with time development. After the jam flow is stable, the amplitude of each mode is preserved.

3.2 Forming Jam in a Periodic Boundary with no Bottleneck

The homogeneous flow solution becomes unstable if the average density exceeds the critical value expressed by the condition $a < 2V'(b)$. In this case we can observe the process of forming jam by the simulation in Fig. 4 (left).

Initially all vehicles move at the same velocity with the same headway. As the homogeneous flow is broken, the fluctuation is enhanced and the small jam clusters appear. After small jam clusters are combined, finally the stable jam flow appears. After the relaxation time, all jam clusters move with the same velocity opposed to the direction of the vehicle.

In the phase space of headway and velocity, all vehicles are initially at the same point. As the homogeneous flow is broken, the vehicle-points gradually scatter and finally the jam flow solution appears as the trajectory in Fig. 4 (right). In the jam flow solution all vehicles are moving along the specific closed curve in the direction of the arrows, so called "hysteresis loop" [2, 3], which is a kind of limit cycle. In a different initial condition, we obtain the

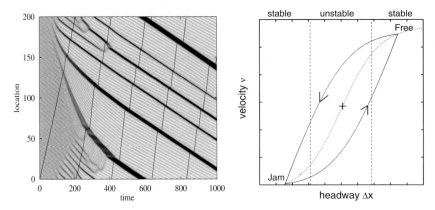

Fig. 4. Space-time plot of all vehicles for the process of forming jam in the periodic boundary. The thin line is the trajectory of a sample vehicle. (Left): The plot in the phase space (headway, velocity) of the jam flow solution. The size of hysteresis loop depends on a. In this paper we fix a as the appropriate value (Right).

different figure of space-time plot for the process of forming jam. But we always recognize the universal profile of jam flow solution in the phase space.

The jam flow solution is obtained on the condition of periodic boundary. The result may seem to be mathematical idealization, and not realistic. But the mechanism of forming jam in periodic boundary obtained in the mathematical model is essential and can be realized. The system in periodic boundary can be naturally identified as the chain of vehicles moving on a circuit with no bottleneck. We check whether the jamming phenomena really occurs in such a mathematical situation.

4 Experiment of Forming Jam on a Circuit

We perform an experiment whether the emergence of jam really occurs on a circuit with no bottleneck. In the situation, the spontaneous enhancement of fluctuation caused by the collective motion of vehicles is observed. The condition for the number of vehicles N and the length of circle L that the homogeneous flow is unstable, can be predictable by the theoretical analysis of the mathematical model. We set $L = 230m$ and $N = 22$. Figure 5 is the snapshot of the experiment. We set the all-around scope mirror built in a video camera at the center of the circle (Fig. 6).

In the beginning of the experiment the vehicles were tuned to move with the same velocity and uniformly distributed on the circle. Then, the performance started. In several minutes, the homogeneous flow was maintained as in Fig. 7 (Top). The small fluctuations of density wave appeared and the fluctuation was gradually enhanced. Finally, a clear stop-and-go wave was observed and stable, which was nothing but a jam cluster, as in Fig. 7 (Bottom).

Fig. 5. The snapshot of the circuit-experiment.

Fig. 6. The all-around scope mirror built in the video camera.

In the experiment we need the relaxation time that the fluctuation is enhanced enough to break down the homogeneous flow and induce the jam. The vehicles in the jam cluster stopped completely and the vehicles outside the jam move smoothly. The top vehicle of the jam starts to accelerate and escapes from the jam, while another vehicle reaches the bottom of the jam. The repetition of such movement preserves the stability of the jam cluster. And the cluster itself propagates in the opposite direction of vehicles.

In the situation of the experiment on the circle, small fluctuation moves around the circle and becomes gradually enhanced, finally the state (the free flow) is broken and the other state (the jam flow) becomes stable. The phenomenon is caused by the effect of collective motion of many-body system. We understand that the mechanism of forming jam is realistic by comparing the experiment on the circle with the simulation of the mathematical model.

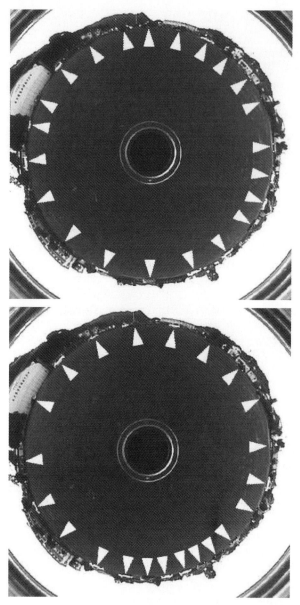

Fig. 7. The pictures taken by the all-around scope mirror of the circuit-experiment. The snapshot at the stage of the homogeneous movement (Top). The snapshot after 5 minutes shows that the jam is formed and stable (Bottom).

5 Forming Jam in Open Boundary with a Bottleneck

In the real traffic flow in highways, we usually observe a jam on the upstream of a bottleneck. A bottleneck restricts the flow-rate of traffic. Then, vehicles slow down and a jam appears. Is the mechanism of forming jam in such a realistic situation is different from the previous experiment?

We think that it is essentially the same as in the periodic boundary with no bottleneck. A bottleneck is not essential for the emergence of jam in the physical point of view. A bottleneck is one of situations to induce the growth of fluctuation. The local increasing of vehicle-density caused by a bottleneck induces the enhancement of fluctuation, and the state of free flow is broken down. This is the effect of collective motion in many-body system just the same as in the circuit. We show the above scenario is true by the simulations together with real data.

5.1 Realistic Simulation in a Bottleneck

We perform the simulations of mathematical model in open system with bottleneck [4, 5]. We apply the OV model to a realistic situation by setting OV function as

$$V(\Delta x) = v_0 \left[\tanh g(\Delta x - d) - \tanh g(l_c - d) \right] \tag{5}$$

We chose the values of the above parameters as follows. They are determined by a data for a car-following measurement by M. Koshi et.al in Chuo Motorway in Japan (1983). a is also determined by a realistic size of the hysteresis loop.

$l_c = 5m$: the length of vehicle
$d = 25m$: the inflection point of OV-function
$g = 0.086$: the gradient at d
v_0 : the maximum velocity
$$v_{max} = v_0 \left[1 - \tanh g(l_c - d) \right]$$
$$= 33.6m/sec = 115km/h$$

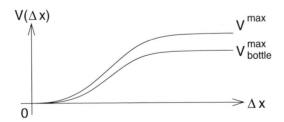

Fig. 8. The OV function in the bottleneck comparing with the original one.

In the region of a bottleneck (Here, we suppose a tunnel as a bottleneck.), the optimal velocity is reduced comparing with that outside a bottleneck,

$V_{bottle}(\Delta x) < V(\Delta x)$, as in Fig. 8. The condition is natural from the view point of psychological effect for drivers as well as the traffic regulations. The suppression-rate of bottleneck is measured by the ratio of the maximum velocities as $S_{bottle} \equiv V_{bottle}^{max}/V^{max} < 1$.

5.2 The Structure of Traffic Flow with a Bottleneck and the Universality of Jam

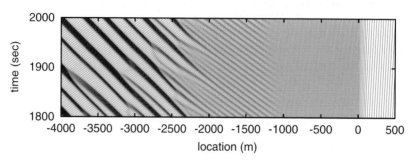

Fig. 9. The three stable patterns of traffic flow upstream of the bottleneck The bottleneck ($100m$) is placed at $0m$-point. Vehicles move from left to right.

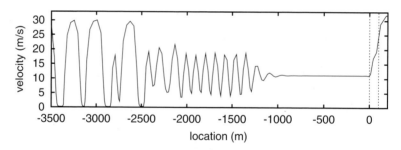

Fig. 10. The snapshot of the velocity-distribution upstream of the bottleneck

Figure 9 is the result of simulation, which shows the stable structure of traffic flow upstream of the bottleneck. After relaxation time the three distinct spatial-temporal patterns of flow are formed. In the immediate upstream of the bottleneck a slow and relatively high-density homogeneous flow appears. We remark that the flow is not a jam flow, but a uniform flow, which is the solution (3) of OV model. The homogeneous flow with such density is not stable on a circuit, but stabilized by a bottleneck. In further upstream, the "oscillatory wave flow" is observed as the fine stripes in Fig. 9, which means that the density is oscillatory changing with the small period. The wave is followed by the jam flow, where jam clusters appear clearly. The patterns are also recognized by the distribution of velocity on the lane in Fig. 10. The similar structure is obtained by the investigation of OV model with localized perturbation as an on-ramp instead of a tunnel [6, 7]. The effect of localized perturbation seems to be essentially the same as the scale-down OV function as a bottleneck. The structure observed upstream of the bottleneck in OV model is quite similar to the real traffic data for the phenomena of so called "synchronized flow" on highways with tunnel or on-ramp [8–10].

Figure 11 is the headway-velocity plot for the jam flow (Left) and the "oscillatory wave flow"(Right) observed upstream of the bottleneck. The profile of the jam flow on the circuit (Fig. 4) and that on the open lane with the bottleneck in Fig. 11 (Left) are just the same. The jam flow is universal either in the system on a circuit or in the open system with a bottleneck.

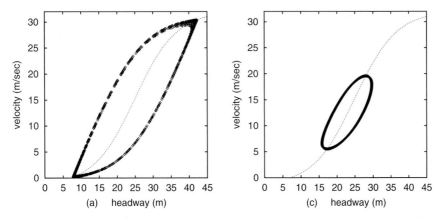

Fig. 11. The phase space (headway-velocity) plots of the flow at -3500m (Left) and -1500m (Right) upstream of the bottleneck.

The small closed loop in Fig. 11 (Right) seems to be a kind of limit cycle solution as the jam flow solution is. The small limit cycle has not been observed in the simulation on a circuit. This implies that this solution is 'originally' unstable. The existence of such an unstable solution is consistent with that the jamming transition in OV model is subcritical Hoph bifurcation [11, 12]. It is very interesting possibility that the unstable solution is convectively stabilized by a bottleneck and related to the phenomena of so called "synchronized flow" [4, 5].

5.3 The Velocity of Jam Cluster Comparing the Simulation Result with Real Data

Figure 12 shows the comparison of the velocity of jam cluster between the result by the simulation of the system with a bottleneck in Fig. 9 and the real data of the aerial photograph in Fig. 2. The velocity obtained by the simulation of the mathematical model is quite similar to the observed real data. The OV model well describes the properties of the jam cluster in real traffic, either in the system of the periodic boundary or in the open boundary with a bottleneck, and explains the universality of jam.

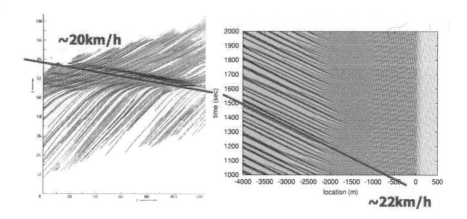

Fig. 12. The velocity of jam cluster upstream of the bottleneck by the simulation (Right) comparing with the real data of the aerial photograph in Fig. 2 (Left).

5.4 Fundamental Diagram by the Simulation in a Bottleneck and Real Data

Figure 13 shows the q-k(flow-density) plots of the flow upstream of a bottleneck by the simulation together with real data measured at the upstream of the tunnel on Tomei Highway by Japanese Highway Cooperation in 1999. The data of the simulation are taken along the sequence of the three stable patterns of the flow for two suppression-rates as $S_{bottle} = 0.7$ and 0.65. In the region between the green lines, the homogeneous flow is originally unstable in the absence of a bottleneck. However, in the case of the existence of a bottleneck, such the homogeneous flow with low-velocity and high-density is stabilized, as in Figs. 9 and 10. The data points of the upstream are spread in the horizontal direction (the same flow rate but the different vehicle-densities) in q-k plane corresponding to the sequence of the three patterns. Moreover, the data points for the different suppression-rates of bottleneck are spread in the vertical direction So, the data points in q-k plane are generally spread in two dimensional space caused by the effect of a bottleneck, which is the specific property for q-k plot observed in real data upstream of a bottleneck.

The flow of the "oscillatory wave" has relatively small amplitude comparing with the jam flow, and it is localized at some distance from the homoge-

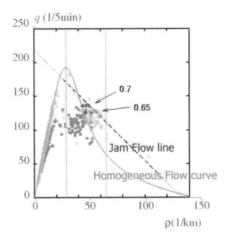

Fig. 13. The plots of 5 min.-averaged data on the upstream of the bottleneck by the simulation data (red points) and the real data (the marks on background). The red curve and the dashed line are the q-k relations for the homogeneous flow and the jam flow solutions, respectively. The region between green lines is the unstable-region for the homogeneous flow in the absence of a bottleneck.

neous flow upstream of the bottleneck. In q-k plot, the average density varies with the same flow-rate. These properties strongly indicate the possibility to understand so called "synchronized flow" as the "oscillatory wave" solution of OV model with a bottleneck [4, 5].

6 Summary and Discussion

A vehicle is a machine controlled by a driver (human). However, the variety of his/her character can be almost ignored in traffic flow and a vehicle plays only a particle with some specific interaction, which leads the softly exclusive effect with each other. Traffic flow should be identified as a many-body system of such kind of interacting particles. In the physical viewpoint, the emergence of jam is essentially the effect of the collective motion. Physical investigations of traffic flow have revealed the essential features to describe the phenomena of freeway traffic. Such studies also can provide a lot of information to the research and development of engineering in real traffic systems.

Acknowledgments

The authors thank very much for the students of Nakanihon Automotive College who cooperate the experiment of forming jam on the circuit as the drivers of vehicles. This work was partly supported by a Grant-in-Aid for Scientific Research (C) (No. 15560051) of the Japanese Ministry of Education, Science, Sports and Culture.

References

1. M. Bando, K. Hasebe, A. Nakayama, A. Shibata, and Y. Sugiyama. Dynamical model of traffic congestion and numerical simulation. *Phys. Rev.* **E**, 51:1035, 1995.
2. M. Bando, K. Hasebe, A. Nakayama, A. Shibata, and Y. Sugiyama. Structure stability of congestion in traffic dynamics. *Japan J. Indust. Appl. Math.*, 11:203, 1994.
3. Y. Sugiyama. Dynamical model for congestion of freeway traffic and its structural stability. In D.E. Wolf, M. Schreckenberg, and A. Bachem, editors, *Traffic and Granular Flow*, page 137. World Scientific, Singapore, 1995.
4. Y. Sugiyama and A. Nakayama. Modeling, simulation and observations for freeway traffic and pedestrian. In H. Emmerich, B. Nestler, and M. Schreckenberg, editors, *Computational physics of transport, and interface dynamics*, page 406. Springer-Verlag, 2003.
5. Y. Sugiyama and A. Nakayama. Understanding "synchronized flow" by optimal velocity model. In P.L. Garrido and J. Marro, editors, *Modeling of Complex Systems*, page 111. American Institute of Physics, 2003.
6. M. Mitarai and H. Nakanishi. Spatiotemporal structure of traffic flow with an open boundary. *Phys. Rev. Lett.*, 85:1766, 2000.
7. M. Mitarai and H. Nakanishi. Convective instability and structure formation in traffic flow. *J. Phys. Soc, Jpn.*, 65:3752, 2000.
8. M. Koshi, M. Iwasaki, and I. Ohkura. Some findings and overview on vehicular flow characteristics. In V. F. Hurdle and G. N. Stewart, editors, *Proceedings of the 8th International Symposium on Transportation and Traffic Flow Theory*, page 403. University of Toronto Press, Toronto, 1983.
9. B.S. Kerner and H. Rehborn. Experimental properties of phase transitions in traffic flow. *Phys. Rev. Lett.*, 79:4030, 1997.
10. B.S. Kerner. The physics of traffic. *Physics World*, 12:25, 1999.
11. Y. Sugiyama and H. Yamada. Aspects of optimal velocity model for traffic flow. In M. Schreckenberg and D.E. Wolf, editors, *Traffic and Granular Flow '97*, page 301. Springer-Verlag, Singapore, 1998.
12. P. Berg and E. Wilson. Wave solutions and metastability of the ov model: a bifurcation viewpoint. In *this volume*. 2005.

Fluctuation in Expressway Traffic Flow

S. Tadaki[1], M. Kikuchi[2], A. Nakayama[3], K. Nishinari[4], A. Shibata[5],
Y. Sugiyama[6], and S. Yukawa[7]

[1] Computer and Network Center – Saga University
   Saga 840-8502 – Japan
   tadaki@cc.saga-u.ac.jp
[2] Cybermedia Center – Osaka University
   Toyonaka 560-0043 – Japan
[3] Department of Administrative Information Sciences – Gifu Keizai University
   Ogaki 503-8550 – Japan
[4] Department of Applied Mathematics and Infomatics – Ryukoku University
   Ohtsu 520-2194 – Japan
[5] Computing Research Center – High Energy Accelerator Research Organization
   (KEK), Oho 1-1, Tsukuba, Ibaraki 305-0801 – Japan
[6] Graduate School of Information Science – Nagoya University
   Nagoya 464-8601 – Japan
[7] Department of Applied Physics – University of Tokyo
   Bunkyo 113-8656 – Japan

Summary. The temporal data of the expressway traffic data are complex mixtures of various time scales. The periodic appearances of the congestion, for example, reflect social activities. We need some filters to extract proper fluctuations from the raw data. We employ the method of detrended fluctuation analysis for studying the time sequence of the traffic flow. We find the long-range correlation from 1 hour to 24 hours.
Keywords: Traffic Flow, Observation, Detrended Fluctuation Analysis

1 Introduction

The traffic flow has been said to bear the $1/f$ fluctuation. The $1/f$ behavior has been observed in simulation results [1–3]. The granular flow in vertical pipes shows the $f^{-4/3}$ fluctuation [4]. Since the early observation by Musha and Higuchi [5], however, the limited number of studies on the $1/f$ fluctuations in the expressway traffic flow has been done [6–9]. It is still unclear which time-scale region the $1/f$ fluctuation appears in.

The observational temporal behavior of the expressway traffic data are complex mixtures of various time scales. The response time of drivers is of order 0.1 second. The phase transition between the free-flow and congestion states takes of order 10 minutes. The periodic appearances of the congestion seem to reflect human social activities. We are interested in the dynamical fluctuations contained in time sequences of the traffic flow. We need some filters to extract proper fluctuations from the raw data.

One of methods for subtracting a trend from raw sequential data is the *detrended fluctuation analysis* (DFA) method [10]. By employing DFA, we study the long-range correlation hidden in the raw data of the traffic flow.

2 Detrended Fluctuation Analysis

The detrended fluctuation analysis (DFA) has been invented for analyzing long-range correlations in DNA sequences [10]. The method has been applied to various time series with non-stationarities.

The method is described as follows: The profile $y(t)$ of the raw temporal data $u(t)$ is defined by

$$y(t) = \sum_{i=0}^{t} [u(i) - \langle u \rangle],$$
(1)

where $\langle u \rangle$ is the average value of $\{u(t)\}$. Then the entire time sequence $y(t)$ of length T is divided into T/l nonoverlapping segments of length l. The *local trend* in each segment is defined by fitting data in the segment with linear least square method. The *detrended sequence* $y_l(t)$ is defined as the difference between the original sequence $y(t)$ and the local trend.

The variance of the detrended sequence is defined by

$$F^2(l) = \frac{1}{T} \sum_{t=0}^{T-1} y_l^2(t).$$
(2)

By analyzing the dependence of the variance $F(l)$ on the length l, we find the correlations in non-stationary time sequences.

3 Observational Data

We analyze the observational data provided by the Japan Highway Public Corporation [11] in this paper. The data was taken at 468km point near Seta East IC in Meishin (Nagoya-Kobe) Expressway. There were two lanes bound for Kobe (West). The time sequence of the flow and the average velocity for 5 minutes on Aug. 11th, 1999, are shown in Figs. 1 and 2. Very sharp traffic congestion in the morning and the evening is found in the time-sequence of the velocity. No significant behavior of congestion, however, can be found in the time-sequence of the flow.

The Detrended Fluctuation Analysis is applied to the one year data of the traffic flow observed there. The profile $y(t)$ is constructed from the raw data

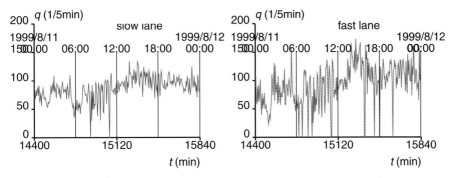

Fig. 1. The time sequence of the traffic flow observed on Aug. 11th, 1999 for the fast lane (right) and the slow lane (left).

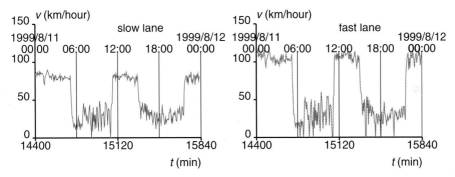

Fig. 2. The time sequence of the average velocity observed on Aug. 11th, 1999 for the fast lane (right) and the slow lane (left).

$q(t)$ of the flow. Figure 3 shows one example of the local trend of length 8 hours.

Figure 4 shows the dependence of the variance $F(l)$ on the segment size l. In the region from one hour to 24 hours, the long range correlation with the exponent $\alpha \sim 1.5$ is observed. The correlation with $\alpha > 1/2$ corresponds to a sequence in which the positive (negative) fluctuations tend to induce the positive (negative) ones [12]. The correlation with $\alpha > 1$ indicates that the correlation is not scale-free.

Another range between one day and 15 days is found to be with $\alpha \sim 0.5$, which corresponds to random walk around the local trend. In the area larger than 15 days, the exponent takes a close value $\alpha = 1$, which corresponds to $1/f$ fluctuation.

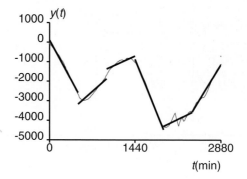

Fig. 3. The profile $y(t)$ of the flow on the slow lane and the local trend. The bold line corresponds to the local trend of 8 hours.

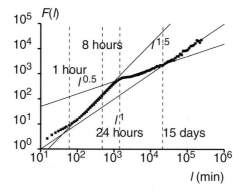

Fig. 4. The dependence of the variance $F(l)$ on the segment size l for the traffic flow data of 1999.

4 Daily Trend Subtraction

There are some possible methods for subtracting *local trend* in the traffic flow data. Here we show another simple method for subtracting the *daily trend*, which is the most fundamental periodic behavior. We call it the *Daily Trend Subtraction*.

The daily trend $q'(t)$ is defined as a average trend given by

$$q'(t) = \frac{1}{D} \sum_{d} q(1440 \times d + t), \tag{3}$$

where $0 \leq d < D$ is the day of year and t runs from 0 to 1440 (one day) minutes. Note that our data are taken every 5 minutes.

The average power spectrum of the fluctuation around the daily trend is given by

$$I_q^2(k) = \frac{1}{D} \sum_d \tilde{q}_d^2(k), \tag{4}$$

where the Fourier transform $\tilde{q}_d(k)$ is defined as

$$\tilde{q}_d(k) = \sum_t [q(1440 \times d + t) - q'(t)] e^{-2\pi kt}. \tag{5}$$

The average power spectrum of the fluctuation around the daily trend, shown in Fig. 5, indicates the existence of the long-range correlation longer than one hour.

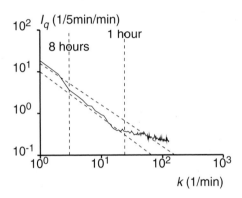

Fig. 5. The power spectrum of the fluctuation around the daily trend.

5 Summary and Discussion

We analyze the time sequence of the traffic flow with the detrended fluctuation analysis (DFA) method and the method of daily trend subtraction. We found some characteristic time scales: one hour, 1 day and 15 days.

The long range correlation is found to be expanding from one hour to 24 hours by DFA. In the area shorter than 30 minutes, we can find other type of correlation. The shortest time of the data is 5 minutes. Therefore the number of data points, however, is too few to analyze such a short range correlation.

The long range correlation is not a result of congestion. Figure 6 is the result for another observational point, which is the 133.11 km point near Taki IC of Ise Expressway. The traffic flow at the point runs smoothly without congestion. Moreover the road has only one lane at the observational point. The features appearing in Fig. 6 are almost the same as those in Fig. 4.

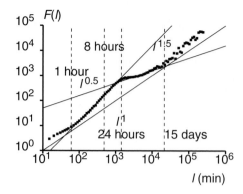

Fig. 6. The dependence of the variance $F(l)$ on the segment size l for the traffic flow data near Taki IC of 1999 .

Acknowledgments

The authors thank the Japan Highway Public Corporation for providing us the observation data. A part of this work is financially supported by Grant-in-aid No. 15607014 from Ministry of Education, Science, Sports and Culture, Japan.

References

1. S. Yukawa and M. Kikuchi. Density fluctuations in traffic flow. *J. Phys. Soc. Jpn.*, 65:916–919, 1996.
2. S. Tadaki, M. Kikuchi, Y. Sugiyama, and S. Yukawa. Coupled map traffic flow simulator based on optimal velocity functions. *J. Phys. Soc. Jpn.*, 67:2270–2276, 1998.
3. S. Tadaki, M. Kikuchi, Y. Sugiyama, and S. Yukawa. Noise induced congested traffic flow in coupled map optimal velocity model. *J. Phys. Soc. Jpn.*, 68:3110–3114, 1999.
4. O. Moriyama, N. Kuroiwa, M. Matsushita, and H. Hayakawa. 4/3 law of granular particles flowing through a vertical pipe. *Phys. Rev. Lett.*, 80:2833–2836, 1998.
5. T. Musha and H. Higuchi. The $1/f$ fluctuation of a traffic current on an expressway. *Jpn. J. Appl. Phys.*, 15:1271–1275, 1976.
6. P. Wagner and J. Peinke. Scaling properties of traffic-flow data. *Z. Naturforsch.*, 52 a:600–604, 1997.
7. L. Neubert, L. Santen, A. Schadschneider, and M. Schreckenberg. Single-vehicle data of highway traffic: A statistical analysis. *Phys. Rev.*, E60:6480–6490, 1999.
8. K. Nishinari and M. Hayashi, editors. *Traffic Statistics in Tomei Express Way*. The Mathematical Society of Traffic Flow, Japan, 1999.
9. M. Kikuchi, A. Nakayama, K. Nishinari, Y. Sugiyama, S. Tadaki, and S. Yukawa. Long-term traffic data from Japanese expressway. In M. Fukui, Y. Sugiyama, M. Schreckenberg, and D.E. Wolf, editors, *Traffic and Granular Flow '01*, pages 257–262. Springer-Verlag, Berlin, 2003.

10. C.-K. Peng, S. V. Buldyrev, S. Havlin, M. Simons, H. E. Stanley, and A. L. Goldberger. Mosaic organization of DNA mucleotides. *Phys. Rev.*, E49:1685–1689, 1994.
11. The Japan Highway Public Corporation, http://www.jhnet.go.jp.
12. S. Havlin, R. Blumberg Selinger, M. Schwartz, H. E. Stanley, and A. Bunde. Random multiplicative processes and transport in structures with correlated spatial disorder. *Phys. Rev. Lett.*, 61:1438–1441, 1988.

Calibration and Validation of Microscopic Traffic Flow Models

E. Brockfeld and P. Wagner

Institute of Transport Research – German Aerospace Center
Rutherfordstr. 2, 10245 Berlin – Germany
Elmar.Brockfeld@dlr.de, Peter.Wagner@dlr.de

Summary. The aim of this paper is to present recent progress in calibrating ten microscopic traffic flow models. The models have been tested using data collected via DGPS-equipped cars (Differential Global Positioning System) on a test track in Japan. To calibrate the models, the data of a leading car are fed into the model under consideration and the model is used to compute the headway time series of the following car. The deviations between the measured and the simulated headways are then used to calibrate and validate the models. The calibration results agree with earlier studies as there are errors of 12 % to 17 % for all models and no model can be denoted to be the best. The differences between individual drivers are larger than the differences between different models. The validation process leads to errors from 17 % to 22 %. But for special data sets with validation errors up to 60 % the calibration process has reached what is known as "overfitting": because of the adaptation to a particular situation, the models are not capable of generalizing to other situations.

Keywords: microscopic models, car following, DGPS, calibration, validation

1 Introduction

Microscopic simulation models are becoming increasingly important tools in modeling transport systems. There is a large number of available models used in many countries. The most difficult stage in the development and use of such models is the calibration and validation of the microscopic sub-models describing the traffic flow, such as the car following, lane changing and gap acceptance models. This difficulty is due to the lack of suitable methods for adapting models to empirical data. The aim of this paper is to present recent progress in calibrating a number of microscopic traffic flow models. By calibrating and validating various models using the same data sets, the models are directly comparable to each other. This sets the basis for a transparent benchmarking of those models. Furthermore, the advantages and disadvantages of each model can be analyzed better to develop a more realistic behavior of the simulated vehicles.

In this work ten microscopic traffic flow models were tested from a very microscopic point of view concerning the car-following behavior and gap-acceptance. This is in contrast to a typical macroscopic analysis which compares aggregated data on links for example. The data used for calibration and validation is from car-following experiments conducted in Japan in October 2001 [1]. The data have been collected by letting nine DGPS-equipped cars follow a lead car driving along a 3 km test track for about 15-30 minutes.

At first the experiments on the test track and the recorded data sets are briefly described and the simulation setup for testing the models is defined. In the following the measurement procedure for calculating the error differences between the recorded data and the data produced by the models is specified. After the tested models are listed and basically described, the calibration and validation results are presented leading to some conclusions.

2 Data and Error Measurement

2.1   The Data and the Simulation set-up

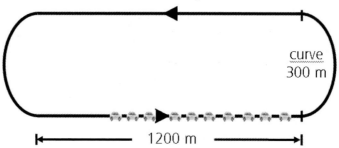

Fig. 1. Sketch of the test track with ten cars driving on the course.

The data sets have been recorded on a test track in Hokkaido, Japan in October 2001 [1]. Eight experiments have been conducted, where nine cars drove on a 3 km test track (2 x 1.2 km straight segments and 2 x 0.3 km curves; see Fig. 1) for about 15-30 minutes in each experiment following a lead car, which performed some driving patterns. These are for example driving with constant speeds of 20, 40, 60 and 80 km/h for some time, varying speeds (regularly increasing/decreasing speed) and emulating many accelerations/decelerations as they are typical at intersections. The regulary increase/decrease of speed is done with different frequencies, the velocity cycles from 20 to 60 km/h being performed one to four times on the straight segments.

To minimize driver-dependent correlations between the data sets, the drivers were exchanged between the cars after each experiment. Having all cars equipped with DGPS (Differential Global Positioning System), the position of each car is stored in 0.1 second intervals throughout each experiment. From these position data other important variables like the speed, the acceleration and the headway between the cars were extracted for simulation purposes. The accuracy of the DGPS is about 1 cm and the appointment of the speeds has got an error of less than 0.2 km/h as described in [1]. Thus, the data sets have got such a high resolution that they are adequate for the analysis of car-following behavior and calibration of car-following models.

In this paper we present analyses concerning four of the eight experiments, namely the patterns mostly with intervals of constant speeds and waveperforming. For the simulation set-up only two cars are considered at a time: the leading car is updated according to the speeds and positions in the recorded data sets and the following car is updated as defined by the equations of the used model.

2.2   Error Measurement

The absolute error a model produces in comparison to a measured data set is calculated via the simple distance between a recorded time series and a simulated time series of gaps. To get a percentage error it is additionally related to the average value of the time series in each particular data set:

$$e = \frac{\dfrac{1}{T}\sum_{t=0}^{T}\left|x^{(sim)}(t)-x^{(obs)}(t)\right|}{\dfrac{1}{T}\sum_{t=0}^{T}x^{(obs)}(t)},\qquad(1)$$

where $x^{(sim)}$ and $x^{(obs)}$ are a simulated and an observed traffic flow variable, which is in this case the gap between two cars. T is the time series over the total time of each experiment.

3   The Models

The models used for the simulations are all microscopic traffic flow models, which describe the behavior of a following car in relation to a leading car. For the vehicle movement, typically equations like the following were used, defining the new speed of a vehicle at time $t+\Delta t$, depending on the values of some variables at time t:

$$v(t+\Delta t) = f(g(t), v(t), V(t), \{p\})$$

$$g(t+\Delta t) = V(t) - v(t), \qquad(2)$$

where v is the speed of the following and V that of the leading car, respectively, and g is the headway between the cars. The symbol $\{p\}$ denotes a set of parameters of the model under consideration.

In the calibration approach the following microscopic traffic flow models of very different kind with 3 to 15 parameters have been tested. Some models are used in commercial simulation programs, which are popular in European countries, the USA and Japan, and some are scientific simulation approaches.

Abbreviation	Description	params
CA0.1	cellular automaton model [2]	4
SK_STAR	model based on the SK-model by S. Krauss [3]	7
OVM	"Optimal Velocity Model", Bando, Hasebe [4]	4
IDM	"Intelligent Driver Model" [5]	7
IDMM	"Intelligent Driver Model with Memory" [6]	7
Newell	Can be understood as a continuous CA with more variable acceleration and deceleration [7,8]	7
GIPPSLIKE	basic model by P.G. Gipps [9]	6
Aerde	used in the simulation package INTEGRATION [10]	6
FRITZSCHE	used in the British software PARAMICS; similar to what is used in the German software VISSIM [11]	13
MitSim	model by Yang and Koutsopulus, used in the software MitSim [12]	15

Tab. 1. List of tested models.

The most basic parameters used by the models are the car length, the maximum speed, an acceleration rate (except for the CA0.1-model) and a deceleration rate (for most models). The acceleration and deceleration rates are specified in more detail in some models depending on the current speed or the current headway to the leading vehicle. Furthermore, some models (CA0.1, SK_STAR and MitSim) use some kind of stochastic parameters describing individual driver behavior. Most models use something like a reaction time of the drivers to the behavior of the leading car. The MitSim and the FRITZSCHE model have got a lot of more parameters defining thresholds concerning the headway and the speed difference to a leading vehicle. Depending on these various driving behaviors are realized like "free driving", "approaching" and "emergency braking" for example.

As the time step for the models is 0.1 seconds according to the recorded data, some models with a traditional time step of 1 second - as for example used for simple cellular automatons - have been modified to adopt for an arbitrarily small time-step.

4   Calibration and Validation

Alltogether 36 vehicle pairs (four experiments, each with nine vehicle pairs) were used as data sets for the analyses of the car following behavior. Each model has been calibrated with each of the 36 different constellations separately gaining optimal parameter sets for each "model - data set" combination. To find the optimal parameter constellations a gradient-free optimization method was used [13] and started many times with different initialization values for each "model - data set" pair. This variation is done to avoid sticking with a local minimum, which of course can occur because getting a global minimum can not be guaranteed by those optimizations.

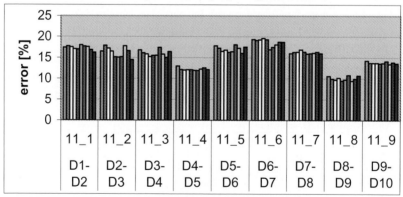

Fig. 4. Some calibration results obtained for one of the four experiments.

As an example Fig. 4 shows the calibration results obtained for the first experiment ("11"). In this case one driver pair ("11_8") can be reproduced well with errors of about 10 %. Other driver pairs like "11_6" or "11_1" are much harder to reproduce with errors up to 17-20 %. In total, for all 36 constellations, the errors mainly range from 12 % to 17 %. In nearly all cases the models do not differ so much when reproducing the behavior of a driver pair, because the average differences between the models reproducing the single driver pairs is about 2.5 percentage points. It is noteworthy that this diversity of the models is much smaller than the differences in the driver behavior (mainly about 5 percentage points), as can be seen in Fig. 4, too.

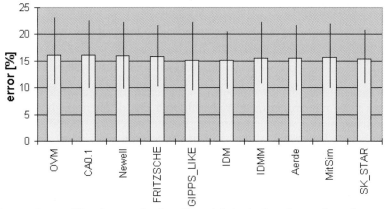

Fig. 5. Mean calibration results for all models including the total result range.

Looking at the average errors each model produces with the 36 data sets, it can be seen in Fig. 5, that, again, the differences of the models are not very big. The best model produces an error of 15.14 %, the worst one of 16.20 %. Thus, no model can be denoted to be the best and especially complex models do not produce better results than simple models.

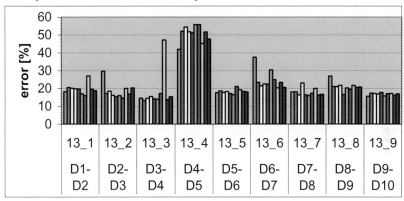

Fig. 6. Validation results using the best parameter sets of one experiment "11" and trying to reproduce the behavior of the drivers in experiment "13".

For validation purposes the optimal parameter results for the data sets in the first experiment "11" were taken to reproduce the data sets in the other three experiments. In Fig. 6 the validation errors are shown exemplarily for the reproduction of experiment "13". Except for some cases, where the parameter sets were not transferable due to very high errors ("13_4"), the validation error over all data sets mainly ranges from 17 % to 22 %, which is for the singular models about 3.2 to 5.5 percentage points higher than in the calibration cases. The average validation errors of the models range from 19.25 % (SK_STAR) to 20.72 % (IDM). Only the model by Aerde (23.13 %) and the OVM model (22.82 %) showed slightly more problems during the validation.

5   Conclusions

The error rates of the models in comparison to the data sets during the calibration for each model reach from 9 % to 24 %. Surprisingly, no model appears to be significantly better than any other model and the average error rates of the models are very close to each other between 15.1 % and 16.2 %. All models share the same problems with certain data sets while other data sets can be reproduced quite well with each model. Interestingly, it can be stated that models with more parameters do not necessarily reproduce the real data better. The results of the validation process give a similar picture. The additional errors in comparison to the calibration are – apart from singular cases of "overfitting" - mainly in the area of 3 to 5 percentage points.

The results after the calibration and the validation agree with results that have been obtained before with a completely different data set taking the travel times on road segments instead of headways for the error measurement [14]. In these studies about 15 % to 27 % were found to be the minimum calibration error and additional validation-errors were found to be about 2 to 5 percentage points. It was found, too, that out of about ten models the differences are not as big as could be expected. However, the results of the validation show, that when calibrating and validating with special data sets, the parameters of a model can be "overfitted" and thus the results can be very unsatisfactory with surprisingly high errors. The calibration tends to optimize the model for a given data-set, thereby sacrificing generality.

There are two conclusions that can be drawn. First, one should call for the development of better models. Additionally, one should think about a different calibration technique which avoids "overfitting" and could produce results which stay more general. The other way to interpret the results is that – from this microscopic point of view – errors of about 15-25 % can probably not be suppressed no matter what model is used. These are due to a really stochastic component in the driver's behavior.

References

1.   G.S. Gurusinghe, T. Nakatsuji, Y. Azuta, P. Ranjitkar, and Y. Tanaboriboon, Transp. Res. Record 1802, 166-180 (2003).
2.   K. Nagel and M. Schreckenberg, J. Physique I, 2, 2221 – 2229 (1992).
3.   S. Krauss, P. Wagner, and C. Gawron, Phys. Rev. E 55, 5597 – 5605 (1997).
4.   M. Bando, K. Hasebe, A. Nakayama, A. Shibata, and Y. Sugiyama, Phys. Rev. E 51, 239-253 (1999).
5.   M. Treiber, A. Hennecke, and D. Helbing, Phys. Rev. E 62, 1805-1824 (2000).
6.   M. Treiber and D. Helbing, Phys. Rev. E 62, 046119 (2003).
7.   G. Newell, J. Op. Soc. Japan 5, 9 – 54 (1962).
8.   G. Newell, Transp. Res. B 36, 195 – 205 (2002).
9.   P.G. Gipps, Transp. Res. B 15, 105 – 111 (1981).
10.   B.C. Crowther, "A Comparison of CORSIM and INTEGRATION ...", Master thesis, Virginia Polytechnic Institute and State University (2001).
11.   H.-T. Fritzsche, Transp. Engin. Contr. 5, 317 – 321 (1994).
12.   Kazi Iftekhar Ahmed, Ph.D. thesis, MIT (1999).
13.   W.H. Press, S.A. Teukolsky, W.T. Vetterling, and B.P. Flannery, Numerical Recipes in C, Chapter 10, Cambridge University Press and references therein (2002).
14.   E. Brockfeld, R.D. Kühne, A. Skabardonis, and P. Wagner. TRR No. 1852, Traffic Flow Theory and Highway Capacity 2003, Washington, D.C. (2003).

Real Highway Traffic Simulations Based on a Cellular Automata Model

E.G. Campari[1,2], M. Clemente[1], G. Levi[1,3], and L. Quadrani[1,3]

[1] Physics Departement – Bologna University
   Viale B. Pichat 6/2, 40127 Bologna – Italy
   clemente@bo.inf.it
[2] Istituto Nazionale per la Fisica della Materia
   Viale B. Pichat 6/2, 40127 Bologna – Italy
[3] Istituto Nazionale di Fisica Nucleare
   Viale B. Pichat 6/2, 40127 Bologna – Italy

Summary. In this paper we present results relative of our cellular automata simulation on Italian highway traffic. Based on the previous two lane highway model that we have realised some time ago this is a three lane model without periodic boundary conditions. In particular this model has on and off-ramps; it provides drivers with an unbound braking ability in order to avoid car accidents and it can take account of road works and lane reductions. It implements the Nagel and Schreckemberg rules and it can reproduce well known experimental phenomena observed in real traffic, such as the typical three regions in the fundamental plot, lane inversion or jams that propagate backward like waves.
Keywords: Traffic, simulation, Italian highway, cellular automata model.

1 Introduction

Ours is a cellular automata model: the highway is divided into cells, and the road status evolves according to simple rules, mainly derived from the Nagel and Schreckemberg model[1]. We have added to our previous two lane highway model the third lane, which is in fact a considerable complication. Thus it is interesting to check for instance the peculiar features of lane inversion that we have obtained or to discuss the creation of some differences in jam nucleation at ramps. One can also compare the different lane usage by cars appearing with the introduction of a third lane, which cannot be used by slow vehicles (trucks). We built our model with the aim of making it capable of simulating real traffic on Italian highways. That is why, for instance, it is unecessary to simulate traffic lights and crossroads, but we have taken into account the presence of on and off-ramps and of roadworks and lane reductions. The model that we have used also has no periodic boundary conditions to reproduce the fact that real highways always have both entrances and exits. In our model we can put entrances and exits in the right cell according to their real position on the part of the highway that we are studying. Our ultimate aim is to compare the forecasts provided by simulation with the results obtained from the analysis of data collected directly by our monitoring system.

In particular we are currently able to reproduce the real traffic on the (two-lane) A14 highway between Rimini and Ancona, but most of all we have used our model to make forecast of traffic conditions on one of the busiest Italian highways, named A1 or "Autostrada del Sole" which has three lanes between Milan and Bologna, and only two lanes between Bologna and Florence. We have chosen this highway because it is between Bologna and Florence that we have installed the BirdEye detector (patented by the University of Bologna) which has been collecting data since March and is still actually working: in this way we can make the necessary comparison between real data and simulation.

Finally, a model such as ours could be used to make traffic forecasts by inserting data coming from detectors along the highway as input data for the model itself.

2 Cellular Automata Model for Highway Traffic

Now we can analyse in some detail the structure of the model. Each lane of the highway is divided into cells each of which corresponds to 5 m, their number can being variable according to the length of the highway under study. The cells can be either empty or occupied by a car. Vehicles can be generated at the beginning of the highway on the first and second lane at the chosen rate. The model is asymmetric in the sense that the first lane is for normal cruising while the second one is used both by cars and trucks to overtake and the third one can only be used by cars to overtake. Data are recorded in a structure: for every cell in every lane, which is identified with a number, one can write for instance car identity, car speed, car type according to the maximum velocity that the vehicle could reach and so on.

At the beginning of each simulation we have to insert some parameters such as car density for each lane, car generation probability, fraction of slow cars, maximum vehicle speed and the chosen simulation length in time ticks (tt). For all on-ramps, at each time tick the program checks if the cell is empty or not. In the latter case a random number in the [0,1] interval is generated and if it is less then the threshold chosen for that simulation, a car is generated in that cell with a speed of 2 cell/tt above all on the first lane but also, with a lower percentage, on the second one. Two kinds of maximum velocity ($maxv$) can be reach according to the type of the generated vehicle (9 cell/tt for car and 6 cell/tt for trucks). Calling $fgap1$, $fgap2$, $fgap3$ and $bgap1$, $bgap2$, $bgap3$ the forward and backward gaps, the number of empty cells between cars for each lane; v, $fv2$ ($fv3$) and $pv1$ ($pv2$) car speed, speed of the following car lane 2 (lane 3) and speed of the preceding car on lane 1 (lane 2), at each time step the system updates as follows [4] :

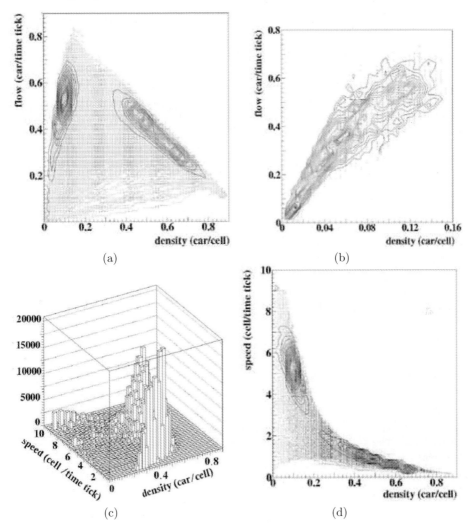

Fig. 1. (a) Fundamental plot obtained with simulated data using general parameters. (b) Simulated data obtained with the specific parameters of the A1 highway and the real traffic conditions of the day of the calibration (15/04/2003). The plot only shows the positive slope which corresponds to free flow that we have observed. (c) Car speed vs density. In this figure, z axis instead of a contour plot is used to put into evidence the distribution points over the bent curve. (d) We have the typical inverse relation between density and velocity known literature.

For every car in lane 1

1) compute gaps (on front and on behind);
2) if $v < maxv : v \to v + 1$;

3) if $v > fgap1$ and there is room enough on lane 2: change lane; else $v = fgap1$.

For every car in lane 2

4) compute gaps;
5) if $v < maxv : v \rightarrow v + 1$;
6) if $fv2$ or $pv1 > v$: return into lane 1 as soon as possible.

For every car in lane 3

7) compute gaps;
8) if $v < maxv : v \rightarrow v + 1$;
9) if $fv3$ or $pv2 > v$: return into lane 2 as soon as possible.

For every car in each lane

10) $v \rightarrow v - 1$ with given probability;
11) compute gaps (to take into account those cars which changed lane as a results of the previous steps);
12) on lane 1: if $v > fgap1 : v = fgap1$;
13) on lane 2: if $v > fgap2 : v = fgap2$;
14) on lane 3: if $v > fgap3 : v = fgap3$;
15) update car positions: cell $\rightarrow$ cell $+ v$.

Note that in such a simulation cars have an unbound braking ability in order to avoid car accidents.

At each time step it is possible to know a number of local and average quantities. For each car position, speed and gaps are recorded. For each cell mean car speed and density are computed, averaging over the 100 cells before and after the given cell. From these quantities the flow is readly computed as density by mean speed.

3 Results

It can be seen from pictures below that our model can reproduce well known global traffic behaviour, as is clear in the fundamental plot realised with the simulated points obtained with the use general parameters (Fig. 1(a)). A population level contour plot superimposed on the same figure reveals that most of the flow (about 99 % of the points) is concentrated in the three region of free, synchronized and congested flow. In particular during the free flow the phenomenon of lane inversion appears, while during the latter kind of flow we have jams propagating backwards like waves as reported in literature. In particular we have simulated the traffic on the first lane of the A1 between Bologna and Florence, where the BirdEye sensor is actually working, to compare simulated with real data collected by the sensors. Simulated data obtained with the specific parameters of the A1 highway and the real traffic

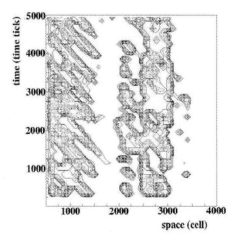

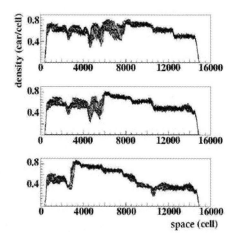

Fig. 2. Regions of space-time where lane inversion in free flow is observed [density of lane 3 > density of lane 2].

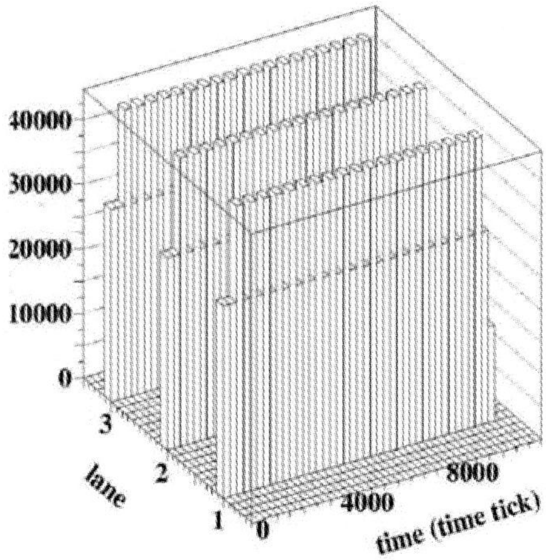

Fig. 3. Third lane occupation.

Fig. 4. Propagation of the wave described in the previous figure. From top to bottom density of lane 2 as a function of space at time 2200, 2350, 2500 tt.

conditions of the day of the calibration (15/04/2003) ideally align along the first curve with positive slope which represents the state of free flow as we have observed. (Fig. 1(b)).

One can also notice that the relations between different important quantities, for instance the inverse relation between car speed and density is well reproduced.

At the end of this short summary of the results obtained with our model which prove its availability we presents some plots tied to the specific use of the third row. In particular lane inversion in free flow can be observed [density of lane 3 > density of lane 2] (Fig. 2(a)). In Fig. 2(b), instead, is represented the growth and propagation of the wave described in the previous figure. From top to bottom density as a function of space at time 2200, 2350, 2500 tt. In this case data of lane 2 are used.

References

1. M. Schreckenberg, A. Schadschneider, K. Nagel, and N. Ito, Phys. Rev E 51, 2939 (1995).
2. K. Nagel, D.E. Wolf, P. Wagner, and P. Simon, Phys. Rev. E 58, 1425 (1998).
3. W. Knospe, L. Santen, A. Schadschneider, and M. Schreckenberg, Physica A 265, 614 (1999).
4. E.G. Campari and G. Levi, Eur. Phys. B 17, 159 (2000).

Traffic Data Collection and Study with the BirdEye System

E.G. Campari[1,2], M. Clemente[1], G. Levi[1,3], and L. Quadrani[1,3]

[1] Physics Departement – Bologna University
     Viale B. Pichat 6/2, 40127 Bologna – Italy
[2] Istituto Nazionale per la Fisica della Materia
     Viale B. Pichat 6/2, 40127 Bologna – Italy
[3] Istituto Nazionale di Fisica Nucleare
     Viale B. Pichat 6/2, 40127 Bologna – Italy
     lucio.quadrani@bo.infn.it

Summary. In this paper we present results obtained with an original traffic data collector named BirdEye, comprising infrared emitters-detectors, microcomputers using a linux operating system and GSM or cable communication systems. It is currently under test, below a portal, along the A1 Italian highway, between Bologna and Florence. Data analysis has revealed the presence of the various phases of traffic; the formation of traffic slowing downs and jams and the highway overcrowding by lorries during working days. In particular we have analysed speed and length spectra, flow and fundamental plot, which reproduces experimental phenomena known from literature. Finally, we present results on the time sequence of vehicles transit which confirm the hypothesis of the existence of a fractal dimension of traffic.
Keywords: traffic, BirdEye detector, fractal dimension

1 Introduction

BirdEye is a traffic monitoring system which allows to acquire data in real time and, through a mathematical model of a given road based on a cellular automata algorithm [1, 2], to make traffic forecasts. This information could be distributed with great benefit to the drivers of that road via information panels or via the web network.

One detector is currently installed along the A1 highway (autostrada del sole) between Bologna and Florence, which is a highway with two lanes per direction and one of the most important (and busiest) road in Italy. Traffic data are collected through a minicomputer based on a linux operating system and are sent to final users via GSM or cable communication systems.

The vehicle sensors are couples of infrared sensors tied by a horizontal bar whose length is 2 meters, and they are designed to be put on top of each highway lane with the bar parallel to the direction of car motion. The detector

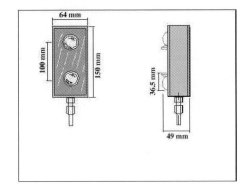

Fig. 1. Schematic picture of BirdEye detection system.

Fig. 2. BirdEye detector sensors.

location can be any portal or bridge along the highway. As with the use of couples of consecutive induction loops, with these sensors it is possible to detect both speed and length of vehicle passing below each detection station, together with the transit time. This alternative to the detection based on inductive loops, allows maintenance of the system and it is not affected by asphalt consumption; finally it is simpler to use and cheaper than cameras.

2 Traffic Data Analysis

The BirdEye system has a good vehicle detection efficiency, a time resolution of one millisecond and therefore also a good space resolution. The system is collecting data without interruption since the beginning of March 2003 and the raw data collected are sent to the central computer via GSM modem. These data consist of binary strings which contain the information relative to the vehicle transit below each one of the detector sensors.

The string length is 32 bits:

- 27 bits record the time of transit in millisecond after midnight;
- 1 bit is for the sensor status (on/off);
- 4 bits are for the detector's identification number (ID).

These raw data are analysed by a program, which reads the data strings and interpret them, providing a double measure of speed and length of each vehicle, together with transit time. From these primary data we can calculate flow, density and mean speed which are the basic quantities for the monitoring and studying of traffic. Moreover the knowledge of the vehicle length allows us to distinguish between cars, trucks and long vehicles.

The detector, installed along the A1 highway, has been tested in order to determine its efficiency. The calibration has been done comparing a movie camera film of traffic with corresponding traffic detected and analysed by the BirdEye system. The results of the calibration test are:

- time duration of test: 1 hour
- total number of vehicles: 425
- vehicles detected: 92%
- vehicles not detected or badly detected: 8%

The data are daily stored on computer files and inserted into a web database, to make them available in our group, then they are analysed. Let l be the vehicle length, v the speed, t_1 and t_2 the transit times below sensor 1 and 2 of the detector respectively, $t_{1,occ}$ and $t_{2,occ}$ the occupation times of the sensors and d the fixed sensors distance. Then, we can determine the vehicle speed $v = d/(t_2 - t_1)$, and the vehicle lengths $l_1 = v * t_{1,occ}$ and $l_2 = v * t_{2,occ}$. As l_1 and l_2 have a good correlation, we assume that the vehicle length is l_1.

The spectra represented in Fig. 3, 4 and 5 show the length distributions of vehicles on A1 highway, near Sasso Marconi. The plot in Fig. 3 refers to a working day and 3 peaks can be seen in the vehicle lengths distribution. A peak centered about 5 m belonging to cars, a second peak between 8 and 15 m (trucks) and a third peak at about 18 m (long vehicles). These peak intensities, with the long vehicles peak more intense than that of the cars, is reliable as the data refer to the first (or right) lane of the highway. The plot in Fig. 4 is that of a Sunday, a day when only cars can circulate. In Fig. 5 it is shown the vehicle lengths distribution on a Saturday. In the plots, the number of events indicates the number of vehicles revealed in a day by the detector. In Fig. 6 it is shown the vehicle speedss distribution during a day in which a jam occurred. The small peak near zero corresponds to a strong slowdown caused by the jam itself. We have analysed mean speed with a minute average and in Fig. 7 the interruption in the flow of vehicles is reported caused by an accident verified on April the 29 between 4.30 and 8 pm.

From the fundamental data (length, speed and time of passage of vehicles) we obtain flow by counting number of events that occur in a time slot (one minute usually, but not necessary). Then we calculate the density by dividing flow by the mean speed (one minute average). Figure 8 shows the behaviour of flow during a whole day. The well known double peak shown in the plot corresponds to rush-hours, while the vehicle flow goes down during the night. The fundamental plot (Fig. 9), obtained from data taken in April 2003 shows how, despite some very busy periods, most of the traffic correspond to free flow.

3 Fractal Analysis and Conclusions

We have tested the hypothesis that the fundamental quantities of highway traffic have a fractal dimension, D. A preliminary fractal analysis has been done on free flow data, of which we have a greater and better amount. The

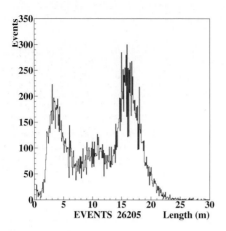

Fig. 3. Vehicles length distribution on a working day.

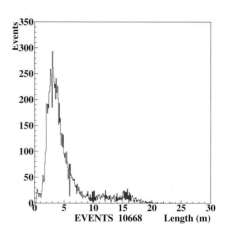

Fig. 4. Vehicles length distribution on a holiday.

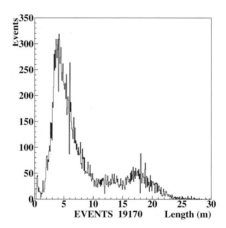

Fig. 5. Vehicles length distribution on a Saturday.

data are analysed with the box counting algorithm [3–6], which is probably the most widely used algorithm to extract a non integer dimension from a data set. We have analysed the time variation during one day of flow and density. The whole time and flow (or density) range of variation has been divided into n^2 boxes of side length $L = 1/n$ of that range. An estimate of the fractal dimension D is obtained from a linear fit of $log(N)$ as a function of $log(1/L)$, where N is, for any value of the side length L, the number of boxes which contain at least one point of the data. As can be seen in Figs. 10a and 10b the points can be fitted by a line whose slope is computed to be $D=(1.542\pm0.004)$ for car density and $D=(1.656\pm0.005)$ for car flow. The fractal dimension spans up to 1.7 decades which is quite good for experimental

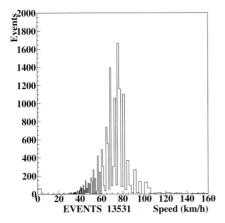

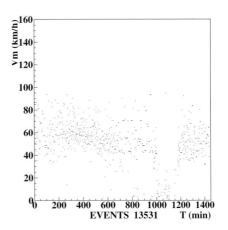

Fig. 6. Vehicles speed distribution on a day when an accident has occurred. The small peak near zero corresponds to a slowdown caused by a jam.

Fig. 7. Vehicles mean speed with one minute average. A slowdown due to an accident is represented.

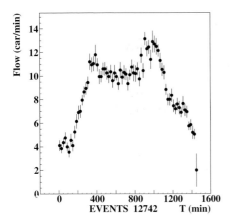

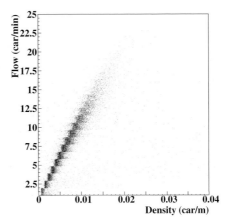

Fig. 8. Vehicles flow during one working day with about twenty minutes average.

Fig. 9. Fundamental plot. Data obtained in April 2003.

data. Other data sets show the same behaviour, supporting the hypothesis of the existence of a fractal dimension. A further analysis of the data is however still in progress to properly compute D as a function of the traffic phases.

From the analysis of the data collected using the BirdEye detector in these eight months we can assert that it even works well for bad weather conditions like snow, rain or wind. Furthermore, up to now the detector has not required

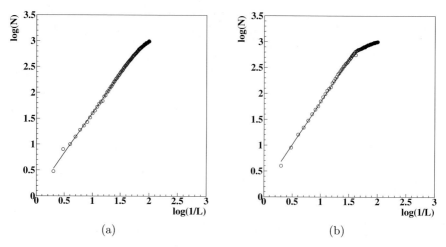

(a)   (b)

Fig. 10. Box counting analysis of car density(a) and car flow(b). The slope changes for the points at the extreme right of the curves occurs when the box side becomes smaller than the number of events of the data plot.

any maintenance. We are waiting for a future installation of other detectors, at different locations along that highway to be able to test the predictive capability of the software related to this detector. In the near future we will also improve the time resolution and the optics of the detector in order to have a better speed resolution and to increase the detector efficiency up to 99%.

References

1. E.G. Campari and G. Levi. *A cellular automata model for highway traffic*, Eur. Phys. J. B 17, 159-166 (2000).
2. M. Schreckenberg, A. Schadschneider, K. Nagel, and N. Ito. *Discrete stochastic models for traffic flow*, Phys. Rev. E 51, 2939 (1995).
3. E.G. Campari and G. Levi. *Self-similarity in highway traffic*, Eur. Phys. J. B 25, 245-251 (2002).
4. P. Addison. *Fractals and Chaos* (IOP, London, 1997).
5. J. Woinowski. *Fractal Frontieres '97*, edited by M.M. Novak and T.G. Dewey, pp. 321-327 (World Scientific, Singapore, 1997).
6. K.T. Alligood, T.D. Sauer, and J.A. Yorke. *Chaos, an introduction to dynamical systems*, Chap. 4 (Springer, New York, 1996).

Traffic Models and Theory

Probabilistic Traffic Flow Breakdown in Stochastic Car Following Models

D. Jost[1] and K. Nagel[2]

[1] Inst. for Computational Science – ETH Zürich
   CH-8092 Zürich – Switzerland
   `jost@inf.ethz.ch`
[2] Inst. for Land and Sea Transport Systems – TU Berlin
   Salzufer 17–19, 10587 Berlin – Germany
   `nagel@vsp.tu-berlin.de`

Summary. There is discussion if traffic displays spontaneous breakdown. This paper presents computational evidence that stochastic car following models can have a control parameter that moves the model between displaying and not displaying spontaneous phase separation for some densities. Those phases can be called "laminar" and "jammed". Models with spontaneous phase separation show three states as a function of density: a first state at low density, where those models are homogeneously laminar; a second state at high density, where they are homogeneously jammed; and a third state at intermediate density, where they consist of a mix between the two phases (phase coexistence). This is the same picture as for a gas-liquid transition when volume of the gas is the control parameter.

Although the gas-liquid analogy to traffic models has been widely discussed, no traffic-related model so far displayed a completely understood *stochastic* version of that transition. Having a stochastic model is important to understand the potentially probabilistic nature of the transition. Most importantly, if indeed models with spontaneous phase separation describe certain aspects correctly, then this leads to an understanding of spontaneous breakdown. Alternatively, if models without spontaneous phase separation describe these aspects better, then there is no spontaneous breakdown (= no breakdown without a reason). Interestingly, even models without spontaneous phase separation can still allow for jam formation on small scales, which may give the impression of having a model with spontaneous phase separation.
Keywords: traffic flow theory, car following models, traffic breakdown, traffic simulation, phase transition, phase separation, critical point

1 Introduction

The capacity of a road is an important quantity. If demand exceeds capacity, queues will form, which represent a cost to the driver and thus to the economic system. In addition, such queues may impact other parts of the system, for example by spilling back into links used by drivers who are on a path that is not overloaded.

This paper discusses freeway capacity. The question concerns the maximum flows that freeways can reach, and if the maximum flows sometimes observed (> 2500 vehicles per hour and lane) are sustainable flows or short-term fluctuations. Let us assume that there is traffic with a fairly high density ρ on a freeway, but vehicles are still able to drive at some fast velocity v. Throughput is $q = \rho v$. The question is what will happen if density is further increased: Can q further increase because ρ increases more than v decreases? Will q gradually decrease because ρ increases but v decreases faster? Or is there a possibility that traffic will break down, leading to stop-and-go traffic?

More technically, the question is if there is, for each density ρ, a velocity $V(\rho)$ and corresponding throughput $Q(\rho) = \rho V(\rho)$ at which traffic flow is smooth and homogeneous. Or is there a density range where that homogeneous traffic flow is unstable, and traffic has a tendency to reorganize into a stop-and-go pattern, with possibly lower throughput?

It is important to note that this paper's focus is on *homogeneous* situations. This concerns both the geometry of the system, which is assumed to be closed (such as a long ring) and spatially uniform (no bottlenecks, no changes in speed limit, no grades, etc.), and the initial condition, which is assumed to be traffic with the same density everywhere along the ring. Clearly, this is a theoretical construct, but the issue is to sort out theoretical questions. Again, the main question is if in such a situation the initially homogeneous traffic has a tendency to reorganize into a stop-and-go pattern; this is what is meant by (spontaneous) "breakdown" in this paper. This is in contrast to induced phases, such as queues upstream of a bottleneck. Induced phases are important, but they are outside the scope of this paper.

There is in fact a long history of publications about breakdown behavior in freeway traffic, sometimes called "reverse lambda shape of the fundamental diagram" [1, 2], "hysteresis" [3], "capacity drop" [4], "catastrophe theory" [5], and the like. From the modeling side, there have since long been discussions about an analogy to a gas-liquid transition [6, 7], and recent work has established traffic models which display deterministic versions of a liquid-gas-like transition [8, 9].

Yet, measurements by Cassidy [10] indicate that there can be stable homogeneous flow at all densities. Many of the "reverse lambda" observations could also be caused by geometrical constraints, in the following way [11]. A bottleneck downstream of a measurement location can cause the following temporal sequence of measurements: (1) The system starts with low flow at low densities. – (2) Both flow and density keep increasing, along the "free flow" branch of the fundamental diagram. – (3) This flow can be larger than what can flow through the bottleneck. Then, a queue starts forming at the bottleneck, but that does not immediately influence the measurement. – (4) Eventually, the queue will have spilled back to the measurement location. At that point in time, data points will move to a much higher density, while the flow value will drop to the bottleneck capacity. It can take up to 20 minutes for the transition zone (transition from free flow to queue) to traverse a fixed detector location, leading to fundamental diagram data points that lie between the free flow and the queue state [11]. This mechanism generates data that looks similar to data that one would expect from a spontaneous breakdown in a homogeneous system, as explained above. Unfortunately, many of the published data sets do not provide enough information about the geometrical layout and the full spatio-temporal picture of the dynamics in order to resolve this question.

Because measurement locations upstream of bottlenecks generate fundamental diagrams that in the past were used to support the spontaneous breakdown hypothesis, at this point few measurements remain that can truly be used to help with the question. The maybe strongest empirical evidence for spontaneous breakdown is an experiment where a number of vehicles drive in a spatially homogeneous circle for an extended period of time [12]. In that experiment, traffic remains laminar for many minutes, but eventually "breaks down" into a stop-and-go pattern. Other evidence is indirect: Assume homogeneous traffic operating at a certain density, and assume the introduction of a strong disturbance, say by stopping one car for several seconds. If the introduced disturbance heals out over time, then homogeneous traffic at that density is stable; if the disturbance grows over time, then the homogeneous solution is unstable at this density. This implies that stable jams, embedded in laminar traffic, support the spontaneous breakdown hypothesis. There are at least three references (Figs. 2 and 3 in [4]; Fig. 3 in [13]; Fig. 4 in [14]) where the data in fact points to the existence of a stable jam, embedded *both upstream and downstream* in free traffic, and where the outflow from the jam is lower than the inflow. In the 2nd and the 3rd of these references, one can in addition see that the jam is remaining compact. In the 1st of these references, the data to decide this question is not sufficient.

This question is not just academic. The correct use of technical devices such as ramp metering [15] or adaptive speed limits [16] depends on the answer. For example, let us assume that the homogeneous solution is unstable in a certain density range, and that the alternative stop-and-go solution has a lower throughput than homogeneous traffic at the same density. In this case, the task of ramp metering might be to keep the density away from the unstable range. If density approaches this value, on-ramp traffic should be reduced.

If, in addition, breakdown is probabilistic, that is, the homogeneous solution can survive for certain amounts of time, then the question becomes which risk of breakdown one would be willing to accept. Accepting higher flow rates in the ramp metering algorithm might increase *average* throughput, but it might also increase the probability of breakdown. There is discussion to include aspects of stochastic transitions into the Highway Capacity Manual [17].

If, in contrast, the homogeneous solution is stable everywhere, then the potentially positive effects of ramp metering need to be derived from something other than breakdown.

Given this state of affairs, it makes sense to look at modeling. The task is to understand which model solutions are possible at all. This understanding will lead to the predictions of additional features that will go along with one mechanism or the other, and it might be possible to measure them, and so the issue will hopefully be eventually resolved.

It is important to note that this paper looks at the issue of spontaneous jam formation in a spatially homogeneous system, e.g. traffic in a long closed ring. In order to be clear about that, the term "spontaneous phase separation" will be used. This is different from boundary-induced phases, such as queues upstream of bottlenecks. Boundary-induced phases are clearly important in traffic, possibly more important than the issue of spontaneous phase separation. Nevertheless, the issue of spontaneous phase separation needs to be understood before conclusive statements on boundary-induced phases can be made.

This paper starts with Sec. 2 which recalls the general idea of a gas-liquid transition. Sec. 3 describes the simulation setup including the car following model that is used, discusses space-time plots of the resulting dynamics, and investigates transients vs. the steady state. Sec. 4 then establishes how a coexistence state can be numerically detected for a given model. Sec. 5 reports similar results for cellular automata (CA) models. Sec. 6 discusses how these results relate to deterministic models; the paper is concluded by a discussion and a summary.

2 Phases and Phase Transitions

The analogy between a gas-liquid transition and the laminar-jammed transition of traffic was pointed out many times (e.g. [7, 9]). The description of traffic in the well-known 2-fluid-model [18] assumes the existence of two phases; and all simulation models which use spatial queues (e.g. [19–21]) will display two phases because of the definition of the dynamics. The two phases in models with queues are however much easier to understand than the phases in more realistic models.

In a gas-liquid transition, one observes the following (Fig. 1):

- In the **gas state**, at low densities, particles are spread out throughout the system. Distances between particles vary, but the probability of having two particles close to each other is small.
- In the **liquid state**, at high densities, particles are close to each other. There is no crystalline structure as in solids, but the density is similar. Because of the fact that the particles are so close to each other, it is difficult to compress the fluid any further.
- In between, there is the so-called **coexistence state**, where gas and liquid coexist. In typical experiments in gravity, the liquid will be at the bottom and the gas will be above it. Without gravity, droplets form within the gas

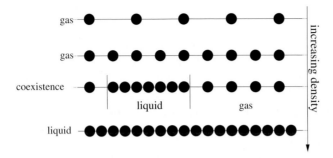

Fig. 1. Schematic representation of the gas-liquid transition in one dimension.

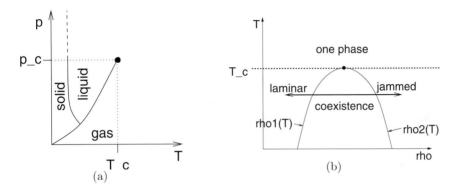

Fig. 2. (a) "Standard" pressure-temperature (PT) diagram with critical point. For $T > T_c$ or $p > p_c$, there is no gas-liquid transition any more. In that case, distinguishing between liquid and gas phases does not really make sense. (b) Phase diagram of the gas-fluid model as a function of the density and the temperature T.

and remain interdispersed. The droplets will slowly merge together into bigger droplets (coagulation). The final state of the system is having one big droplet of liquid, surrounded by gas.

If a system in the coexistence state is compressed, more droplets form and/or existing ones grow, but the density both inside and outside the droplets remains constant. That is, the system reacts by allocating more space to the liquid, but *not* by changing the density either of the gas or the liquid. Let us call those two densities ρ_1 and ρ_2, with $\rho_1 < \rho_2$. Eventually, all the space is used up by the liquid. At this point, the system will be homogeneous again and remain so if density is increased further.

The above picture is probably known to many people, e.g. from high school or undergraduate physics. Still, it is important to be clear about the details. The above description refers to a view where temperature is kept constant and volume is controlled. With regards to pressure, one should recall that pressure does not change in the coexistence state. If one uses pressure instead of volume as the control parameter, then there is no finite range of control parameter values where the system is in coexistence (Fig. 2(a)). For the traffic analogy, it will be important to use volume as the control parameter. The inverse of volume is system-wide density, which is a more common variable for traffic systems. It is, however, important to distinguish system-wide density, which we will denote by $\rho_L = 1/V$, from any local density.

Also note that the above description refers to *two phases*, called "liquid" and "gas", but to *three states*, called "liquid", "gas", and "coexistence". The first two states are homogeneous states, since they contain only one phase and are thus spatially homogeneous. The coexistence state contains both phases together.

The above picture is correct in equilibrium, which essentially means after waiting "long enough" while the system is at a fixed ρ_L. If one compresses the system rather quickly beyond ρ_1, then the system is not able to immediately re-organize into droplets: Some time is necessary to achieve this.

The kinetics of the droplet formation (e.g. [22]) is ruled by a balance between surface tension and vapor pressure. Since surface tension pulls the droplet together, it increases the pressure inside the droplet. This interior pressure pushes water molecules out of the droplet. Vapor pressure outside the droplet is the balancing force – it pushes particles into the droplet.

Surface tension and thus interior pressure depend on the droplet radius – the smaller the droplet, the larger the surface tension and thus the interior pressure. The result is that slightly above ρ_1 large droplets are stable, but small droplets are not. Stable and unstable droplets are separated by a critical radius $r_c(\rho)$: Droplets smaller than r_c in the average shrink and thus in the average eventually dissolve; droplets larger than r_c in the average grow.

When ρ_L comes from a low density, the homogeneous phase can survive for some time even slightly above ρ_1, because small droplets are suppressed, while large droplets are not (yet) there. This super-critical gas is thus *meta-stable*. Only after some waiting time one or more droplets will become, by a fluctuation, large enough to go beyond r_c, at which point these droplets will continue to grow until they have swallowed up enough molecules to reduce the gas density outside the droplets to ρ_1. A direct consequence of meta-stability is *hysteresis*: When coming from low densities, it is possible to have $\rho_L > \rho_1$ and still remain in the gas phase.

The description so far refers to a constant temperature. However, ρ_1 and ρ_2 depend on the temperature (Fig. 2(b)). With increasing temperature the densities approach each other, meaning that the densities inside and outside the droplets become more similar. Eventually, there is a temperature T_c where $\rho_1(T_c) = \rho_2(T_c)$. At this point, the densities inside and outside the droplets become the same, which means that they become indistinguishable. In other words: for $T \geq T_c$ there is no coexistence state any more; the system is homogeneous at every density ρ_L.

Said again differently: Depending on the temperature T, our system will either display spontaneous transitions between gas and coexistence and between coexistence and liquid, *or there will be no spontaneous phase separation at all.* (In that latter case, boundary-induced phase separation is still a possibility.)

We will now move on to describe the supporting evidence for the claim that traffic models can show a similar behavior. As is typical in computational science, our evidence is based on computer simulations. It is backed up by generic knowledge about phase transitions as they are well understood in physics.

3 Simulations

In this paper, we will start by using the model by Krauß [23]. As one will see in Sec. 5, the precise details of the model do not really matter. Nevertheless, they are given for technical completeness. The velocity update of the Krauß model reads as follows:

$$v_{\text{safe}} = \tilde{v}(t) + \frac{\frac{g(t)}{\tau} - \tilde{v}(t)}{\bar{v}(t)/(b\,\tau) + 1} \tag{1}$$

$$v_{\text{des}} = \min\{v(t) + a\,\Delta t, v_{\text{safe}}, v_{\text{max}}\} \tag{2}$$

$$v(t + \Delta t) = \max\{0, v_{\text{des}} - \varepsilon\,a\,\eta\} . \tag{3}$$

g is the gap (front-bumper-to-front-bumper distance minus space a vehicle uses in a jam), $\tilde{v}$ is the speed of the car in front, $\bar{v} = (v + \tilde{v})/2$ is the average velocity of the two cars involved, v_{max} is the maximum velocity, a is the maximum acceleration of the vehicles, b their maximum deceleration for $\varepsilon = 0$, ε is the noise amplitude, and η is a random number in $[0, 1]$. The meaning of the terms is as follows:

- Eq. 1: Calculation of a "safe" velocity. This is the maximum velocity that the follower can drive to be sure to avoid a crash [23]. The equation states that the follower tries to have the same velocity as the leader, with a gap proportional to the leader's velocity: $g = \tau\,\tilde{v}$. If the gap is larger than that, v_{safe} is larger than the velocity of the leader; if the gap is smaller than that, then v_{safe} is smaller than the velocity of the leader.
- Eq. 2: The desired velocity is the minimum of: (a) current velocity plus acceleration, (b) safe velocity, (c) maximum velocity (e.g. speed limit).
- Eq. 3: Some randomness is added to the desired velocity.

After the velocities of all vehicles are updated, all vehicles are moved.

The Krauß model has been proven to be free of crashes for numerical time steps Δt smaller than or equal to the reaction time, τ [23]. We will use $\Delta t = \tau = 1$ as has conventionally been used for the Krauß model. We further use $a = 0.2$, $b = 0.6$, $v_{\text{max}} = 3$ for all simulations.

The model as defined above is free of units. A reasonable calibration is: one time unit corresponds to one second, and one space unit correspond to 7.5 meters, which is the space that a vehicle occupies in a jam. The reaction time is then 1 second, and $v_{\max} = 3$ corresponds to 22.5 m/s or 81 km/h. $a = 0.2$ means a maximum acceleration of 1.5 m/s (5.4 km/h) per second. $b = 0.6$ corresponds to a maximum deceleration of 16.2 km/h per second.

All simulations are done in a 1-lane system of length L with periodic boundary conditions (i.e. the road is bent into a ring). Let N be the number of cars on the road. The (global) density is $\rho_L = N/L$.

Before analysing the Krauß model quantitatively, it is instructive to look at space-time plots (Fig. 3). The following refers to the subfigures (i)–(vi) of Fig. 3. They are arranged so that they correspond to Fig. 2(b). The *bottom* row corresponds to a smaller noise amplitude $\varepsilon = 1.0$. One recognizes

(iv) The laminar state: All cars drive at high speed. The available space is shared evenly among the cars. The traffic is homogeneous.

(v) The coexistence state: The slow cars are all together in one big jam. On the rest of the road, the cars drive at high speed. In consequence, the traffic is very inhomogeneous.

(vi) The jammed state: The density is so high that no single car can drive fast. As in (iv), the traffic is homogeneous.

In contrast, the top row (i)–(iii) corresponds to a larger noise amplitude $\varepsilon = 1.8$. Here, many small jams are distributed over the whole system. There is neither a larger area of free flow, nor a major jam. The traffic is homogeneous at all densities. Note that "homogeneous" here means "homogeneous on large scales". In (i) and (ii), there is structure, i.e. small jams and laminar flow, but these are not visible when looking at the plots from a distance. In contrast, the coexistence state, as in (v), will never look homogeneous (see Sec. 4 for a more technical version of this).

For many parameters of the Krauß model, there is a unique equilibrium state, which the system will attain after a finite time t_{relax}, no matter how it was started. Deciding when the equilibrium is reached is not trivial. Our criterion was to look at the number of jams in the system (Fig. 4). The system was once started with equidistant vehicles (maximally homogeneous) and once with all vehicles in a "mega-jam" (maximally inhomogeneous). Initially, the number of jams in the system shows very different behavior in those two simulations. However, eventually that number becomes the same in both simulations, at which point it was assumed that equilibrium was reached. A *jam* here is defined as a sequence of adjacent cars driving with speed less or equal $v_{\max}/2$. This definition of a jam is used nowhere else in this paper; it is only used to decide how long a simulation needs to run until one can assume that it has reached equilibrium.

(i) $\varepsilon = 1.8$, $\rho = 0.2$ (ii) $\varepsilon = 1.8$, $\rho = 0.3$ (iii) $\varepsilon = 1.8$, $\rho = 0.75$

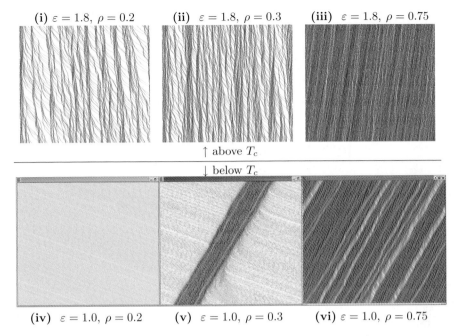

↑ above T_c

↓ below T_c

(iv) $\varepsilon = 1.0$, $\rho = 0.2$ (v) $\varepsilon = 1.0$, $\rho = 0.3$ (vi) $\varepsilon = 1.0$, $\rho = 0.75$

Fig. 3. Space-time plots for different parameters. Space is horizontal; time increases downward; each line is a snapshot; vehicles move from left to right; fast cars are green, slow cars red. $L = 600$ for all plots.

4 Establishment of a Phase Diagram via a Measure of Inhomogeneity

One needs to establish a criterion that distinguishes homogeneous from co-existence states. As pointed out before, coexistence states, for example at $\varepsilon = 1.0$ and $\rho = 0.3$ in our model, see Fig. 3(v), are characterized by the coexistence of laminar and jammed traffic. Inside the coexistence regime, the phases coagulate, leading to one large laminar and one large jammed section in the system. When approaching the boundaries of the coexistence regime, this characterization will become less clear-cut, and it may be possible to have more than one jam. Typically, there will be one major jam and many small ones, and for many measurement criteria this will cause enough problems to no longer be able to differentiate between the coexistence and a homogeneous state. This is particularly true for criteria that attempt a binary classification into homogeneous or not. In contrast, our criterion will show a gradual transition.

The criterion is defined as follows: Partition the road into segments of length ℓ (for simplicity let ℓ divide L without remainder). For each segment the local density ρ_ℓ can be computed as the number of cars in that segment divided by ℓ. An interesting value is the variance of the local density (see, e.g.,

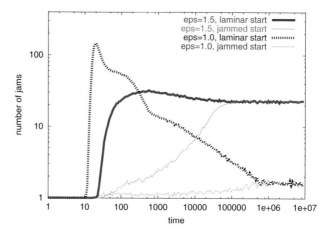

Fig. 4. Time evolution of the number of jams. All four curves are for 1000 cars and $\rho = 0.3$. Each curve is an average over at least 80 realizations, each with a different random seed.

[24]):

$$\text{Var}(\rho_\ell) = \frac{\ell}{L} \sum_{i=1}^{L/\ell} (\rho_\ell(i) - \rho_L)^2 \; , \qquad (4)$$

where ρ_L is the systemwide average density. Note that since density values always lie within $[0, 1]$, the variance cannot exceed $1/4$.

   This value picks up how much each individual measurement segment of length ℓ deviates, in terms of its density, from the average density. Assume a system consisting of jammed and laminar traffic. If there is a jam in one segment, then the segment's density will be much higher than the average density. Conversely, if there is only laminar traffic in a segment, then the segment's density will be much lower than the average density. $\text{Var}(\rho_\ell)$ takes the average over the square of these deviations.

   Fig. 5 shows this value as a function of the global density ρ and the noise parameter ε. Each gridpoint is the result of a computer simulation. The simulations run until the average number of jams over the last 100'000 time steps is (almost) equal for a system started with a big jam and a system started with laminar flow (recall Fig. 4). The variance of the local density is averaged over those same 100'000 time steps.

   Look at Fig. 5 for fixed noise ε, say $\varepsilon = 1$. One sees that at densities up to $\rho \approx 0.2$, the value of $\text{Var}(\rho_\ell)$ is close to zero, indicating a homogeneous state, which is in this case the laminar state. Similarly, for densities higher than 0.8, $\text{Var}(\rho_\ell)$ is again close to zero, indicating a homogeneous state, which is in this case the jammed state. In between, for $0.2 \leq \rho \leq 0.8$, the value of $\text{Var}(\rho_\ell)$ is significantly larger than zero, indicating a coexistence state.

   Now slowly increase ε. We see that the two critical densities approach each other (see Fig. 5). At $\varepsilon \approx 1.7$, the coexistence phase goes away; for larger ε,

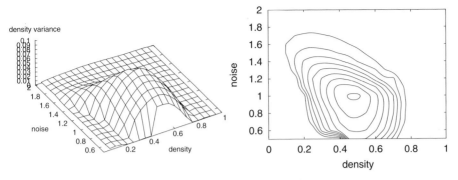

Fig. 5. 3d-plot and contour plot of the density variance for the Krauß model. Both plots show the same data. The outermost isoline is $\text{Var}(\rho_\ell) = 0.01$, the innermost $\text{Var}(\rho_\ell) = 0.09$. $L = 1000$ and $\ell = 62.5$

we do not pick up any inhomogeneity at *any* density (look at the contour plot in order to get information about behavior not visible in the 3d plot). Compare this to the theoretical expectation in Fig. 2(b), where for increasing T the two densities eventually merge and thus the different phases go away. Note that close to but slightly above the critical ε, the system still looks like it possesses different phases (see Fig. 3(i) and locate the corresponding $\varepsilon = 1.8$ and $\rho = 0.2$ in Fig. 5). These structures exist, however, *on small scales only*. This means that for system size $L \to \infty$ and measurement interval $\ell \to \infty$ (but $\ell \ll L$), all intervals of size ℓ will eventually return the same density value. A segment length of $\ell = 62.5$, as used for Fig. 5, is already sufficient in order to not measure any inhomogeneity for the states in Fig. 3(i) and (ii). This will not be the case for coexistence states: In coexistence state, there will always be segments with different densities, unless $\ell \approx L$. This is because droplets will coagulate so that they will eventually show up on all possible length scales ℓ.

Remember again that ε is a model parameter while ρ is a traffic observable. That is, once one has settled for an ε, the model behavior is fixed, and one has decided if one can encounter spontanoues phase separation (= spontaneous jam formation) or not. *If* one can encounter spontaneous phase separation, it will come into existence through changing traffic demand throughout the day – traffic can move from the laminar into the coexistence and potentially into the jammed state and back.

As a side remark, let us note that there is also another regime without spontaneous phase separation for $\varepsilon \to 0$. Albeit potentially interesting, this is outside the scope of this paper.

In summary, one obtains, for the above traffic model and a spatially homogeneous geometry, a phase diagram as in Fig. 2(b), which is the schematic phase diagram for a gas-liquid transition in fluids. Again, the important feature of this phase diagram is that there are three states for low temperatures (small T or small ε): gas/laminar; coexistence; liquid/jammed. For higher temperatures, the coexistence range becomes more and more narrow, while the density of the gas phase and the density of the liquid phase in the coexistence state approach each other. Eventually, these densities become equal, and the coexistence state dies out. The only notable difference is that for our traffic model the phase diagram is bent to the left with increasing ε.

There are other criteria which can be used to understand these types of phase transitions. In particular, one can look at the gap distribution between jams, and one would expect a fractal structure at the critical point, i.e. at $\rho \approx 0.2$ and $\varepsilon \approx 1.7$. This is indeed the case but goes beyond the scope of this paper; see [25] for further information.

5 Cellular Automata Models

Many of the arguments regarding the nature of a stochastic and possibly critical phase transition [26–30] have been made using so-called cellular automata (CA) models. CA models use coarse spatial, temporal, and state space resolution. For traffic, a standard way is to segment a 1-lane road into cells of length l_c, where l_c is the length a vehicle occupies in the average in a jam, i.e. $l_c = 1/\rho_{jam} \approx 7.5$ m. Cells are either occupied by exactly one car, or are empty. Vehicles move by jumping from one cell to another. As with the Krauß model, the time step for the CA models is best selected similar to the reaction time; a time step of 1 second works well in practice. Taking this time step together with l_c, one finds that a speed of 135 km/h corresponds to five cells per time step; this is often taken as maximum velocity v_{max}.

A possible CA velocity update rule is [31]:

- Deterministic car driving:

$$v_{t+\frac{1}{2}} = \min[g_t, v_t + 1, v_{max}] \,, \tag{5}$$

  where g_t is the gap (number of empty cells ahead) at time t.
- Randomization:

$$v_t = \begin{cases} \max[0, v_{t+\frac{1}{2}} - 1] & \text{with probability } p_{slow}(v_t) \\ v_{t+\frac{1}{2}} & \text{else} \end{cases} \,, \tag{6}$$

  where $p_{slow}(v)$ is a velocity-dependent randomization. Often-selected values are

$$p_{slow}(v) = \begin{cases} p_0 = 0.5 & \text{if } v_t = 0 \\ p_{>0} = 0.01 & \text{if } v_t > 0 \end{cases} \,.$$

  which models that drivers, once stopped, are a bit sloppy in re-starting again. $p_{>0} = p_0 = 0.5$ returns the CA of [32].

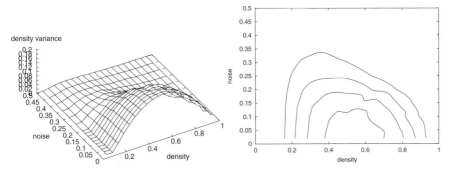

Fig. 6. 3d-plot and contour plot of the density variance for the CA model. Both plots show the same data. Instead of the noise parameter ε, the randomness-parameter $p_{>0}$ is varied from zero to one.

With this family of models, one can again plot the density variance (Fig. 6). Instead of the noise amplitude ε, the parameter $p_{>0}$ is used. $p_{>0} = 0$ means deterministic driving except when accelerating from zero; increasing $p_{>0}$ means increasingly more randomness when moving. In Fig. 6, one finds a behavior similar to Fig. 5: For small $p_{>0}$, the system displays three states (laminar, coexistence, and jammed). For $p_{>0} \to 0.5$, the system becomes eventually a system without spontaneous phase separation.

From this plot, it is impossible to decide exactly at which $p_{>0}$ the transition from a model with to a model without spontaneous phase separation takes place. Nevertheless, this plot makes clear why there was so much discussion about possible fractals for the original model [32] in which $p_{>0} = 0.5$: That model is indeed close to the critical point, and in consequence one should expect fractals up to a certain cut-off length scale. That cut-off length scale should depend on the distance to the critical point; further investigations are necessary to exactly determine the correct value of the critical point.

6 Phase Transitions in Deterministic Models

Only stochastic models can display *spontaneous* transitions between homogeneous and coexistence states. The nature of the transition can however also become clear in deterministic models. We will discuss these similarities first for a deterministic car following model and then for deterministic fluid-dynamical models.

A possible **car-following model** is [33]

$$a(t) = \alpha \cdot (V(g(t)) - v(t)) \ , \quad \text{with } V(g) = v_f \cdot (\tanh(g + l_c) - \tanh(l_c)) \ , \quad (7)$$

where a is the acceleration, g is again the gap, $V(.)$ is a desired velocity, l_c is the space a vehicle occupies in a jam (7.5 m in the previous models), and

v_f is the free speed. For this model, it was shown [9] that the homogeneous solution of the model is linearly unstable for densities where $dV/dg > \alpha/2$. The instability sets in for intermediate densities; for low and high densities *all* models are stable in the homogeneous (laminar or jammed) state. One can thus select the curve $V(g)$ and the parameter α such that the model either has unstable ranges, or not. If all parameters, including the density, are such that the homogeneous solution is not stable, then the system rearranges itself into a pattern of stop-and-go traffic, corresponding to the coexistence state. The density of the laminar and the jammed phase in the coexistence state are independent from the average system density, that is, if in that state system density goes up, it is reflected in the jammed phase using up a larger fraction of space.

Fluid-dynamical theory, of the type

$$\partial_t \rho + \partial_x(\rho v) = 0 \tag{8}$$

$$\partial_t v + v\, \partial_x v = \frac{1}{\tau} \left(V(\rho) - v \right) + \alpha(\rho)\, \partial_x \rho + \nu(\rho)\, \partial_x^2 v \tag{9}$$

can, depending on the choice of parameters including the $V(\rho)$-curve, either display or not display spontaneous phase separation [34]. For example, the homogeneous solution of the model with $\alpha(\rho) = c_0^2/\rho$ and $\nu(\rho) = \nu_0$ is linearly unstable at densities where $|dV/d\rho| > c_0/\rho$ [34]. This is similar to the instability condition for the car following model above; note that $V'(\rho)$ and $V'(g)$ are, albeit related, not the same.

As pointed out before, these models are deterministic. In no situation will these models display *stochastic* transitions.

7 Discussion

As mentioned in the introduction, there is discussion in the literature if traffic shows spontaneous jam formation, or if all jams are caused by geometrical constraints such as bottlenecks. That discussion was in the past hampered by the fact that no clear picture for spontaneous jam formation in stochastic models was available: The introduction of the slow-to-start CA (s2s-CA) models was guided by the observation that the original CA [32] model did not display true meta-stability, but no convincing overall picture emerged. In particular, it was never clarified if or why the origial CA displayed fractal properties, and how these fractal properties change when moving towards s2s-CA models. In contrast, the present paper allows, for people sufficiently versed in the theory, a clear prediction: The original CA should display fractal properties up to a certain cut-off; that cut-off should become larger and eventually diverge with decreasing $p_{>0}$; it should then become smaller again, until eventually one cannot speak of fractals any more.

In addition, better understanding allows to make better predictions for properties besides spontaneous breakdown. For example, one would predict that a traffic queue, when operating at an average density between ρ_1 and ρ_2, would show phase separation; those phases should have the densities ρ_1 and ρ_2, and they should coagulate with increasing distance from the bottleneck. Unfortunately, coagulation is a slow process, and for that reason once more the issue cannot be resolved easily.

8 Summary

This paper shows, via computational evidence, that a specific stochastic car following model can either display or not display spontaneous phase separation, depending on the choice of parameters. The two phases are: "laminar", and "jammed". Models with spontaneous phase separation possess two homogeneous states, which correspond to the phases. They also possess a third state, at intermediate densities, which is a coexistence state. It consists of sections with jammed and sections with laminar traffic.

With respect to cellular automata (CA) models, it turns out that one of the early CA models for traffic [32] is a model without spontaneous phase separation, but close to such a model, which explains the near-fractal structures which have been observed. In contrast, the so-called slow-to-start models [31] display clear phase separation.

Some of these findings can also be understood by looking at deterministic models for traffic, either car-following or fluid-dynamical. However, the stochastic elements of the transition cannot be explained by deterministic models. An important stochastic element is meta-stability, which means that a "super-critical" homogeneous state can survive for long times before it "breaks down" and reorganizes into stop-and-go traffic.

It is important to understand this possibility of stochastic models to be in different regimes if one considers to enter the notion of traffic breakdown probabilities into the Highway Capacity Manual. If traffic is best described by a model without spontaneous phase separation, then there is, in our view, no theoretical justification for (spontaneous) breakdown probabilities. If, however, traffic is best described by a model with spontaneous phase separation, then such breakdown probabilities make sense.

Acknowledgments

F. Wagner, D. Stauffer, and M. Droz have significantly helped in defining computational criteria for the detection of multiple phases. P. Wagner, C. Daganzo, and M. Cassidy have helped indirectly (and maybe inadvertently), via the discussion of a separate but related paper. The paper has also benefited from discussions with P. Nelson, although the consequences of those discussions have not yet completely sunk in. Several weeks of computing time on

192-CPU cluster Xibalba and on 44-CPU cluster Linneus were used for the computational results.

Note added – Gray, Levine, Mukamel and Ziv have made considerable progress with respect to theoretical results in the same area ([35] and references therein). However, none of their results so far completely explains stochastic cases which simultaneously have "slow-to-start" and $v_{max} > 1$.

References

1. M. Koshi, M. Iwasaki, and I. Ohkura. Some findings and an overview of vehicular flow characteristics. In *Proceedings of the 8th International Symposium on Transportation and Traffic Theory*, pages 403–426. Elsevier, Amsterdam, Toronto, 1983.
2. B. S. Kerner. Phase transitions in traffic flow. In D. Helbing, H. J. Hermann, M. Schreckenberg, and D. E. Wolf, editors, *Traffic and Granular Flow '99*, pages 253–284. Springer, Berlin, 1999.
3. J. Treiterer and J.A. Myers. The hysteresis phenomenon in traffic flow. In D.J. Buckley, editor, *Proc. 6th Internat. Sympos. on Transportation and Traffic Theory*, pages 13–38, Sydney, Australia, 1974. Elsevier, New York.
4. B. S. Kerner and H. Rehborn. Experimental features and characteristics of traffic jams. *Phys. Rev. E*, 53(2):R1297–R1300, 1996.
5. J.A. Acha-Daza and F.L. Hall. Graphical comparison of predictions for speed given by catastrophe theory and some classic models. *Transportation Research Record*, 1398:119–124, 1993.
6. I. Prigogine and R. Herman. *Kinetic theory of vehicular traffic*. Elsevier, New York, 1971.
7. H. Reiss, A.D. Hammerich, and E.W. Montroll. Thermodynamic treatment of nonphysical systems: Formalism and an example (single-lane traffic). *Journal of Statistical Physics*, 42(3/4):647–687, 1986.
8. B. S. Kerner and P. Konhäuser. Structure and parameters of clusters in traffic flow. *Phys. Rev. E*, 50(1):54–83, 1994.
9. M. Bando, K. Hasebe, A. Nakayama, A. Shibata, and Y. Sugiyama. Structure stability of congestion in traffic dynamics. *Japan Journal of Industrial and Applied Mathematics*, 11(2):203–223, 1994.
10. M.J. Cassidy. Bivariate relations in nearly stationary highway traffic. *Transportation Research B*, 32B:49–59, 1998.
11. J. C. Muñoz and C. F. Daganzo. Structure of the transition zone behind freeway queues. *Transportation Science*, 37(3):312–329, 2003.
12. Y. Sugiyama, A. Nakayama, M. Fukui, M. Kikuchi, K. Hasebe, K. Nishinari, S.-i. Tadaki, and S. Yukawa. Observations, theories, and experiments for freeway traffic as physics of many-body system, this volume.
13. B. S. Kerner and H. Rehborn. Experimental properties of complexity in traffic flow. *Phys. Rev. E*, 53(5):R4275–R4278, 1996.
14. B. S. Kerner and H. Rehborn. Experimental properties of phase transitions in traffic flow. *Phys. Rev. Letters*, 79(20):4030–4033, 1997.
15. B. Persaud, S. Yagar, D. Tsui, and H. Look. Breakdown-related capacity for freeway with ramp metering. *Transportation Research Record*, 1748:110–115, 2001.

16. H. Zackor, R. Kühne, and W. Balz. Untersuchungen des Verkehrsablaufs im Bereich der Leistungsfähigkeit und bei instabilem Fluß. Forschung Straßenbau und Straßenverkehrstechnik 524, Bundesminister für Verkehr, Bonn–Bad Godesberg, 1988.

17. L. Elefteriadou. Personal communication, Nov 2001.

18. R. Herman and I. Prigogine. A two-fluid approach to town traffic. *Science*, 204:148–151, 1979.

19. DYNAMIT www page. See its.mit.edu and dynamictrafficassignment.org, accessed 2003.

20. DYNASMART. See www.dynasmart.com and dynamictrafficassignment.org, accessed 2003.

21. C. Gawron. An iterative algorithm to determine the dynamic user equilibrium in a traffic simulation model. *International Journal of Modern Physics C*, 9(3):393–407, 1998.

22. E.M. Lifschitz and L.P. Pitajewski. Statistische Physik, Teil 1. Lehrbuch der Theoretischen Physik. Akademie-Verlag, 1987.

23. S. Krauß. *Microscopic modeling of traffic flow: Investigation of collision free vehicle dynamics*. PhD thesis, University of Cologne, Germany, 1997. See www.zaik.uni-koeln.de/~paper.

24. D. Helbing and T. Platkowski. Drift- or fluctuation-induced ordering and self-organization in driven many-particle systems. *Europhysics Letters*, 60(2):227–233, 2002.

25. D. Jost. Breakdown and recovery in traffic flow models. Master's thesis, ETH Zurich, Switzerland, 2002. See e-collection.ethbib.ethz.ch.

26. K. Nagel. Life-times of simulated traffic jams. *International Journal of Modern Physics C*, 5(3):567–580, 1994.

27. K. Nagel and M. Paczuski. Emergent traffic jams. *Phys. Rev. E*, 51:2909–2918, 1995.

28. M. Sasvári and J. Kertész. Cellular automata models of single lane traffic. *Phys. Rev. E*, 56(4):4104–4110, 1997.

29. L. Roters, S. Lübeck, and K.D. Usadel. Critical behavior of a traffic flow model. *Phys. Rev. E*, 59(3):2672–2676, 1999.

30. D. Chowdhury, J. Kertész, K. Nagel, L. Santen, and M. Schadschneider. Comment on: "Critical behavior of a traffic flow model". *Phys. Rev. E*, 61(3):3270–3271, 2000.

31. R. Barlovic, L. Santen, A. Schadschneider, and M. Schreckenberg. Metastable states in CA models for traffic flow. *European Physical Journal B*, 5(3):793–800, 1998.

32. K. Nagel and M. Schreckenberg. A cellular automaton model for freeway traffic. *Journal de Physique I France*, 2:2221–2229, 1992.

33. M. Bando, K. Hasebe, A. Nakayama, A. Shibata, and Y. Sugiyama. Dynamical model of traffic congestion and numerical simulation. *Phys. Rev. E*, 51(2):1035–1042, 1995.

34. R.D. Kühne and R. Beckschulte. Non-linearity stochastics of unstable traffic flow. In C.F. Daganzo, editor, *Proc. 12th Int. Symposium on Theory of Traffic Flow and Transportation*, page 367. Elsevier, Amsterdam, The Netherlands, 1993.

35. E. Levine, D. Mukamel, and G. Ziv. Condensation transition in zero-range processes with diffusion. *J. Stat. Mech.: Theor. Exp.*, P05001, 2004.

A Wave-Based Resolution Scheme for the Hydrodynamic LWR Traffic Flow Model

V. Henn

LICIT (INRETS–ENTPE)
Rue Maurice Audin, 69 518 Vaulx-en-Velin cedex – France
vincent.henn@entpe.fr

Summary. Resolution of LWR model is considered. The method proposed in this paper differs from (continuous) characteristic based analytical resolutions or from (discretized) finite difference schemes. It is based on an approximation of the flow-density relationship which yields a solution with piecewise constant density. This solution is then exactly calculated by tracking waves and handling their collisions. Extensions for incorporating boundary conditions such as traffic signals or discontinuity of the flow-density diagram are considered and provide an illustration of the potentiality of the method.
Keywords: Continuum traffic model; shock wave; rarefaction wave

1 Introduction

The LWR model was introduced separately by Lighthill, Whitham [1] and Richards [2]. It is based on a hydrodynamic analogy, supposing that the discrete stream of vehicles could be represented by a continuous flow. This model has received a great interest among scientists in particular because of its simplicity which makes it possible to draw analytical solutions.

But actually such analytical solutions are generally restricted to some isolated cases or simplistic scenarios where they can be derived so that numerical methods have been developed for a wider use. Those methods are mainly based on finite difference principle and more precisely on Godunov scheme which proved to be very efficient and enables a lot of extensions.

However, the numerical solutions provided by such a scheme are built in a very different way than analytical ones. More precisely, they are not made of waves which are the fundamental elements of the LWR theory. This is not a problem as long as only density and flow are of interest (since they are correctly approximated) but it may become one when trying to "understand" a numerical solution by the way of the LWR theory or, conversely, to illustrate LWR theory with numerical solutions.

This latter case was the initial reason why the author was involved in this research: the desire for a small computer program which would be able to calculate automatically analytical solutions in a pedagogical purpose, in order to illustrate LWR theory on some small examples for traffic courses.

When developing such a program, it appears that a sound numerical wave based method could be of use for this purpose: the *Wave Tracking method*. This method is based on the approximation of the fundamental flow-density relationship and on an explicit tracking of waves. It was initiated by Dafermos in the 70's [3] and received more mathematical foundations in the late 80's and in the 90's by Holden, Risebro, Bressan and others (see the historical notes of [4] or the web page by Lie http://www.math.ntnu.no/~andreas/fronttrack/).

When applying this method to the LWR resolution case it comes that it could be used for the initial educational purpose, but also as a full numerical method and that it could even be of a help for solving some of the limits or problems of the Godunov scheme.

In this paper, only the single road case will be presented without any network consideration. After a review of the LWR model and its "classical" resolution methods, we will present the basic features and a formal application of the Wave Tracking method to an infinite road with initial density conditions. We will then see how boundary conditions can be incorporated in order to represent more realistic roads with finite length, traffic signals, capacity reduction, etc. Finally, a small theoretical example will illustrate the potentiality of the method.

2 The LWR Traffic Flow Model

2.1 Model Hypothesis and Equations

The LWR model is a macroscopic model which is based on a description of the traffic flow rather than on a representation of the trajectory of vehicles. It considers traffic as a homogeneous and continuous stream characterized by the following three variables:

- The density K (number of vehicles by space unit)
- The flow Q (number of vehicles passing through a position by time unit)
- The flow speed V (speed of vehicles composing the stream)

Those variables are linked by the following relation:

$$Q = K \times V \tag{1}$$

The main hypothesis of the model is that traffic is always and everywhere in an equilibrium state such that traffic flow and speed only depend on the density of traffic at the same position. So there exists a *fundamental relation* defining flow and speed as functions of density:

$$V(x,t) = V_E(K(x,t)) \tag{2}$$
$$Q(x,t) = V_E(K(x,t)) \times K(x,t) = Q_E(K(x,t)) \tag{3}$$

Q_E is supposed to be concave and reaches its maximum value at a critical density: $Q_{max} = Q_E(K_{crit}) = \max Q_E(K)$; it is null for null density and for K_{max}, the maximum density, which corresponds to traffic jam.

In addition to the above equations, the conservation of vehicles has to be considered on a portion of road (excluding intersections). This conservation law can be written (as classically in fluid dynamic):

$$\frac{\partial K(x,t)}{\partial t} + \frac{\partial Q(x,t)}{\partial x} = 0 \tag{4}$$

The LWR model can thus be written into a single partial differential equation in K:

$$\frac{\partial K(x,t)}{\partial t} + \frac{\partial Q_E(K(x,t))}{\partial x} = 0 \tag{5}$$

and flow and speed are calculated afterwards using Eq. 2–3.

2.2 Analytical Resolution

It is interesting to note that Eq. 5 belongs to the family of hyperbolic conservation laws which are quite well known (interested readers may refer to some reference books such as [4–6]). The general method for analytically derive its solutions will be briefly reminded hereafter: it is based on the study of the propagation of density along *characteristic lines* where, by construction it is constant.

Characteristic Lines – Kinematic Waves

Let us define a *characteristic* curve (sometimes also called a *kinematic wave*) to be a continuous set of points with the same density. Such a curve can be parameterized by t such that it is the set of points of positions $(\chi(t), t)$. By definition, density must be constant along this curve:

$$\frac{\partial K(\chi(t), t)}{\partial t} = 0$$

From Eq. 5, it comes easily that $\chi'(t) = Q'_E(K_0)$ where K_0 is the value of the density along the curve. In particular, this means that the slope is constant so that characteristic curves are straight lines. Let us note $\mathcal{W}(K)$ the speed of the kinematic wave propagating density K:

$$\mathcal{W}(K) = Q'_E(K) \tag{6}$$

Such a property makes the resolution of the LWR model quite easy since in order to know the density profile $K(\cdot, t)$ at a time t, we only have to consider the propagation of the initial profile $K(\cdot, t_0)$ at time t_0 along characteristic lines.

It is to be noted that since Q_E is a concave function, its derivative is decreasing (everywhere it is defined). In particular, this means that density propagates all the slower when it is greater.

Solution of a Riemann Problem

In a *Riemann problem* the initial conditions are a single density discontinuity at a given position (x_0, t_0):

$$\mathcal{R}(x_0, t_0) : \quad \begin{cases} \forall x < x_0, \ K(x, t_0) = K_{\text{up}} \\ \forall x > x_0, \ K(x, t_0) = K_{\text{down}} \end{cases} \tag{7}$$

When solving a Riemann problem, the objective is to calculate the density profile at times $t > t_0$.

Two cases may arises:

- If $K_{\text{up}} < K_{\text{down}}$ (increasing density jump)
  Then the upstream density will propagate faster than the downstream density and characteristic lines will collide. Such a collision corresponds to a density discontinuity and the set of all those collisions at different times can be seen as the propagation of the discontinuity observed at t_0.
  It is easy to show from Eq. 5 that the speed W of propagation of the discontinuity point is given by the Rankine-Hugoniot formula:

$$W(K_{\text{up}}, K_{\text{down}}) = \frac{Q_E(K_{\text{up}}) - Q_E(K_{\text{down}})}{K_{\text{up}} - K_{\text{down}}} \tag{8}$$

  In particular, this means that the discontinuity is stable over time and travels along a linear *shock wave* (see Fig. 1).
- If $K_{\text{up}} > K_{\text{down}}$ (decreasing density jump)
  Then, on the contrary, in this case, the characteristics emanating from the upstream and downstream zones will not collide since they diverge.
  In order to finish solving the Riemann problem, we have to add a condition to know what happens between those characteristics. This condition corresponds to some entropic criterion. It can be interpreted as if at position (x_0, t_0) not only K_{up} and K_{down} densities were present but also all densities $K \in [K_{\text{up}}; K_{\text{down}}]$.
  This implies that such densities also propagate along their characteristic lines and finally form a continuous fan, called a *rarefaction wave*, as can be seen on Fig. 2.

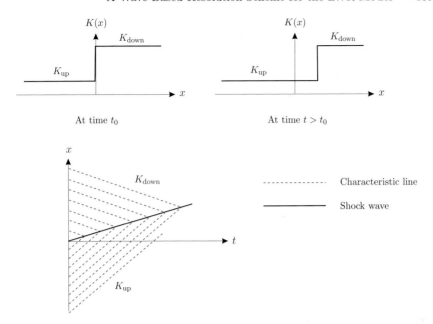

Fig. 1. Resolution of a Riemann problem with a increasing density jump which results in a shock wave.

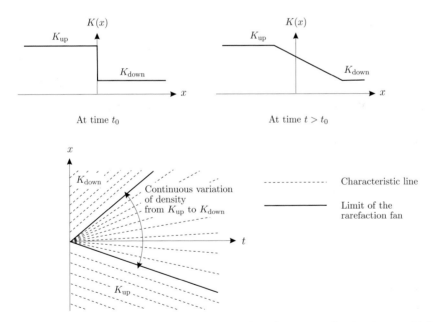

Fig. 2. Resolution of a Riemann problem with a decreasing density jump which results in a rarefaction fan.

Solution of the General Problem

In order to solve the general problem of the LWR model on a road, initial conditions have to be considered. At time $t = 0$, the state of the road is given by:

$$K(x,0) = k_0(x) \tag{9}$$

Where $k_0(\cdot)$ is the initial density profile of the road.

If this function is piecewise constant, it implies a discrete number of density discontinuities located at $\{x_1, x_2, \dots\}$. Since the speed of density propagation (that is the speed of characteristic lines) is given by the derivative of the flow-density diagram $(Q_E{}')$, which is supposed to be concave, it is in particular bounded by $Q_E{}'(0)$ which is equal to the *free flow speed* V_f, that is the speed a vehicle would have it it was alone on the road. This speed is of course finite, so that two adjacent discontinuities located at x_i, x_{i+1} cannot interact instantaneously.

As a consequence, each initial discontinuity can be solved independently as an isolated Riemann problem $\mathcal{R}(x_i, 0)$, generating a shock wave or a rarefaction fan. Later, those different generated waves may intersect and imply other discontinuities that can also be solved as Riemann problems, etc. It is much more complicated to consider the case where a rarefaction fan intersect with a shock wave or even with another fan. Indeed, because of the continuous nature of fans, this implies a continuous set of discontinuities to be considered.

Then it appears that, in the general case, the LWR model is not easy to solve analytically. Solutions have been drawn but restricted to some particular situations such as for example traffic signals with constant demand [7, 8] or isolated incident [9, 10] or to some particular flow-density relation, such as Newell [11] using triangular shaped relationship or Wong and Wong [12] using Greenshield parabolic relationship. For this reason, numerical solutions are needed and will be presented in the next section.

2.3 Space–Time Discretization for a Numerical Resolution

For numerically solving the LWR model in the general case, different solutions have been proposed in the literature (see for example [13, 14]). The main idea is to discretize space and time in order to transform the conservation equation from a partial derivative equation (Eq. 4) into a finite difference one.

Presentation of the General Scheme

If we note:

- Δx and Δt the time and space discretization steps
- Q_x^t the flow at position $x\Delta x$, supposed to be constant during period $[t\Delta t; (t+1)\Delta t]$

- K_x^t the density at time $t\Delta t$, supposed to be uniform on a discretization cell $[x\Delta x; (x+1)\Delta x]$

then the flow conservation writes:

$$K_x^{t+1} = K_x^t + \frac{\Delta t}{\Delta x}(Q_x^t - Q_{x+1}^t) \tag{10}$$

While density is calculated through this conservation equation, the remaining issue is to compute the flow from cell to cell.

The Godunov Discretization Scheme

An interesting method to be considered is the one proposed by the Godunov scheme. Indeed, at each time $t\Delta t$, it takes into account the density jumps at each position $x\Delta x$ and solves the Riemann problems $\mathcal{R}(x\Delta x, t\Delta t)$.

If the space step Δx is sufficiently big regarding the time step Δt, two adjacent Riemann problems do not interfere before a time step has elapsed. This condition (known as the Courant-Friedrich-Lewy condition) ensures the stability of the numerical scheme and then its convergence towards the analytical solution when time and space steps tend to zero [6, p. 110].

This scheme is very efficient since it has been proved that the flow is constant during a time step and it could be easily calculated by the following formula [15]:

$$Q_x^t = \min\left\{\Delta(K_{x-1}^t); \Sigma(K_x^t)\right\} \tag{11}$$

where Δ and Σ are demand and supply functions defined by:

$$\Delta(K) = \begin{cases} Q_E(K) & \text{if } K < K_{crit} \\ Q_{max} & \text{if } K \geq K_{crit} \end{cases} \qquad \Sigma(K) = \begin{cases} Q_{max} & \text{if } K < K_{crit} \\ Q_E(K) & \text{if } K \geq K_{crit} \end{cases} \tag{12}$$

Boundary conditions on the flow are almost straightforward to introduce in such a scheme. Indeed, only the demand or supply functions are to be modified. For example capacity reduction (due for example to a change of the number of lanes), traffic lights or incidents can be modeled by limiting the supply locally (see for example [16]).

3 Application of the Wave Tracking Resolution Scheme to the LWR Model

3.1 Motivation and Principles of the Wave Tracking Method

As we have just seen, the efficiency of the Godunov scheme is great. However, it would be interesting to have at our disposal an approximated solution of the LWR model which would be based on its fundamental elements (namely shock and rarefaction waves).

In particular this means that we are searching for a *continuous* solution in the space-time diagram in order for example to get rid of the rigidity of the discretization grid which may not necessarily correspond to some boundary conditions for example.

The problem is that such continuous functions are much more complicated to handle in a practical point of view than functions presenting discontinuity especially if they are constant or linear between discontinuity points (even if, from a theoretical point of view, mathematicians often prefer continuous or even smooth functions which ensure proper behaviour).

More precisely, the difficulty of handling a continuous representation of density is caused by rarefaction fans which are continuous, whereas shock waves are discrete and, furthermore have constant speed. So the question is: can we build an approximated solution to the LWR model which would only be made of (linear) shock waves limiting constant density zones?

The answer to this question is (fortunately) positive and the *Wave Tracking method* is based on this idea. Actually the solution it provides is based on a piecewise linear approximation of the flow-density relationship. We will see that a solution of such a type exactly responds to the objectives we have just fixed and verifies the following statements:

- It verifies the LWR model everywhere (ensuring proper behaviour).
- It is only made of linear shock waves separating constant density zones, hence it is easy to handle.

The presentation of the method is quite tricky. Indeed the description of the algorithm is based on this peculiarity of the piecewise linear flow-density relationship to generate only discrete shock waves (and no continuous fan). But in order to show this interesting behaviour, we have to make use of the resolution scheme. We decided to solve this difficulty by the following steps:

- We will first show that, if the flow-density relation is piecewise linear, then a Riemann problem results in the generation of one or several shock waves (§3.2). This will help the reader understand the special behaviour introduced by this piecewise linearity property.
- After this, we will describe formally the method (§3.3).
- We will then explain why it can effectively be applied (§3.4): discontinuities can actually be solved independently by the algorithm (as *independent* Riemann problems) and no combinatorial issue will be encountered.
- Finally, we will show that the method is efficient, that is it gives a good approximation of the real solution (§3.5).

3.2 Resolution of Riemann Problems with Piecewise Linear Flow-Density Diagram

The idea for getting rid of continuous rarefaction waves and to keep only discrete shock waves is to use a piecewise linear flow-density relationship.

Indeed as we are going to see, in such a case, a decreasing density discontinuity does not result in continuous rarefaction fan but in a discrete set of constant density zones separated by linear shock waves. [1]

For demonstrating this property, let us consider a piecewise linear fundamental relationship Q_E. Let us note $\{K_0, K_1 \ldots K_n\}$ the (ordered) densities where it presents angular points and $Q_i = Q_E(K_i)$ the flow at those densities.

By construction, the slope of Q_E is constant between two consequent densities K_{i-1} and K_i. It is equal to:

$$W_i = \frac{Q_i - Q_{i-1}}{K_i - K_{i-1}} \tag{13}$$

We can note that this slope is equal to the speed of propagation of any shock wave separating two zones of density K_{i-1} and K_i (Eq. 8):

$$W_i = W(K_{i-1}, K_i) \tag{14}$$

Furthermore it is also equal to the speed of propagation of any density in the interval $[K_{i-1}, K_i]$ given by Eq. 6:

$$\forall K \in [K_{i-1}, K_i], \quad \mathcal{W}(K) = W_i \tag{15}$$

Since Q_E is not differentiable at its angular points K_i, its derivative is not strictly defined at those points. Nevertheless, we can consider that the speed of propagation of densities K_i can take any value between W_i and W_{i+1}.

Let us consider a *decreasing* density Riemann problem $\mathcal{R}(x_0, t_0)$ and let us first suppose that K_{up} and K_{down} are "around" an angular point of the flow-density relation, that is there exists $i \in \{0, \ldots n\}$ such that:

$$K_{i-1} < K_{\text{down}} < K_i < K_{\text{up}} < K_{i+1}$$

If we suppose that at point (x_0, t_0) all the densities of $[K_{\text{down}}, K_{\text{up}}]$ are present (that is to say that the discontinuity is actually the limit of a continuous density profile), they will all propagate along their own characteristic or kinematic wave at speed $\mathcal{W}(K)$. But Eq. 15 implies that there are only two different propagation speeds (except for K_i), namely W_i and W_{i+1}.

As for density K_i, it also propagates along its characteristics which speed can take any value between W_i and W_{i+1}, so that any point between the two lines of speeds W_i and W_{i+1} can be reached by a characteristic of density K_i.

This implies that the $(K_{\text{up}}; K_{\text{down}})$ discontinuity will actually split into two discontinuities $(K_{\text{up}}; K_i)$ and $(K_i; K_{\text{down}})$ propagating at speeds W_i and W_{i+1} respectively (and verifying the Rankine Hugoniot condition Eq. 8). So that finally the Riemann problem results in two shocks separating the initial densities K_{up} and K_{down} and an intermediate state of density K_i. In particular, there is no continuous rarefaction wave (see Fig. 3).

---

[1] This property is nothing really new and original (see for example [17, §4.4] or [18]), but we present here a quite detailed analysis for sake of completeness of the present paper.

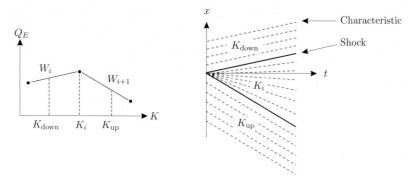

Fig. 3. Resolution of a Riemann problem near an angular point with a piecewise linear flow-density relationship.

It is easy to generalize such a reasoning to any K_{up} and K_{down} verifying:

$$K_{i-1} < K_{\text{down}} < K_i < \cdots < K_j < K_{\text{up}} < K_{j+1}$$

and to conclude that it results in a discrete set of zones of constant densities $K_i, \ldots, K_j$, limited by classical shock waves of speeds $W_i, \ldots, W_{j+1}$ (see Fig. 4). [2]

As for a Riemann problem with an *increasing* density jump $K_{\text{up}} < K_{\text{down}}$ it results classically in a shock wave of speed $W(K_{\text{up}}, K_{\text{down}})$.

So, in conclusion, when the flow-density is piecewise linear, any Riemann problem generates one or several waves depending on the position of the up- and down-stream initial densities.

3.3 Formal Description of the Wave Tracking Method

Definition of the Method

Given an initial problem $\mathcal{P}$:

$$\mathcal{P}: \begin{cases} \dfrac{\partial K(x,t)}{\partial t} + \dfrac{\partial Q_E(K(x,t))}{\partial x} = 0 \\ K(x,0) = k_0(x) \end{cases} \tag{16}$$

---

[2] It is interesting to note that those shock waves inside a rarefaction fan correspond to *decreasing* density jumps and hence do not respect the Ansorge's "ride impulse" principle for entropic solution [19]. But this does not mean that the constructed solution does not verify the maximum entropy condition Indeed, as Velan and Florian [18] noticed, this principle is only valid for smooth flow-density, which is not the case here, by construction.

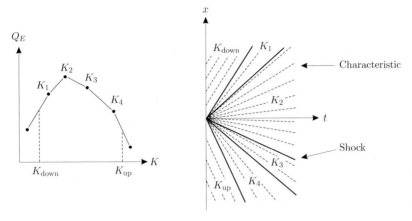

Fig. 4. General resolution of a Riemann problem with a piecewise linear flow-density relationship.

the Wave Tracking method approximates the solution $K(x,t)$ by building the *exact* solution $\widetilde{K}(x,t)$ of the following approximated problem:

$$\widetilde{\mathcal{P}}: \quad \begin{cases} \dfrac{\partial K(x,t)}{\partial t} + \dfrac{\partial \widetilde{Q}_E(K(x,t))}{\partial x} = 0 \\ K(x,0) = \widetilde{k}_0(x) \end{cases}$$

where $\widetilde{Q}_E$ is a piecewise linear approximation of the flow-density relationship and $\widetilde{k}_0$ is a piecewise constant approximation of the initial density profile of the road.

We have just seen that this linear property implies discrete rather than continuous solutions. We can thus focus on those discrete discontinuities for building such solutions, which will be far easier than constructing a continuous solution in the general case when using the characteristic based method we described earlier (§ 2.2).

The main issue is to effectively build this solution $\widetilde{K}$. This construction is based on two points: the propagation of waves and their collisions.

Propagation and Intersection of Waves

While propagation is straightforward since waves have constant speed, collision is neither very difficult. Indeed, when two adjacent waves $w_{ab} = W(K_a, K_b)$ and $w_{bc} = W(K_b, K_c)$ collide at (x,t), this yields a Riemann problem $\mathcal{R}(x,t)$ with $K_{\text{up}} = K_a$ and $K_{\text{down}} = K_c$. Since wave w_{ab} has reached w_{bc}, it means it goes faster and it is easy to see that this necessarily implies $K_a < K_c$. In particular this implies that $\mathcal{R}(x,t)$ will result in only one shock wave $w_{ac} = W(K_a, K_c)$. Then it is quite easy to build the entire solution by sequentially generating waves at new discontinuity points (= collision points) and tracking them till they collide.

Formal Algorithm

More formally, the solution of the approximated problem can be calculated by the following algorithm:

1. At time t the density profile $\widetilde{K}(\cdot, t)$ is considered. Each density discontinuity is solved locally as a Riemann problem and results in the generation of one shock wave separating two constant density zones.
2. The waves are tracked till the next possible collision.
3. A collision between two shock waves at time t' results in a new density discontinuity in the density profile $\widetilde{K}(\cdot, t')$.
4. If end of simulation is reached or if there is no more density discontinuity (that is $\widetilde{K}(\cdot, t')$ is constant), then algorithm stops and resolution is finished. Else, go back to item 1 with $t := t'$.

Of course, such an algorithm is a bit more complicated to implement than the Godunov scheme since it does not use a fixed set of data (with fixed space and time steps). However, evolved computer languages such as C++ give powerful tools to implement ordered lists with deletion or insertion capabilities, so that it is not too difficult to program this algorithm (see [20] for an example).

3.4 The Approximated Solution can be Effectively Calculated by Wave Tracking

We will see in this section that the proposed algorithm can effectively be applied. For this, we have to verify that (1) density discontinuities can be solved separately and (2) that the proposed algorithm effectively finds the solution in a finite number of operations (without falling down into infinite loops for example).

Density Discontinuities can be Considered Separately

We have noticed, when dealing with the analytical resolution, that since propagation speeds were finite, density discontinuities could not interact immediately. This was the reason for solving discontinuity locally as Riemann problems.

In our case this property is still valid and it means that the proposed algorithm can effectively solve each discontinuity by generating one new shock wave. Furthermore, we are certain that starting from a piecewise constant density profile and propagating discontinuities, we will never have a continuously varying density profile which would prevent us from handling independently discontinuities.

So, in conclusion, the problem $\mathcal{P}$ is discrete and it remains so, so that we can treat it in a discrete way.

Effectiveness of the Algorithm

Another question that may arise is: how many waves will have to be considered by the algorithm in order to construct the entire solution? Indeed, for example, we could be afraid of some combinatorial explosion when the number of waves is high or we could even fear that the algorithm do not converge in a finite time.

The first point is quite direct to solve. Actually, even if a huge number of waves had to be treated, the number of collisions to be considered is just equal to this number. Indeed, two waves can only collide if they are adjacent, otherwise it would mean that one would have jumped over another one, which is impossible without collision.

In order to illustrate this issue, the previous formal algorithm can be detailed.

- *Initialization:*
  1. Waves are created at each discontinuity of $\widetilde{K}_0$.
     They are ordered so that the positions $x_i(t)$ of wave w_i at time $t = 0$ respect: $x_1(t) \leq x_2(t) \leq \cdots \leq x_n(t)$.
     A wave w_i separates two constant density zones $Z_{\mathrm{up}}(w_i)$ and $Z_{\mathrm{down}}(w_i)$. It is clear that two adjacent waves share a same zone: $Z_{\mathrm{down}}(w_i) = Z_{\mathrm{up}}(w_{i+1})$.
  2. Calculate the potential intersections dates t_i between adjacent waves w_i and w_{i+1}.
- *Repeat:*
  3. Find $t_i = \min_j t_j$. That is find w_i and w_{i+1} which will be the first ones to collide.
  4. Generate the new wave w^* between $Z_{\mathrm{up}}(w_i)$ and $Z_{\mathrm{down}}(w_{i+1})$.
  5. Replace w_i and w_{i+1} by w^* in the wave list and suppress $Z_{\mathrm{down}}(w_i) = Z_{\mathrm{up}}(w_{i+1})$ in the zone list.
  6. Update the potential intersections dates t_{i-1} between w_{i-1} and w^* and t_{i+1} between w^* and w_{i+2}.
  Until end of the simulation is reached

Moreover, it is clear from this algorithm that the number of waves decrease at each iteration: two colliding waves are replaced by one new wave w^*. If we note m the number of initially generated waves at step 1, due to initial discontinuities in the density profile $\widetilde{k}_0$, then the number of iterations (steps 3–6) is equal to m. Step 3 takes $O(m)$, while steps 4–6 only takes $O(1)$, so that the number of operations of the algorithm is $O(m^2)$ (which can be even reduced by a proper implementation of data in an ordered structure).

3.5 Efficiency of the Method

Holden *et al.* have demonstrated [21] that the wave tracking method was an efficient numerical scheme that is:

- It is *stable*: the numerical errors introduced by the method do not increase over time.
- It is *convergent*: the approximated solution $\widetilde{K}$ converges towards the real solution K when the approximation of the flow-density relationship $\widetilde{Q}_E$ tends to the initial relationship Q_E.

(Pedagogical presentations of the needs for a numerical scheme have been made by Zhang [22] or by LeVeque [6].)

In other words, this means that when precision (defined by the distance between the initial data Q_E and k_0 and the approximated ones $\widetilde{Q}_E$ and $\widetilde{k}_0$) is getting higher, the approximated solution is getting closer to the real one and, furthermore, for a given precision, the distance between those two solutions is "acceptable".

It could be argued against the method that the approximated solution is less physically relevant than the real one since it introduces some discontinuities in the speed field (located at the density discontinuities) hence some problems for the definition of acceleration. Indeed, while acceleration might be defined (since speed is smooth) in rarefaction fans if the flow-density is smooth, in the approximated solution it is either null (in constant density zones) or undefined (on the shocks waves). But it is to note that such a smooth speed does not imply a physically admissible acceleration. Furthermore, the problem is rather the admissibility of kinematic profiles than the definition of acceleration, and this could be obtained by approximating (by wave tracking) the solution of an extension of LWR model considering physical bounds for acceleration (see [23, 24] or [25] for such type of extension).

4 Introducing Boundary Conditions

Till now, we have only considered the LWR model given initial conditions $k_0(x) = K(x, 0)$. This means that we are now able to represent the traffic on an infinite road, given its initial state. In order to represent more realistic situations, we need to incorporate some boundary conditions such as a traffic demand at the entrance of the road, a capacity at the exit of the road, some traffic signals, etc.

All those conditions are conditions on the flow rather than on the density and they are all located at a given place. We can consider them as special waves which are stationary and whose behaviour in case of collision with another wave is particular. Such special waves can appear at the beginning of the simulation and vanish at the end (such as the waves representing the entrance and exit of the road) or they can appear and vanish during the simulation, at some fixed times (such as the wave corresponding to the red phase of a traffic signal).

Three types of events have to be considered: (1) the appearance or (2) the vanishing of a special wave and (3) the collision of a shock wave with a special wave. Each of those events can be represented by a special discontinuity problem which will be called an *extended Riemann problem*.

4.1 Extended Riemann Problems

An *extended Riemann problem* $R^*(x_0, t_0, Q^*)$ is defined by a density profile at time t_0 presenting a (possible) discontinuity at position x_0 and by a boundary condition at position x_0 constraining the flow by a function Q^* (depending on the nature of the boundary condition to be represented):

$$R^*(x_0, t_0, Q^*): \quad \begin{cases} \forall x < x_0, & K(x, t_0) = K_{\text{up}} \\ \forall x > x_0, & K(x, t_0) = K_{\text{down}} \\ \forall t > t_0, & Q(x_0, t) = Q^* \end{cases} \tag{17}$$

The flow has to be conserved through the spatial discontinuity, so that for times $t > t_0$, the density just upstream K_{up}' and just downstream K_{down}' must verify:

$$Q_E(K_{\text{up}}') = Q_E(K_{\text{down}}') = Q^* \tag{18}$$

The problem is then to find K_{up}' and K_{down}' to verify such equation. Indeed, given the shape of the flow-density relation two density values are eligible for this equation. In order to fix the proper one, we need to know whether the traffic is congested or not. Let us note λ_{up} and λ_{down} the initial state of the traffic and λ_{up}' and λ_{down}' the state of traffic at time $t' > t$, just upstream and downstream the discontinuity. We further need to define the inverse equilibrium flow-density relationship Q_E^{-1}. It is defined such that for $K = Q_E^{-1}(Q, \lambda)$: $Q_E(K) = Q$ and, if $\lambda =$ "uncongested", then $K < K_{\text{crit}}$, else, if $\lambda =$ "congested", then $K > K_{\text{crit}}$.

So finally, defining:

$$K_{\text{up}}' = Q_E^{-1}(Q^*, \lambda_{\text{up}}') \tag{19}$$

$$K_{\text{down}}' = Q_E^{-1}(Q^*, \lambda_{\text{down}}') \tag{20}$$

ensures that the flow is conserved through the discontinuity.

The next step is to consider the discontinuities introduced by K_{up}' and K_{down}'. Indeed, they may differ from the initial K_{up} and K_{down} and the two following (classical) Riemann problems have to be solved at position (x_0^-, t_0) just upstream the discontinuity and (x_0^+, t_0) just downstream:

$$R(x_0^-, t_0): \quad \begin{cases} \forall x < x_0^-, & K(x, t_0) = K_{\text{up}} \\ \forall x > x_0^-, & K(x, t_0) = K_{\text{up}}' \end{cases}$$

$$R(x_0^+, t_0): \quad \begin{cases} \forall x < x_0^+, & K(x, t_0) = K_{\text{down}}' \\ \forall x > x_0^+, & K(x, t_0) = K_{\text{down}} \end{cases}$$

Those problems may (classically) generate one or several waves originated at (x_0, t_0), depending on the respective positions of the densities.

So finally, when considering boundary conditions, we only have to incorporate special (stationary) waves propagating the Q^* information and to handle collisions with such waves as extended Riemann problems. In particular, the previously proposed algorithm is still valid as soon as we add the appearance times of those special waves in the list of events to be considered. We also have to keep in mind that now step 4 may correspond to an *extended* Riemann problem and thus may generate several waves instead of only one. In particular, this means that the complexity of the algorithm is the same as previously but considering the number of initially generated waves plus the number of waves generated by each special wave as it appears, vanishes or is collided.

4.2 Application to Specific Boundary Conditions

In order to be complete, we still need to express the Q^* constrain as well as the λ_{up}' and λ_{down}' states of traffic for some particular boundary conditions.

Traffic signals

A red traffic signal compels the flow to be equal to zero. So that in this case, $Q^* = 0$.

The beginning of the red phase leads to the creation of two densities $K_{up}' = K_{max}$ and $K_{down}' = 0$ and the classical Riemann problems yield the generation of two shock waves.

On the other hand, the end of a red phase corresponds to the vanishing of the constrain Q^*, so that a classical Riemann problem can be solved between $K_{up} = K_{max}$ and $K_{down} = 0$, generating several shock waves.

Demand at the Road Entrance

Let us note $\Delta_0(t)$ the demand of traffic at the entrance of the road. Let us suppose it is piecewise constant (or approximated by a piecewise constant function).

Each variation of this demand as well as each intersection of a shock wave with the special wave corresponding to the entrance of the road introduces an extended Riemann problem at $x = 0$ where only the downstream part is of interest but which is similar to other ones with the flow constrains $Q^* = \min\{\Delta_0; \Sigma(K_{down})\}$.

Several cases may appear:

- If $\lambda_{down} =$ "uncongested", then it is easy to see that $Q^* = \Delta_0(t)$ and $\lambda_{down}' =$ "uncongested" so that $K_{down}' = Q_E^{-1}(\Delta_0(t), \text{"uncongested"})$.

Solving the classical Riemann problem between $K_{down}{}'$ and K_{down} may result in the generation of one or several waves depending if $K_{down}{}'$ is greater or lesser than K_{down}.

- If λ_{down} = "congested", then:
  - if $Q_E(K_{down}) > \Delta_0(t)$, then $Q^* = \Delta_0(t)$ and $\lambda_{down}{}'$ = "uncongested" and $K_{down}{}' = Q_E^{-1}(\Delta_0(t),$ "uncongested"). Solving the Riemann problem will lead to the generation of one shock wave (since $K_{down}{}' < K_{crit} < K_{down}$).
  - if $Q_E(K_{down}) < \Delta_0(t)$ then $Q^* = Q_E(K_{down})$ and $K_{down}{}' = K_{down}$ and in particular there is no shock.

Spatial Discontinuity of Flow-Density Profile

In a general case, it can be assumed that the flow density-relation is only piecewise constant and that it changes at a given position (where a capacity reduction or a road enlargement is observed).

While shock waves do not intersect the stationary wave representing this discontinuity, their behaviour is unchanged and when they collide that special wave, we have to study an extended Riemann problem which constrain is given by [15]: $Q^* = \min\{\Delta(K_{up}); \Sigma(K_{down})\}$ with the demand and supply functions defined earlier (Eq. 12) and calculated respectively with the upstream and downstream flow-density relationships.

Many different cases may appear depending on the positions of the densities, they are detailed in [15]. In appears from their study that the following rule could be used for determining the upstream and downstream states $\lambda_{up}{}'$ and $\lambda_{down}{}'$:

$$(\lambda_{up}{}'; \lambda_{down}{}') = \begin{cases} (\text{"congested"}; \lambda_{down}) & \text{if } \Delta(K_{up}) > \Sigma(K_{down}) \\ (\lambda_{up}; \text{"uncongested"}) & \text{otherwise} \end{cases} \quad (21)$$

Once those states are known, the densities just upstream and downstream the discontinuity can be calculated thanks to Eq. 19–20 and classical Riemann problems can be solved, generating one or several waves (or none). It is interesting to note that the proposed scheme is quite direct and without considering explicitly the different possible cases it is perfectly able to reconstruct them.

4.3 General Illustration

As an illustration of the method, we propose a small theoretical example. We consider a 3 lane road with a 2 lane capacity reduction and a traffic signal pulsing the flow. Figure 5 depicts the approximated solution $\widetilde{K}(x,t)$ with a 50 segment approximation of the flow-density relationship. It looks rather continuous than piecewise constant and shock waves look really smooth while they are not.

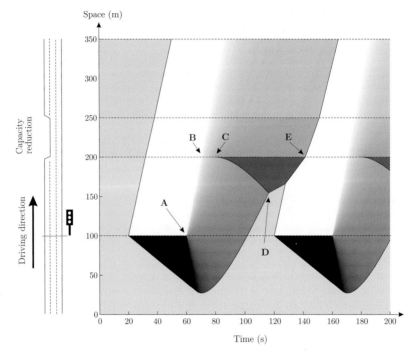

Fig. 5. Density and shock waves on a road with traffic signal at $x = 100\,\mathrm{m}$ and capacity reduction for $x \in [200\,\mathrm{m}; 250\,\mathrm{m}]$ (the darker the color, the denser the traffic).

Different phases can be seen on the figure: (**A**) The classical rarefaction wave appearing at the start of the green phase, with its continuous looking (even they are approximated here). (**B**) At the beginning, the platoon moving off after the signal turns green is sufficiently light to pass through the capacity reduction. (**C**) But as times goes up, it becomes denser and cannot drive through anymore thus generating congestion moving backward. (**D**) Once the platoon has gone, this congestion can diminish and even disappear (**E**).

5 Conclusion and Perspectives

We have presented in this paper a method for solving the LWR model. It is based on the iterative construction of the exact solution of an approximated problem with a piecewise linear flow-density relationship.

Such a method turned out to be quite efficient when incorporating extensions of the LWR model in order to represent different boundary conditions. Actually, other extensions have been studied elsewhere such as the representation of a bus moving inside a traffic flow and acting like a moving bottleneck

[26] or the consideration of network aspects in order to calculate dynamic traffic assignment [27].

All those extensions towards a more complete model have lead us much further than our initial objective of a simple pedagogical tool. The issue could be now to wonder if such a method could be used as a practical method for solving LWR model in real applications. But actually, whatever might be the answer, it already provides a tool for calculating (almost) exactly the solution of LWR problems and such a tool could be very interesting for other numerical methods, for validation purposes but also for the design of such methods.

For example, Godunov like methods, as we have seen, are based on the exact resolution of Riemann problems at each time step and on the boundary of each cell. But in some cases, the close formulation proposed by Lebacque (Eq. 11) cannot be applied anymore because of the presence of a bus in the cell for example (see [24]). In such cases, the Wave Tracking method could be used locally for calculating the exact solution of the LWR model.

References

1. M.J. Lighthill and G.B. Whitham. On kinematic waves II. A theory of traffic flow on long crowded roads. *Proc. of the Royal Soc.*, A(229):317–345, 1955.
2. P.I. Richards. Shocks waves on the highway. *Oper. res.*, 4:42–51, 1956.
3. C.M. Dafermos. Polygonal approximations of solutions of the initial value problem for a conservation law. *J. Math. Anal. Appl.*, 38:33–41, 1972.
4. H. Holden and N.H. Risebro. *Front tracking for hyperbolic conservation laws*, volume 152 of *Applied mathematical sciences*. Springer, 2002.
5. E. Godlewski and P.-A. Raviart. *Hyperbolic systems of conservation laws*, volume 3/4 of *Mathématiques et applications*. Ellipses, 1990.
6. R.J. LeVeque. *Numerical methods for conservation laws*. Lectures notes in mathematics. Birkhäuser, Basel, 2nd edition, 1992.
7. S.C. Wirasinghe. Determination of traffic delays from shock-wave analysis. *Transpn. Res.*, 12:343–348, 1978.
8. P.G. Michalopoulos, G. Stephanopoulos, and V.B. Pisharody. Modeling of traffic flow at signalized links. *Transpn. Science*, 14:9–41, 1980.
9. B. Heydecker. Incidents and interventions on freeways. PATH Research report UCB-ITS-PRR-94-5, University of California, Berkeley, 1994.
10. H. Mongeot and J.-B. Lesort. Analytical expressions of incident-induced flow dynamic perturbations using the macroscopic theory and an extension of the lighthill and whitham theory. *Transpn. Res. Record*, 1710, 2000.
11. G.F. Newell. A simplified theory of kinematic waves in highway traffic; part I: general theory; part II: queueing at freeway bottlenecks. *Transpn. Res.-B*, 27(4):281–303, 1993.
12. S.C. Wong and G.C.K. Wong. An analytical shock-fitting algorithm for LWR kinematic wave model embedded with linear speed-density relationship. *Transpn. Res.-B*, 36(8):683–706, 2002.
13. C.F. Daganzo. A finite difference approximation of the kinematic wave model of traffic flow. *Transpn. Res.-B*, 29(4):261–276, 1995.

14. Ch. Buisson, J.-P. Lebacque, and J.-B. Lesort. Strada, a discretized macroscopic model of vehicular traffic flow in complex networks based on the Godunov scheme. In *CESA'96 IMACS Multiconference. Computational Engineering in Systems Applications*, pages 976–981, Lille, France, July 1996.

15. J.-P. Lebacque. The Godunov scheme and what it means for first order flow models. In J.-B. Lesort, editor, *Proc. of the 13th International Symposium on the Theory of Traffic Flow and Transportation*. Pergamon, 1996.

16. H. Mongeot. *Traffic incident modelling in mixed urban networks*. PhD thesis, Faculty of Engineering and applied science, Southampton, 1998.

17. C.F. Daganzo. *Fundamentals of transportation and traffic operations*. Pergamon, 1997.

18. S. Velan and M. Florian. A note on the entropy solutions of the hydrodynamic model of traffic flow. *Transpn. Science*, 36(4):435–446, 2002.

19. R. Ansorge. What does the entropy condition mean in traffic flow theory? *Transpn. Res.-B*, 24(2):133–143, 1990.

20. J.O. Langseth. On an implementation of a front tracking method for hyperbolic conservation laws. *Advances in engineering software*, 26:45–63, 1996.

21. H. Holden, L. Holden, and R. Høegh-Krohn. A numerical method for first order nonlinear scalar conservation laws in one dimension. *Comput. Math. Applic.*, 15:595–602, 1988.

22. H.M. Zhang. A finite difference approximation of a non-equilibrium traffic flow model. *Transpn. Res.-B*, 35:337–365, 2001.

23. L. Leclercq. *A traffic flow model for dynamic estimation of noise*. PhD thesis, INSA, Lyon, 2002. In french.

24. F. Giorgi, L. Leclercq, and J.-B. Lesort. A traffic flow model for urban traffic analysis: extensions of the LWR model for urban and environmental applications. In M. Taylor, editor, *Proc. of the 15th International Symposium on Transportation and Traffic Theory*, pages 393–415. Pergamon, July 2002.

25. J.-P. Lebacque. Two-phase bounded acceleration traffic flow model: analytical solutions and applications. *Transpn. Res. Record*, 1852:220–230, 2003.

26. V. Henn and L. Leclercq. Wave tracking resolution scheme for bus modelling inside the LWR traffic flow model. In *The Fifth Triennial Symposium on Transportation Analysis*, Le Gosier, Guadeloupe, French West Indies, June 2004.

27. V. Henn. Wave tracking resolution for the LWR model in a dynamic assignment perspective. 2004. In preparation.

Various Scales for Traffic Flow Representation: Some Reflections

J.-B. Lesort, E. Bourrel, and V. Henn

LICIT-INRETS/ENTPE
Rue Maurice Audin, 69518 Vaulx-en-Velin CEDEX – France
jean-baptiste.lesort@inrets.fr emmanuel.bourrel@entpe.fr,
vincent.henn@entpe.fr

Summary. This paper is an analysis of the various scales at which traffic flow can be represented, from vehicular to continuum flow models, with various time and space resolutions. The paper first investigates the question of scales and scale separation in traffic flow, both in modelling and measurement, using analogy with other disciplines. It then evaluates the representation of vehicles heterogeneities at various scales. It concludes on the interest of developing multiscale modeling.
Keywords: traffic flow, model, microscopic, continuum, scales

1 Introduction

From the origins of traffic science, traffic flow has been considered from two points of view: a microscopic one, analysing vehicle trajectories and interactions, and a continuum one, analysing traffic flow as a whole. Basically, these two viewpoints correspond to two different scales for analysis. Intermediate scales have also been considered (platoons).

Very early too, some analysis has been made of the relationships between models representing traffic at various scales. In [1] it has been shown that, under steady-state conditions, some car-following models resulted in a particular shape of the flow density equilibrium relationship. More recently, continuum models have been derived from microscopic considerations [2,3], and vice versa [4,5]. Furthermore, a whole branch of traffic theory is devoted to continuum models based on a statistical description of individual vehicles behaviour, following the work of Hermann and Prigogine [6]. Finally some models have tried to combine several approaches, for instance by embedding a microscopic model into a continuum one [7].

What can be found by examining this literature is that the relationships between the various model classes are more complex than it can appear on first thought. Particularly, most comparisons are made under steady-state conditions and are no longer valid under dynamic ones. Derivations of continuum models from microscopic ones or vice–versa include additional assumptions or simplifications which generally make the final model quite different from the initial one. On the other hand, there has been little consideration of the scale problem itself as it has been analysed in other disciplines such as statistical mechanics. For instance, the scale separations has not been studied, nor homogenisation techniques have been investigated.

The objective of this paper is to make a step towards a better understanding of the various aspects of the influence of the representation scale on traffic description, and on the relationships between the various scales. These reflec-

tions have mostly been made while studying hybrid models coupling vehicular and flow representations [8,9,9]: coupling models was a good opportunity
to investigate their relationships and the differences in the traffic flow representation they provide.

The first part of the paper is an analysis of the scale problem in other fields
such as solid or fluid mechanics, and of the application of the scale concept
to traffic flow. The second part present some variables and indicators used at
each scale and the influence of the scale on their definitions and properties,
and the third part shows, through the question of taking account of vehicles
heterogeneity, the different properties of different scale models. Finally, the
paper concludes on the interest of integrating multiple scales into consistent
models.

2   Various Scales for Traffic Flow Analysis and Modelling

2.1   Scales and Scale Separation

The multiplicity of scales to be considered in traffic flow analysis has several
causes:
- the nature of traffic phenomena (traffic flow phenomena occur and interact
  on a wide range of space and time scales [11]);
- application domains (relevant space and time scales for road managers
  range from a few tens of meters and seconds in the case of an isolated traffic signal for example to several kilometres over a few hours for a motorway).

The diversity of the models used is a direct consequence of this multiplicity.

This scale mutiplicity is not specific to the case of traffic flow theory. It exists in many other scientific domains, such as fluid mechanics, civil engineering, ... For example, several scales exist to model the behaviour of a gas. At a
microscopic level, molecules constitutive of the gas are considered, the objective being to represent the interaction phenomena between them. Conversely, at a macroscopic level, aggregate variables of flow (like flow or density) are considered. In this last case, the number of considered particles is
very large (to give an example, there are $6,023 \times 10^{23}$ molecules in a mole,
Avogadro's number).

Such examples are numerous and show that the same object can be seen
differently according to the considered scale. The choice of this scale is of
primary importance since it conditions the phenomena which will be represented. In the previous example, collisions phenomena between molecules
are studied in the microscopic approach, whereas more global phenomena are
considered in the macroscopic approach, such as mass transfer.

In some of these fields, the link between the different scales of modelling is
well known. For example, in fluid mechanics, it is possible to obtain the macroscopic laws of the flow from the kinetic theory (for example, it is possible
to derive the Navier-Stokes equations from the Boltzmann one in the case of
a low Knudsen number). In the same way, in civil engineering, some methods make it possible to deduce the macroscopic behaviour of a material from
the microscopic one. In the case of heterogeneous periodic materials (like
concrete), the homogenization method [12] makes it possible to give a global

description of the material behaviour, defining a fictitious homogeneous material whose behaviour is equivalent to that of the real heterogeneous one.

This link between different modelling scales is made possible thanks to scale separation. Indeed, in these domains, it is feasible to differentiate the different scales of the variation of a phenomenon. For example, in the case of heterogeneous periodic materials, a phenomenon ϕ can be observed either at a microscopic level (characterised by the length l in Fig. 1.a) or at a macroscopic level (characterised by the length L in Fig. 1.b).

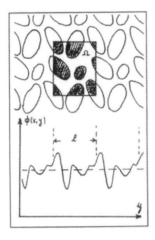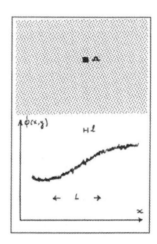

Fig. 1. Variation of ϕ at the microscopic (a) and the macroscopic level (b), from [12].

The main difference between traffic flow theory and the other scientific domains is that, in the case of traffic, separation between the microscopic and the macroscopic scales is hardly possible, as is mentioned in [13]. According to Lebacque [14], even on the scale of one kilometer, which is a relevant scale for applications in traffic, as the maximum concentration is lower than 200 veh/km per lane, a difference of one vehicle is not completely negligible in the definition of the average concentration.

To illustrate that lack of scale separation in traffic, we take the example of the choice of the aggregation period when measuring flows on a road [15]. Fig. 2 represents the distribution of flow values measured at one point according to the aggregation period used to compute the flows. Aggregation periods range from 10 seconds to 7 minutes, values usually used in traffic. If there was a clear separation between the local and the global variation of flow, one should observe a curve similar to the solid lines on the figure, that is:

- a large dispersion of flow values for small aggregation periods (corresponding to the microscopic variations of traffic phenomena);
- the presence of a "critical point" where the envelope narrows;
- a widening of the envelope (corresponding to the macroscopic variations of traffic phenomena).

If such a curve is observed, it would show that a scale separation exists. In this case, the critical point is the good scale to observe the variations at the macroscopic level. In our case, the envelope of data is quite far from this type of curve; this shows that there is no clear separation between the microscopic and the macroscopic scales in traffic.

We will see in the following that this lack of a scale separation makes it more difficult to derive macroscopic models from microscopic ones than in other domains.

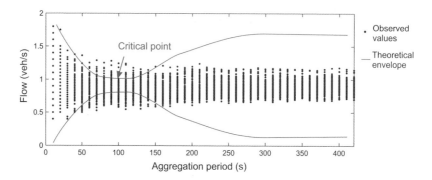

Fig. 2. Distribution of flows according to the aggregation period.

2.2   Scales in Traffic Flow Modelling

Since there is no real scale separation in traffic flow, it is difficult to properly define which traffic variations are microscopic and which ones are macroscopic. On the other hand, microscopic (vehicular) and macroscopic (flow) representations do exist. The question is then to examine the relationships between these various types of models and the consistency of both approaches.

2.2.1   From microscopic to continuum models

The link between microscopic and continuum models has early been studied. In [1], a pioneer paper in this field, and later in [16,17] the authors show that, under steady-state conditions, flow density equilibrium relationships corresponding to the global behaviour of microscopic models are similar (if not identical) to those used for the macroscopic models. However, this link is not sufficient since it is only valid under steady-state conditions (the objective of traffic flow models being to study dynamic situations). When considering dynamic situations, it is much more difficult to derive rigorously a continuum model from a microscopic one, due to the lack of a scale separation. In the literature, some higher order macroscopic models are presented by their authors as being derived from a microscopic one [2,3,18,19]. However, the assumptions of those derivations are valid only for slow variations of traffic states (rigorously, the derivation of the Payne model holds only for a stationary and homogeneous situation [14]). In a very dynamic situation, those approximations (in particular the one supposing that the concentration K is the inverse of the distance s between two vehicles) are no longer valid because speed and spacing between vehicles strongly vary; thus, higher order terms must not be neglected (cf. the Appendix C of [20]). It is one of the criticisms formulated by Daganzo against higher order models [21].

More rigorous methods have been proposed by some authors in order to improve the derivation of a continuum model from a microscopic one. In [13], the field of flow speeds in (x,t) is obtained by linear interpolation between the speeds of the two vehicles whose position verify $x_i \geq x \geq x_{i+1}$ (x_i being the position of vehicle i). In [20], the suggested method differs from that proposed by Payne the definition of the concentration. Observing that $\int_{x_{i+1}}^{x_i} K(x)\,dx = 1$ for all i , the authors propose to link concentration and spacing by:

$$ s \approx \frac{1}{K} - \frac{1}{2K^3}\frac{\partial K}{\partial x} - \frac{1}{6K^4}\frac{\partial^2 K}{\partial x^2} + \frac{1}{2K^5}\left(\frac{\partial K}{\partial x}\right)^2 $$

However, in both methods, the authors still suppose a slow variation of traffic flow conditions between two vehicles (and thus that the strong variations of traffic occur on the level of several vehicles), which is not necessarily the case, in particular when traffic jams propagate.

Thus, although the link between models is well established in stationary situation, no method of derivation of a continuum model from a microscopic one is really satisfactory.

Actually, continuum models have been developed without any link to a microscopic model, but rather by analogy with fluid mechanics [21] (it is besides a reason why traffic has an isotropic behaviour in the first generation of models as those proposed in [2,3]).

Derivation from microscopic models has been realized afterwards, in order to show that the continuum models proposed integrate the microscopic behaviour of vehicles. This clearly appears when considering higher order macroscopic models of new generation as those proposed in [18,19,22]. The general form of these models has been proposed in order to eliminate the main flaws of the analytical properties of higher order macroscopic models. In particular, in [23,5] the link between the higher order model and a microscopic one has been studied afterwards.

This could put some doubt on the validity of continuum models. However, if the link between microscopic and continuum models is fundamental in other scientific domains, it is of less importance in traffic. The main difference is that microscopic models of traffic flow are strictly empirical, with a weak connection to the experimental reality. Because of the nature of traffic, traffic flow models cannot have the same precision as in other domains [24]. Thus, even if it were possible to derive a continuum model in a rigorous way from a microscopic model, this would not be a sufficient argument to validate this model.

2.2.2   *From continuum to microscopic models*

It also exists some microscopic models that are derived from continuum ones. For example, the microscopic models proposed in [4], in [25] or in [9] are based on the LWR model [26,27]. Those models can bee seen as a particle discretization of the continuum model. A similar approach can be found in [28]; the authors propose a particle resolution of a gas-kinetic model for pedestrian flows. In the same way, in [5], the authors conclude that the microscopic model can be seen as a semi-discretization of the continuum one.

However it is quite difficult to consider these models as *real* microscopic models. Although they represent the trajectory of individualized vehicles, which is a characteristic of microscopic models, they describe the collective behaviour of drivers, which is a characteristic of the continuum models.

2.2.3   An intermediate level: the gas-kinetic models

Gas-kinetic traffic flow models are mesoscopic models that describe the dynamics of the phase-space density [6]. This type of model allows to integrate at an aggregated level (vehicles are not described individually) distributions of individual behavioural laws (such as speed distribution).

Some of these models are derived from a microscopic one (see [29] for an example of such a derivation). Conversely, some authors [29,30] derived higher order macroscopic models from gas-kinetic models using the method of moments. Thus, those models are generally considered as an intermediate level between microscopic and continuum models.

2.3   Modelling and Representation

The previous analysis shows that the usual micro/macro classification is not sufficient to clearly identify the characteristics of the various models.

Thus, we propose a new classification in order to characterize more precisely the models and to clarify the links that exist between them. This classification, presented in [9], is founded onto two criteria: the behavioural law and the representation scale. The behavioural law makes a distinction between the models that try to reproduce the individual behaviour of a driver in its environment (individual behavioural law) and the models that are based on the analysis of the propagation of concentrations or flows (collective behavioural law).

The second criterion is the representation scale. It differentiates the models according to the type of representation they use: vehicle or flow representation. These two criteria define a classification grid where all existing models can be classified (Fig. 3). That classification has the advantage to highlight the relationships between the models. However, it is quite difficult to define precisely the limits between the different classes. Indeed, some models, as gas-kinetic models, can be considered as intermediate models between vehicle / flow representation and collective / individual behavioural law.

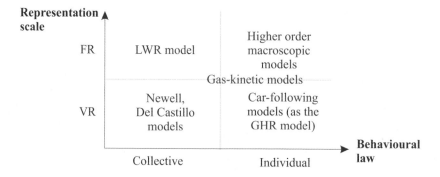

Fig. 3. A new classification of traffic flow models.

3   Variables and Indicators at Various Scale

The vertical axis of Fig. 3, which corresponds to the representation scales, also corresponds to various sets of variables characterizing traffic flow. At the vehicular level, the basic variables are the position $x_i(t)$ of each vehicle, its speed $\dot{x}_i(t)$, possibly its acceleration $\ddot{x}_i(t)$, the spacing $s_i = x_{i+1} - x_i$ and others. At the flow level, classical variables will be the flow $Q(x,t)$, the density $K(x,t)$ and the flow speed $V(x,t)$. The intermediate representation consists of using individual variables, but only under a distributed form. These variables may in turn be used to compute individual or global indicators.

In many cases, it is necessary to compute some indicators at another scale than the original one of the model. For example one may need to calculate some "macroscopic" indicator from the results of a microscopic model or vice-versa. Such scale changes can be more or less easy to define and to calculate (from straightforward to impossible changes), but moreover it nearly always lead to questions on the relevancy of looking at some results at a different scale than the original one.

3.1   Macroscopic Indicators Derived From Vehicular Results

The first way to try to derive macroscopic indicators from vehicular results is to derive locally at each vehicle position variables such as flow or density. The density, e.g., is linked directly to the space "around" a vehicle and a possible definition thus could be: $K_i = 1/s_i$, linking density to spacing.

The question is then to know where such a density is valid except at the position of the vehicle i. Several possibilities could be used such as a constant density between successive vehicles or any interpolation between the densities defined at vehicles positions. But the main point is to notice that no definitive solution exists.

Another way to derive macroscopic indicators from vehicular results is to aggregate individual variables into continuum ones. The main problem is then similar to the classical problem of measuring traffic variables. Indeed, if the aggregation is made on a large scale, every small variation will be erased. Conversely, if the aggregation scale is too small, there is a risk of artificial variations only due to the size of the aggregation step.

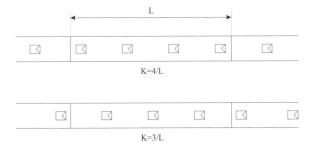

Fig. 4. Different results for density measurement due to individual vehicle positions.

For example, let us consider a steady-state traffic flow along a road. Density can be evaluated for a section of length L as the number of vehicles present on this section divided by its length. But depending on the time this density is evaluated, a one vehicle variation may appear and lead to a significant variation of density (Fig. 4): density may then oscillate even under stationary conditions.

3.2   Microscopic Indicators Derived From Continuum Results

On the other hand, microscopic indicators can also be derived from a continuum model. Of course, since the scale of continuum traffic flow is larger than vehicles, some information will be missing and cannot be reconstituted. This will generally lead to non unique microscopic representation.

However the notion of particle can be defined properly from a continuum representation. Indeed, if density and flow are defined simultaneously, a flow speed can be derived which is equal to $V = Q/K$. We can define the notion of a particle by analogy to a leaf floating on the flow and moving with the flow speed. This gives us an individual tracer for the movement of the flow. It is important to note that such a particle is not exactly a vehicle and has no size. It just helps to see the trajectory a vehicle would have if it passed at point x at time t.

Given this notion of particle, other indicators can be derived such as travel time for example. It is defined as the difference between the time when a particle reaches point B and the time it had passed point A. This definition makes it possible to continuously compute a travel time for any entry (or exit) time.

But some problems may arise because the continuum model does not necessarily ensure a proper vehicular behaviour (which is not in its objectives). To illustrate this, let us consider for example the numeric discretization of the LWR model based on the Godunov scheme [31]. Such a scheme is proved to tend to the continuous solution when the size of cells reduces and is empirically known to give good estimations of flows and densities if the size of cells is "reasonable". However, if we try to calculate travel times with such a scheme, the errors introduced by the discretization (that is the distance between the discretized and the continuous solutions of the model) are no more negligible.

In the discretization scheme, density and flow are not calculated at the same positions. Flow speed thus cannot be defined as above. Various solutions have been proposed in order to estimate the speed, but they are only estimations. The method to determine the travel time from A to B uses cumulative flows and gives a representation of vehicles which is consistent with the continuum one. Travel time is continuously calculated for particle passing by point A at time t_A by:

$$\tau_{AB}(t_A) = \tau_{AB}(t_A^0) + \int_{t_A^0}^{t_A} \frac{Q_A(t)}{Q_B(t + \tau_{AB}(t))} dt$$

where $\tau_{AB}(t_A^0)$ is the travel time from an initial vehicle passing by point A at time t_A^0. The problem is that, though this definition is correct, the microscopic "reality" hidden behind the continuum model is not necessarily correct. The Godunov scheme does not consider at all vehicles; the way densities and flows are calculated implies that some vehicles may have unphysical trajectories (jumping instantaneously from a position to another one and even stepping back). This is not a problem as far as only density and flow are computed. It may lead to difficulties when indicators based on trajectories are computed. This is the case for travel times, and the results obtained with a Godunov computation are significantly different from the ones calculated with the continuous solution of LWR-model. The influence of the discretization is obvious and much more important than purely numerical effects (numerical viscosity) which can be observed in flows and densities [32] (Fig. 5).

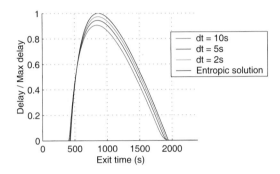

Fig. 5. Influence of the time step on delay.

4   Scales and Heterogeneity

4.1   Introducing Vehicles Heterogeneities Into Models

There are many kinds of heterogeneities to be introduced in traffic models. The simplest case is to introduce characteristics which do not interfere with traffic flow: these are for instance characteristics connected to assignment, such as destination, route and assignment method (vehicles equipped or not with on board guidance systems). Introducing these distinctions is straightforward in vehicular models, and quite simple in flow models by using partial flows [33]. There is not much influence of the modeling scale: The only point is that numerical solutions of these models do not always ensure a FIFO (First-In-First-Out) flow behaviour, and that FIFO violations are in the range of the resolution scale.

The next step is to introduce, into a uniform flow, a specific vehicle (typically a bus) whose behaviour is exogenous to the model. This is again straightforward in vehicular models, and has been shown for some flow models [34,35]. That type of model is by nature a combination of scales, with a single identified vehicle inside a continuum flow. From a mathematical view point it is not a problem, the vehicle being only a moving boundary condition. From the traffic representation viewpoint this combination is more questionable: What is the significance level of the local phenomenon observed in the vicinity of the obstacle? Is there any interest in representing traffic flow along the bus, or is the punctual bus model more relevant?

Introducing differences in the vehicle kinematics or sizes raises other questions. Basically, the choice is simple: a high scale representation, using classes (or distributions) for the parameters of a vehicular model, or a low scale using partial flows with different behaviours. The former representation is classical in vehicular models; the latter one has been studied recently by several authors [36,37]. The vehicular representation is necessarily based on a distribution of the various types of vehicles in space or time. This leads to all the questions related to stochastic models, which will be examined in the next paragraph. Multiclass flow models present the same limits than other flow models, with the additional one of taking no account of local vehicle/vehicle interactions (a slow vehicle passing a slower one…).

A third and intermediate approach consists of describing parameters distributions and integrating these distributions to come up with a unique flow model. This is what is done with kinetic models, resulting in models similar to higher order flow models. Zhang [36] observed that there was some kind of convergence between multiclass first order models and higher order ones. The kinetic approach is thus not much different, from the scale viewpoint, from the basic flow one.

The conclusion is that introducing vehicle heterogeneities into flow models, or even introducing and integrating distributions such as in kinetic models, does not change drastically the question of scale significance. On the other hand, introducing distributions into a vehicular model has a strong influence on the model, which has been little studied till now.

4.2   Stochastic Models

Another possibility to take heterogeneity of traffic into account is to introduce statistical distributions into the models.

This approach is classical in vehicle represented simulations. Generally, stochastic aspects are introduced when generating vehicles at the entry of the network, distributing headways between consecutive vehicles and some characteristic parameters of the behaviour of vehicles, as desired speed (in particular, this makes it possible to introduce several classes of vehicles). In some simulations, stochastic aspects are also introduced in the behaviour law of vehicles, but this remains relatively rare.

The objective of these distributions is to reproduce the vehicle feature on the network taking the effect of heterogeneity of traffic into account, especially for some special elements of the network which global behaviour is directly related to the random nature of traffic (intersections for example).

However, little attention has been paid in the literature concerning the effects of these distributions. It is possible to show through very simple examples that these distributions do not have the same effects, and that they modify the model.

First, it is to notice that distributions introduced at the entry of the network can be quite different from that observed at a certain distance of the entry, due to the propagation of vehicles. To illustrate that point, we use Newell's vehicular model [25] and we only introduce a headway distribution at the entry of the network. If one observes the headway distribution at a certain distance of the entry, it appears that the propagation of vehicles modifies the headway distribution, and that the distribution vanishes rapidly, as it can be seen in Fig. 6 (the distribution used here is an Erlang distribution, but another one would have led to the same result).

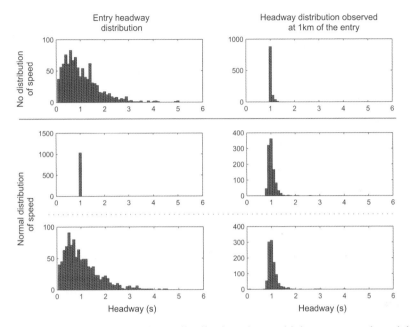

Fig. 6. Propagation of headway distributions in a vehicle represented model according to the distributions introduced when generating vehicles.

The farther headways are observed from the entry, the more homogeneous are they. This result is quite different if we introduce a distribution of desired speeds. The observed headways remain distributed. However, the observed distribution is quite different from the one introduced at the entry.

This very simple example shows that it is necessary when introducing distributions into a model to wonder on the effects of the introduced distributions and the way they propagate in the network. If the objective is for example to generate random arrivals of vehicles at in intersection following a given distribution, it is necessary to ensure that the introduced distributions at the entry of the network make it possible to obtain the desired result. It is our opinion that this type of problems are crucial in the use of stochastic vehicle represented simulations: however, such an analyze is relatively rare in the literature!

The introduced distributions also have an effect on the model itself. If we compute the flow density equilibrium relationship corresponding to the global behaviour of the stochastic vehicle represented model, it appears that this relationship is different from that obtained using the same model with a deterministic generation process of vehicles, as can be seen in Fig. 7 (this is due to the distribution of desired speeds since lower vehicles generate platoons).

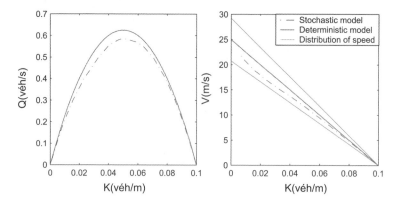

Fig. 7. Flow density equilibrium relationship.

In addition, the use of distributions raises the question of the representativeness of the results of the model. Each run of the simulation is a possible realization and is different from the preceding due to stochastic elements (provided that the root of the random generator is not fixed). Thus, it is necessary to answer to the following question: how many runs (or replications) are necessary so that the whole results are representative for the behaviour of the studied network? However, this question has been little studied in the literature.

Thus, although it is relatively simple to introduce distributions into a vehicle represented model (what makes it possible to take the heterogeneity of traffic into account), this generates some problems in which few authors were interested.

Conversely, several ways exist to introduce stochastic aspects in a flow represented model. One can imagine to distribute the boundary conditions, to introduce some stochastic terms in the equations of the model (as it is done in [38] to distribute the parameters of the model (e.g., those of the fundamental diagram) on a part of the network). If, in the case of models based on a collective behavioural rule, the mean effect of randomness can be considered as being ever present in the model, the introduction of distribution can be seen as a good way to take the effect of randomness into account when considering specific elements of a network like intersections. However, to the authors' knowledge, few models of this type have been developed (see [38] for a rare example). Besides, the introduction of distributions leads to the same problems as above.

As already explained, gas-kinetic models constitute another way to introduce stochastic aspects by using distribution models. It has been seen that the results are basically similar to higher order models, and might be similar to multiclass first order models.

Integrating stochastic aspects this way thus seems a good way of introducing heterogeneities, and it may improve the quality of the flow description on freeways. On the other hand, distributions are also used to introduce queuing models into traffic flow ones, which is necessary to properly describe toll gates, roundabouts and other particular locations. Continuum models have a little ability to answer this question, even if some possibilities do exist to integrate queuing models inside continuum ones. Vehicular models may give some answer to this question for they give some kind of interval distributions. However, it has been seen that the actual interval distribution is by no means straightforward and has little reason to be similar to the one used for vehicle generation: it depends on speed distributions but certainly also on other parameters such as car following laws, passing models etc. It is thus certainly difficult to properly calibrate a vehicular model in order to obtain a specific interval distribution.

There are two conclusions to be drawn from this short examination of the question of heterogeneity. The first one is that taking account of heterogeneities makes even larger the differences between the various representation scales. The second one is that, again, the good scale (or scales) does not exist.

5   Conclusion

There are several reasons why it is difficult to deal properly with the scale question in traffic flow modelling. The first one is that traffic itself includes several scales and that there no scale separation: traffic particles (vehicles) are quite large and traffic variations occur at a scale similar to the vehicle size. On the other hand, measuring or observing traffic at various scales has a great influence over the results which can be obtained and, even for a specific application, a unique appropriate scale seldom exists. There is thus a major interest in finding a consistent way to combine several representation scales into a unique model. This can be done by embedding a representation into another one as in [7]or by coupling consistent vehicular and continuum models as in [9] in order to use the best suited model for each location. One could also imagine real multiscale models, explicitly combining various consistent representations of traffic flow and making it possible to compute composite indicators.

References

1. D.C. Gazis, R. Herman, and R.B. Potts. Car-following theory of steady state flow. Operations Research, 1959, vol. 7, n° 4, pp. 499-505.
2. H.J. Payne. Models of freeway traffic and control. Mathematical Models of Public Systems, Bekey, G.A., vol. 1, La Jolla, California: Simulation Council, 1971, pp. 51-61.
3. D. Helbing. From microscopic to macroscopic traffic models. In J. Parisi, S.C. Müller and W. Zimmermann, editors., A Perspective Look at Nonlinear Media. From Physics to Biology and Social Sciences, Berlin: Springer, 1998, pp. 122-139.
4. J.M. Del Castillo. A car following model based on the Lighthill-whitham theory. In J. B. Lesort editor, Transportation and Traffic Theory, proceedings of the 13th ISTTT, Oxford: Pergamon, 1996, pp. 517-538.
5. A. Aw, A. Klar, T. Materne, and M. Rascle. Derivation of continuum traffic flow models from microscopic follow-the-leader models. SIAM Journal on Applied Mathematics, 2002, vol. 63, n° 1, pp. 259-278.
6. I. Prigogine and R. Herman. Kinetic theory of vehicular traffic. New York: Elsevier, 1971, p. 100.
7. C.M.J. Tampère, B. Van Arem, and S.P. Hoogendoorn. Gas kinetic traffic flow modelling including continuous driver behaviour models, 82nd Trsp. Res. Board, 2003, Washington, D.C.
8. E. Bourrel. Modélisation dynamique de l'écoulement du trafic routier: du macroscopique au microscopique. Phd thesis in civil engineering, INSA de Lyon, Lyon, France, 2003, p. 261.

9. E. Bourrel and J.B. Lesort. Mixing micro and macro representations of traffic flow: a hybrid model based on the LWR theory. 82nd Trsp. Res. Board, 2003, Washington, D.C., Trsp. Res. Board editor, 2003, 16 p.

10. E. Bourrel and V. Henn. Mixing micro and macro representations of traffic flow: a first theoretical step. Proceedings of the 9th meeting of the Euro Working Group on Transportation, 10-13 June 2002, Bari, Italy, Polytechnic of Bari, 2002, pp. 610-616.

11. G. Lerner, A. Hochstaedter, R. Kates, C. Demir, J. Meier, and A. Poschinger. The interplay of multiple scales in traffic flow: coupling of microscopic, mesoscopic, and macroscopic simulation. Proceedings of the 7th World Congress on Intelligent Transport Systems, 6-9 November 2000, Turin, Italy, 2000, p. 5.

12. C. Boutin. Comportement macroscopique de matériaux hétérogènes. HDR thesis, Université Joseph Fournier de Grenoble, Grenoble, 1994, p. 188.

13. A. Hennecke, M. Treiber, and D. Helbing. Macroscopic simulation of open systems and micro-macro link. In D. Helbing, H.J. Herrmann, M. Schreckenberg et D.E. Wolf, editors., Traffic and Granular Flow '99, Springer, Berlin, 1999, pp. 383-388.

14. J.P. Lebacque. Les modèles d'écoulement du trafic. Actes du groupe de travail Modélisation du trafic, Arcueil, France, INRETS, 1995, pp 51-75.

15. Ch. Buisson, Une note sur la relation fondamentale, internal note, 2003.

16. D.C. Gazis, R. Herman and R. W. Rothery. Nonlinear follow-the-leader models of traffic flow. Operations Research, 1961, vol. 9, n° 4, pp. 545-567.

17. A.D., May and H.E.M. Keller. Non-integer car-following models. Highway Research Records, 1967, vol. 199, pp. 19-32.

18. H.M. Zhang. A non-equilibrium traffic model devoid of gas-like behaviour. Trsp. Res. Part B, 2002, vol. 36, n° 3, pp. 275-290.

19. R. Jiang, Q.S. Wu, and Z.J. Zhu. A new continuum model for traffic flow and numerical tests. Trsp. Res. Part B, 2002, vol. 36, n° 5, pp. 405-419.

20. P. Berg. Optimal-velocity models of motorway traffic. Phd Thesis in Mathematics, University of Bristol, Bristol, 2001, 194 p.

21. C.F. Daganzo. Requiem for second-order fluid approximations of traffic flow. Trsp. Res. Part B, 1995, vol. 29, n° 4, pp. 277-286.

22. A. Aw and M. Rascle. Resurrection of "second order" models of traffic flow? Journal of Applied Mathematics, 2000, vol. 60, n° 3, pp. 916-938.

23. R. Jiang and Q.S. Wu. Study on propagation speed of small disturbance from a car-following approach. Trsp. Res. Part B, 2003, vol. 37, n° 1, pp. 85-99.

24. M. Papageorgiou. Some remarks on macroscopic traffic flow modelling. Trsp. Res. Part A, 1998, vol. 32, n° 5, pp. 323-329.

25. G.F. Newell. Nonlinear effects in the dynamics of car following. Operations Research, 1961, vol. 9, pp. 209-229.

26. M.J. Lighthill and G.B. Whitham. On kinematic waves II. A theory of traffic flow on long crowded roads. Proceedings of the Royal Society A, 1955, vol. 229 , pp. 317-345.

27. P.I. Richards. Shockwaves on the highway. Operations research, 1956, vol. 4, pp. 42-51.

28. S.P. Hoogendoorn, P.H.L. Bovy, and H. Van Lint. Normative behaviour theory and modelling. In M.A.P. Taylor editor, Transportation and Traffic Theory in the 21st Century, proceedings of the 15th ISTTT, Pergamon, Oxford, 2002, pp. 625-651.

29. A. Klar and R. Wegener. A hierarchy of models for multilane vehicular traffic I&II: Modeling. SIAM Journal of Applied Mathematics, 1998, vol. 59, n° 3, pp. 983-1001.

30. W. Leutzbach. Introduction to the theory of traffic flow. 1988, Springer-Verlag, Berlin

31. J.P Lebacque. The Godunov scheme and what it means for first order traffic flow models. Transportation and Traffic Theory, proceeding of the 13th ISTTT, J.B.Lesort ed., 1996, Pergamon, Oxford, pp. 647-677.

32. V. Henn, Macroscopic traffic flow models for traffic assignment : which definition for travel time? May 2003, internal note (in French).

33. M. Papageorgiou. Dynamic modelling, assignment and road guidance on traffic networks, Trsp. Res. part B, 1990, Vol 24 N°6, pp. 471-495.

34. G.F. Newell. A moving bottleneck. Trsp. Res. B, 1998, Vol32 N°8, pp 531-538.

35. J.P Lebacque, J.B. Lesort, and F. Giorgi Introducing buses into   first order macroscopic traffic flow models. Trsp. Res. Rec, 1998, N°1664, pp. 70-79.

36. H.M. Zhang and W.L. Jin. Kinematic wave traffic flow model for mixed traffic, Trsp. Res. Records, 2002, N°1804, pp. 197-204.

37. S. Chanut and C. Buisson. Godunov discretization of a two flow macroscopic model for mixed traffic with distinguished speeds and lengths. 82nd Trsp. Res. Board, 12-16 january 2003, Washington, D.C., 2003, p. 20 (accepted for publication in Trsp. Res. Records).

38. S. Smulders. Control of freeway traffic flow by variable speed signs. Trsp. Res. Part B, 1990, vol. 24, n° 2, pp. 111-132.

Comparison of Congested Pattern Features at Different Freeway Bottlenecks

B.S. Kerner[1] and S.L. Klenov[2]

[1] DaimlerChrysler AG
    RIC/TS, HPC: T729, 70546 Stuttgart – Germany
[2] Department of Physics – Moscow Institute of Physics and Technology
    141700 Dolgoprudny – Russia

Summary. Based on a microscopic theory of spatial-temporal congested traffic patterns at freeway bottlenecks, diagrams and features of the congested patterns at bottlenecks due to on-ramps, merge bottlenecks (a reduction in freeway lanes), and off-ramps are studied and compared.

1 Introduction

Traffic flow in freeways can be either free or congested (e.g., [1–29]). Congested patterns usually occur near freeway bottlenecks (e.g., [2]), which act like as defects for pattern emergence in physical systems.

The empirical features of congested traffic flow patterns at bottlenecks have been understood only recently [17–19]. In particular, it has been found that moving jams do *not* emerge spontaneously in free flow. Instead, first a transition from free flow to synchronized flow (F→S transition for short) occurs at a bottleneck [18, 19]. If a vehicle speed in the synchronized flow is high enough then wide moving jams do *not* necessarily emerge in the synchronized flow. In this case, a synchronized flow pattern (SP) consisting of synchronized flow only can occur at the bottleneck [19]. Wide moving jams can spontaneously emerge only in synchronized flow (S→J transition) when the density in the synchronized flow is high enough (the pinch effect [18]). As a result, a general pattern (GP) occurs at the bottleneck [19]: an GP consists of synchronized flow upstream of the bottleneck and wide moving jams that spontaneously emerge in the synchronized flow and propagate upstream. Thus, two main types of congested patterns, SP and GP, can occur at an isolated bottleneck (the bottleneck that is far from other bottlenecks) [19].

To explain the empirical results, Kerner introduced a three-phase traffic theory [7, 18, 20–22]. In this theory, there are two qualitative different traffic phases that should be distinguished in congested traffic: The "synchronized flow" traffic phase and "wide moving jam" traffic phase. The three-phase traffic theory is a qualitative theory that gives some objective (empirical) criteria for traffic phases and explain main features of phase transitions and congested patterns in traffic flow [7, 21, 22]. Based on the three-phase traffic theory, several microscopic traffic models have recently been introduced [24–28], to describe a diverse variety of empirical congested pattern features at freeway bottlenecks.

In this article, a further development of a microscopic theory of phase transitions and congested patterns at freeway bottlenecks of Ref. [24] is presented. In particular, diagrams and features of congested patterns at isolated bottlenecks due to on- and off-ramps and at a merge bottleneck (a reduction in freeway lanes) are examined and compared.

2 A Microscopic Traffic Flow Model for Spatial-Temporal Congested Patterns

2.1 Vehicle Motion Rules

In a single-lane traffic flow model the general rules of the vehicle motion first introduced in [25] are used:

$$v_{n+1} = \max(0, \min(v_{\text{free}}, v_{c,n}, v_{s,n})), \tag{1}$$

$$x_{n+1} = x_n + v_{n+1}\tau, \tag{2}$$

$$v_{c,n} = \begin{cases} v_n + \Delta_n & \text{at } x_{\ell,n} - x_n \le D_n \\ v_n + a_n\tau & \text{at } x_{\ell,n} - x_n > D_n, \end{cases} \tag{3}$$

where

$$\Delta_n = \max(-b_n\tau, \min(a_n\tau, \; v_{\ell,n} - v_n)). \tag{4}$$

In (1)–(4), v_n and x_n are the speed and space coordinate of the vehicle; the index n corresponds to the discrete time $t = n\tau$, $n = 0, 1, 2, ..$; τ is the time step; v_{free} is the maximum speed in free flow that is constant; $v_{s,n}$ is the safe speed; the lower index ℓ marks values and variables related to the preceding vehicle; all vehicles have the same length d; values $a_n \ge 0$ and $b_n \ge 0$ in (3), (4) restrict changes in vehicle speed when the vehicle accelerates or adopts its speed to the speed of the preceding vehicle.

2.2 Speed Adaptation Effect Within Synchronization Distance

Equations (3), (4) describe the adaptation of the vehicle speed to the speed of the preceding vehicle. The vehicle speed adaptation takes place within the synchronization distance D_n: at $x_{\ell,n} - x_n \le D_n$ the vehicle tends to adjust the vehicle speed to the speed of the preceding vehicle. This means that the vehicle decelerates if $v_n > v_{\ell,n}$, and accelerates if $v_n < v_{\ell,n}$ [25].

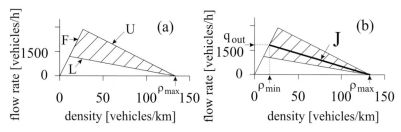

Fig. 1. Model steady states in the flow-density plane (**a**) and the line J (**b**) related to the downstream front of a wide moving jam. In (**a**), (**b**) the two-dimensional region in the flow-density plane (dashed region) is limited by three boundaries U, L, and F, which are related to the safe speed, the synchronization distance, and the maximum free flow speed, respectively, as has been explained in [25, 26]. At chosen model parameters (Sect. 2.7) the flow rate in the wide moving jam outflow is $q_{\mathrm{out}} = 1810$ vehicles/h and the velocity of the downstream jam front is $v_{\mathrm{g}} = -15.5$km/h, the maximum flow rate in free flow is $q_{\mathrm{max}}^{(\mathrm{free})} \approx 2400$ vehicles/h.

This speed adaptation effect within the synchronization distance is related to the fundamental hypothesis of the three-phase traffic theory: hypothetical steady states of synchronized flow cover a 2D region in the flow–density plane (Fig. 1((a)).

In the general rules (1)–(4), the synchronization distance D_n depends on the vehicle speed v_n and on the speed of the preceding vehicle $v_{\ell,n}$

$$D_n = d + G(v_n, v_{\ell,n}), \tag{5}$$

where the function $G(u, w)$ is chosen as

$$G(u, w) = \max(0, k\tau u + \beta a^{-1} u(u - w)), \tag{6}$$

$k > 1$ and β are constants, a is the constant acceleration. If $v_n = v_{\ell,n}$, the synchronization distance D_n is $d + kv_n\tau$. This corresponds to a fixed time gap $k\tau$. If $v_n > v_{\ell,n}$, the distance D_n increases and vice versa.

2.3 Motion State Model for Random Acceleration and Deceleration

Random Acceleration and Deceleration. At the first step, the preliminary speed of each vehicle is found as $\tilde{v}_{n+1} = v_{n+1}$ where v_{n+1} is calculated based on (1)–(4). *At the second step,* a noise component ξ_n is added to the calculated speed $\tilde{v}_{n+1}$ and then the final value of the speed v_{n+1} at time step $n + 1$ is found from the condition introduced in [26]

$$v_{n+1} = \max(0, \min(v_{\mathrm{free}}, \tilde{v}_{n+1} + \xi_n, v_n + a\tau, v_{\mathrm{s},n})). \tag{7}$$

In the motion state model [24], a random deceleration and acceleration are applied depending on whether the vehicle decelerates or accelerates or else maintains its speed:

$$\xi_n = \begin{cases} -\xi_b & \text{if } S_{n+1} = -1 \\ \xi_a & \text{if } S_{n+1} = 1 \\ 0 & \text{if } S_{n+1} = 0, \end{cases} \tag{8}$$

where ξ_b and ξ_a are random sources for deceleration and acceleration, respectively; S in (8) denotes the state of motion ($S_{n+1} = -1$ is related to a deceleration, $S_{n+1} = 1$ to an acceleration and $S_{n+1} = 0$ to the motion at nearly constant speed)

$$S_{n+1} = \begin{cases} -1 & \text{if } \tilde{v}_{n+1} < v_n - \delta \\ 1 & \text{if } \tilde{v}_{n+1} > v_n + \delta \\ 0 & \text{otherwise}, \end{cases} \tag{9}$$

where δ is a constant ($\delta \ll a\tau$).

The random components in (8) are given as "impulsive":

$$\xi_b = a\tau\Theta(p_b - r), \tag{10}$$
$$\xi_a = a\tau\Theta(p_a - r). \tag{11}$$

In (10), (11), p_b and p_a are probabilities of random deceleration and acceleration, respectively. $r = \text{rand}(0,1)$, i.e., this is an independent random value uniformly distributed between 0 and 1. The function $\Theta(z)$ has the following meaning: $\Theta(z) = 0$ at $z < 0$ and $\Theta(z) = 1$ at $z \geq 0$.

Random Time Delays. To simulate a time delay either in vehicle acceleration or in vehicle deceleration, a_n and b_n in (4) are taken as the following stochastic functions

$$a_n = a\Theta(P_0 - r_1), \tag{12}$$
$$b_n = a\Theta(P_1 - r_1), \tag{13}$$

$$P_0 = \begin{cases} p_0 & \text{if } S_n \neq 1 \\ 1 & \text{if } S_n = 1, \end{cases} \tag{14}$$

$$P_1 = \begin{cases} p_1 & \text{if } S_n \neq -1 \\ p_2 & \text{if } S_n = -1, \end{cases} \tag{15}$$

where $r_1 = \text{rand}(0,1)$, i.e., this is a random number uniformly distributed between 0 and 1. The function P_0 in (14) determines the probability $\psi_a = 1 - P_0$ of a random delay in vehicle acceleration at a time step $n+1$, whereas the function P_1 (15) determines the probability $\psi_b = 1 - P_1$ of a random delay in vehicle deceleration at the time step $n+1$.

Safe Speed. In the model, the safe speed $v_{s,n}$ in (1) is chosen in the form:

$$v_{s,n} = \min(v_n^{(\text{safe})}, g_n/\tau + v_\ell^{(a)}), \tag{16}$$

where $v_n^{(\text{safe})} = v^{(\text{safe})}(g_n, v_{\ell,n})$ is the safe speed in the model of Krauß et al. [15], $g_n = x_{\ell,n} - x_n - d$ is the space gap, $v_\ell^{(a)}$ is an 'anticipation' speed of the preceding vehicle at the next time step.

The safe speed $v^{(\text{safe})}(g_n, v_{\ell,n})$ in the model of Krauß et al. [15] ensures collisionless vehicle motion if $g_n \geq v_{\ell,n}\tau$. In the model under consideration, it is assumed that in some cases, namely due to lane changing or merging of vehicles onto the neighboring lane within merging regions of bottlenecks, the gap g_n can become lower than $v_{\ell,n}\tau$ for a short time. In these critical situations, collisionless vehicle motion in the model is due to the second term in (16) in which some prediction of the speed of the preceding vehicle $(v_\ell^{(a)})$ at the next time step is used. The 'anticipation' speed $v_\ell^{(a)}$ is chosen as

$$v_\ell^{(a)} = \max(0, \min(v_{\ell,n}^{(\text{safe})} - a\tau, v_{\ell,n} - a\tau, g_{\ell,n}/\tau)), \tag{17}$$

where $v_{\ell,n}^{(\text{safe})}$ is the safe speed for the preceding vehicle, $g_{\ell,n}$ is the space gap in front of the preceding vehicle. Simulations show that (16), (17) ensure collisionless vehicle motion in a wide range of parameters of merging regions of bottlenecks (Sect. 2.6) and at the chosen lane changing rules (Sect. 2.4).

2.4 Lane Changing Rules

Lane changing rules in a two-lane model are based on the well-known incentive and security conditions (see Nagel et al. [16]). However, these conditions are adjusted to take the effect of the synchronization distance into account.

The following incentive conditions for lane changing from the right lane to the left lane $(R \to L)$ and a return change $(L \to R)$ have been used in the model:

$$R \to L: \ v_n^+ \geq v_{\ell,n} + \delta_1 \text{ and } v_n \geq v_{\ell,n}, \tag{18}$$

$$L \to R: \ v_n^+ > v_{\ell,n} + \delta_2 \text{ or } v_n^+ > v_n + \delta_2, \tag{19}$$

where $\delta_1 \geq 0$, $\delta_2 \geq 0$ are constants.

The security conditions by lane changing are given by inequalities:

$$g_n^+ > \min(v_n\tau, \ G_n^+), \quad g_n^- > \min(v_n^-\tau, \ G_n^-). \tag{20}$$

Here

$$G_n^+ = G(v_n, v_n^+), \quad G_n^- = G(v_n^-, v_n), \tag{21}$$

where the function $G(u, w)$ is given by formula (6).

Superscripts $+$ and $-$ in variables, parameters, and functions denote the preceding vehicle and the trailing vehicle in the "target" (neighboring) lane, respectively. For example, g_n^+ is the space gap between the vehicle that wants to change the lane and the preceding vehicle in the target lane. Similar to lane changing rules in other models [16], the speed v_n^+ or the speed $v_{\ell,n}$ in (18), (19) is set to ∞ if the space gap g_n^+ or the space gap g_n exceeds a given look-ahead distance L_a, respectively. If the conditions (18)–(20) are satisfied, then as in [29] the vehicle changes the lane with probability p_c.

2.5 Boundary and Initial Conditions

In the model, open boundary conditions are applied. At the start of each lane new vehicles are generated one after another at equal time intervals $\tau_{in} = 1/q_{in}$, q_{in} is the flow rate in the incoming boundary flow per lane. A new vehicle appears only if the distance from the start of the road ($x = x_b$) to the position $x = x_{\ell,n}$ of the farthest upstream vehicle in the lane is not smaller than the safe distance $\ell_{safe} = v_{\ell,n}\tau + d$, i.e., $x_{\ell,n} - x_b \geq \ell_{safe}$. The speed v_n and the coordinate x_n of the new vehicle are $v_n = v_{\ell,n}$, $x_n = \max(x_b, x_{\ell,n} - \max(v_n\tau_{in}, \ell_{safe}))$.

In the initial state ($n = 0$), all vehicles have the maximum speed $v_n = v_{free}$ and they are positioned at space intervals $x_{\ell,n} - x_n = v_{free}\tau_{in}$. After a vehicle has reached the end of the road it is removed; before this, the farthest downstream vehicle maintains its speed and lane.

2.6 Models of Bottlenecks

A bottleneck due to the on-ramp, a merge bottleneck, where two lanes are reduced to one lane, and a bottleneck due to the off-ramp (Fig. 2) are considered.

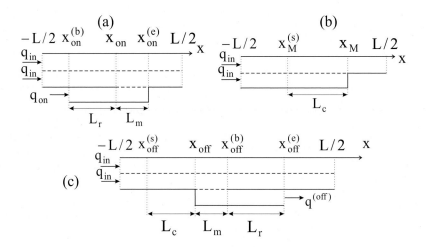

Fig. 2. Models of bottlenecks. **(a)** An on-ramp, **(b)** a merge bottleneck, **(c)** an off-ramp.

The on-ramp bottleneck consists of two parts (Fig. 2 (a)): (i) The merging region of length L_m where vehicle can merge onto the main road from the on-ramp lane. (ii) A part of the on-ramp lane of length L_r upstream of the merging region where vehicles move in accordance with the model (1)– (17)

with the maximal speed $v_{\text{free}} = v_{\text{free on}}$. At the beginning of the on-ramp lane $(x = x_{\text{on}}^{(b)})$ the flow rate to the on-ramp q_{on} is given like as q_{in}.

At the merge bottleneck (Fig. 2(b)), within the merging region of length L_c upstream of the merge point $x = x_M$ vehicles have to change from the right lane to the left lane.

The off-ramp bottleneck consists of two parts (Fig. 2(c)): (i) A merging region of length L_m where vehicle can merge from the main road onto the off-ramp lane. (ii) A part of the off-ramp lane of length L_r downstream of the merging region where vehicles move in accordance with the model (1)–(17). The maximal speed in the off-ramp lane is $v_{\text{free}} = v_{\text{free off}}$. Within a second merging region of length $L_m + L_c$ that is on the main road ($x_{\text{off}}^{(s)} \le x \le x_{\text{off}}^{(b)}$ in Fig. 2(c)) vehicles going to the off-ramp have to change from the left lane to the right lane of the main road. The flow rate of vehicles that go to the off-ramp is given as a percentage η of the flow rate q_{in}.

Vehicle Motion Rules in Merging Region. There are some peculiarities of vehicle motion for vehicles that move within the merging regions of these bottlenecks (Fig. 2(a–c)). This is related only to those vehicles that intend to merge onto the target lane. These vehicles move either in the on-ramp lane, or in the right lane of the merge bottleneck, or else these are the vehicles that want to leave the main road to the off-ramp.

Each of these vehicles moves in accordance with the model rules (1)–(17). However, in (3)–(5) the coordinate $x_{\ell,n}$ and the speed $v_{\ell,n}$ of the preceding vehicle are replaced by values x_n^+ and $\hat{v}_n^+$, respectively, where

$$\hat{v}_n^+ = \max(0, \ \min(v_{\text{free}}, \ v_n^+ + \Delta v_r^{(2)})), \tag{22}$$

x_n^+ and v_n^+ are the coordinate and the speed of the preceding vehicle in the target lane, $\Delta v_r^{(2)}$ is a constant. However, the safe speed in the vehicle motion equations (1) is further determined by the safe speed $v_{s,n}$ (16) related to the preceding vehicle in the lane where the vehicle currently moves.

Models of Vehicle Merging. The lane changing rules of Sect. 2.4 are used beyond all bottleneck merging regions. The following rules for vehicle merging within the merging regions are assumed to be the same for all three types of bottlenecks.

The rule (∗): A speed $\hat{v}_n$ is calculated corresponding to

$$\hat{v}_n = \min(v_n^+, \ v_n + \Delta v_r^{(1)}), \tag{23}$$

where $\Delta v_r^{(1)}$ is a constant that describes the maximum possible increase in speed after merging. Then the speed $\hat{v}_n$ is used instead of v_n in the security lane changing rules (20). If these conditions are satisfied, then the vehicle merges within the merging region of a bottleneck. In this case, the vehicle speed v_n is set to $\hat{v}_n$ (23) and the vehicle spatial coordinate in the new lane does not change in comparison with the spatial coordinate in the old lane.

If the rule (∗) is not satisfied, the rule (∗∗) is applied: The gap between two neighboring vehicles in the target lane exceeds some value $g_{\text{on}}^{(\min)}$

$$x_n^+ - x_n^- - d > g_{\mathrm{on}}^{(\mathrm{min})}, \tag{24}$$

$$g_{\mathrm{on}}^{(\mathrm{min})} = \lambda^{(\mathrm{on})} v_n^+ + d, \tag{25}$$

$\lambda^{(\mathrm{on})}$ is a constant. In addition, the condition that the vehicle passes the midpoint

$$x_n^{(\mathrm{m})} = (x_n^+ + x_n^-)/2 \tag{26}$$

between two neighboring vehicles in the target lane for time step n should be satisfied, i.e.,

$$\begin{aligned} & x_{n-1} < x_{n-1}^{(\mathrm{m})} \quad \text{and} \quad x_n \geq x_n^{(\mathrm{m})} \\ & \text{or} \\ & x_{n-1} \geq x_{n-1}^{(\mathrm{m})} \quad \text{and} \quad x_n < x_n^{(\mathrm{m})}. \end{aligned} \tag{27}$$

After merging, the coordinate of the merging vehicle is set to $x_n = x_n^{(\mathrm{m})}$ and the vehicle speed v_n is set to $\hat{v}_n$ (23).

If neither the rule (*) nor the rule (**) is satisfied, the vehicle does not merge onto the target lane. In this case, the vehicle moves in the old lane until it comes to a stop at the end of the merging region.

2.7 Model Parameters

In simulations, the length of the main road is $L = 40$ km, the point $x = 0$ is at the distance $L/2 = 20$ km from the end of the road so the road starts at $x = -L/2 = -20$ km. The model parameters are: $\tau = 1$ s, $v_{\mathrm{free}} = 30$ m/s (108 km/h), $d = 7.5$ m, $a = 0.5$ m/s^2, $k = 3$, $\beta = 1$, $p_1 = 0.3$, $p_a = 0.17$, $p_b = 0.1$, $\delta = 0.01$, $p_0(v) = 0.575 + 0.125 \min(1, v/10)$, $p_2(v) = 0.48 + 0.32\theta(v - 15)$. Lane changing parameters are $\delta_1 = \delta_2 = 1$ m/s, $p_c = 0.2$. The parameters of the merging region are: $\lambda = 0.75$ ($\lambda = 0.6$ for the merging region in the off-ramp lane), $\Delta v_r^{(1)} = 10$ m/s, $\Delta v_r^{(2)} = 5$ m/s at the on-ramp and $\Delta v_r^{(2)} = -2$ m/s at the off-ramp and at the merge bottleneck. $x_{\mathrm{on}} = x_{\mathrm{M}} = x_{\mathrm{off}} = 16$ km. $L_m = 500$ m, $L_r = 2$ km for the on-ramp bottleneck; $L_m = 0.6$ km, $L_r = 2$ km, $L_c = 1.6$ km for the off-ramp bottleneck; $L_c = 0.6$ km for the merge bottleneck.

3 Congested Patterns at Bottlenecks on Two-lane Freeway

3.1 Types of Congested Patterns and Their Diagram at On-Ramp Bottlenecks

In the model, the first-order F→S transition is governed by a competition between a tendency towards free flow due to vehicle over-acceleration and a tendency towards synchronized flow due to the speed adaptation effect [24]

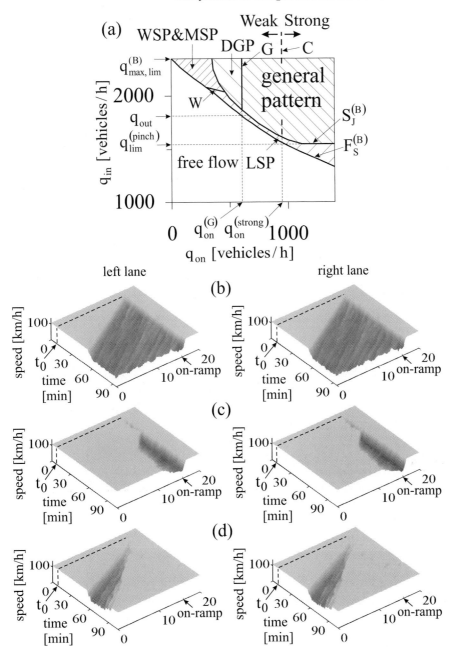

Fig. 3. The diagram of congested patterns at an on-ramp bottleneck (**a**) on a two-lane road and speed on the main road in space and time within SPs for the left and right freeway lanes (**b–d**). (**b**) The widening SP (WSP), (**c**) the localized SP (LSP), (**d**) the moving SP (MSP). In (**b–d**) the flow rates (q_{on}, q_{in}) are: (**b**) (300, 2250), (**c**) (550, 1846), and (**d**) (25, 2323), vehicles/h. In (**a**) criteria for the boundaries $F_S^{(B)}$ and $S_J^{(B)}$ are the same as those in [26]. $q_{max,lim}^{(B)} \approx 2350$ vehicles/h.

(Sect. 2.2). This competition explains the physics of the onset of congestion in free flow [21, 24].

Types of congested patterns at an on-ramp bottleneck on a two-lane freeway are qualitative the same (Figs. 3 and 4) as found in [24] for a single-lane road. This is because the vehicle over-acceleration that is modeled by acceleration noise in (8) is described in the model as a "collective effect", which occurs *on average* in traffic flow [24].

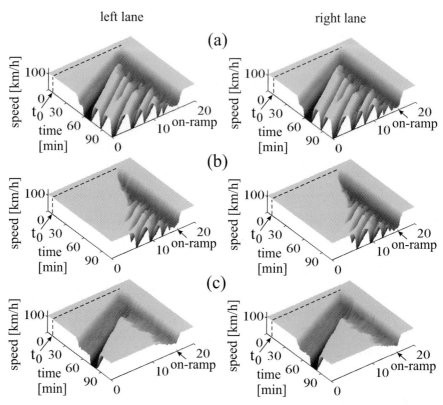

Fig. 4. GPs related to the diagram in Fig. 3(**a**). Speed on the main road for the left and right lanes in space and time. (**a**) GP at $q_{in} > q_{out}$, (**b**) GP at $q_{in} < q_{out}$, (**c**) Dissolving GP (DGP). The flow rates (q_{on}, q_{in}) are: (**a**) (1000, 2308), (**b**) (1200, 1698), and (**c**) (450, 2308) vehicles/h.

Diagram of congested patterns at the on-ramp bottleneck on the two-lane road, i.e., the regions of the spontaneous occurrence of the patterns in the flow-flow plane with coordinates q_{in} and q_{on}, is presented in Fig. 3 (a). Different SPs (Figs. 3 (b–d)) occurs between the boundaries $F_S^{(B)}$ and $S_J^{(B)}$ in the diagram (Fig. 3 (a)). The boundaries $F_S^{(B)}$ and $S_J^{(B)}$ are related to *spontaneous* F→S and S→J transitions, respectively. The downstream front of a widening SP (WSP) is fixed at the on-ramp (Fig. 3 (b)). The upstream front of the WSP is continuously widening upstream. The WSP occurs above the boundary W in Fig. 3 (a). Below the boundary W a localized SP (LSP) occurs. As in WSPs, the downstream front of the LSP is fixed at the on-ramp.

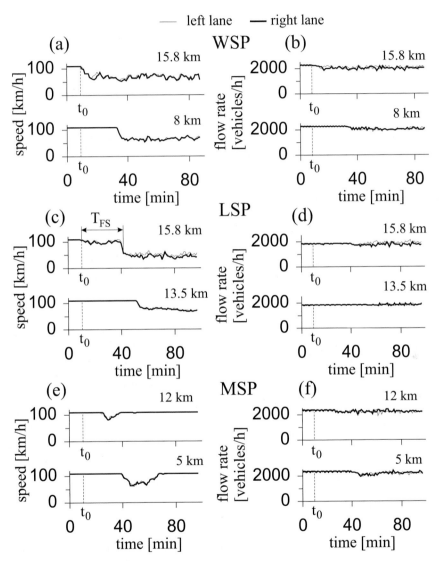

Fig. 5. Vehicle speed (**a, c, e**) and flow rate (**b, d, f**) for WSP (**a, b**), LSP (**c, d**), and MSP (**e, f**) shown in Fig. 3 (b), (c), and (d) respectively. 1-min data of vitual detectors.

However, the upstream front of the LSP is localized at a some distance L_{LSP} upstream of the on-ramp (Fig. 3 (c)). At higher q_{in} and low q_{on} a moving SP (MSP) can occur (Fig. 3 (d)) rather than an WSP appears. Since an F→S transition is a first-order phase transition, SPs often emerge after some time delay (Figs. 3 (c), 5 (c)). Qualitatively the same SPs appear spontaneously in the left and right lanes of the road (Figs. 3(b–d), 5).

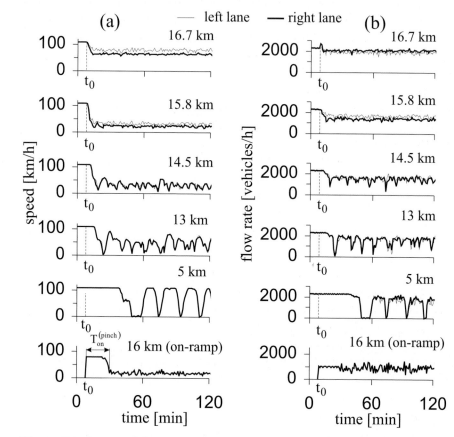

Fig. 6. Vehicle speed (**a**) and flow rate (**b**) at virtual detectors for GP in Fig. 4 (a).

In the diagram (Fig. 3 (a)), right of the boundary G and right of the boundary $S_J^{(B)}$ GPs occur (Fig. 4 (a, b)). In synchronized flow of an GP the pinch effect, i.e., the self-compression of synchronized flow upstream of the on-ramp, occurs [25]. In the related pinch region of synchronized flow in the GP the speed and flow rate decrease (Fig. 6, locations labeled "15.8 km" and "14.5 km"), however the decrease in flow rate is much smaller than the one in speed. In the pinch region of the GP, narrow moving jams spontaneously emerge and grow (Fig. 6, locations labeled "14.5 km" and "13 km"). Some of these narrow moving jam transform into wide moving jams. As a result, upstream of the pinch region of the GP a sequence of wide moving jams appears (Figs. 4 (a, b), and 6). The mean time between wide moving jams, $T_{\mathrm{J}}^{(\mathrm{wide})}$, is considerably greater than the mean time between narrow moving jams in the pinch region T_{J} (in Fig. 6 $T_{\mathrm{J}}^{(\mathrm{wide})} \approx 18$ min at $x = 5$ km and $T_{\mathrm{J}} \approx 6$ min at $x = 14.5$ km). In the GP shown in Fig. 6, the pinch region also occurs after some delay $T_{\mathrm{on}}^{(\mathrm{pinch})}$ in the on-ramp lane (Fig. 6, location labeled "16 km (on-ramp)"). Right of the boundary $S_J^{(B)}$ and left of the boundary G in the diagram in Fig. 3 (a) a dissolving GP (DGP) occurs (Fig. 4 (c)).

Right to the line C in Fig. 3 (a), i.e., at $q_{\mathrm{on}} > q_{\mathrm{on}}^{(\mathrm{strong})}$, the strong congestion occurs in the pinch region of GP. In the case of strong congestion, the flow rate $q^{(\mathrm{pinch})}$ in the pinch region reaches its limit value $q^{(\mathrm{pinch})} = q_{\mathrm{lim}}^{(\mathrm{pinch})}$, and the frequency of the moving jam emergence $f_{\mathrm{narrow}} = 1/T_{\mathrm{J}}$ reaches a maximum [24]. In contrast, left of the boundary C, i.e., at $q_{\mathrm{on}} < q_{\mathrm{on}}^{(\mathrm{strong})}$, and right of the boundary $S_J^{(B)}$ the weak congestion occurs in GP. In this case the pinch region characteristics depend on traffic demand, i.e., on q_{on} [24].

3.2 Congested Patterns at Merge Bottlenecks

At the merge bottleneck (Fig. 2 (b)), the following peculiarities of congested patterns occur: (i) independent of traffic demand q_{in} *only* strong congestion occurs in the pinch region of an GP (Figs. 7 (a–c) and 8). (ii) In contrast to a GP under the strong congestion condition at the on-ramp bottleneck, the average speed between moving jams in an GP at a merge bottleneck has a *saturation* to a limit (maximum) speed $v_{\mathrm{max}}^{(\mathrm{M})} < v_{\mathrm{free}}$. The actual speed $v^{(\mathrm{M})}$ between the moving jams shows only a small amplitude oscillations around $v_{\mathrm{max}}^{(\mathrm{M})}$ (Fig. 7 (b)). (iii) The limit flow rate per freeway lane at the merge bottleneck $q_{\mathrm{lim\ M}}^{(\mathrm{pinch})}$ is considerably lower than $q_{\mathrm{lim}}^{(\mathrm{pinch})}$ at the on-ramp bottleneck. (iv) At the same model parameters the mean time between narrow jams $T_{\mathrm{J}}^{(\mathrm{M})}$ is lower at the merge bottleneck ($T_{\mathrm{J}}^{(\mathrm{M})} \approx 3.5$ min in Fig. 8 at $x = 14$ km) than at the on-ramp bottleneck ($T_{\mathrm{J}} \approx 6$ min in Fig. 6 at $x = 14.5$ km). The distances between the downstream fronts of wide moving jams in an GP, $T_{\mathrm{J}}^{(\mathrm{wide\ M})}$, are lower at the merge bottleneck ($T_{\mathrm{J}}^{(\mathrm{wide\ M})} \approx 7$ min in Fig. 8 at $x = 5$ km) than at the on-ramp bottleneck ($T_{\mathrm{J}}^{(\mathrm{wide})} \approx 18$ min in Fig. 6 at $x = 5$ km). Nevertheless, after speed saturation has occurred between the looking-like narrow moving jams in the GP at the merge bottleneck, the downstream jam fronts propagate with the characteristic velocity v_{g} through other upstream bottlenecks and through other states of traffic flow while *maintaining* the velocity of the downstream front v_{g}, i.e., the jams are wide moving jams.

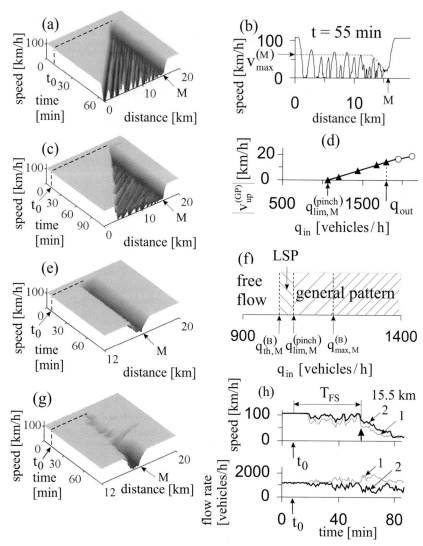

Fig. 7. Features of congested patterns at a merge bottleneck. (**a–c**) Speed distributions averaged across both lanes in an GP at $q_{in} > q_{out}$ (**a, b**) and at $q_{in} < q_{out}$ (**c**). (**d**) The absolute value of the average velocity $v_{up}^{(GP)}$ of the upstream boundary of the GP versus q_{in}. (**e**) Speed distributions in an LSP. (**f**) The diagram of patterns at the merge bottleneck. (**g, h**) The critical perturbation development that leads to spontaneous GP formation (**g**) and the related data of virtual detector (**h**). q_{in} is: (**a, b**) 1946, (**c**) 1440, (**e**) 1050, and (**g, h**) 1180 vehicles/h. The vehicle motion in the right lane upstream of the merge bottleneck is switched at $t = t_0 = 8$ min. The location of the merge bottleneck is labeled M. Curves 1 and 2 in (**h**) are related to the left and right lanes, respectively.

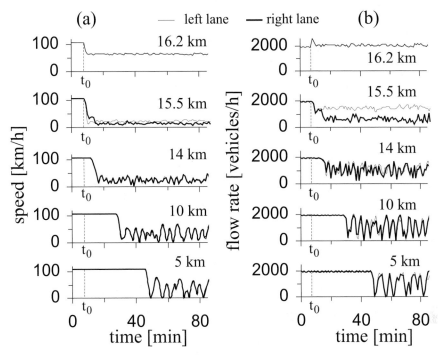

Fig. 8. Vehicle speed (**a**) and flow rate (**b**) at virtual detectors for the GP shown in Fig. 7 (a).

Similarly to an GP at the on-ramp bottleneck, in an GP at merge bottlenecks, the region of wide moving jams is expanded upstream. At $q_{in} > q_{out}$ the width (in the longitudinal direction) of the farthest upstream jam increases over time (Fig. 7 (a)), whereas at $q_{in} < q_{out}$ jams at the upstream boundary of the GP dissolve one after another (Fig. 7 (c)). The average velocity of the upstream boundary of the GP $v_{up}^{(GP)}$ decreases as q_{in} decreases (Fig. 7 (d)), and $v_{up}^{(GP)}$ becomes zero at $q_{in} = q_{lim\ M}^{(pinch)}$. At $q_{in} < q_{lim\ M}^{(pinch)}$ only LSPs can occur at a merge bottleneck (Fig. 7 (e)).

Pattern emergence at the merge bottleneck shows the following pattern diagram (Fig. 7 (f)): (i) If $q_{in} \geq q_{max,\ M}^{(B)}$, an GP usually can occur spontaneously within a given time interval T_0. At $T_0 = 30$ min the maximum flow rate $q_{max,\ M}^{(B)} \approx q_{max}^{(free)}/2$ (the factor $1/2$ appears because at the merge bottleneck *two*-lane road transforms into *single*-lane road at $x \geq x_M$). There is a random time delay T_{FS} for GP formation after traffic flow has been switched: Many local perturbations can appear spontaneously, grow, and then decay at the merge bottleneck before the perturbation appears, which growth indeed leads to GP formation (Fig. 7 (g, h)). (ii) If $q_{th\ M}^{(B)} \leq q_{in} < q_{max,\ M}^{(B)}$ free flow is metastable: (1) At $q_{lim\ M}^{(pinch)} \leq q_{in} < q_{max,\ M}^{(B)}$ an GP can be excited by a

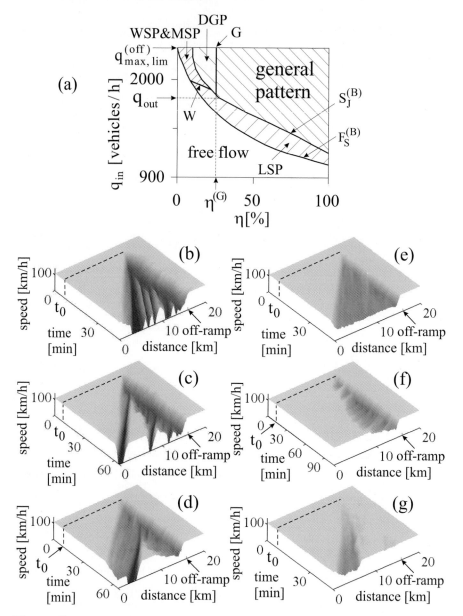

Fig. 9. Congested patterns at an isolated off-ramp bottleneck. (**a**) The diagram of congested patterns. (**b**) GP at high η. (**c**) GP at low η. (**d**) DGP. (**e**) WSP. (**f**) LSP. (**g**) MSP. η (in %) and the flow rate q_{in} (in *vehicles/h*) are: (**b**) (75, 2250), (**c**) (30, 2000), (**d**) (14.5, 2182), (**e**) (9.5, 2279), (**f**) (14, 1925), and (**g**) (0.5, 2308). In (**b-g**) speed distributions averaged over both lanes are shown. Vehicles start to leave the main road to the off-ramp at $t = t_0 = 8\ min$.

high amplitude short-time external perturbation. (2) At $q_{\text{th M}}^{(B)} \leq q_{\text{in}} \leq q_{\text{lim M}}^{(\text{pinch})}$ rather than an GP only an LSP can be excited. At $q_{\text{in}} < q_{\text{th M}}^{(B)}$ free flow is stable.

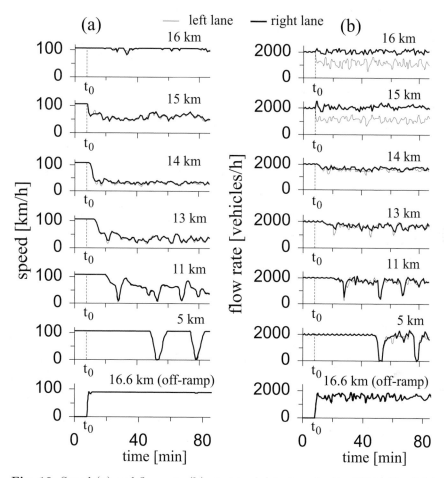

Fig. 10. Speed (**a**) and flow rate (**b**) at virtual detectors for the GP in Fig. 9 (c).

3.3 Congested Patterns at Off-ramp Bottlenecks

A study of pattern formation at an off-ramp bottleneck shows the following pattern features: (i) In contrast to the on-ramp and merge bottlenecks, at the off-ramp bottleneck (Fig. 2 (c)) a GP *only* under weak congestion condition can be formed at all $\eta < 100\%$. In this case, all characteristics of the pinch region in the GP do depend on η [24]. (ii) The pattern diagram at the off-ramp bottleneck in the $(\eta,\ q_{in})$ plane (Fig. 9 (a)) qualitative resembles the pattern diagram at the on-ramp bottleneck (Fig. 3 (a)). In particular, there are different SPs (MSP, WSP, and LSP), DGPs, and GPs (Fig. 9 (b-g)), which look like the related patterns at the on-ramp bottleneck (Figs. 3 (b-g), 4). However, there is no saturation of the boundary $S_J^{(B)}$ at higher η. This is related to the result that only weak congestion occurs in GPs at the off-ramp bottleneck at all $\eta < 100\%$. (iii) In an GP, wide moving jams are often formed at a considerable distance (more than 5 km) upstream of the off-ramp (Fig. 10, "11 km"). The downstream fronts of GPs and SPs are located at some distance about $1 - 1.5$ km for GPs (Fig. 10, "15 km") and about 2 km for WSPs and LSPs (Fig. 11, "14 km") upstream of the off-ramp. (iv) At the same model parameters, the values $q^{(pinch)}$ and T_J in an GP at the off-ramp bottleneck at high η can be considerably *lower* than $q_{lim}^{(pinch)}$ and $T_{J,\ lim}$, respectively, in a GP under strong congestion condition at the on-ramp bottleneck: The off-ramp bottleneck forces vehicles at higher η to much stronger traffic compression. (v) At $\eta = 100\%$ an off-ramp bottleneck transforms into a merge bottleneck. We find qualitatively the same GP features at $\eta = 100\%$ as for the merge bottleneck at the same model parameters.

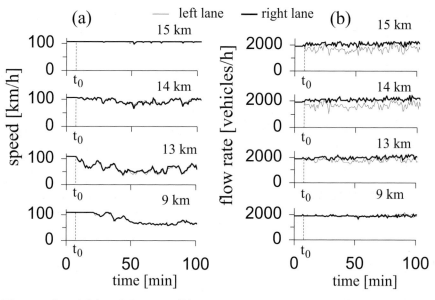

Fig. 11. Speed (**a**) and flow rate (**b**) at virtual detectors for the LSP in Fig. 9 (f).

4 Conclusion

The microscopic traffic flow theory presented in [24] and in this paper, which is based on three-phase traffic theory [18, 20, 21], can explain and reproduce the main empirical features of spatial-temporal traffic patterns at different freeway bottlenecks [18, 19]: (i) the metastability of free flow and the first order-phase transition from free flow to synchronized flow, (ii) spontaneous emergence of wide moving jams in synchronized flow due to S→J transition, (iii) three types of SPs at on- and off-ramp bottlenecks: WSP, MSP, and LSP, (iv) features of GPs of different types at bottlenecks, (v) the cases of strong and weak congestions and diverse transformations between different congested patterns under weak congestion.

References

1. G.B. Whitham: *Linear and Nonlinear Waves* (Wiley, New York 1974).
2. A.D. May: *Traffic Flow Fundamentals* (Prentice Hall Inc., New Jersey 1990).
3. D. Chowdhury, L. Santen, and A. Schadschneider: Physics Reports **329**, 199 (2000).
4. D. Helbing: Rev. Mod. Phys. **73**, 1067 (2001).
5. T. Nagatani: Rep. Prog. Phys. **65**, 1331 (2002).
6. K. Nagel, P. Wagner, and R. Woesler: Operation Res. **51**, 681 (2003).
7. B.S. Kerner: *The Physics of Traffic* (Springer, Berlin, New-York, Tokio 2004).
8. M. Schreckenberg and D.E. Wolf (editors): *Traffic and Granular Flow' 97* (Springer, Singapore 1998).
9. D. Helbing, H.J. Herrmann, M. Schreckenberg, and D.E. Wolf (editors): *Traffic and Granular Flow' 99* (Springer, Heidelberg 2000).
10. M. Fukui, Y. Sugiyama, M. Schreckenberg, and D.E. Wolf (editors): *Traffic and Granular Flow'01* (Springer, Berlin 2003).
11. D. Helbing, A. Hennecke, and M. Treiber: Phys. Rev. Lett. **82**, 4360 (1999); M. Treiber, A. Hennecke, D. Helbing: Phys. Rev. E **62**, 1805 (2000).
12. H.Y. Lee, H.-W. Lee, and D. Kim: Phys. Rev. E **59**, 5101 (1999); Physica A **281**, 78 (2000) Phys. Rev. E **62**, 4737 (2000).
13. R. Barlovic, L. Santen, A. Schadschneider, and M. Schreckenberg: Eur. Phys. J. B. **5**, 793 (1998).
14. W. Knospe, L. Santen, A. Schadschneider, and M. Schreckenberg: J. Phys. A **33**, L477 (2000) Phys. Rev. E **65**, 015101 (R) (2001).
15. S. Krauß, P. Wagner, and C. Gawron: Phys. Rev. E **55**, 5597 (1997).
16. K. Nagel, D.E. Wolf, P. Wagner, and P. Simon: Phys. Rev. E **58**, 1425 (1998).
17. B.S. Kerner and H. Rehborn: Phys. Rev. E **53**, R4275 (1996).
18. B.S. Kerner: Phys. Rev. Lett. **81**, 3797 (1998).
19. B.S. Kerner: Phys. Rev. E **65**, 046138 (2002).
20. B.S. Kerner: 'A Theory of Congested Traffic Flow'. In: *Proceedings of the 3rd Symposium on Highway Capacity and Level of Service*, ed by R. Rysgaard, Vol. 2 (Road Directorate, Ministry of Transport – Denmark 1998) pp. 621–642.
21. B.S. Kerner: Transportation Research Record **1678**, 160 (1999); Physics World **12**, 25 (August 1999); J. Phys. A: Math. Gen. **33**, L221 (2000).

22. B.S. Kerner: Networks and Spatial Economics **1**, 35 (2001).
23. B.S. Kerner: Physica A **333**, 379 (2004).
24. B.S. Kerner and S.L. Klenov: Phys. Rev. **68**, 036130 (2003).
25. B.S. Kerner and S.L. Klenov: J. Phys. A: Math. Gen. **35**, L31 (2002).
26. B.S. Kerner, S.L. Klenov, and D.E. Wolf: J. Phys. A: Math. Gen. **35**, 9971 (2002).
27. L.C. Davis: Phys. Rev. E **69**, 016108 (2004).
28. H.K. Lee, R. Barlović, M. Schreckenberg, and D. Kim: Phys. Rev. Lett. **92**, 238702 (2004).
29. M. Rickert, K. Nagel, M. Schreckenberg, and A. Latour: Physica A **231**, 534 (1996).

Congestion Due to Merging Roads: Predictions of Three-Phase Traffic Theory

B.S. Kerner[1], S.L. Klenov[2], and D.E. Wolf[3]

[1] DaimlerChrysler AG
    RIC/TN, HPC: T729, 70546 Stuttgart – Germany
[2] Department of Physics – Moscow Institute of Physics and Technology
    141700 Dolgoprudny, Moscow Region – Russia
[3] Institute of Physics – University Duisburg-Essen
    47048 Duisburg – Germany

Summary. Congested traffic patterns at a symmetric merger of two single-lane roads into one road are studied in the KKW-model, a cellular automaton that traces all observed congestion patterns back to two basic concepts: A typical distance below which a driver decides to adjust his speed to the one of the car in front, and the minimal distance between cars that is compatible with secure driving. The diagram of congested patterns in the flow–flow plane whose co-ordinates are the inflow rates onto the two merging roads is determined. Depending on the parameters, every congestion pattern can occur on either road, but not all combinations of patterns on both roads are realized. There is no indication of spontaneous symmetry breaking: Equal demand on both roads leads to the same type of pattern on both sides.
Keywords: three-phase traffic theory, cellular automata, congested patterns

1 Introduction

Traffic flow theory is a quickly developing and very controversial field of statistical and non-linear physics. There are two qualitatively different theoretical approaches for the description of phase transitions and spatial-temporal patterns of traffic: (i) The fundamental diagram approach, and (ii) the three-phase traffic theory approach.

Almost all earlier traffic flow theories and models belong to the fundamental diagram approach (see references in reviews [1–7]). In the fundamental diagram approach, hypothetical steady state solutions lie on a curve(s) in the flow–density plane. This curve is called the fundamental diagram. Steady states are model solutions, which are related to a hypothetical unperturbed and noiseless vehicle motion. In a steady state, all vehicles move with the same time-independent speed at the same distance to one another. Obviously fluctuations destroy steady states. However, dynamical features of steady states determine to a large degree both the features of possible phase transitions and the dynamics of model solutions far away from the steady states.

In particular, models based on the fundamental diagram approach show the spontaneous emergence of wide moving jams in free flow [3–6]), if some critical density is exceeded. This is in contradiction with empirical observations [8]: In real traffic flow, first the phase transition from free flow to synchronized flow occurs (the F→S transition) and later and at another freeway location a wide moving jam can spontaneously emerge in this synchronized flow (the S→J transition). In other words, wide moving jams emerge due to the sequence of the F→S→J transitions.

To explain this and other empirical pattern features, Kerner introduced a three-phase traffic theory in 1996-1999. The three phases are: 1. Free flow, 2. Synchronized flow, 3. Wide moving jam. In this qualitative traffic flow theory, there is no fundamental diagram for hypothetical steady states of synchronized flow: Steady states of the "synchronized flow" traffic phase cover a two-dimensional region in the flow–density plane. Fluctuations lead to non-homogeneous spatial-temporal states of synchronized flow with special properties concerning the phase transitions in traffic flow. These dynamical features of the phase transitions in three-phase traffic theory were first formulated as hypotheses [8–10], which allowed to conjecture also some general features of spatial-temporal congested patterns at a freeway bottleneck and the pattern diagram [11]. These predictions of three-phase traffic theory have been confirmed with models, which simulate individual cars in a continuous space model [12] or as KKW-cellular automata (KKW-CA) models [13]. These mathematical results of three-phase traffic theory explain and predict all known features of phase transitions and congested traffic patterns, which have been observed in empirical studies. Recently, some new microscopic traffic flow models based on the three-phase traffic theory have been introduced [14, 15] that confirm features of congested patterns at an on-ramp bottleneck earlier found in [12, 13].

A detailed consideration of empirical features of freeway congested traffic patterns, traffic flow theory and microscopic models, and their engineering applications have recently been made in the book [7].

Different types of congested patterns can occur at a bottleneck, depending on the traffic demand. So far diagrams of congested patterns for three-phase traffic theory were only analyzed for asymmetric vehicle priority regulations in the vicinity of the freeway bottleneck [11–16]. Specifically, vehicles on the main road had the right of way in comparison with vehicles, which merge onto the main road from an on-ramp. Then it was natural to focus on the congested pattern forming on the main road, although the global situation is characterized by a *combination of patterns* forming on the main road and on the merging road upstream of the on-ramp bottleneck.

It is the purpose of this article to investigate the mutual coupling of traffic patterns forming on two merging roads. Priority rules will influence this coupling. Therefore, the symmetric case is of special interest, where two identical roads merge into one road and vehicle priority regulating rules are symmetric. This is the case investigated in the following.

2 Model of a Symmetric Merger Bottleneck

2.1 KKW-2 CA Traffic Flow Model

For the simulation we use the KKW-2 CA model [13], the definition of which will be briefly recalled in this subsection. The motion rules for a given vehicle are as follows:

$$v_{n+1} = \max(0, \min(v_{\text{free}}, v_{\text{s},n}, v_{\text{c},n})), \tag{1}$$

$$x_{n+1} = x_n + v_{n+1}\tau, \tag{2}$$

$$v_{\text{c},n} = \begin{cases} v_n + a\tau & \text{for } x_{\ell,n} - x_n > D_n, \\ v_n + a\tau \ \text{sgn}(v_{\ell,n} - v_n) & \text{for } x_{\ell,n} - x_n \leq D_n. \end{cases} \tag{3}$$

$$v_{\text{s},n} = g_n/\tau, \tag{4}$$

where v_n and x_n denote the speed and the space co-ordinate of the vehicle, respectively, at discrete time $t = n\tau$, $n = 0, 1, 2, ...$ Eq. (1) means that the speed cannot be negative nor larger than any of the three velocities v_{free}, $v_{\text{s},n}$ and $v_{\text{c},n}$, where the latter two depend on time. v_{free} is the maximum speed in free flow, which is assumed to be the same for all vehicles, $v_{\text{s},n}$ is the safe speed and $v_{\text{c},n}$ determines the rule of "speed change", which are specified by Eq. (4) and Eq. (3), respectively. a denotes the acceleration as well as a comfortable deceleration (no emergency braking). $g_n = x_{\ell,n} - x_n - d$ is the space gap between vehicles. The lower index ℓ marks functions (or values) related to the vehicle in front of the one at x_n, the preceding vehicle, and d is the vehicle length assumed to be the same for all vehicles. D_n is a "synchronization distance" [12, 13]: At $x_{\ell,n} - x_n \leq D_n$ the driver tends to adjust his speed to the one of the vehicle in front, i.e. he decelerates if $v_n > v_{\ell,n}$, and accelerates if $v_n < v_{\ell,n}$. In the KKW-2 CA model [13] the synchronization distance D_n in (3) is a nonlinear function of the vehicle speed:

$$D(v_n) = d + v_n\tau + \beta v_n^2/2a, \tag{5}$$

where β is a positive constant. That a driver changes his/her behaviour from speed maximation to speed adaption, when the driver comes closer than a typical distance, the synchronization distance, to the preceding vehicle and the driver cannot pass the preceding vehicle, provides a *microscopic reason* for the fundamental ingredient of three-phase traffic theory [12, 13]: Hypothetical steady states of synchronized flow in the model cover a 2D region in the flow density plane (Fig. 1 (a)) [8–10].

As in the NaSch CA-models [17–22], acceleration and deceleration in the cellular automata model KKW-2 [13] are stochastic functions. *At a first step*, a preliminary speed $\tilde{v}_{n+1}$ is calculated for each vehicle,

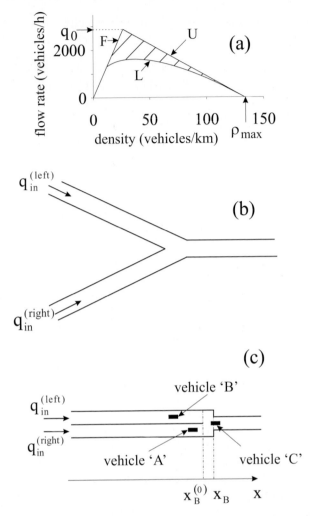

Fig. 1. (a) Steady states of the KKW2-CA model. (b,c) Symmetric merger bottleneck.

$$\tilde{v}_{n+1} = v_{n+1}, \tag{6}$$

according the system of the dynamical equations (1) – (4). *At a second step,* a fluctuation $a\tau\eta_n$ (to be specified below) is added. *Finally,* the speed v_{n+1} at the time $n + 1$ is calculated according to

$$v_{n+1} = \max(0, \min(\tilde{v}_{n+1} + a\tau\eta_n, v_n + a\tau, v_{\text{free}}, v_{s,n})). \tag{7}$$

This means, that the stochastic contribution $a\tau\eta_n$ may neither lead to a speed smaller than zero, nor to a speed larger than what the deterministic acceleration a would give, taking the limitations by v_{free} and $v_{s,n}$ into account.

We implement the fluctuation η_n in (7) as

$$\eta_n = \begin{cases} -1 \text{ if } r < p_b, \\ 1 \text{ if } p_b \leq r < p_b + p_a, \\ 0 \text{ otherwise} \end{cases} \tag{8}$$

where r denotes a random number uniformly distributed between 0 and 1. As in a NaSch CA-model in the fundamental diagram approach [19], the probability p_b in (8) is taken as a decreasing function of the vehicle speed v_n:

$$p_b(v_n) = \begin{cases} p_0 \text{ if } v_n = 0 \\ p \ \text{ if } v_n > 0. \end{cases} \tag{9}$$

where p and $p_0 > p$ are constants. The probability p_a of the random acceleration in (8) is taken as a constant in the model KKW-2 [13].

2.2 Implementation of the Symmetric Merger

In this article, traffic phenomena are considered, which occur at a symmetric merger bottleneck of two identical single-lane roads (Fig. 1 (b)). In our model (Fig. 1 (c)), the two roads merge at a point $x_B^{(0)}$. Downstream of another point $x_B \geq x_B^{(0)}$ vehicles move on a single-lane road. Within the merging region, $x_B^{(0)} < x < x_B$, the vehicles merge onto the single-lane road from one of the two roads upstream of the bottleneck. Within this merging region vehicles move like on a single-lane road. There is no priority for vehicles from either side upstream to get onto the road downstream. To simulate the vehicle motion at this bottleneck, the following symmetric rules are used:

(i) At each time $t = n\tau$ the vehicles, which are nearest to the point $x_B^{(0)}$, are determined on the right and left roads, respectively. Then the space coordinates of these two vehicles are compared. The vehicle with a larger coordinate is considered vehicle A (see Fig. 1 (c)), the other one is called vehicle B. If the coordinates of the two vehicles are equal, vehicle A is chosen randomly with probability $1/2$ among them.

(ii) The first vehicle downstream of $x_B^{(0)}$ is considered the preceding vehicle for A (vehicle C in Fig. 1 (c)). Vehicle A accelerates if it is slower than the preceding vehicle:

$$v_{n+1} = \max(0, \min(v_n + \Delta v, v_{\text{free}}, v_{s,n})), \tag{10}$$

provided it is already close enough to the merger, i.e., its coordinate x_n satisfies the condition

$$x_n \geq x_B^{(0)} - L, \tag{11}$$

where Δv and L are constants. If the condition (11) is not fulfilled, vehicle A moves according to the rules (1)-(5), and the "jump" in the speed (10) occurs after it reached the point $x = x_B^{(0)} - L$.

(iii) For the safe speed of vehicle B the following formula is used instead of (4):

$$v_{s,n} = \lfloor (x_B^{(0)} - x_n)/(2\tau) \rfloor, \tag{12}$$

where $\lfloor \ldots \rfloor$ denotes the integer part. In addition, the vehicle B adjusts its speed to the one of A: The values $v_{\ell,n}$ and $x_{\ell,n}$ in (3) are the speed and the coordinate of vehicle A. Note that according to (12), vehicle B decelerates, when approaching the bottleneck, i.e. the point $x_B^{(0)}$, and it can come to a stop there. This guarantees a collision-free motion during the changing to the road downstream of the bottleneck i.e., within the region $x_B^{(0)} < x < x_B$. Often, however, vehicle A will pass the bottleneck before B stops. Then B may become the new vehicle A and start to accelerate again according to (ii).

(iv) If at time $t = (n+1)\tau$ the co-ordinate of vehicle A,

$$x_{n+1} = x_n + v_{n+1}\tau \geq x_B^{(0)} + d, \tag{13}$$

then the vehicle A plays the role of the preceding vehicle C for another vehicle A on one of the two roads upstream of the point $x = x_B^{(0)}$. This new vehicle A is chosen in accordance with the rules of point (i).

2.3 Parameters of Discretization and Simulation

The time and space discretization and simulation parameters are the same as those in the CA model KKW-2 in [13]: The time step τ is $\tau = 1s$, the length of cells is chosen equal to $\delta x = 0.5\ m$, the corresponding speed discretization is $\delta v = 1.8\ km/h$. The other parameters are $a = 0.5\ m/s^2$, $v_{\text{free}} = 108\ km/h = 60\ \delta v$, $d = 7.5\ m = 15\ \delta x$, $p = 0.04$, $p_a = 0.052$, $\beta = 0.05$. The probability p_0 is chosen equal to $p_0 = 0.425$. This corresponds to a velocity $v_g = -15.5\ km/h$ of the downstream front of a wide moving jam and an outflow $q_{\text{out}} = 1810$ vehicles/h from a wide moving jam. The parameters for the dynamics at the bottleneck are $\Delta v = 10$ m/s, $L = v_{\text{free}}\tau$.

The congested patterns are simulated for a system of 100 km length (200000 cells) with open boundary conditions. The reference point $x = 0$ is placed at the distance 20 km from the end of the road, so that the beginning of the roads is at the coordinate $x_0 = -80$ km. The merger bottleneck is at $x_B^{(0)} = 16$ km (32000 cells). Note that the coordinate x_B was only introduced for illustrative purposes. It is not used in the rules (10) - (13) for vehicle motion at the bottleneck and therefore needs not be specified. What matters for the kinetics of a merger bottleneck is the point, downstream of which two vehicles can no longer drive side by side, but must arrange such that the x-intervals they cover, do not overlap. This point is $x_B^{(0)}$.

Open boundary conditions are used: At the beginning of each of the two roads new vehicles are generated one by one independently of one another at rates which depend on the traffic demand, characterized by asymptotic incoming flow rates $q_{\text{in}}^{(\text{left})}$ and $q_{\text{in}}^{(\text{right})}$, and the local congestion. For example, if the time which passed since the previous vehicle was generated at the beginning of the left road is greater than or equal to $\tau_{in}^{(left)} = 1/q_{in}^{(left)}$, and if its current position $x_{\ell,n}$ and velocity $v_{\ell,n}$ are such that $x_{\ell,n} - x_0 \geq \ell_{safe}$, where $\ell_{safe} = v_{\ell,n}\tau + d$ is the safe distance, a new vehicle appears on the left road. The speed of this vehicle v_n and the coordinate x_n are set to $v_n = v_{\ell,n}$ and $x_n = \max(x_0, \; x_{\ell,n} - \max(v_n\tau_{in}^{(left)}, \ell_{safe}))$, respectively. This boundary condition applies also for the right road, with $\tau_{in}^{(left)}$ replaced by $\tau_{in}^{(right)} = 1/q_{in}^{(right)}$.

3 Results of Simulations and Their Discussion

Simulations of the model show that different types of synchronized flow patterns (SP) and general patterns (GP) can spontaneously occur at the symmetric bottleneck depending on the flow rates $q_{in}^{(left)}$ and $q_{in}^{(right)}$ at the beginning of the left and right roads, respectively. Recall that a SP is a congested pattern which consists of the "synchronized flow" traffic phase only. A GP is a congested pattern where synchronized flow occurs and wide moving jams spontaneously emerge in this synchronized flow: The GP consists of the "synchronized flow" and "wide moving jam" traffic phases [11–13].

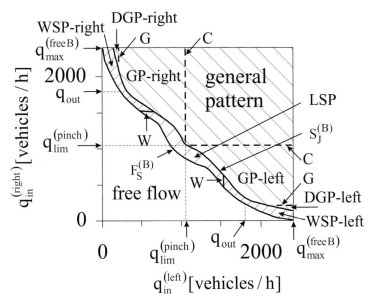

Fig. 2. Diagram of congested patterns at the symmetric bottleneck.

The diagram of congested patterns, Fig. 2, shows the regions in the $(q_{in}^{(left)}, \; q_{in}^{(right)})$-plane, where the different congested patterns occur. For a symmetric merger bottleneck this diagram must be symmetric with respect to the initial flow rates $q_{\text{in}}^{(\text{left})}$ and $q_{\text{in}}^{(\text{right})}$, in contrast to the diagram of congested patterns at an on-ramp bottleneck [12, 13].

The types of congested patterns at the symmetric bottleneck are qualitatively the same as those at an on-ramp bottleneck [12, 13]. In particular, different SPs (Fig. 3) occur between the boundaries $F_{\text{S}}^{(\text{B})}$ and $S_{\text{J}}^{(\text{B})}$ in the diagram (Fig. 2). The boundaries $F_{\text{S}}^{(\text{B})}$ and $S_{\text{J}}^{(\text{B})}$ are related to the *spontaneous* F→S and S→J transitions, respectively. The downstream front of all SPs is fixed at the bottleneck.

For medium traffic demands on both roads one finds the localized SP (LSP) on both sides. Its upstream front is localized at some distance from the bottleneck (Fig. 3 (d,e)). If, however, the traffic demand on the two roads is very different (upper left and lower right corner of the diagram, Fig. 2) one finds the widening SP (WSP) (Fig. 3 (a), right, and (b), left), which close to the boundary $F_{\text{S}}^{(\text{B})}$ in the diagram, is repeatedly interrupted by free flow (so-called alternating SP (ASP)) (Fig. 3 (c), right). The upstream front of these patterns is continuously moving upstream. On the other road, one finds free flow or a localized SP (LSP) in this case, but never a WSP or a GP: If one increases $q_{\text{in}}^{(\text{left})}$ for fixed, large $q_{\text{in}}^{(\text{right})} > q_{\text{out}}$ (q_{out} is the flow rate out of a wide moving jam) one moves on a horizontal line in the upper part of Fig. 2. Firstly, one finds free flow on both roads; then an alternating SP or a widening SP on the right road and free flow or a LSP on the left road; for still larger $q_{\text{in}}^{(\text{left})}$ one obtains a general pattern (GP) on the right, but still LSP on the left (cf. Fig. 3 (a)). Only if $q_{\text{in}}^{(\text{left})}$ exceeds $q_{\text{lim}}^{(\text{pinch})}$ (line C in Fig. 2), a GP is also found on the left road, i.e., wide moving jams will also form in the synchronized flow on the left road. Whether or not this sequence depends on the rules at the merger bottleneck (see Sec. 2.2), remains to be investigated. Our example only shows, that *in general not all combinations of congested patterns on the two merging roads are realized.*

In the diagram (Fig. 2), there are two lines G, a vertical and a horizontal one. In the upper right corner of the diagram, i.e. right and above the boundaries G and $S_{\text{J}}^{(\text{B})}$, GPs occur. Within the GPs, the pinch effect is realized, i.e., the self-compression of synchronized flow occurs [12]. This leads to the spontaneous formation of narrow moving jams, some of which may grow into wide moving jams. As a result, upstream of the pinch region of a GP a sequence of wide moving jams appears (Fig. 4 (a), right). Between the boundary $S_{\text{J}}^{(\text{B})}$ and the lines G a dissolving GP (DGP) occurs (Fig. 4 (d), right), combined with a LSP on the other road.

If traffic demand is equal on both roads, we find another remarkable result: The pattern on both roads is always the same. *There is no indication of spontaneous symmetry breaking.* This is nontrivial, because the flow that reaches the bottleneck is lower than q_{in} on a road, where a GP (that is widening upstream over time) forms. One could imagine, that such a GP forms first

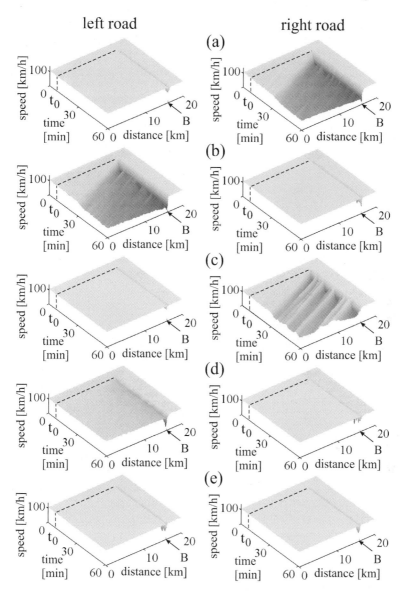

Fig. 3. Synchronized patterns (SP) at the symmetric bottleneck: (**a**) WSP on the right road. (**b**) WSP on the left road. (**c**) WSP with alternating regions of free and synchronized flow (ASP). (**d**) LSP at $q_{in}^{(right)} < q_{in}^{(left)}$ near the boundary "W" between WSP and LSP in the diagram (Fig.2). (**e**) LSP at $q_{in}^{(right)} = q_{in}^{(left)}$. The flow rates ($q_{in}^{(right)}$, $q_{in}^{(left)}$) are: (**a**) (2300, 72), (**b**) (72, 2300), (**c**) (2300, 45), (**d**) (663, 1510), (**e**) (1030, 1030) vehicles/h. The position of the bottleneck ($x_B^{(0)} = 16$ km) is shown by the point "B". At $t_0 = 7$ min flow on the roads is switched on. Single vehicle data are averaged over a space interval of 20 m and a time interval of 1 min. Each figure consists of two parts: speed distribution on the left (right) road upstream of the bottleneck (the point "B") and speed distribution on the common single-lane road downstream of the bottleneck.

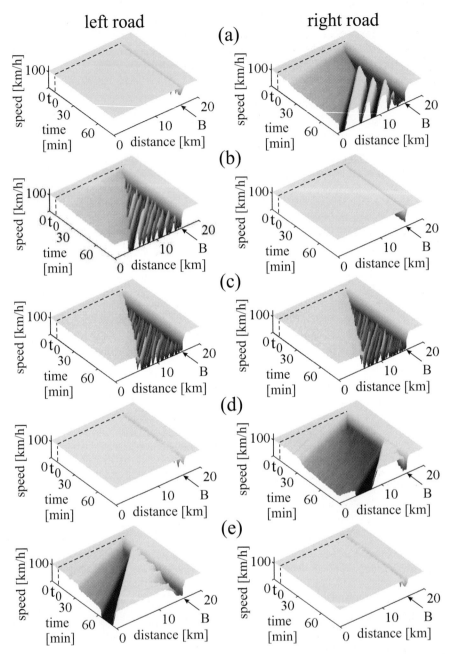

Fig. 4. General patters (GP) at the symmetric bottleneck: (**a**) GP on the right road for $q_{in} > q_{out}$: Moving jams grow. (**b**) GP on the left road at $q_{in} < q_{out}$: Moving jams die out. (**c**) GPs on both roads at $q_{in}^{(right)} = q_{in}^{(left)}$. (**d**, **e**) Dissolving GPs (DGP) on the right road (**d**) and on the left road (**e**). The flow rates $(q_{in}^{(right)}, q_{in}^{(left)})$ are: (**a**) (1964, 432), (**b**) (720, 1660), (**c**) (1440, 1440), (**d**) (2400, 144), (**e**) (144, 2400) vehicles/h.

on the right road, for example. Then the reduced flow reaching the merger from the right might drop below the value needed to trigger the same congested pattern on the left. This is not the case. Moreover, near the boundary $S_J^{(B)}$ it may take quite long until wide moving jams form. However, once they occurred on one road, they occur on the other one almost immediately, as well. In other words, the pattern emergence on one road induces the pattern emergence on the other road without any significant time delay.

The reason for this might be the following: In the case $q_{in}^{(right)} = q_{in}^{(left)} = q_{in}$ GPs are only formed at $q_{in} \geq q_{lim}^{(pinch)}$, where $q_{lim}^{(pinch)}$ is the limit flow rate in the pinch region of GP. In the case of the symmetric bottleneck, $q_{lim}^{(pinch)}$ is also the maximal flow rate on one road upstream of the merger when a GP exists on the other road (this value is called $q_{on}^{(strong)}$ in the case of an on-ramp bottleneck [16]). Thus, if a GP starts to form on one road, the flow rate on the other road drops down to $q_{lim}^{(pinch)}$ without any delay. Because of the condition $q_{in} \geq q_{lim}^{(pinch)}$ this causes GP emergence on the other road.

In accordance with empirical observations [11] and a microscopic three-phase traffic theory [16], we found both weak and strong congestion conditions in the pinch region of GPs. However, in contrast to the case of an on-ramp bottleneck [16], at the symmetric bottleneck there are two lines C in the diagram, which separate the weak congestion condition in the pinch region of GPs from the strong congestion condition (Fig. 2). Right of the vertical line C and above the horizontal line C strong congestion conditions in GPs are realized. By contrast, left of the vertical line C and below the horizontal line C and also right of the boundary $S_J^{(B)}$ weak congestion conditions in GPs occur.

From the above consideration it also follows that only the strong congestion condition in a GP occurs in the case $q_{in}^{(right)} = q_{in}^{(left)} = q_{in}$. Indeed, in the case of weak congestion in a GP at an on-ramp bottleneck all vehicles from the on-ramp lane can merge to the main road, i.e., there is no restriction of the flow rate from the on-ramp lane to the main road ($q_{on} < q_{on}^{(strong)}$). In contrast, the strong congestion condition occurs in the opposite case $q_{on} > q_{on}^{(strong)}$. At the symmetric bottleneck one road can be considered as an on-ramp for the other one. Since in a GP $q_{in} \geq q_{lim}^{(pinch)}$, only GPs with the strong congestion condition occur.

In summary, we have shown that congested patterns on two equal roads that merge into one depend on each other: Not all pattern combinations occur and there is no spontaneous symmetry breaking. We believe that these results are generally valid. However, the precise form of the diagram of congested patterns may depend on the conditions at the merger bottleneck reflected by the way it is implemented. Another interesting question, which remains for further studies, concerns the dependence of the pattern combinations on the asymmetry, if the roads do not have equal priority.

Acknowledgements

BSK acknowledges funding by BMBF within project DAISY.

References

1. N. Gartner, C. Messer, and A. Rathi (eds.): *Special Report 165: Revised Monograph on Traffic Flow Theory* (Transportation Research Board, Washington, D.C. 1997).
2. D.E. Wolf: Physica A **263**, 438 (1999).
3. D. Chowdhury, L. Santen, and A. Schadschneider: Physics Reports **329**, 199 (2000).
4. D. Helbing: Rev. Mod. Phys. **73**, 1067 (2001).
5. T. Nagatani: Rep. Prog. Phys. **65**, 1331 (2002).
6. K. Nagel, P. Wagner, and R. Woesler: Operation Res. **51**, 681 (2003).
7. B.S. Kerner: *The Physics of Traffic* (Springer, Berlin, New York, Tokio 2004).
8. B.S. Kerner: Phys. Rev. Lett. **81**, 3797 (1998).
9. B.S. Kerner: In: *Proceedings of the 3rd Symposium on Highway Capacity and Level of Service*, edited by R. Rysgaard, Vol. 2 (Road Directorate, Ministry of Transport - Denmark 1998) 621–642; B.S. Kerner: In: *Transportation and Traffic Theory*, Proceedings of the 14th International Symposium on Transportation and Traffic Theory, edited by A Ceder (Elsevier Science Ltd, Oxford 1999) p. 147.
10. B.S. Kerner: Transportation Research Record **1678**, 160 (1999).
11. B.S. Kerner: Phys. Rev. E **65**, 046138 (2002).
12. B.S. Kerner and S.L. Klenov: J. Phys. A: Math. Gen **35**, L31 (2002).
13. B.S. Kerner, S.L. Klenov, and D.E. Wolf: J. Phys. A: Math. Gen. **35**, 9971 (2002).
14. L.C. Davis: Phys. Rev. E **69**, 016108 (2004).
15. H.K. Lee, R. Barlović, M. Schreckenberg, and D. Kim: Mechanical restriction versus human overreaction triggering congested traffic states, cond-mat/0404315 (2004); Phys. Rev. Lett. **92**, 238702 (2004).
16. B.S. Kerner and S.L. Klenov: Phys. Rev. E **68**, 036130 (2003).
17. K. Nagel and M. Schreckenberg: J Phys. I France **2**, 2221 (1992).
18. M. Schreckenberg, A. Schadschneider, K. Nagel, and N. Ito: Phys. Rev. E **51**, 2939 (1995).
19. R. Barlovic, L. Santen, A. Schadschneider, and M. Schreckenberg: Europ. J. Phys. B **5**, 793 (1998).
20. W. Knospe, L. Santen, A. Schadschneider, and M. Schreckenberg: J. Phys. A: Math. Gen **33**, L477 (2000).
21. W. Knospe, L. Santen, A. Schadschneider, and M. Schreckenberg: Phys. Rev. E **65**, 015101(R) (2002).
22. K. Nagel, D.E. Wolf, P. Wagner, and P. Simon: Phys. Rev. E **58**, 1425 (1998).
23. M. Takayasu and H. Takayasu: Fractals **1**, 860 (1993).

Production, Supply, and Traffic Systems: A Unified Description

D. Helbing

Institute for Economics and Traffic – Dresden University of Technology
Andreas-Schubert-Str. 23, 01062 Dresden – Germany
helbing@trafficforum.org

Summary. The transport of products between different suppliers or production units can be described similarly to driven many-particle and traffic systems. We introduce equations for the flow of goods in supply networks and the adaptation of production speeds. Moreover, we present two examples: The case of linear (sequential) supply chains and the case of re-entrant production. In particular, we discuss the stability conditions, dynamic solutions, and resonance phenomena causing the frequently observed "bullwhip effect", which is an analogue of stop-and-go traffic. Finally, we show how to treat discrete units and cycle times, which can be applied to the description of vehicle queues and travel times in freeway networks.
Keywords: stop-and-go traffic, supply network, production system, bullwhip effect, resonance, convective instability, re-entrant production, cycle time, travel time, queueing network

Supply chain management is a major subject in economics, as it significantly determines the efficiency of production processes [1]. Many related studies focus on subjects like optimum buffer sizes and stock levels, but a stable dynamics and optimal network structure are also important subjects [2, 3]. Therefore, this scientific field is connected with the statistical physics of networks [4].

In this paper, we will investigate the dynamic properties and linear stability of supply chains by a generalization of "fluid-dynamic" production models, which have been inspired by traffic models [5–10]. Fluid-dynamic models take into account non-linear interactions and are suitable for on-line control, as they are numerically much more efficient than event-driven (Monte-Carlo) simulations. Moreover, unlike queueing theory, they are not mainly restricted to the treatment of stationary situations, but can reflect variations in the consumption rate and effects of machine breakdowns or changes in the production schedule.

The organization of this contribution is as follows: Section 1 focusses on the description of supply and production systems, in particular linear supply chains and re-entrant production. Section 2 treats freeway traffic as a queueing network and applies the formula for the cycle time to the determination of travel times.

1 Modeling Production Systems as Supply Networks

Our production model assumes u production units or suppliers j which deliver d_{ij} products of kind $i \in \{1, \ldots, p\}$ per production cycle to other suppliers and consume c_{kj} goods of kind k per production cycle. The coefficients c_{kj} and d_{ij} are determined by the respective production process, and the number of production cycles per unit time (e.g. per day) is given by the production speed $Q_j(t)$. That is, supplier j requires an average time interval of $1/Q_j(t)$ to produce or deliver d_{ij} units of good i. The temporal change in the number $N_i(t)$ of goods of kind i available in the system is given by the difference between the inflow

$$Q_i^{\text{in}}(t) = \sum_{j=0}^{u} d_{ij} Q_j(t) \tag{1}$$

and the outflow

$$Q_i^{\text{out}}(t) = \sum_{j=1}^{u+1} c_{ij} Q_j(t). \tag{2}$$

In other words, it is determined by the overall production rates $d_{ij}Q_j(t)$ of all suppliers j minus their overall consumption rates $c_{ij}Q_j(t)$:

$$\frac{dN_i}{dt} = Q_i^{\text{in}}(t) - Q_i^{\text{out}}(t) = \sum_{j=1}^{u} (d_{ij} - c_{ij})Q_j(t) - Y_i(t). \tag{3}$$

Herein, the quantity

$$Y_i(t) = \underbrace{c_{i,u+1}Q_{u+1}(t)}_{\text{consumption and losses}} - \underbrace{d_{i0}Q_0(t)}_{\text{inflow of resources}} \tag{4}$$

comprises the consumption rate of goods i, losses, and waste (the "export" of material), minus the inflows into the considered system (the "imports"). In the following, we will assume that the quantities are measured in a way that $0 \le c_{ij}, d_{ij} \le 1$ (for $1 \le i \le p$, $1 \le j \le u$) and the "normalization conditions"

$$d_{i0} = 1 - \sum_{j=1}^{u} d_{ij} \ge 0, \qquad c_{i,u+1} = 1 - \sum_{j=1}^{u} c_{ij} \ge 0 \tag{5}$$

are fulfilled. Equations (3) can then be interpreted as conservation equations for the flows of goods.

1.1 Adaptation of Production Speeds

The production speeds $Q_j(t)$ may be changed in time in order to adapt to varying consumption rates $Y_i(t)$. We will assume that it takes a typical time period T to adapt the actual production speeds $Q_j(t)$ to the desired production ones, W_j:

$$\frac{dQ_j}{dt} = \frac{1}{T}\left[W_j(\{N_i(t)\}, \{dN_i/dt\}, \{Q_k(t)\}) - Q_j(t)\right].\tag{6}$$

Here, the curly brackets indicate that the so-called management or control function $W_j(\dots)$ may depend on all inventories $N_i(t)$ with $i \in \{1,\dots,p\}$, their derivatives dN_i/dt, and/or all production speeds $Q_k(t)$ with $k \in \{1,\dots,u\}$. The resulting dynamics of the supply network can be investigated by means of a linear stability analysis, which has been carried out for linear supply chains in Ref. [2] and for supply networks with $d_{ij} = \delta_{ij}$ in Ref. [3]. Some of the main results will be discussed in the following.

1.2 Modelling one-Dimensional Supply Chains

For simplicity, let us investigate a model of one-dimensional supply chains, here, which corresponds to $d_{ij} = \delta_{ij}$ and $c_{ij} = \delta_{i+1,j}$, where $\delta_{ij} = 1$ for $i = j$ and $\delta_{ij} = 0$ otherwise. This implies $Q_i^{\text{in}}(t) = Q_i(t)$ and $Q_i^{\text{out}}(t) = Q_{i+1}(t) = Q_{i+1}^{\text{in}}(t)$. The assumed model consists of a series of u suppliers i, which receive products of kind $i-1$ from the next "upstream" supplier $i-1$ and generate products of kind i for the next "downstream" supplier $i+1$ at a rate $Q_i(t)$ [7,9]. The final products are delivered at the rate $Q_u(t)$ and removed from the system with the consumption rate $Y_u(t) = Q_{u+1}(t)$. The consumption rate is typically subject to perturbations, which may cause a "bullwhip effect" [11–18], i.e. growing variations in the stock levels and deliveries of upstream suppliers. This is due to delays in the adaptation of their delivery rates.

The inventory of goods of kind i changes in time t according to

$$\frac{dN_i}{dt} = Q_i^{\text{in}}(t) - Q_i^{\text{out}}(t) = Q_i(t) - Q_{i+1}(t),\tag{7}$$

while the temporal change of the delivery rate will be assumed as

$$\frac{dQ_i}{dt} = \frac{1}{T}\left[W_i(N_i(t), dN_i/dt, Q_i(t), Q_{i+1}(t)) - Q_i(t)\right].\tag{8}$$

The management strategy

$$\frac{dQ_i}{dt} = \frac{1}{T}\left\{\frac{N_i^0 - N_i(t)}{\tau} - \beta\frac{dN_i}{dt} + \epsilon[Q_i^0 - Q_i(t)]\right\}\tag{9}$$

appears to be appropriate to keep the inventories $N_i(t)$ stationary, to maintain a certain optimal inventory N_i^0 (in order to cope with stochastic variations due to machine breakdowns etc.), and to operate with the equilibrium production rates $Q_i^0 = Y_u^0$, where Y_u^0 denotes the average consumption rate. The supplier-independent parameter values τ, β, and ϵ can be justified with suitable scaling arguments [3].

1.3 Dynamic Solution and Resonance Effects

In the vicinity of the stationary state characterized by $N_i(t) = N_i^0$ and $Q_i(t) = Q_i^0$, it is possible to calculate the dynamic solution of the one-dimensional supply chain model [3, 8]. For this, let $n_i(t) = N_i(t) - N_i^0$ be the deviation of the inventory from the stationary one, and $q_i(t) = Q_i(t) - Q_i^0$ the deviation of the delivery rate. The linearized model equations read

$$\frac{dn_i}{dt} = q_i(t) - q_{i+1}(t) \tag{10}$$

and

$$\frac{dq_i}{dt} = -\frac{1}{T}\left(\frac{n_i}{\tau} + \beta\frac{dn_i}{dt} + \epsilon q_i\right). \tag{11}$$

Deriving Eq. (11) with respect to t and inserting Eq. (10) results in the following set of second-order differential equations:

$$\frac{d^2 q_i}{dt^2} + \underbrace{\frac{(\beta+\epsilon)}{T}}_{=2\gamma}\frac{dq_i}{dt} + \underbrace{\frac{1}{T\tau}}_{=\omega^2} q_i(t) = \underbrace{\frac{1}{T}\left[\frac{q_{i+1}(t)}{\tau} + \beta\frac{dq_{i+1}}{dt}\right]}_{=f_i(t)}. \tag{12}$$

This corresponds to the differential equation for the damped harmonic oscillator with damping constant γ, eigenfrequency ω, and driving term $f_i(t)$. The two eigenvalues of this system of equations are

$$\lambda_{1,2} = -\gamma \pm \sqrt{\gamma^2 - \omega^2} = -\frac{1}{2T}\left[(\beta+\epsilon) \mp \sqrt{(\beta+\epsilon)^2 - 4T/\tau}\right], \tag{13}$$

i.e. for $(\beta + \epsilon) > 0$ their real parts are always negative, corresponding to a stable behavior in time. Nevertheless, we will identify a convective instability below, i.e. the oscillation amplitude can grow from one supplier to the next one upstream.

The set of equations (12) can be solved successively, starting with $i = u$ and progressing to lower values of i. For example, assuming periodic oscillations of the form $f_u(t) = f_u^0 \cos(\alpha t)$, after a transient time much longer than $1/\gamma$ we find

$$q_u(t) = f_u^0 F \cos(\alpha t + \varphi) \tag{14}$$

with

$$\tan \varphi = \frac{2\gamma\alpha}{\alpha^2 - \omega^2} = \frac{\alpha(\beta + \epsilon)}{\alpha^2 T - 1/\tau} \tag{15}$$

and

$$F = \frac{1}{\sqrt{(\alpha^2 - \omega^2)^2 + 4\gamma^2\alpha^2}} = \frac{T}{\sqrt{[\alpha^2 T - 1/\tau]^2 + \alpha^2(\beta + \epsilon)^2}} , \qquad (16)$$

where the dependence on the eigenfrequency ω is important to understand the occuring resonance effect. Equations (12) and (14) imply

$$f_{u-1}(t) = \frac{1}{T}\left[\frac{q_u(t)}{\tau} + \beta\frac{dq_u}{dt}\right] = f_{u-1}^0 \cos(\alpha t + \varphi + \delta) \qquad (17)$$

with

$$\tan\delta = \alpha\beta\tau \quad\text{and}\quad f_{u-1}^0 = f_u^0 \frac{F}{T}\sqrt{(1/\tau)^2 + (\alpha\beta)^2} . \qquad (18)$$

1.4 "Bull-Whip Effect" and Stop-and-Go Traffic

The oscillation amplitude increases from one supplier to the next upstream one, if

$$\frac{f_{u-1}^0}{f_u^0} = \left\{1 + \frac{\alpha^2[\epsilon(\epsilon + 2\beta) - 2T/\tau] + \alpha^4 T^2}{(1/\tau)^2 + (\alpha\beta)^2}\right\}^{-1/2} > 1 . \qquad (19)$$

One can see that this resonance effect can occur for $0 < \alpha^2 < 2/(T\tau) - \epsilon(\epsilon + 2\beta)/T^2$. Therefore, variations in the consumption rate are magnified under the instability condition

$$T > \epsilon\tau\left(\beta + \epsilon/2\right) . \qquad (20)$$

Supply chains show this "bullwhip effect" (which corresponds to the phenomenon of convective, i.e. upstream moving instability), if the adaptation time T is too large, if there is no adaptation to the equilibrium production speed Q_i^0, corresponding to $\epsilon = 0$, or if the management reacts too strong to deviations of the actual stock level N_i from the desired one N_i^0, corresponding to a small value of τ. The latter is very surprising, as it implies that the strategy

$$\frac{dQ_i}{dt} = \frac{1}{T\tau}[N_i^0 - N_i(t)] , \qquad (21)$$

which tries to maintain a constant work in progress $N_i(t) = N_i^0$, would ultimately lead to an undesireable bullwhip effect. In contrast, the management strategy

$$\frac{dQ_i}{dt} = \frac{1}{T}\left\{-\beta\frac{dN_i}{dt} + \epsilon[Q_i^0 - Q_i(t)]\right\} \qquad (22)$$

would avoid this problem, but it would not maintain a constant work in progress. The control strategy (9) with a sufficiently large value of τ would fulfill both requirements.

The "bullwhip effect" has, for example, been reported for beer distribution [22, 23], but similar dynamical effects are also known for other distribution or transportation chains with significant adaptation times T. It has, for example, some analogy to stop-and-go traffic [3, 10], where delayed adaptation also leads to an unstable behavior. In the case $\beta = 0$ and $\epsilon = 1$, the stability condition (20) agrees exactly with the one of the optimal velocity model [24], which is a particular microscopic traffic model. This car-following model assumes an acceleration equation of the form

$$\frac{dv_i(t)}{dt} = \frac{V_{\mathrm{opt}}(d_i(t)) - v_i(t)}{T} \tag{23}$$

and the complementary equation

$$\frac{dd_i(t)}{dt} = -[v_i(t) - v_{i+1}(t)]. \tag{24}$$

In contrast to the above supply chain model, however, the index i represents single vehicles, $v_i(t)$ is their actual velocity of motion, V_{opt} the so-called optimal (safe) velocity, which depends on the distance $d_i(t)$ to the next vehicle ahead. T denotes again an adaptation time. Comparing this equation with Eq. (9), the velocities v_i would correspond to the delivery rates Q_i, the optimal velocity V_{opt} to the desired delivery rate $W_i(N_i) = Q_i^0 + (N_i^0 - N_i)/\tau$, and the inverse vehicle distance $1/d_i$ would approximately correspond to the stock level N_i (apart from a proportionality factor). This shows that the analogy between supply chain and traffic models concerns only their mathematical structure, but not their interpretation, although both relate to transport processes. Nevertheless, this mathematical relationship can give us hints, how methods, which have been successfully applied to the investigation of traffic models before, can be generalized for the study of supply networks. Compared to traffic dynamics, supply networks and production systems have some interesting new features: Instead of a continuous space, we have discrete production units j, and the management strategy (6) is generally different from the velocity adaptation (23) in traffic. With suitable strategies, in particular with large values of τ and β, the oscillations can be mitigated or even suppressed. Moreover, production systems are frequently supply networks with complex topologies rather than one-dimensional supply chains, i.e. they have additional features compared to (more or less) one-dimensional freeway traffic. They are more comparable to street networks of cities [25].

1.5 Calculation of the Cycle Times

Apart from the productivity or throughput Q_i of a production unit, production managers are highly interested in the cycle time T_i, i.e. the time interval between the beginning of the generation of a product and its completion. Let us assume that the queue length $L_i(t)$ of products waiting to be processed by production unit i is given by the inventory $N_{i-1}(t)$ of product $i-1$. The change of the queue length $L_i(t)$ in time is then determined by the difference between the arrival rate $Q_i^{\mathrm{arr}}(t)$, which corresponds to the inflow $Q_{i-1}^{\mathrm{in}} = Q_{i-1}(t)$ from production unit $i - 1$, and the departure rate $Q_i^{\mathrm{dep}}(t)$, which corresponds to the outflow $Q_{i-1}^{\mathrm{out}}(t) = Q_i(t)$ to production unit i:

$$\frac{dL_i}{dt} = \frac{dN_{i-1}}{dt} = Q_{i-1}(t) - Q_i(t) = Q_i^{\mathrm{arr}}(t) - Q_i^{\mathrm{dep}}(t). \tag{25}$$

On the other hand, the waiting products move forward $Q_i^{\mathrm{dep}}(t) = Q_i(t)$ steps per unit time, as $Q_i(t)$ is the processing rate (production speed). For this reason, the overall time $T_i(t)$ until having been processed is given by the implicit equation

$$N_{i-1}(t) = L_i(t) = \int\limits_{t}^{t+T_i(t)} dt'\, Q_i^{\mathrm{dep}}(t') = \int\limits_{-\infty}^{t+T_i(t)} dt'\, Q_i^{\mathrm{dep}}(t') - \int\limits_{-\infty}^{t} dt'\, Q_i^{\mathrm{dep}}(t'),$$

$$\tag{26}$$

if the queue of length $L_i(t)$ was joined at time t. Accordingly, high inventories imply long cycle times, which favours just-in-time production with small stock levels. From Eqs. (25) and (26), one can finally derive a delay-differential equation for the waiting time under varying production conditions [10]:

$$\frac{dT_i}{dt} = \frac{Q_i^{\mathrm{arr}}(t)}{Q_i^{\mathrm{dep}}(t + T_i(t))} - 1 = \frac{Q_{i-1}(t)}{Q_i(t + T_i(t))} - 1. \tag{27}$$

This equation can be solved numerically as a function of the production rates $Q_i(t)$, since the production initially starts with a cycle time of $T_i(0) = 1/Q_i(0)$, corresponding to the average processing time $1/Q_i(0)$ when the production unit i is started. In this way, it is possible to determine the process cycle times T_i and the overall production time as the sum of the cycle times of all single production steps, taking into account the respective time delays T_i: When a specific product enters the queue before production unit i at time $t = t_i$, it enters the queue before production unit $i+1$ at time $t_{i+1} = t_i + T_i(t_i)$. The time of delivery to the customer is given by $t_{u+1} = t_u + T_u(t_u)$.

1.6 Modeling of Discrete Units

The above "fluid-dynamic" model equations can not only be used to represent approximate mean values of large numbers of products. They can also be transfered to the treatment of discrete units (such as single units of a product), if their dynamics is sufficiently deterministic. For example, one could represent the time interval ΔT, during which a discrete unit occupies a certain production unit i by a step function. However, as step functions are not differentiable everywhere, we will replace them by smooth functions.

One possible specification uses a Fourier approximation of the step functions. To lowest order, one may take

$$Q_i(t) = A \sum_k \{1 - \cos[\pi(t - t_i^k)/\Delta T]\} \tag{28}$$

where t_i^k denotes the starting time of occupation by object k. The cosinus function is set to zero for $t < t_i^k$ and $t > t_i^k + \Delta T$. The prefactor A is determined in a way that satisfies the normalization condition,

$$\int_{t_i^k}^{t_i^k + \Delta T} dt' \, Q_i(t') = 1 \,, \qquad (29)$$

which implies $A = 1/\Delta T$. Under this condition, the integral

$$N_i^{\text{tot}}(t) = \int_0^t dt' \, Q_i(t') \qquad (30)$$

counts the number of units that have passed the cross section or production unit under consideration.

Another possible specification assumes

$$Q_i(t) = B \sum_k (t - t_i^k)^2 (t - t_i^k - \Delta T)^2 \,, \qquad (31)$$

where we set the fourth order polynomial to zero for $t < t_i^k$ and $t > t_i^k + \Delta T$. The normalization condition (29) implies $B = 30/(\Delta T)^5$.

Together with (28) or (31), the relationship (30) can also be applied to situations where several units occupy a production unit at the same time. As before, the management function can be chosen as a function of the inventories $N_i(t)$, but in many cases, it would be reasonable to replace a reaction to temporal changes dN_i/dt in the inventories by exponentially smoothed values.

1.7 Re-Entrant Production

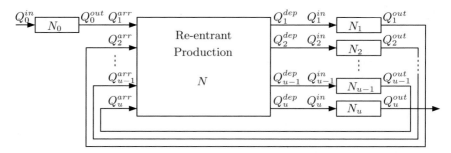

Fig. 1. Illustration of the model of re-entrant production and its variables (see main text for details).

Semiconductor production [19] relies on some extremely expensive (lithographic) production units, which are therefore used for many similar subsequent production steps i, namely the production of the various layers of a chip. This is the reason for re-entrant production [20, 21] (see Fig. 1.7), posing particular control problems not only for the overall arrival rate $Q^{\text{arr}}(t)$ in the re-entrant production area, but also concerning the fractions $p_i(t)$ of products in the different production stages i fed into it. If $N_i(t)$ denotes the stock level of chips after the ith entry (production step), $Q_i^{\mathrm{in}}(t)$ the respective inflow, and $Q_i^{\mathrm{out}}(t)$ the outflow, the related balance equation is again

$$\frac{dN_i}{dt} = Q_i^{\mathrm{in}}(t) - Q_i^{\mathrm{out}}(t). \tag{32}$$

Assuming also a buffer before the first and a buffer after the last (uth) production step, this equation applies to $i \in \{0, 1, \dots, u\}$, where Q_u^{out} corresponds to the removal rate of products that have completed the re-entrant (lithographic) production steps. The outflow $Q_{i-1}^{\mathrm{out}}(t)$ determines the arrival rate $Q_i^{\mathrm{arr}}(t)$ for the ith re-entrant production step, while the departure rate $Q_i^{\mathrm{dep}}(t)$ determines the inflow $Q_i^{\mathrm{in}}(t)$ into the subsequent buffer. The overall stock level $N(t)$ of chips at various production stages in the re-entrant production area changes in time according to

$$\frac{dN}{dt} = \sum_{i=1}^{u} [Q_{i-1}^{\mathrm{out}}(t) - Q_i^{\mathrm{in}}(t)] = \sum_{i=1}^{u} [Q_i^{\mathrm{arr}}(t) - Q_i^{\mathrm{dep}}(t)] = Q^{\mathrm{arr}}(t) - Q^{\mathrm{dep}}(t). \tag{33}$$

The overall departure rate $Q^{\mathrm{dep}}(t)$ from the re-entrant production area depends on the number $N(t)$ of products processed at the same time. Moreover, if $p_i(t)$ denotes the fraction of chips that have entered the re-entrant area for the ith production step at time t and $T^0(t)$ is the related cycle time of these chips in the re-entrant area, their departure rate is given by

$$Q_i^{\mathrm{dep}}(t) = p_i(t - T^0(t))Q^{\mathrm{dep}}(N(t)) = Q_i^{\mathrm{in}}(t). \tag{34}$$

As before, the delay-differential equation for the temporal change of the cycle time is

$$\frac{dT^0}{dt} = \frac{Q^{\mathrm{arr}}(t)}{Q^{\mathrm{dep}}(t + T^0(t))} - 1. \tag{35}$$

The overall arrival rate may be adapted according to

$$\frac{dQ^{\mathrm{arr}}}{dt} = \frac{1}{T}[W(N(t), dN/dt, Q^{\mathrm{arr}}(t), Q^{\mathrm{dep}}(t), N_u(t), dN_u/dt, Q_u^{\mathrm{out}}(t)) - Q^{\mathrm{arr}}(t)], \tag{36}$$

where the management function $W(\dots)$ depends not only on the variables $N(t)$ and $Q^{\mathrm{dep}}(t)$ characterizing the re-entrant production area, but also on the (desired) removal rate $Q_u^{\mathrm{out}}(t)$ and the stock level $N_u(t)$ of the final buffer. Finally, the specific arrival rates of chips for the $(i+1)$st re-entrant production step are

$$Q_{i+1}^{\mathrm{arr}}(t) = p_{i+1}(t)Q^{\mathrm{arr}}(t) = Q_i^{\mathrm{out}}(t), \tag{37}$$

where their relative percentages

$$p_{i+1}(t) = \frac{W_i(N_i(t), dN_i/dt, Q_i^{\mathrm{in}}(t), Q_i^{\mathrm{out}}(t), Q_u^{\mathrm{out}}(t))}{\sum_j W_j(N_j(t), dN_j/dt, Q_j^{\mathrm{in}}(t), Q_j^{\mathrm{out}}(t), Q_u^{\mathrm{out}}(t))} \tag{38}$$

are controlled by another management function $W_i(\ldots)$, which depends on the stock levels $N_i(t)$ of the respectively preceding buffers and on the removal rate $Q_u^{\text{out}}(t)$. The management functions $W_i(\ldots)$ allow one to specify different priorities such as push or pull strategies. Specifications of the management function, which can avoid bullwhip (resonance) effects and long cycle times will be presented in a forth-coming paper. Generalizations to the simultaneous production of various products are possible, requiring the consideration of an individual number of re-entrant steps with potentially different production speeds and cycle times ("overtaking") in the re-entrant area. The problem is similar to the treatment of heterogeneous multi-lane traffic [26]. In the following, we will present a simpler traffic model for uniform vehicles, which shows how to transfer the above approach from production processes to transportation processes in street networks.

2 Queueing Model of Vehicle Traffic in Freeway Networks

In the following, we will propose a traffic model [25], which was inspired by the above model of supply networks. Here, the elements to be served are the vehicles. The formula for the queue length determines the density in a road section, and the cycle time corresponds to the travel time.

When we now specify road traffic as a queueing system, we will take into account essential traffic characteristics such as the flow-density relation or the properties of extended congestion patterns at bottlenecks. In fact, traffic congestion is usually triggered by spatial inhomogeneities of the road network [27], and queueing effects are normally not observed along sections of low capacity, but upstream of the beginning of a bottleneck. Therefore, we will subdivide roads into sections i of homogeneous capacity and length l_i^{max}, which start at place x_i and end with some kind of inhomogeneity (i.e. an increase or decrease of capacity) at place $x_{i+1} = x_i + l_i^{\text{max}}$. In other words, the end of a road section i is, for example, determined by the location of an on- or off-ramp, a change in the number of lanes, or the beginning or end of a gradient.

The model can be derived from the "fluid-dynamic" continuity equation

$$\frac{\partial \rho(x,t)}{\partial t} + \frac{\partial Q_i(x,t)}{\partial x} = \text{source terms} \qquad (39)$$

describing the conservation of the number of vehicles. Here, $\rho(x,t)$ denotes the vehicle density per lane at place x and time t, and $Q_i(x,t)$ the traffic flow per lane. The source terms originate from ramp flows $Q_i^{\text{ramp}}(t)$, which enter the road at place x_{i+1}. Let us define the arrival rate at the upstream end of road section i (the "inflow") by $Q_i^{\text{arr}}(t) = I_i Q_i(x_i + dx, t)$, where dx is a differential space interval and I_i the number of lanes of road section i.

Analogously, the departure rate from the downstream end of this section is defined by $Q_i^{\text{dep}}(t) = I_i Q_i(x_{i+1} - dx, t)$. (Note that this "outflow" from section i is to be distinguished from the outflow Q_{out} per lane from congested traffic). The conservation of the number of vehicles implies that the departure rate plus the ramp flow determine the arrival rate in the next downstream section $i + 1$:

$$Q_{i+1}^{\text{arr}}(t) = Q_i^{\text{dep}}(t) + Q_i^{\text{ramp}}(t).$$ (40)

In order to guarantee non-negative flows, we will demand for the ramp flows that the consistency condition $-Q_i^{\text{dep}}(t) \leq Q_i^{\text{ramp}}(t) \leq Q_{i+1}^{\text{arr}}(t)$ is always met.

Integrating the continuity equation over x with $x_i < x < x_{i+1}$ provides a conservation equation for the number $N_i(t) = \int_{x_i}^{x_{i+1}} dx\, I_i \rho(x, t)$ of vehicles in road section i. It changes according to

$$\frac{dN_i(t)}{dt} = Q_i^{\text{arr}}(t) - Q_i^{\text{dep}}(t) = Q_{i-1}^{\text{dep}}(t) + Q_{i-1}^{\text{ramp}}(t) - Q_i^{\text{dep}}(t).$$ (41)

In the terminology of queueing theory, equation (41) reflects the change in the number $N_i(t)$ of vehicles waiting to be served by the downstream end of the section, and the number I_i of lanes corresponds to the number of "channels" serving in parallel. Moreover, $N_i(t)$ corresponds to the "queue length". However, this does not necessarily mean that we have congested traffic as, in the terminology of traffic theory, "queue length" refers to something else, namely the spatial extension $l_i(t) > 0$ of a congested road section. In the following, we will try to express the traffic dynamics and the travel times only through the flows at the cross sections x_i, taking into account the features of traffic flow in a simplified way.

For free flow, i.e. below some critical vehicle density ρ_{cr} per lane, the relation between the traffic flow Q_i per lane and the density ρ per lane can be approximated by an increasing linear relationship, while above it, a falling linear relationship is consistent with congested flow-density data (in particular, if the average time gap T is treated as a time-dependent, fluctuating variable) [28]. This implies $Q_i(x, t) \approx Q_i(\rho(x, t))$ with

$$Q_i(\rho) = \begin{cases} Q_i^{\text{free}}(\rho) = \rho V_i^0 & \text{if } \rho < \rho_{\text{cr}} \\ Q_i^{\text{cong}}(\rho) = (1 - \rho/\rho_{\text{jam}})/T & \text{otherwise.} \end{cases}$$ (42)

Here, V_i^0 denotes the average free velocity, T the average time gap, and ρ_{jam} the density per lane inside of traffic jams. Moreover, we define the free and congested densities by

$$\rho_i^{\text{free}}(Q_i) = Q_i/V_i^0 \quad \text{and} \quad \rho_i^{\text{cong}}(Q_i) = (1 - TQ_i)\rho_{\text{jam}}.$$ (43)

The quantity $Q_{\text{out}} = (1 - \rho_{\text{cr}}/\rho_{\text{jam}})/T$ corresponds to the outflow per lane from congested traffic [29]. Depending on the parameter specification, the model describes a continuous flow-density relation (for $\rho_{\text{cr}} V_i^0 = Q_{\text{out}}$) or a capacity drop at the critical density ρ_{cr} and high-flow states immediately before (if $\rho_{\text{cr}} V_i^0 > Q_{\text{out}}$).

According to shock wave theory [30], density variations at place x propagate with velocity $C(t) = [Q_i(x+dx,t) - Q_i(x-dx,t)]/[\rho(x+dx,t) - \rho(x-dx,t)]$. Accordingly, the propagation velocity is $C = V_i^0$ in free traffic, and $C = -c = -1/(T\rho_{\max})$ in congested traffic. Therefore, it takes the time period $T_i^{\text{free}} = l_i^{\max}/V_i^0$ for a perturbation to travel through free traffic, while it takes the time period $T_i^{\text{cong}} = l_i^{\max}/c$, when the entire road section i is congested.

Now, remember that congestion in section i starts to form upstream of a bottleneck, i.e. at place x_{i+1}. Let $l_i(t)$ denote the length of the congested area and $x(t) = x_{i+1} - l_i(t) = x_i + l_i^{\max} - l_i(t)$ the location of its upstream front. Then, we have free traffic between x_i and $x_i + l_i^{\max} - l_i(t)$, i.e. $Q_i(x-dx,t) = Q_i^{\text{arr}}(t - (x-x_i)/V_i^0)$ (considering $dx \to 0$), and congested traffic downstream of $x(t)$, i.e. $Q_i(x+dx,t) = Q_i^{\text{dep}}(t - (x_{i+1}-x)/c)$. With $dx/dt = -dl_i/dt = C(t)$ and Eq. (43) we find

$$\frac{dl_i}{dt} = -\frac{Q_i^{\text{dep}}(t - l_i(t)/c)/I_i - Q_i^{\text{arr}}(t - [l_i^{\max} - l_i(t)]/V_i^0)/I_i}{\rho_i^{\text{cong}}(Q_i^{\text{dep}}(t - l_i(t)/c)/I_i) - \rho_i^{\text{free}}(Q_i^{\text{arr}}(t - [l_i^{\max} - l_i(t)]/V_i^0)/I_i)}. \tag{44}$$

The capacity of a *congested* road section i is approximated as the outflow $Q_{\text{out}} = (1 - \rho_{\text{cr}}/\rho_{\text{jam}})/T$ from congested traffic per lane times the number I_i of lanes, minus the maximum bottleneck strength at the end of this section. This may be given by an on-ramp flow $Q_i^{\text{ramp}}(t) > 0$ or analogously by $(I_i - I_{i+1})Q_{\text{out}}$ in case of a reduction $I_{i+1} - I_i < 0$ in the number of lanes, or in general by some time-dependent value $\Delta Q_i(t)$ in case of another bottleneck such as a gradient:

$$Q_i^{\text{cap}}(t) = I_i Q_{\text{out}} - \max[Q_i^{\text{ramp}}(t), (I_i - I_{i+1})Q_{\text{out}}, \Delta Q_i(t), 0]. \tag{45}$$

Analogously, the maximum capacity $Q_i^{\max}(t)$ of the road section i under free flow conditions is given by the maximum flow $I_i\rho_{\text{cr}}V_i^0$ minus the reduction by bottleneck effects:

$$Q_i^{\max}(t) = I_i\rho_{\text{cr}}V_i^0 - \max[Q_i^{\text{ramp}}(t), (I_i - I_{i+1})\rho_{\text{cr}}V_i^0, \Delta Q_i(t), 0]. \tag{46}$$

Moreover, we have to specify the departure rate $Q_i^{\text{dep}}(t)$ as a function of the respective traffic situation. Focussing on the cross section at location x_{i+1} and considering the directions of information flow (i.e. the propagation direction of density variations), we can distinguish three different cases:

1. If we have free traffic in the upstream section i and free or partially congested traffic in the downstream section $i+1$, density variations propagate downstream and the departure rate $Q_i^{\text{dep}}(t)$ at time t is given as the arrival rate $Q_i^{\text{arr}}(t - T_i^{\text{free}}) = Q_{i-1}^{\text{dep}}(t - T_i^{\text{free}}) + Q_{i-1}^{\text{ramp}}(t - T_i^{\text{free}})$, since the vehicles entering section i at time $t - T_i^{\text{free}}$ leave the section after an average travel time T_i of T_i^{free}.

2. In the case of partially or completely congested traffic upstream and free or partially congested traffic downstream, the departure rate $Q_i^{\text{dep}}(t)$ is given by the capacity $Q_i^{\text{cap}}(t)$ of the congested road section i.

3. In the case of congested traffic on the entire downstream road section $i + 1$, the departure rate $Q_i^{\text{dep}}(t)$ is given by the departure rate $Q_{i+1}^{\text{dep}}(t - T_{i+1}^{\text{cong}})$ from the downstream section at time $t - T_{i+1}^{\text{cong}}$ minus the ramp flow $Q_i^{\text{ramp}}(t)$ entering at location x_{i+1}.

Summarizing this, we have

$$Q_i^{\text{dep}}(t) = \begin{cases} Q_i^{\text{arr}}(t - T_i^{\text{free}}) & \text{if } l_{i+1}(t) < l_{i+1}^{\max} \text{ and } l_i(t) = 0 \\ Q_i^{\text{cap}}(t) & \text{if } l_{i+1}(t) < l_{i+1}^{\max} \text{ and } l_i(t) > 0 \quad (47) \\ Q_{i+1}^{\text{dep}}(t - T_{i+1}^{\text{cong}}) - Q_i^{\text{ramp}}(t) & \text{if } l_{i+1}(t) = l_{i+1}^{\max}. \end{cases}$$

A numerical solution of the above defined section-based queueing-theoretical traffic model is carried out as follows: First, calculate the new arrival and departure rates by means of Eqs. (40) and (47), taking into account the boundary conditions for the flows at the open ends of the road network. Second, determine the queue lengths $l_i(t)$ in all road sections i: (i) If traffic in the road section flows freely ($l_i(t) = 0$) and the maximum capacity $Q_i^{\max}(t)$ is not reached, i.e. $Q_i^{\text{arr}}(t - T_i^{\text{free}}) < Q_i^{\max}(t)$, we have $dl_i(t)/dt = 0$ and the traffic flow in the road section remains free. (ii) If the road section is completely congested ($l_i(t) = l_i^{\max}$) and the arrival rate $Q_i^{\text{arr}}(t)$ is not below the departure rate at time $t - T_i^{\text{cong}}$, i.e. $Q_i^{\text{dep}}(t - T_i^{\text{cong}}) \leq Q_i^{\text{arr}}(t)$, the road section i stays fully congested and $dl_i/dt = 0$. (iii) In other cases, we have partially congested traffic in road section i and the length $l_i(t)$ of the congested area changes according to Eq. (44). Next, one continues with the first step for the new time $t + dt$, and so on. It is obvious, that this numerical solution is significantly more simple and robust than the numerical solution of the Lighthill-Whitham model, as shock waves (i.e. the interfaces between free and congested traffic) are treated analytically and the propagation velocities of perturbations within the free and congested regions are constant.

The travel time $T_i(t)$ of a vehicle that enters road section i at time t can be calculated analogously to Eq. (27) [25], i.e. via the delay-differential equation

$$\frac{dT_i(t)}{dt} = \frac{Q_i^{\text{arr}}(t)}{Q_i^{\text{dep}}(t + T_i(t))} - 1 = \frac{Q_{i-1}^{\text{dep}}(t) + Q_{i-1}^{\text{ramp}}(t)}{Q_i^{\text{dep}}(t + T_i(t))} - 1. \quad (48)$$

According to this, the travel time $T_i(t)$ increases with time, when the arrival rate Q_i^{arr} at the time t of entry exceeds the departure rate Q_i^{dep} at the leaving time $t + T_i(t)$, while it decreases when it is lower. It is remarkable that this formula does not explicitly depend on the velocities on the road section, but only on the arrival and departure rates. The calculation of the travel time based on the velocity is considerably more complicated: Let $v(t)$ be the velocity and $x(t) = x_i + \int_{t_0}^{t} dt'\, v(t')$ the location of a vehicle at time t, when it enters section i at time t_0. Its travel time $T_i(t_0)$ on section i is given by the implicit equation $x(t + T_i(t_0)) = x_{i+1}$, which says that this vehicle reaches place $x_{i+1} = x_i + l_i^{max}$ at time $t_0 + T_i(t_0)$. The vehicle speed $v(t)$ is also difficult to determine, as it depends on the (free or congested) traffic state at its respective location $x(t)$: It is $v(t) = V_i^0$ in free flow, i.e. for $V_i^0(t - t_0) < l_i^{max} - l_i(t)$. In congested flow, i.e. for $t_0 + [l_i^{max} - l_i(t)]/V_i^0 \le t \le t_0 + T_i(t_0)$, it is determined via $v(t) = Q_i(x(t), t)/\rho(x(t), t) = [1/\rho(x(t), t) - 1/\rho_{max}]/T$ with $\rho(x(t), t) = \rho(x_{i+1}, t - [x_{i+1} - x(t)]/c) = \{1 - TQ_i^{dep}(t - [x_{i+1} - x(t)]/c)/I_i\}\rho_{jam}$.

3 Summary and Conclusions

In this contribution, traffic and production systems have been treated in a uniform way as dynamic queueing networks, since it does not matter whether one treats the transport of goods from one production unit to the next one or of vehicles from one cross section of the road network to the next. Consequently, the stability conditions for supply chains looked similar to the ones of some specific traffic models. Moreover, we have presented a delay-differential equation for the determination of cycle (production) times, which can be also applied to calculate the travel times of vehicles. Interestingly enough, this formula did not require to calculate the vehicle speed on freeway sections, but only the in- and outflows at certain cross sections of the street network, namely, where the road capacity changed.

The derived instability conditions allow one to choose appropriate management strategies which can avoid the well-known bullwhip effect. This effect describes an amplification of variations in the delivery rate and inventory from one supplier to the next one upstream. Such a convective instability can occur despite of a stable behavior in time because of the possibility of resonance effects. Apart from this, we have sketched the treatment of re-entrant production processes and of discrete units.

The advantage of "fluid-dynamic" models of traffic, supply, and production networks is their great numerical efficiency and their consideration of non-linear interaction effects. Therefore, in contrast to most classical queueing theoretical approaches and to event-driven (Monte Carlo) simulations, they are suitable for on-line control and the treatment of variations in the consumption rate or in the production program.

Present research focusses on the effect of the topology of supply networks on their dynamics [3]. The dynamics of production processes can even be chaotic [22, 31–33]. Other studies concentrate on the subject of optimal control (including chaos control) [33–39], which is particularly challenging for re-entrant production [20, 21]. Finally, the presented traffic model is now being applied to the simulation of city networks with adaptive traffic light control and to dynamic assignment problems.

Acknowledgements

The author would like to thank Stefan Lämmer, Martin Treiber, and Thomas Seidel for their valuable comments. This work was partially supported by the German Research Foundation (DFG), grant no. He 2789/5-1.

References

1. W. J. Hopp and M. L. Spearman, *Factory Physics* (McGraw-Hill, Boston, 2000).
2. T. Nagatani and D. Helbing, Stability analysis and stabilization strategies for linear supply chains, *Physica A*, in print (2004).
3. D. Helbing, S. Lämmer, P. Šeba, and T. Płatkowski, Physics, stability and dynamics of supply networks, preprint (2004).
4. R. Albert and A.-L. Barabási, Statistical mechanics of complex networks, *Rev. Mod. Phys.* **74**, 47–97 (2002).
5. U. Witt and G.-Z. Sun, Myopic behavior and cycles in aggregate output, in *Jahrbücher f. Nationalökonomie u. Statistik* (Lucius & Lucius, Stuttgart, 2002), Vol. 222/3, pp. 366–376.
6. C. Daganzo, *A Theory of Supply Chains* (Springer, New York, 2003).
7. D. Armbruster, Dynamical systems and production systems, in G. Radons and R. Neugebauer (eds.) *Nonlinear Dynamics of Production Systems* (Wiley, New York, 2004).
8. D. Helbing, Modeling and optimization of production processes: Lessons from traffic dynamics, in G. Radons and R. Neugebauer (eds.) *Nonlinear Dynamics of Production Systems* (Wiley, New York, 2004).
9. E. Lefeber, Nonlinear models for control of manufacturing systems in G. Radons and R. Neugebauer (eds.) *Nonlinear Dynamics of Production Systems* (Wiley, New York, 2004).
10. D. Helbing, Modelling supply networks and business cycles as unstable transport phenomena, *New Journal of Physics* **5**, 90.1–90.28 (2003).
11. F. Chen, Z. Drezner, J. K. Ryan, and D. Simchi-Levi, Quantifying the bullwhip effect in a simple supply chain: The impact of forecasting, lead times, and information, *Management Science* **46**(3), 436–443 (2000).
12. H. Lee, P. Padmanabhan, and S. Whang, The bullwhip effect in supply chains, *Sloan Management Rev.* **38**, 93–102 (1997).
13. H. L. Lee, V. Padmanabhan, and S. Whang, Information distortion in a supply chain: The bullwhip effect, *Management Science* **43**(4), 546–558 (1997).
14. R. Metters, Quantifying the bullwhip effect in supply chains, in *Proc. 1996 MSOM Conf.*, pp. 264–269 (1996).
15. J. Dejonckheere, S. M. Disney, M. R. Lambrecht, and D. R. Towill, Transfer function analysis of forecasting induced bullwhip in supply chains. *International Journal of Production Economics* **78**, 133–144 (2002).
16. J. Dejonckheere, S. M. Disney, M. R. Lambrecht, and D. R. Towill, Measuring and avoiding the bullwhip effect: A control theoretic approach, *European Journal of Operational Research* **147**, 567–590 (2003).
17. K. Hoberg, U. W. Thonemann, and J. R. Bradley, Analyzing the bullwhip effect of installation-stock and echelon-stock policies with linear control theory, in *Proceedings of the Operations Research Conference 2003, Heidelberg* (2003).

18. B. Faisst, D. Arnold, and K. Furmans, The impact of the exchange of market and stock information on the bullwhip effect in supply chains, in *Proceedings of the Operations Research Conference 2003, Heidelberg* (2003).
19. D. Marthaler, D. Armbruster, and C. Ringhofer, A mesoscopic approach to the simulation of seminconductor supply chains, in G. Mackulak *et al.* (eds.) *Proc. of the Int. Conf. on Modeling and Analysis of Semiconductor Manufacturing*, pp. 365–369 (2002).
20. I. Diaz-Rivera, D. Armbruster, and T. Taylor, Periodic orbits in a class of re-entrant manufacturing systems, *Math. and Op. Res.* **25**, 708–725 (2000).
21. D. Armbruster, D. Marthaler, and C. Ringhofer, Modeling a re-entrant factory, preprint 1/2003, submitted to *Operations Research*.
22. E. Mosekilde and E. R. Larsen, Deterministic chaos in beer production-distribution model, *System Dynamics Review* **4**(1/2), 131–147 (1988).
23. J. D. Sterman, *Business Dynamics* (McGraw-Hill, Boston, 2000).
24. M. Bando, K. Hasebe, A. Nakayama, A. Shibata, and Y. Sugiyama, Dynamical model of traffic congestion and numerical simulation. *Phys. Rev. E* **51**, 1035–1042 (1995).
25. D. Helbing, A section-based queueing-theoretical traffic model for congestion and travel time analysis in networks, *J. Phys. A: Math. Gen.* **36**, L593–L598 (2003).
26. D. Helbing, A. Hennecke, V. Shvetsov, and M. Treiber, MASTER: Macroscopic traffic simulation based on a gas-kinetic, non-local traffic model, *Transportation Research B* **35**, 183–211 (2001).
27. D. Helbing, A. Hennecke, and M. Treiber, Phase diagram of traffic states in the presence of inhomogeneities, *Phys. Rev. Lett.* **82**, 4360–4363 (1999).
28. K. Nishinari, M. Treiber, and D. Helbing, Interpreting the wide scattering of synchronized traffic data by time gap statistics, *Phys. Rev. E* **68**, 067101 (2003).
29. B. S. Kerner and H. Rehborn, Experimental features and characteristics of traffic jams, *Phys. Rev. E* **53**, R1297–R1300 (1996).
30. M. J. Lighthill and G. B. Whitham, On kinematic waves: II. A theory of traffic on long crowded roads, *Proc. Roy. Soc. London, Ser. A* **229**, 317–345 (1955).
31. T. Beaumariage and K. Kempf, The nature and origin of chaos in manufacturing systems, in *Proceedings of the 5th IEEE/SEMI Advanced Semiconductor Manufacturing Conference*, pp. 169–174 (1994).
32. I. Katzorke and A. Pikovsky, Chaos and complexity in simple models of production dynamics, *Discrete Dynamics in Nature and Society* **5**, 179–187 (2000).
33. T. Ushio, H. Ueda, and K. Hirai, Controlling chaos in a switched arrival system, *Systems & Control Letters* **26**, 335–339 (1995).
34. B. Rem and D. Armbruster, Control and synchronization in switched arrival systems, *Chaos* **13**, 128-137 (2003).
35. L. A. Bunimovich, Controlling production lines, in H. Schuster (ed.) *Handbook of Chaos Control*, pp. 324–343 (Wiley-VCH, Berlin, 1999).
36. J. J. Bartholdi and D. D. Eisenstein, A production line that balances itself, *Operations Research* **44**, 21–34 (1996).
37. E. Zavadlav, J. O. McClain, and L. J. Thomas, Self-buffering, self-balancing, self-flushing production lines, *Management Science* **42**, 1151–1164 (1996).
38. P. H. Zipkin, *Found. of Inventory Management* (McGraw-Hill, Boston, 2000).
39. J. A. Hołyst, T. Hagel, G. Haag, and W. Weidlich, How to control a chaotic economy?, *Journal of Evolutionary Economics* **6**, 31–42 (1996).

Capacity Funnel Explained Using the Human-Kinetic Traffic Flow Model

C. Tampère[1,3], S.P. Hoogendoorn[2], and B. van Arem[1]

[1] TNO Inro
   PO box 6041, 2600 JA Delft – The Netherlands
   `b.vanarem@inro.tno.nl`
[2] Delft University of Technology
   PO box 5048, 2600 GA Delft – The Netherlands
   `s.hoogendoorn@citg.tudelft.nl`
[3] Katholieke Universiteit Leuven
   Kasteelpark Arenberg 40, 3001 Heverlee – Belgium
   `chris.tampere@bwk.kuleuven.ac.be`

Summary. We present a macroscopic traffic flow model that explicitly builds on continuous individual driving behaviour. Not only do we start from classical car-following rules (like the kind that are used in microscopic simulators), the model also explicitly accounts for the finite reaction times of drivers, anticipation behaviour, anisotropy in driver responses and the finite space requirement of drivers in the stream. Moreover we allow variations in driver psychology, which lets drivers adopt different 'driving styles' dependent on traffic conditions, like the presence of a merging zone.
We illustrate the potential of such model by simulating a busy highway with an on-ramp. Plausible assumptions about driver psychology allow us to reproduce the so-called 'capacity funnel', i.e. the onset of congestion typically occurs some distance downstream of the merge area.
Keywords: traffic flow model, gas-kinetic traffic flow model, driver behaviour, capacity funnel, micro-macro link

1 Introduction: Individual Driver Behaviour in Macroscopic Traffic Flow Models

In this article we present some recent developments of traffic flow models in which special attention is given to the representation of individual driver behaviour. The motivation for this work was raised by criticism heard in our daily co-operation with human factors' experts on the role of individual driver behaviour in traffic flow models (e.g. [1]) and by the desire to explore the impact of (semi-) automated driving on traffic dynamics. The latter would require a macroscopic traffic flow model that does not build on any empirical macroscopic 'equilibrium' or 'fundamental' relationship.

In the remainder of this chapter we summarise our findings with respect to the state-of-the-art on this topic. In the next chapters we show how we combine various elements that we found in the literature into a new traffic flow model, based on the gas-kinetic approach but with explicit 'human' behaviour of the vehicles in the flow. For this reason we refer to our model as the *'human kinetic'* traffic flow model.

A common component of all macroscopic traffic flow models is the equation for the density dynamics. With k for the density, q for the flow and V for the speed, it writes:

$$\frac{\partial k(t,x)}{\partial t} + \frac{\partial q(t,x)}{\partial x} = 0 \quad \text{with} \quad q(t,x) = k(t,x)V(t,x) \tag{1}$$

This equation is a purely physical law of 'conservation of vehicles'. All reference to individual driver behaviour is therefore found in the equations accompanying this density equation. Traditionally a dominant role is played by the so-called equilibrium speed or fundamental relationship $V^e(k(t,x))$ between the macroscopic speed and the density. This holds equally for the first order model – in which V is set equal to V^e at all times – and for higher order models in which the speed V is allowed to vary dynamically around V^e. The equilibrium speed implicitly contains all information on individual driver behaviour, since it is the result of individual actions and interactions between drivers, which is either obtained empirically or (theoretically) through calculations in the gas-kinetic theory. We found four approaches in literature that aim at refining the representation of individual driver behaviour in macroscopic traffic flow models – mostly through the equilibrium speed:

- *Extensions in analogy to hydrodynamics*:
  Some authors propose extra terms in the speed dynamics equation (of second or higher order models) to represent anticipation of drivers or the tendency to go along with the flow (viscosity term); these extensions are seldom built on individual driver behaviour specifications, but are inspired by qualitative behavioural interpretations of terms that are introduced to obtain better dynamic behaviour or numerical performance of the model. Contributions of this kind are numerous, among others: [6,8,9,11,12,20].

- *'Direct' micro-macro links*:
  A microscopic car-following rule is directly 'scaled up' to the macroscopic level by a Taylor expansion (in analogy to [16], e.g. [21]) or by an appropriate transition from discrete to continuous variables ([7]). Interestingly the latter author adds non-local behaviour to this model, i.e. it is numerically reflected that drivers respond to conditions some distance downstream.

- *Direct refinement of the equilibrium speed*:
  Inspired by behavioural considerations or by macroscopic empirical findings, the equilibrium relationship is refined, for instance with different branches for accelerating or decelerating traffic ([15,22]) or different branches for distinct classes of drivers (e.g. the rabbits and slugs in [4] and [5]). Interestingly, the equilibrium speed of the rabbits in the latter example has a special reversed-lambda shape that is explained by assuming a collective loss of motivation of drivers to follow their predecessor closely.

- *(Gas-) kinetic approaches*:
  This theory accounts for the fact that the flow consists of individual vehicles, but their behaviour is often modelled only coarsely, e.g. with instant deceleration. This theory provides an analytical foundation of the equilibrium speed as a macroscopic balance between the tendency of the

driver population to relax to its desired speed and the gross effect of interactions with other vehicles. The basic assumptions ([17] and [18]) were explicitly refined with respect to driver behaviour modelling in [13,14,6] and [10]. Still, interactions between vehicles in these models have a discrete nature and are little refined.

In our human-kinetic traffic flow model we use the mathematical framework of the gas-kinetic theory to scale up microscopic car-following relationships (Chapt. 2), we use non-local and even non-temporal interactions inspired by the idea of Hennecke [7] (Chapt. 3) and we allow changes in driver psychology as an extension to what is suggested by Daganzo [4] (Chapt. 4).

2   Human-Kinetic Traffic Flow Model With Smooth Accelerations and Decelerations Balanced on the Microscopic level

Let us summarise our application of the gas-kinetic theory as follows (more details can be found in [19]). We interpret V and k as the moments of a probability distribution of individual vehicle-driver combinations in a road section around x. The density k is the inverse of the average of the probability distribution of the rear-bumper to rear-bumper headway s; V is the average of the individual speeds v. The combined (mesoscopic) generalized density function $\rho(t,x,v,s)$ is interpreted as the probability distribution function describing the probability of finding at time t at location x a vehicle with speed v and distance s to its predecessor. The predecessor has an individual speed w drawn from a probability distribution with average V(t, x+s). For each of these predecessor-follower pairs we can apply a microscopic longitudinal or car-following model acc = CF(v,s,w) to calculate the individual acceleration or deceleration of the follower. After averaging this individual acceleration – i.e. multiplying it with the probability of occurrence of the vehicle pair and integration of all possible vehicle pairs whose follower is located at x – we find the expected acceleration or macroscopic acceleration at x. This procedure is illustrated in Fig. 1.

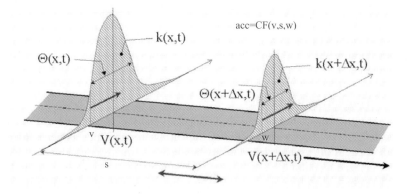

Fig. 1. Calculation of the expected macroscopic acceleration by integration of the car-following relation CF weighed with the probabilities of its inputs v, s and w.

Note how we balance the individual decisions to either accelerate or decelerate (dependent on the outcome of the CF function) on the microscopic scale so that a positive or negative macro acceleration can occur.

The gas-kinetic theory now states that we can transform a generalized law for the conservation of probability:

$$\frac{\partial \rho(t,S)}{\partial t} + \nabla_S \cdot \left(\rho(t,S) \frac{dS(t,S)}{dt} \right) = \left(\frac{d\rho(t,S)}{dt} \right)_{event} \tag{2}$$

with $\nabla_S = \left(\frac{\partial}{\partial s_1}, ..., \frac{\partial}{\partial s_n} \right)$ and S the state vector (e.g. S = (x,v)) by applying the so-called 'method of moments'. This method turns Eq. 2 into dynamic equations for the density dynamics (Eq. 3) and for the speed dynamics (Eq. 4, in which we have used the notation $\left\langle \frac{dv}{dt} \right\rangle_v$ as the outcome of the integration described above and in [19] to obtain the macro acceleration and θ for the speed variance):

$$\frac{\partial k}{\partial t} + \frac{\partial kV}{\partial x} = \left(\frac{dk}{dt} \right)_{event} \tag{3}$$

$$\frac{\partial kV}{\partial t} + \frac{\partial (kV^2 + k\theta)}{\partial x} = k \left\langle \frac{dv}{dt} \right\rangle_v + \left(\frac{dkV}{dt} \right)_{event} \tag{4}$$

With Eq. 3 and 4 we now have a macroscopic traffic flow model that is directly derived from a microscopic car-following model. Since the car-following model typically aims at a velocity dependent following distance and we only look to cars in front our model implicitly obeys the finite space requirement (no point-sized vehicles) and is anisotropic (not sensitive to traffic conditions upstream but only downstream).

3   Refinements to the Basic Human-Kinetic Traffic Flow Model

We now introduce finite reaction times and anticipation behaviour as typical behavioural aspects and show how they can be accounted for in the human-kinetic model.

3.1   Finite Reaction Times

We know that traffic flows are not always stable in the sense that local small perturbations of the speed or density can grow in amplitude under certain conditions. From asymptotic stability analysis in microscopic models we know that the reaction time of drivers is an important contributor to instability that should therefore also be part of our human-kinetic model. We implement a finite reaction time T explicitly in the numerical scheme – a 'non-temporal' term analogously to the non-local term in [7]. For this purpose we calculate the expected acceleration at time t and location x as described in the previous chapter, but only implement it at time t+T, when traffic is at location x(t+T) = x(t)+T·V(t). It turns out that this short delay has little effect in smoothly changing traffic conditions; but can cause instabilities in dense and rapidly changing traffic conditions, as is the case in real traffic.

3.2   Anticipation

It is known that finite reaction times cannot result in collision free traffic flows unless it is compensated by some sort of anticipation. In the model of Payne ([16]) for instance an anticipation term is present in the speed equation, and stability can be mathematically proven to depend on the ratio of the anticipation coefficient and the reaction time.

We model anticipation by taking a subjective or perceived speed w of the predecessor in the integration procedure for the expected acceleration of Chapt. 2. That is: instead of the actual average speed $V(t,x+s)$ we take the speed of the predecessor as if it came from a distribution with perceived average $V^{perc}(t,x+s)$. Although other definitions are equally possible, we define the perceived speed here as follows. We assume that the driver scans some range $(0,\Delta x^{anticip})$ for the steepest speed drop that can be perceived in this interval: $\left(\dfrac{dV}{dx}\right)^{perceived}$; instead of the actual average speed $V(t,x+s)$, we take

$$V^{perc}(t,x+s) = V(t,x)+s\left(\frac{dV}{dx}\right)^{perceived}.$$

Note that this specification models anticipation of drivers with an asymmetric sensitivity for accelerating or decelerating spatial speed profiles. It models anticipation, since the driver already accounts for the fact that soon his predecessors will have to adapt their speeds to the decreasing speeds further ahead. The model is asymmetrically sensitive for deceleration or acceleration because only the steepest (lowest) speed gradient is taken into account. If speeds are lower at the maximum anticipation range than immediately in front then the driver will anticipate to this trend. If speeds far ahead are higher but the immediate predecessors have not yet accelerated, then neither will do the driver since the minimum speed gradient is then the one over the nearby range $\Delta x \ll \Delta x^{anticip}$. Indeed, it would not be realistic in this case to assume long-range anticipation, because nearby traffic restricts the acceleration possibility of the driver.

4   Modelling (Temporal) Driving Style Variations

Just like other macro or micro models we have assumed so far that driver behaviour is rigid, i.e. drivers respond in an identical manner to the same actual traffic conditions, no matter what has happened before. In reality the driving style of an individual might change over time, for instance due to increased motivation, awareness or attention level. This might be modelled through an adapted reaction time, shorter headways, longer anticipation range etc. We now illustrate how attention level dynamics can be modelled in the human-kinetic traffic flow model and apply this to the case of a freeway on-ramp in the next chapter.

As an extension to Chapt. 2 and 3, we characterise an individual driver's state not only by the individual speed v (macro equivalent V) and the distance s tot the predecessor (macro Eq. k), but also by the attention level a of the driver (macro Eq. A). Before we can simulate we have to mathematically describe:

- how the macro property A flows with traffic; (see next sections)
- how the attention level of an individual driver affects the driving behaviour as specified so far; (see next sections)
- how the attention level varies as a result of autonomous processes of the driver and of (changes in) traffic conditions (see next chapter).

We mentioned in Chapt. 2 that the speed dynamic Eq. 4 is obtained directly from the basic Eq. 2 by applying the method of moments. We do the same here but with respect to the attention level instead of the speed. For that purpose we multiply both sides of Eq. 2 by the attention level a and integrate over a and v. One can verify that this operation yields Eq. 5 as a complement to Eq. 3 and 4:

$$\frac{\partial A}{\partial t} + V \frac{\partial A}{\partial x} = \left\langle \frac{da}{dt} \right\rangle_{v,a} + \frac{1}{k} \iint a \cdot \left(\frac{d\rho}{dt} \right)_{events} dv\, da - \frac{A}{k} \left(\frac{dk}{dt} \right)_{event} \tag{5}$$

In this equation, the first term on the right hand side stands for the gross effect of autonomous (driver induced) changes of A, the second term is the gross effect due to events in the flow and the last term redistributes the total attention level A over the new population k in case the density does not remain constant (nett in- or outflow $\left(\frac{dk}{dt} \right)_{event}$ of traffic).

Next we specify the effect of the attention level on the individual driver behaviour. Among many possible specifications we choose – for illustrative purposes – the simple assumption that the space required by a driver for car-following at a certain speed decreases with increasing attention level. Macroscopically speaking this means that an increase in the density cannot only be accommodated by a decrease of the average speed (as always in traffic flow models) but possibly also by an increased attention level, so that less space for the same driving speed suffices and speeds can (temporarily) remain unaffected.

5   The Attention Level Based Human-Kinetic Model Explaining the Capacity Funnel

In this chapter we illustrate the potential of the newly developed traffic flow model by means of a case study of a bottleneck due to merging traffic. It is well-known that near merges the so-called 'capacity funnel' phenomenon can occur. This means that the on-set of congestion is located some distance after the actual merging zone. This phenomenon was first described in [2]. In [3] a well-documented example is presented of a capacity funnel in Toronto where congestion sets in 1000 m or more after the end of the slip lane.

5.1   Specification of the Attention Level Dynamics Near a Merge zone

A final step in the specification is to describe how the attention level changes in time. We assume (a) that the event of other vehicles merging in front of a driver is primarily handled by raising the attention level to a_{main} (and not by immediate braking due to the suddenly reduced headway), (b) that merging drivers from an on-ramp enter with the same high attention level a_{merge}, which enables them to merge into relatively small gaps, and (c) in absence of any of the previous causes for a raised attention level the drivers tend towards a comfortable low attention level a_{min}. Without going into the details of the specification and the mathematical derivation we here immediately present the rewritten Eq. 5 which – together with Eq. 3 and 4 – forms our human-kinetic traffic flow model:

$$\frac{\partial A}{\partial t} + V \frac{\partial A}{\partial x} = \underbrace{\frac{a_{min} - A}{\tau_a}}_{(c)\ \ relaxation} + \underbrace{\left(\frac{dk}{dt}\right)_{merge} \frac{a_{merge} - A}{k}}_{(b)\ \ nett\ effect\ of\ inflow} + \underbrace{\left(\frac{dk}{dt}\right)_{merge} \frac{a_{main} - A}{k}}_{(a)\ \ interaction} \qquad (5')$$

5.2 Effect of Attention Level Variations on Traffic Flow Near a Merge: the Capacity Funnel

Let us now investigate the effect of attention level dynamics on traffic flows near a merge zone. As a case study we look at a circular road of length 3000 m. We start with homogeneously distributed speeds and density. Starting from x_0=1375 m a 250m long merging zone is located. The inflow increases linearly from 0 at t=0 to 100 veh/h at t=100s and remains constant thereafter. Moreover we assume that the speed of merging traffic is always equal to the current average speed on the main lanes adjacent to the middle of the merging zone. In Fig. 2 we show the solution of three experiments for t∈[0,500].

(a)

(b)

(c)

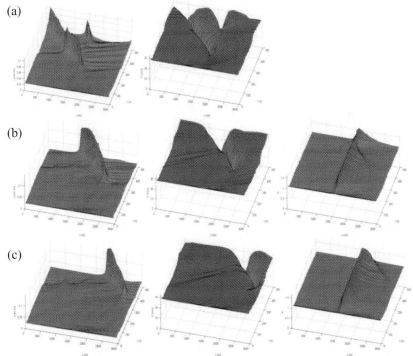

Fig. 2. Effect of the attention level dynamics near a merge (at x=1500m). From left to right: density, speed and the attention level (if applicable). In (a) no attention level variation is modelled and congestion sets in at the location of the merge. In (b) and (c) the merging traffic causes an increase of the attention level that relaxes with time constant τ_a=12 in (b) and τ_a=18 in (c). Congestion sets in further after the actual merge as τ_a increases.

6   Conclusions

Challenged by human factors experts and inspired by 50 years of macro-scopic traffic flow modeling and the creative solutions proposed by various authors to represent valid driver behaviour in these models, we developed the human-kinetic traffic flow model. The model borrows the mathematical framework from the gas-kinetic traffic flow modeling to build macroscopic equations based on a microscopic car-following rule. The model is refined with finite reaction times and a specification of anticipation behaviour and accounts for (temporal) variations of the driving style by taking the attention level as a dynamically varying index for the driving style. The result is a flexible traffic flow simulation the parameters of which all have a specific microscopic meaning directly related to individual driver behaviour. The po-tential of such a model is illustrated by simulating delayed onset of conges-tion near a merge (so-called capacity funnel) by purely making assumptions on the microscopic level.

Acknowledgement

The authors wish to thank the TNO-TRAIL programming committee T3 for its financial support of this research.

References

1. Boer, E.R. (1999), Car following from the driver's perspective, Tr. Res. F Vol. 2, No. 4, pp.201-206.
2. Buckley, D.J. & S. Yagar (1974), Capacity funnels near on-ramps, Proceedings of the 6th International Symposium on Transportation and Traffic Theory, Sydney, Australia.
3. Cassidy, M.J. & R.L. Bertini, (1999), Observations at a Freeway Bottleneck, Proc. of the 14th International Symposium on Transportation and Traffic Theory, Jeru-salem 1999, pp.107-124.
4. Daganzo, C.F. (2002a), A behavioral theory of multi-lane traffic flow. Part I: Long homogeneous freeway sections, Transportation Research B vol 36B, issue 2, pp.131-158 (available at www.path.berkeley.edu).
5. Daganzo, C.F. (2002b), A behavioral theory of multi-lane traffic flow. Part II: Merges and the onset of congestion, Transportation Research B vol 36B, issue 2, pp. 159-169 (available at www.path.berkeley.edu).
6. Helbing, D., (1997), Verkehrsdynamik, Neue physikalische Modellierungskon-zepte, Springer Verlag, Berlin.
7. Hennecke, A., M. Treiber & D. Helbing, (2000), Macroscopic Simulation of Open Systems and Micro-Macro Link, Traffic and Granular Flow '99, edited by: Helbing, D., H.J. Herrmann, M. Schreckenberg & D.E. Wolf (available at: www.tu-dresden.de/vkiwv/vwista/sta1_hp.html).
8. Hoogendoorn, S.P., (1999), Multiclass Continuum Modelling of Multilane Traffic Flow, Dissertatoin thesis, Delft University of Technology, Faculty of Civil Engi-neering and Geosciences.
9. Kerner, B.S. & P. Konhäuser, (1993), Cluster effect in initially homogeneous traffic flow, Phys. Rev. E 48(4), pp.2335-2338.
10. Klar, A. & R. Wegener (1999), A hierarchy of models for multilane vehicular traffic I: modeling, SIAM Journal of Applied Mathematics, Vol. 59, No. 3, pp.983-1001   (available   through: http://www.mathematik.tu-darmstadt.de/~klar/veroeffentlichungen.html)

11. Kühne, R.D., (1984), Macroscopic freeway model for dense traffic – stop-start waves and incident detection, Proc. of the 9th International Symposium of Transportation and Traffic Theory, pp.21-42, ed. by: I. Volmuller & R. Hamerslag.

12. Liu, G., A.S. Lyrintzis & P.G. Michalopoulos (1998), Improved higher-order model for freeway traffic flow. Tr. Res. Rec. 1644, pp.37-46.

13. Nelson, P. (1995), A kinetic model of vehicular traffic and its associated bimodal equilibrium solutions, Transport Theory and Statistical Physics, Vol. 24(1-3), pp.383-409.

14. Nelson, P., D.D. Bui & A. Sopasakis (1997), A novel traffic stream model deriving from a bimodal kinetic equilibrium, Proceedings of the 1997 IFAC meeting, Chania, Greece, pp.799-804.

15. Newell, G.F. (1965), Instability in dense highway traffic, a review, In: J. Almond, ed., Proceedings of the second International Symposium on the Theory of Traffic Flow, pp. 73-83.

16. Payne, H.J., (1971), Models of freeway traffic and control, In: Mathematical models of public systems, volume 1 of simulation councils proc. ser., ed. by: Bekey, G.A., pp, 51-60.

17. Prigogine, I. & R. Herman (1971), Kinetic Theory of Vehicular Traffic, American Elsevier, New York.

18. Prigogine, I. (1961), Boltzmann-like approach to the statistical theory of traffic flow, In: Theory of Traffic Flow (Ed.: R. Hermand).

19. Tampère, C.M.J., S.P. Hoogendoorn & B. Van Arem (2003), Gas kinetic traffic flow modelling including continuous driver behavior model, Preprints of the 82nd Annual Meeting of the Transportation Research Board.

20. Zhang, H.M. (1998), A theory of nonequilibrium traffic flow, Tr. Res. B, Vol. 32B, No. 7, pp.485-498.

21. Zhang, H.M. (2003), Driver memory, traffic viscosity and a general viscous traffic flow model, Tr. Res. B, Vol. 37, No. 1, pp.27-41.

22. Zhang, H.M., (1999), A mathematical theory of traffic hysteresis, Tr. Res. B 33. pp.1-23.

A Comparison of a Cellular Automaton and a Macroscopic Model

S. Maerivoet[1], S. Logghe[2], B. De Moor[1], and B. Immers[3]

[1] Department of Electrical Engineering ESAT-SCD (SISTA) –
Katholieke Universiteit Leuven
Kasteelpark Arenberg 10, B-3001 Leuven – Belgium
sven.maerivoet@esat.kuleuven.ac.be
[2] Transport & Mobility Leuven
Tervuursevest 54 bus 4, B-3000 Leuven – Belgium
steven.logghe@tmleuven.be
[3] Department of Civil Engineering (Transportation Planning and Highway
Engineering) – Katholieke Universiteit Leuven
Kasteelpark Arenberg 40, B-3001 Leuven – Belgium

Summary. In this paper we describe a relation between a microscopic stochastic traffic cellular automaton model (i.e., the STCA) and the macroscopic first-order continuum model (i.e., the LWR model). The innovative aspect is that we explicitly incorporate the STCA's stochasticity in the construction of the fundamental diagram used by the LWR model. We apply our methodology to a small case study, giving a comparison of both models, based on simulations, numerical, and analytical calculations of their tempo-spatial behavior.

Keywords: STCA model, cellular automaton, LWR model, hydrodynamic model, stochasticity

1 Introduction

Dating back to the mid '50s, Lighthill, Whitham, and Richards introduced their macroscopic first-order continuum model (i.e., the LWR model) [1, 2]. It is based on a fluid dynamics analogy, in which the collective behaviour of infinitesimally small particles is described, using aggregate quantities such as flow q, density k and (space) mean speed $\overline{v}_s$. Models like this, can be solved using cell-based numerical schemes (e.g., using the Godunov scheme [3, 4]).

Later, microscopic traffic flow models have been developed that explicitly describe vehicle interactions at a high level of detail. During the early nineties, these models were reconsidered from an angle of particle physics: cellular automata models were applied to traffic flow theory, resulting in fast and efficient modelling techniques for microscopic traffic flow models [5]. These cellular automata models can be looked upon as a particle based discretisation scheme for macroscopic traffic flow models.

It is from this latter point of view that our paper addresses the common structure between the seminal STCA model and the first-order LWR model. Our main goal is to provide a means for explicitly incorporating the STCAs stochasticity in the LWR model. After explaining the methodology of our approach, we present an illustrative case study that allows us to compare the tempo-spatial behavioural results obtained with both modelling techniques.

2 Methodology

Already, relations between both types of models (i.e., STCA and LWR) have been investigated (e.g., [6]). Our approach is however different, in that it provides a practical methodology for specifying the fundamental diagram to the LWR model. Assuming that a stationarity condition holds on the STCA's rules, this allows us to to incorporate the STCA's stochasticity directly into the LWR's fundamental diagram.

We assume that we have the ruleset of the STCA available, as well as the maximum allowed speed v_{max} and the stochastic noise term p (i.e., the *slowdown probability*). Furthermore, a discretisation is given, expressed by the cell length $\Delta X = 7.5$ m, the time step $\Delta T = 1$ s, and its coupled speed increment $\Delta V = \Delta X \div \Delta T = 27$ km/h.

Relating both the STCA and the LWR models is done using a simple two-step approach, in which we first rewrite the STCA's rules (assuming a stationarity condition holds), and then convert these new rules into a space gap/speed diagram (which is equivalent to a stationary density/flow fundamental diagram).

2.1 Rewriting the STCA's Rules

Starting from the ruleset of the STCA, we rewrite it using a min-max formulation. Instead of having several individual rules that give a discrete speed, we now have one rule that returns a continuous speed (for an individual vehicle):

$$v(t + \Delta T) = p \cdot \min \{v(t), g_s(t) - 1, v_{max} - 1\}$$
$$+ (1 - p) \cdot \min \{v(t) - 1, g_s(t), v_{max}\}, \tag{1}$$

with $v(t + \Delta T) = \max \{0, v(t + \Delta T)\}$. The stationarity condition previously mentioned, asserts that the speed $v(t)$ of a vehicle at time t is the same as its speed at time $(t + \Delta T)$:

$$v(t + \Delta T) = v(t). \tag{2}$$

This allows us to reformulate equation (1) as a set of linear inequalities that express constraints on the relations between $v(t)$, $v_{\max}$, p and the space gap g_s.

2.2 Deriving the Fundamental Diagram

The linear inequalities derived in section 2.1 together form a set of boundaries that can be plotted in a diagram that shows the space gap g_s of a vehicle versus its speed v. Knowing that the space headway h_s equals the vehicle's length L plus its space gap g_s, we can plot a stationary (h_s,v) diagram as can be seen in the left part of Fig. 1. Because the space headway h_s is inversely proportional to the density k, we can derive an equivalent *triangular* (k,q) fundamental diagram, corresponding to the right part of Fig. 1.

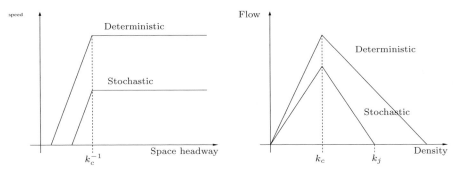

Fig. 1. Deriving stationary (h_s,v) (left) and (k,q) (right) fundamental diagrams for the LWR model, after incorporation of the STCA's stochastic noise.

Considering the previously derived constraints and the diagrams in Fig. 1, the following important observations can be made:

- The stochastic effects from the STCA, are now incorporated in a stationary fundamental diagram, which can then be specified as a parameter to the LWR model.
- The stochastic diagrams lie lower than their deterministic counterparts, so the capacity flow is lower for the stochastic variants.
- The jam density for stochastic systems is different from that for deterministic systems, but the critical density remains unchanged.

3 An Illustrative Case Study

As a toy example, we apply our methodology to a case study, in which we model a single lane road that has a middle part with a reduced maximum speed (e.g., an elevation, or a speed limit, ...). The road consists of three consecutive segments A, B, and C. The first segment A consists of 1500 cells (11.25 km), while the second and third segments B and C each consist of 750 cells (i.e., each approximately 5.6 km long). We consider a time horizon of 2000 s. The maximum speed for segments A and C is 5 cells/s, whereas it's 2 cells/s for segment B. The stochastic noise p was set to 0.1 for all three segments. Vehicles enter the road at segment A, travel through segment B, and exit at the end of segment C.

This road is simulated using both the STCA and the LWR model. As for the boundary conditions, we assume an overall inflow of $q_c^B \div 2$ (q_c is the capacity flow), except from $t = 200$ s to $t = 600$ s, where we create a short traffic burst with an inflow of $(q_c^{A,C} + q_c^B) \div 2$. Fig. 2 shows the individual vehicle trajectories in a time/space diagram: heavy congestion sets in and flows upstream into segment A, where it starts to dissolve at the end of the traffic burst.

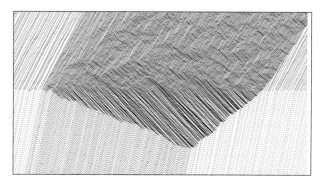

Fig. 2. A time/space diagram after simulation of the STCA: each vehicle is represented by a single dot (the time and space axes are oriented horizontally, respectively vertically). At the end of segment A, we can see the formation and dissolution of an upstream growing congested region, related to the short traffic burst.

Applying our previously discussed methodology, we construct a stationary (k,q) fundamental diagram, and numerically solve the LWR model. The result can be seen in the left part of Fig. 3. Comparing this spatio-temporal behaviour of the LWR model with the microscopic system dynamics from the STCA model (i.e., the right part of Fig. 3), we find a qualitatively good agreement between the two approaches.

Even more interesting, is the fact that the STCA model reveals a higher-order effect that is not visible in the LWR model: there exists a fan of forward propagating density waves in segment B. Furthermore, in its tempo-spatial diagram, the STCA seems to be able to visualise the characteristics that constitute the solution of the LWR model.

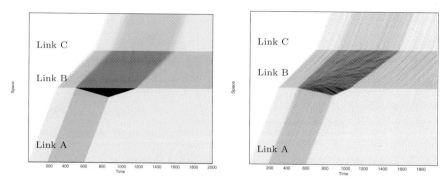

Fig. 3. Time/space diagrams showing the propagation of densities during 2000 s for the road in the case study. The left part shows the results for the LWR model, while the right part shows the microscopic system dynamics from the STCA model (note that darker regions correspond to congested traffic).

4 Conclusions

The novel approach taken in our research, allows us to incorporate the STCA's stochasticity directly in the first-order LWR model. This is accomplished by means of a stationarity condition that converts the STCA's rules into a set of linear inequalities. In turn, these constraints define the shape of the fundamental diagram that is specified to the LWR model.

Our methodology sees the STCA complementary to the LWR model and vice versa, so the results can be of great assistance when interpreting the traffic dynamics in both models. Nevertheless, because the LWR model is only a coarse representation of reality, there are still some mismatches between the two approaches. One of the main concerns the authors discovered, is the fact that using a stationary fundamental diagram (i.e., an equilibrium relation between density and flow), always overestimates the practical capacity of a cellular automaton model (see e.g., Fig. 4, where the true capacity for $v_{max} = 5$ cells/s lies somewhere near 2400 vehicles/h, which is a rather low value).

Further research will focus on the dynamics of multi-lane traffic, on the heterogeneity of the traffic stream (using a heterogeneous LWR model), and on the relation between the capacity of a cellular automaton and the level of stochastic noise in the system.

Acknowledgements

Our research is supported by: **Research Council KUL**: GOA-Mefisto 666, GOA-AMBioRICS, several PhD/postdoc & fellow grants, **FWO**: PhD/postdoc grants, projects, G.0240.99 (multilinear algebra), G.0407.02 (support vector machines), G.0197.02 (power islands), G.0141.03 (identification and cryptography), G.0491.03 (control for intensive care glycemia), G.0120.03 (QIT),

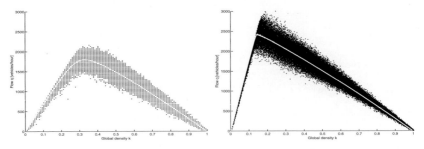

Fig. 4. The (k,q) phase space diagrams of the STCA for $v_{\text{max}} = 2$ cells/s (left) and $v_{\text{max}} = 5$ cells/s (right). The stochastic noise was $p = 0.1$ in both diagrams. The small points denote individual measurements, whereas the white curves represent long-time averages.

G.0452.04 (new quantum algorithms), G.0499.04 (robust SVM), research communities (ICCoS, ANMMM, MLDM), **AWI**: Bil. Int. Collaboration Hungary/Poland, **IWT**: PhD Grants, GBOU (McKnow), **Belgian Federal Science Policy Office**: IUAP P5/22 ('Dynamical Systems and Control: Computation, Identification and Modelling', 2002-2006), PODO-II (CP/40: TMS and Sustainability), **EU**: FP5-Quprodis, ERNSI, Eureka 2063-IMPACT, Eureka 2419-FliTE, **Contract Research/agreements**: ISMC/IPCOS, Data4s, TML, Elia, LMS, Mastercard.

References

1. M.J. Lighthill and G.B. Whitham. On kinematic waves : II. A theory of traffic flow on long crowded roads. In *Proceedings of the Royal Society*, volume A229, pages 317–345, 1955.
2. P.I. Richards. Shockwaves on the highway. *Operations Research*, (4):42–51, 1956.
3. Carlos F. Daganzo. A finite difference approximation of the kinematic wave model of traffic flow. *Transportation Research 29B*, pages 261–276, 1995.
4. J.P. Lebacque. The Godunov scheme and what it means for first order traffic flow models. In J.B. Lesort, editor, *Transportation and Traffic Theory, Proceeding of the 13th ISTTT*. Pergamon, Oxford, November 1995.
5. Kai Nagel and Michael Schreckenberg. A cellular automaton model for freeway traffic. *Journal de Physique I France*, 2:2221–2229, 1992.
6. Kai Nagel. Particle hopping models and traffic flow theory. *Physical Review E*, 53(5):4655–4672, May 1996.

Stochastic Description of Traffic Breakdown: Langevin Approach

R. Mahnke[1], J. Kaupužs[2], and J. Tolmacheva[3]

[1] Fachbereich Physik – Universität Rostock
   18051 Rostock – Germany
   `reinhard.mahnke@physik.uni-rostock.de`
[2] Institute of Mathematics and Computer Science – University of Latvia
   29 Rainja Boulevard, LV, 1459 Riga – Latvia
[3] Institute of Single Crystals – NASU
   61 Lenin Ave., 61001 Kharkov – Ukraine

Summary. From *probabilistic point of view* we investigate a quite classical dynamical system given by stochastic differential equations, i. e. ordinary differential equations driven by multiplicative noise. Based on this Langevin approach the probability density distributions of vehicular velocities as well as headway distances are calculated and discussed.

Our work is a continuation of a stochastic theory of freeway traffic based on a Master equation approach presented first at TRAFFIC AND GRANULAR FLOW '97 as the one–cluster model. The extension to our multi–cluster model can be found at TRAFFIC AND GRANULAR FLOW '99.

Keywords: transportation science, noise, phase diagram.

1 Introduction

The formation and growth of clusters is a widely known phenomenon in physics, e. g. condensation of liquid droplets in a supersaturated vapour [1]. The formation of car clusters (jams) at overcritical densities in traffic flow is an analogous phenomenon in the sense that cars can be considered as interacting particles [2, 3], and the clustering process can be described by similar equations. In particular, the probability that the system has a given cluster distribution at a certain moment in time can be described by the stochastic Master equation in both cases. The transition probabilities depend on the specific physical model under consideration. The spontaneous emergence of car clusters has been studied by different authors (see Proceedings TGF [4, 5]) based on different models and approaches. In spite of the complexity of real traffic, we believe that some general properties of traffic flow, such as headway and velocity distributions, exist which can be described and understood by relatively simple models.

2 Equations of Motion: Langevin Approach

Recently, the theoretical and empirical foundations of *physics of traffic flow* have come into the focus of the physical community. Different approaches like deterministic and stochastic nonlinear dynamics as well as statistical physics of many-particle systems have been very successful in understanding empirically observed structure formation on roads like jam formation in freeway traffic. The motion of an individual vehicle has many peculiarities, since it is controlled by motivated driver behaviour together with physical constraints. Nevertheless, on macroscopic scales the car ensemble displays phenomena like phase formation (nucleation), widely met in different physical systems. Although the cooperative behaviour of cars treated as active particles seems to be rather complex, we concentrate at first on a well-investigated approximation known as safety distance or optimal velocity (OV) model, first proposed by Bando et al. [6, 7].

Considering the problem of a car following a leading vehicle our starting point is the following set of equations of motion

$$m \frac{dv_i}{dt} = F_{det}(v_i, \Delta x_i) + F_{stoch} . \tag{1}$$

This can be written as random dynamical system [8] with multiplicative Gaussian white noise

$$m \, dv_i(t) = F_{det}(v_i, \Delta x_i) \, dt + \sigma v_i \, dW_i(t) \tag{2}$$

$$dx_i(t) = v_i \, dt \tag{3}$$

for a point-like particle (vehicle i) of mass m with speed $v_i(t)$ at location $x_i(t)$. Here σ is the noise amplitude and the fluctuations $dW_i = Z\sqrt{dt}$ are given by the increment of a Wiener process, where Z is a $\mathcal{N}(0,1)$ standard normal-distributed random number. Now we split the deterministic force into two parts

$$F_{det} = F_{acc}(v_i) + F_{dec}(\Delta x_i) \tag{4}$$

with an acceleration and deceleration ansatz

$$F_{acc}(v_i) = m \frac{v_{max} - v_i}{\tau} \geq 0 , \quad F_{dec}(\Delta x_i) = m \frac{v_{opt}(\Delta x_i) - v_{max}}{\tau} \leq 0 \tag{5}$$

taking into account the optimal velocity function proposed by Mahnke et al. [9–13]

$$v_{opt}(\Delta x) = v_{max} \frac{(\Delta x)^2}{D^2 + (\Delta x)^2} . \tag{6}$$

The conservative force term is due to interaction between cars

$$F_{dec}(\Delta x) = v_{max} \frac{m}{\tau} \left(\frac{(\Delta x)^2}{D^2 + (\Delta x)^2} - 1 \right) \tag{7}$$

and is always negative, starting from $F_{dec}(\Delta x = 0) = -v_{max}m/\tau$, and approaching zero at infinite distances. Using the definition of the conservative potential U_{cons} by $F_{dec}(\Delta x) = -dU_{cons}(\Delta x)/d\Delta x$, we obtain

$$U_{cons}(\Delta x) = -v_{max}\frac{D\,m}{\tau}\arctan\left(\frac{\Delta x}{D}\right) + U^0_{cons}. \tag{8}$$

Having discussed the conservative force $F_{dec}(\Delta x)$ in our model, we now turn our attention to the dissipative force in the equation of motion. In order to investigate moving cars on a circle, the dissipative or friction function $F_{acc}(v)$ has to be characterized as an active one which can be understood as an external energy supply. Engine equipped vehicles are active particles which move on the road in one direction. Taking into account fluctuations, these vehicles are called active Brownian particles or motors. Usually active Brownian particles are able to take up energy from the environment which can be stored in a depot. The transfered internal energy can be converted into kinetic energy which results in acceleration of the Brownian particle in the direction of motion. By introducing a dissipative potential U_{diss} via $F_{acc}(v) = -dU_{diss}(v)/dv$ we find

$$U_{diss}(v) = \frac{m}{\tau}\left(\frac{v^2}{2} - v_{max}v\right) + U^0_{diss}. \tag{9}$$

It is convenient to write our equations of motion dimensionless. By introducing the dimensionless coordinates $y_i = x_i/D$ and velocities $u_i = v_i/v_{max}$ of cars, the dimensionless optimal velocity function $u_{opt}(\Delta y)$ can be written as

$$u_{opt}(\Delta y) = \frac{v_{opt}}{v_{max}} = \frac{(\Delta y)^2}{1 + (\Delta y)^2}. \tag{10}$$

This normalization is particularly suitable in the case of point–like cars with $\ell \to 0$, where the headway (distance) to the next car is given by $\Delta y_i = y_{i+1} - y_i$. Introducing the dimensionless time $T = t/\tau$, our equations of motion (2) and (3) now become

$$du_i = (u_{opt}(\Delta y_i) - u_i)\,dT + a\,u_i\,dW_i(T) \tag{11}$$

$$dy_i = \frac{1}{b}u_i\,dT, \tag{12}$$

where $a = \sigma\sqrt{\tau}/m$ (the dimensionless noise intensity) and $b = D/(v_{max}\tau)$ are dimensionless control parameters. Other relevant parameters are the total number of cars N and the dimensionless car density $c = N/\mathcal{L}$, where $\mathcal{L} = L/D$ is the length of the road L measured in units of the interaction distance D.

3 Discussion of Results

The system of $2N$ stochastic differential equations (11) and (12) has been solved numerically by an algorithm called *explicit 1.5 order strong scheme* [14].

We have fixed the noise intensity $a = 0.1$ and the number of cars $N = 60$ and have studied the distribution of vehicular velocities and headway distances depending on the dimensionless car density c and parameter b. In Fig. 1 (left)

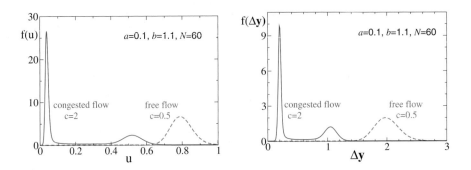

Fig. 1. The probability density distribution of velocities (left) and headway distances (right) for two different car densities $c = 0.5$ and $c = 2$ at fixed dimensionless noise amplitude $a = 0.1$ and parameter $b = 1.1$.

the velocity distribution function $f(u)$ is shown at $b = 1.1$ for a relatively small density $c = 0.5$ as well as for higher density $c = 2$. In the first case we have only one maximum located near $u = 0.8$ which is the steady state velocity of the homogeneous free traffic flow without noise. The noise only smears out the delta–like distribution to yield a smooth maximum seen in Fig. 1. The traffic flow is homogeneous with small fluctuations in velocities and also headway distances, as shown in Fig. 1 (right). The headway–distance distribution function $f(\Delta y)$ has a maximum near $\Delta y = \Delta y_{hom} = 1/c = 2$ which is the average headway distance in a homogeneous flow. In the second case ($c = 2$) the velocity distribution function as well as the headway distribution function have two maxima. It indicates the coexistence of two phases – free flow with relatively large headways and velocities and jam with small headways and velocities.

The coexistence of two phases is possible within a certain region of b and c values. In the thermodynamic limit $N \to \infty$ the homogeneous flow, described by the deterministic equations without noise, becomes unstable [6, 11] and develops into a heterogeneous situation with two coexisting phases when

$$\frac{2\Delta y_{hom}}{[1 + (\Delta y_{hom})^2]^2} > \frac{b}{2} \tag{13}$$

holds. For a finite system of size $N = 60$, the corresponding region is that below the dashed curve in the phase diagram shown in Fig. 2. The homogeneous traffic flow becomes unstable when entering the region below the dashed curve (at $c = c'_{1,2}$) and develops into the heterogeneous solution known as limit cycle. Taking into account the hysteresis effect which is a property of first order

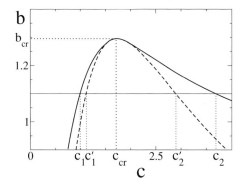

Fig. 2. Phase diagram as b-c–plane for a system of $N = 60$ cars with vanishing noise intensity $a = 0$. The homogeneous traffic flow becomes unstable when entering the region below the dashed curve (at $c = c'_{1,2}$). The heterogeneous traffic flow becomes unstable when exiting the region below the solid curve (at $c = c_{1,2}$). Both situations are possible (depending on the initial conditions) in between of these curves. The maximum corresponds to the critical point $c_{cr} \simeq 1.714$ and $b_{cr} \simeq 1.295$.

phase transitions, the heterogeneous traffic flow becomes unstable when exiting the region below the solid curve (at $c = c_{1,2}$). Both states of the system are coexist within a narrow range of densities $c_1 < c < c'_1$ between the solid and dashed curves in Fig. 2. An analogous situation takes place at large densities within $c'_2 < c < c_2$. These regions are characterised by a bistability, where the system must go through an unstable state or an unstable limit cycle to switch from one stable state to another. This picture is analogous to the nucleation phenomena in physical systems like supersaturated vapour, where the formation of liquid droplets can proceed only by overcoming a certain nucleation barrier, i. e. by going through an unstable or critical cluster (droplet) size.

Although the behavior of a car system is affected by noise, the condition (13) makes sense as a first approximation for small noise intensities $a \ll 1$. This diagram implies that the homogeneous flow is always stable if b exceeds certain critical value b_{cr} corresponding to the maximum of both curves located at the critical density $c = c_{cr}$. In this sense, the parameter b plays the role of temperature and the critical point $c = c_{cr}$, $b = b_{cr}$ is analogous to that of the liquid–gas transition. The critical point of a finite system of $N = 60$ cars at vanishing noise amplitude $a = 0$ has been evaluated: $c_{cr} \simeq 1.714$ and $b_{cr} \simeq 1.295$. In our example, where b is fixed at $b = 1.1 < b_{cr}$ (horizontal solid line in Fig. 2), we observe free flow as a gaseous phase at small densities ($c < c_1$), and heavy dense traffic as a liquid phase at large densities ($c > c_2$). The coexistence of both phases is observed at intermediate densities ($c_1 < c < c_2$) known as congested traffic or stop–and–go regime.

The heavy traffic at $c = 3.5$, as well as the critical situation at $c = c_{cr}$ and $b = b_{cr}$ are illustrated in Fig. 3 with the same notation as in Fig. 1. In both cases the distributions have only one maximum, as consistent with the

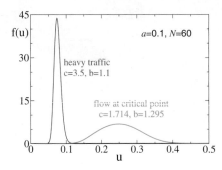
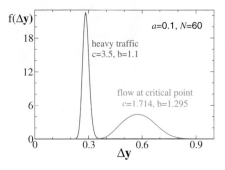

Fig. 3. The probability density distribution of velocities (left) and headway distances (right) in the case of heavy traffic with large density of cars $c = 3.5$ (at $b = 1.1$) and at the critical point $c = c_{cr} \simeq 1.714$, $b = b_{cr} \simeq 1.295$. In both cases the dimensionless noise intensity is $a = 0.1$.

existence of only one phase. A distinguishing feature of the critical point is that the distributions over the headway distances and velocities are relatively broad.

The simulation of the stochastic equations (11) and (12) in the coexistence region allows us also to find the probability $P(n, t)$ that just n cars are involved in the jam. This can be compared with the results of the stochastic master equation approach developed in [10–12].

References

1. J. Schmelzer, G. Röpke, R. Mahnke: Aggregation Phenomena in Complex Systems, Wiley–VCH, Weinheim, 1999.
2. I. Prigogine, R. Herman: Kinematic Theory of Vehicular Traffic, Elsevier, New York, 1971.
3. D. Helbing: Rev. Mod. Phys. **73**, 1067–1141, 2001.
4. D. E. Wolf, M. Schreckenberg, A. Bachem (Eds.): Traffic and Granular Flow, World Scientific Publ., Singapore, 1996.
5. M. Schreckenberg, D. E. Wolf (Eds.): Traffic and Granular Flow '97, Springer, Singapore, 1998.
6. M. Bando, K. Hasebe, A. Nakayama, A. Shibata, Y. Sugiyama: Japan J. Indust. and Appl. Math. **11**, 203, 1994; Phys. Rev. E **51**, 1035, 1995.
7. M. Bando, K. Hasebe, K. Nakanishi, A. Nakayama, A. Shibata, Y. Sugiyama: J. Phys. I France **5**, 1389, 1995.
8. L. Arnold: Random Dynamical Systems, Springer, Berlin, 1998.
9. R. Mahnke, J. Kaupužs: Presentation at second workshop on *Traffic and Granular Flow*, Duisburg, October 1997, see [5], p. 439; 447.
10. R. Mahnke, N. Pieret: Phys. Rev. E **56**, 2666, 1997.
11. R. Mahnke, J. Kaupužs: Phys. Rev. E **59**, 117, 1999.
12. R. Mahnke, and J. Kaupužs: Networks and Spatial Economics **1**, 103–136, 2001.
13. R. Mahnke, J. Kaupžs, V. Frishfelds: Atmospheric Research **65**, 261–284, 2003.
14. P. E. Kloeden, E. Platen: Numerical Solution of Stochastic Differential Equations, Springer, Berlin, 1992.

Towards Noised-Induced Phase Transitions in Systems of Elements with Motivated Behavior

I. Lubashevsky[1], M. Hajimahmoodzadeh[2], A. Katsnelson[2], and P. Wagner[3]

[1] General Physics Institute – Russian Academy of Sciences
Vavilov str., 38, 119991 Moscow – Russia
`ialub@fpl.gpi.ru`
[2] Faculty of Physics – M. V. Lomonosov Moscow State University
Moscow 119992 – Russia
`albert@solst.phys.msu.su`
[3] Institute of Transport Research – German Aerospace Center (DLR)
Rutherfordstr. 2, 12489 Berlin – Germany
`peter.wagner@dlr.de`

Summary. A new type of noised-induced phase transitions that should occur in systems of elements with motivated behavior is considered. By way of an example, a simple oscillatory system $\{x, v = \dot{x}\}$ with additive white noise is analyzed numerically. A chain of such oscillators is also studied in brief.
Keywords: motivated behavior, noise-induced phase transitions

1 Introduction

Systems of elements with motivated behavior (systems with motivation), e.g., fish and bird swarms, car ensembles on highways, stock markets, *etc.* often display noise-induced phase transitions (for a review see Ref. [3]). The ability of noise to induce phase transitions is now well established (see, e.g., Refs [1, 2]). However, in systems with motivation there is a special mechanism endowing the corresponding noise-induced phase transitions with distinctive properties.

For example, people as elements of a certain system cannot individually control all the governing parameters. Therefore one chooses a few crucial parameters and focuses on them the main attention. When the equilibrium with respect to these crucial parameters is attained the human activity slows down retarding, in turn, the system dynamics as a whole. For example, in driving a car the control over the relative velocity v is of prime importance in comparison with the correction of the headway distance x. So, under normal conditions a driver should eliminate the relative velocity between her car and a car ahead first and only then correct the headway.

These speculations lead us to the concept of dynamical traps, a certain "low" dimensional region in the phase space where the main kinetic coeffi-

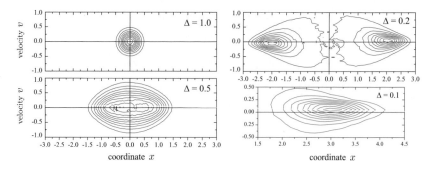

Fig. 1. Evolution of the distribution function $\mathcal{P}(x, v)$ (shown by level contours) as the parameter $\triangle$ decreases. The parameters used are $\sigma = 0.1$ and $\epsilon = 0.1$. The lower right window depicts only one maximum of the distribution function.

cients specifying the characteristic time scales of the system dynamics become sufficiently small in comparison with their values outside the trap region [4, 5]. The present paper analyzes the effect of noise on such a system and demonstrates that additive noise in a system with dynamical traps is able to give rise to new phases.

2 Noised Oscillatory System with Dynamical Traps

By way of example, the following dimensionless system typically used to describe the oscillatory dynamics is considered:

$$\frac{dx}{dt} = v \,, \quad \frac{dv}{dt} = -\Omega(v)\left[x + \sigma v\right] + \epsilon\xi(t) \,, \tag{1}$$

where σ is the damping decrement and the term $\epsilon\xi(t)$ is a random Langevin "force" of intensity ϵ proportional to the white noise $\xi(t)$ with unit amplitude. The function $\Omega(v)$ describes the dynamical trap effect arising in the vicinity of $v = 0$. For this function, the following simple *Ansatz*

$$\Omega(v) = \frac{v^2 + \triangle^2}{v^2 + 1} \,. \tag{2}$$

is used. In the chosen scales the thickness of the trap region is equal to unity and the parameter $\triangle \leq 1$ measures the trapping efficacy. When $\triangle = 1$ the dynamical trap effect is ignorable, for $\triangle = 0$ it is most effective.

System (1), (2) was analyzed numerically using the algorithm described in [6]. Figure 1 shows the distribution function $\mathcal{P}(x, v)$ of the system on the phase plane $\{x, v\}$, depending on the parameter $\triangle$. As can be seen this system undergoes a second order phase transition manifesting itself in the change of the shape of the phase space density $\mathcal{P}(x, v)$ from unimodal to bimodal as the trap parameter $\triangle$ decreases.

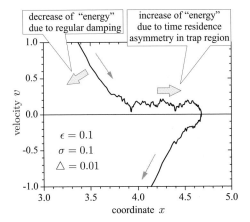

Fig. 2. A typical fragment of the system path going through the trap region. The parameters $\sigma = 0.1$, $\epsilon = 0.1$, and $\triangle = 0.01$ were used in numerical simulations in order to make the trap effect more pronounced.

Mechanism of the Phase Transition

Figure 2 illustrates the mechanism of this phase transition. It depicts a typical fragment of the system motion through the trap region for $\triangle \ll 1$. When the path goes into the trap region $\mathcal{Q}_t$ ($|v| \ll 1$) the regular "force" $\Omega(v)(x + \sigma v)$ is depressed. So inside this region the system dynamics is mainly random due to the remaining weak Langevin "force" $\epsilon \xi(t)$. Crucial is the fact that the boundaries $\partial_+ \mathcal{Q}_t$ (where $v \sim 1$) and $\partial_- \mathcal{Q}_t$ (where $v \sim -1$) are not identical in properties with respect to the system motion. At the boundary $\partial_+ \mathcal{Q}_t$ the regular "force" leads the system inwards the trap region $\mathcal{Q}_t$, whereas at the boundary $\partial_- \mathcal{Q}_t$ it causes the system to leave the region $\mathcal{Q}_t$. Outside the trap region $\mathcal{Q}_t$ the regular "force" is dominant. Thereby, from the standpoint of the system motion inside the region $\mathcal{Q}_t$, the boundary $\partial_+ \mathcal{Q}_t$ is "reflecting" whereas the boundary $\partial_- \mathcal{Q}_t$ is "absorbing".

As a result the distribution of the residence time at different points of the region $\mathcal{Q}_t$ should be asymmetric as shown in Fig. 1 (it is most clear in the lower right window). Therefore, during location inside the trap region the mean velocity must be positive, causing the system to go away from the origin. Outside the trap region the system motion is damping. So, when the former effect becomes sufficiently strong the distribution function $\mathcal{P}(\eta, u)$ becomes bimodal.

3 Chain of Oscillators

A similar noise-induced phase transition for an ensemble of oscillators with dynamical traps is analyzed. The following one-dimensional model is considered, N balls can move along x-axis interacting with the nearest neighbors, which is described by the equations for $i = 1, 2, \ldots N - 1$

$$\frac{dx_i}{dt} = v_i, \quad \frac{dv_i}{dt} = -\Omega_s(v_{i-1}, v_i, v_{i+1})\left[\eta_i + \sigma\vartheta_i\right] + \epsilon\xi_i(t). \tag{3}$$

Here x_i is the coordinate of ball i, v_i is its velocity, the variables η_i and ϑ_i are given by the expression:

$$\eta_i = x_i - \tfrac{1}{2}(x_{i-1} + x_{i+1}), \quad \vartheta_i = v_i - \tfrac{1}{2}(v_{i-1} + v_{i+1}), \tag{4}$$

and $\{\xi_i(t)\}$ is the collection of white noise sources of unit amplitude and being mutually independent. The damping decrement σ and the noise intensity ϵ are assumed to be the same for all the oscillators. The boundary balls ($i = 0$ and $i = N$) are let to be fixed to prevent the ball system from moving as a whole. The function $\Omega_s(v_{i-1}, v_i, v_{i+1})$ measuring the trap effect due to the nearest neighbor interaction is given by the *Ansatz*

$$\Omega_s(v_{i-1}, v_i, v_{i+1}) = \frac{(v_{i-1} - v_i)^2 + (v_i - v_{i+1})^2 + \triangle^2}{(v_{i-1} - v_i)^2 + (v_i - v_{i+1})^2 + 1}, \tag{5}$$

where the parameter $\triangle$ has the same meaning as previously, it measures the intensity of trapping. The balls are assumed to be either mutually permeable or impermeable. In the latter case the absolutely elastic collision approximation is used.

The system of equations (3)–(5) was analyzed numerically. Below, the results are presented for the impermeable balls. Again integration of the stochastic differential equations was performed with the algorithms described in [6]. We analyzed an ensemble of 500 balls initially spaced 5 units apart and counted the number of balls falling into 100 identical intervals covering the system location region in the corresponding space.

Figure 3 actually depicts the ball distribution along the spatial axis and the velocity space at a fixed moment of time. As seen, the trap effect leads to the formation of a sufficiently inhomogeneous spatial distribution of balls whereas for the same system but without trapping, $\triangle = 1$, the balls are actually uniformly distributed over the space region. In the velocity space the trap effect causes the balls to spread over a much wider domain.

To make the trap effect in the velocity space more clear we averaged the velocity distribution over 500 time units. The result is presented in Fig. 4. As seen, the strong trap effect causes a nonanalytic behavior of the velocity distribution at zero value, it takes the form of a cusp.

Finally, Fig. 5 illustrates the time pattern of the system dynamics. It is clear that the shown spatial structure is characterized by a long life time, because, we recall, the period of individual ball oscillations without traps is about 2π.

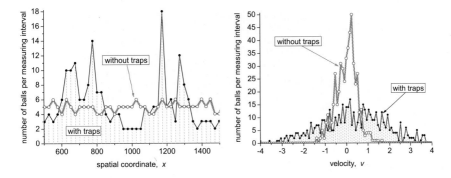

Fig. 3. The ball distribution along the spatial axis and the velocity distribution at a fixed moment of time. These distributions are represented in terms of number of balls falling into measuring interval. The measuring intervals of number 100 cover the whole region of the system location in the corresponding space. The thick grey line and the boundary of dashed region match the cases where the trap effect is absent, $\triangle = 1$, and strong, $\triangle = 0.1$, respectively. The other parameters used are $\sigma = 0.1$ and $\epsilon = 0.1$.

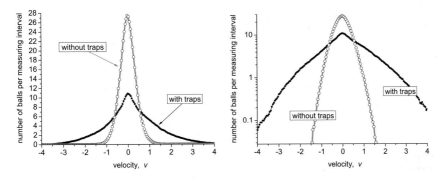

Fig. 4. The time averaged velocity distribution presented in the same way as in Fig. 3. Here, however, the individual length of the measuring interval was decreased so their total number was 250. The parameters used are $\triangle = 0.1$, $\sigma = 0.1$ and $\epsilon = 0.1$, the averaging time interval is 500 unit. The left and right windows depict the same data in linear and logarithmic scales, respectively.

4 Conclusion

A new type of noise-induced phase transitions in systems of elements with motivated behavior is considered. Dynamics of such systems, we think, should exhibit a number of anomalies due to dynamical traps. The dynamical traps form a low dimensional region in the phase space where the kinetic coefficients become sufficiently small and, as a result, the system spends a long time in it. The cause of the dynamical traps is, e.g., the inability of people or animals to control all the system governing parameters. So they have to focus

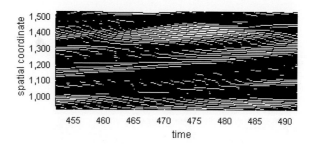

Fig. 5. A fragment of the system time pattern. Black regions correspond to increased number of balls located in them.

the main attention on a few crucial ones and the intensity of their activity decreases when the system attains a local quasi-equilibrium with respect to these parameters.

To illustrate this effect a simple oscillatory system $\{x, v = \dot{x}\}$ is studied when the trap region is located in the vicinity of the x-axis and without noise the stationary point $\{x = 0, v = 0\}$ is absolutely stable. For this system as shown numerically an additive white noise can cause the phase-space density to take a bimodal shape. For the chain of such oscillators the trap effect gives rise to a substantially nonuniform spatial distribution and leads to nonanalytic behavior of the velocity distribution near zero value.

It should be underlined that a possible phase state that could be ascribed in this case to a maximum of the distribution function in the phase space does not match any stationary point of the "regular" or "random" forces.

Acknowledgements

These investigations were supported by RFBR Grant 02-02-16537, Moscow Government Grant 1.1.133, and Russian Program "Integration", Project B0056.

References

1. W. Horsthemke and R. Lefever, *Noise-Induced Transitions* (Springer, Berlin, 1984).
2. *Stochastic Dynamics*, Lutz Schimansky-Geier and Thorsten Pöschel (eds.), Lecture Notes in Physics, Vol. 484 (Springer-Verlag, Berlin, 1998).
3. D. Helbing, Rev. Mod. Phys. **73**, 1067 (2001).
4. I. Lubashevsky, R. Mahnke, P. Wagner, and S. Kalenkov, Phys. Rev. E **66**, 016117 (2002).
5. I. Lubashevsky, M. Hajimahmoodzadeh, A. Katsnelson, and P. Wagner, e-print arXiv:cond-mat/0304300.
6. K. Burrage and P. M. Burrage, Appl. Numer. Math. **22**, 81 (1996).

Using the Road Traffic Simulation "SUMO" for Educational Purposes

D. Krajewicz, M. Hartinger, G. Hertkorn, P. Mieth, C. Rössel, J. Zimmer, and P. Wagner

Institute for Transportation Research – German Aerospace Centre
Rutherfordstr. 2, 12489 Berlin – Germany
Daniel.Krajewicz@dlr.de

Summary. Since the year 2000, the Centre of Applied Informatics and the Institute for Transport Research at the German Aerospace Centre develops a microscopic road traffic simulation package named "SUMO" – an acronym for "Simulation of Urban MObility". Meanwhile, the simulation is capable to deal with realistic scenarios such as large cities and is used for these purposes within the institute's projects. The idea was to support the traffic research community with a common platform to test new ideas and models without the need to reimplement a framework that handles road data, vehicle routes, traffic light steering etc. To achieve this goal, the simulation code is available as open source. Within this publication, we would like to demonstrate how most attributes of traffic flow can be simulated. This should be mainly interesting for educational purposes.

Keywords: traffic simulation, road traffic, car following, microscopic, continuous, multi-modal, open source, car-driver model, traffic research, education

1 Introduction

SUMO is a road traffic simulation package based on the microscopic car-following model developed by Stefan Krauß. A detailed description of this model's behaviour and fitting to real-world car behaviour can be found in [4], [7] and [8], a short description will follow.

SUMO by now consists of a set of applications, needed to perform the different steps of preparing a road traffic simulation. By now, the following modules exist: a network conversion utility, capable to import networks from other simulation packages such as Visum, Vissim or ARTEMIS but also from ArcView databases and from native XML descriptions, an OD-matrix to trip converter, a router (needed for traffic assignment) and two different versions of the simulation - one completely without a visualisation, used to perform fast simulations in a loop and a second one, slower, but with a graphical user interface written using openGL which is nevertheless surprisingly fast. We will not go into detail about the package itself, herein, as it has been described in some other places, such as [5].

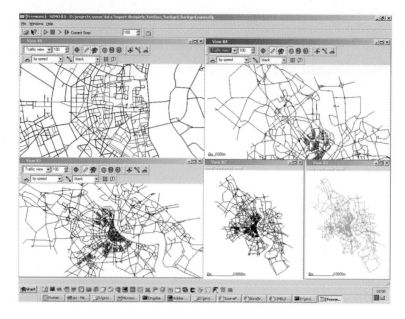

Fig. 1. A screenshot from a simulation of the city of Cologne, the Centre of Applied Information Science is located at, showing different zooming scales and colouring schemes (down right: streets coloured by the maximum speed allowed).

2 The Model

The model developed by Krauß in 1998 is a microscopic, space-continuous, car-following model based on the safe speed – paradigm: a driver tries to stay away from the driver in his front at a distance and a safe speed that allows him to adapt this leader's deceleration. The model assumes the driver to have a reaction time tau of about one second. The model uses the following parameter:

a: the maximum acceleration of the vehicle (in m^2)
b: the maximum deceleration of the vehicle (in (in m/s^2)
vmax: the maximum velocity of the vehicle   (in m/s)
l: the length of the vehicle (in m)
e: the driver's imperfection in holding the wished speed (between 0 and 1)

This safe velocity is computed using the following equation:

$$v_{safe}(t) = v_l(t) + \frac{g(t) - v_l(t)\tau}{\frac{\overline{v}}{b(\overline{v})} + \tau} \qquad (1)$$

Where:

vl(t):   speed of the leading vehicle in time t
g(t):   gap to the leading vehicle in time t
tau: the driver's reaction time (usually 1s).

As vsafe may be larger than the maximum speed allowed on the road he uses or larger than the vehicle is capable to reach until the next step due to his acceleration capabilities, the minimum of these values is computed as next. The resulting speed is called the "desired" or "wished" speed.

$$v_{des}(t) = \min\{v_{safe}(t), v(t-1) + a, v_{max}\} \tag{2}$$

Assuming the driver is not able to perfectly adapt the desired velocity, the "driver's imperfection" value multiplied with the car's acceleration ability and a random number is substracted from the desired velocity. Finally, one must assure, the vehicle is not driving backwards. Due to this, the last of the model's equation is:

$$v(t) = \max\{0, \text{rand}[v_{des}(t) - \epsilon a, v_{des}(t)]\} \tag{3}$$

The velocity, multiplied with the simulation step duration, which is constantly equal to one second, here, is added to the vehicle's current position to achieve the position for the next time step. This model is collision-free and due to the small number of equations very fast – on a 1GHz computer one can easily perform the simulation step for about 1.000.000 vehicles in per second – each describing a real vehicle's movement within one second.
It is known within the traffic research that the traffic flow is a function of the traffic density. This is the fundamental diagram. As Fig. 2 shows, the model by Krauß is capable to replicate the flow function well.

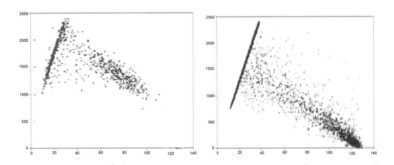

Fig. 2. The traffic flow as a function of the traffic density; left: original highway data, right: simulation results using the Krauß-model.

3 Possible Traffic Scenarios for Educational Purposes

3.1   Traffic Assignment

One important part of a traffic simulation of a large region is the dynamic assignment. In this process, the demand (O/D-matrix) is distributed onto the network. This is done by approaching a so-called dynamic user equilibrium (DUE), or in the case here, a stochastic dynamic UE (SDUE). This means that for any O/D-pair the flows on the various routes connecting O and D are distributed in such a manner, that the travel times (travel costs) are equal on all used routes.

This state is computed by an algorithm that has been described in Gawron [3]. SUMO uses this algorithm within its routing module. The simulation package includes two basic examples to demonstrate this approach. We will discuss only one here, which takes place within a network with only three streets. Driving from bottom to the top of the network, vehicles may use one of the side arms or the centre road. Within the first step, all vehicles use the centre road, as it seems to be the shortest one. Due to the arising high density and the fact, a traffic light is positioned at the road's end it's not. This is adapted by the drivers within the next iterations where each driver adapts the now fastest route with a certain probability. The plot within the next picture displays the overall number of simulated seconds until the last vehicle has left the simulated area. As one can see, drivers are gaining almost half of the time they needed if all use the path assumed to be the shortest one at the beginning.

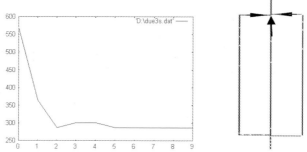

Fig. 3. The reduction of the simulation time until the last vehicle has ended his route over DUE-iterations as a measurement of the equilibrium-quality; left: the "network" used.

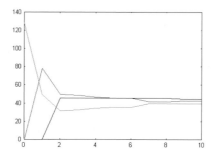

Fig. 4. Plot of the route usages over iterations; green: vehicles using the centre road, red: vehicles using the way to the right, blue vehicles using the way to the left.

3.2   Jam Dynamics

Jams are known to grow as long as the inflow stays high and wanders upwards to the stream. To visualise this behaviour, the following scenario was chosen: vehicles are emitted into a lane with a high density. A traffic light is placed at the end of this lane and with a very low frequency. This traffic light turns red causing a temporary jam. After it turns back green, one can see how the jam is moving backwards and grows. This scenario is included in the application's distribution, too. Still, the visualisation within a single plot is not done, yet.

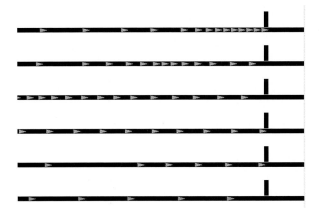

Fig. 5. The development of a jam – from the begin (top) back to free flow (bottom); the time difference between the pictures is 25s, the bluer a vehicle the faster it is, red vehicles are standing.

3.3   Model Calibration

Investigations on microscopic models of traffic flow are one of the main research areas of the Institute of Transport Research and so many approaches are made to validate and calibrate these models (see [1] or [6]). It is assumed that drivers behave differently when different areas are regarded and so, the ability to being calibrated to a given set of traffic data is a traffic simulation's increasingly wished feature. The approach we use most often is to use real-world detector data and emit (release) vehicles on the begin of the road each time one of the real-world detectors was passed by a vehicle. Mostly, when using data from the I880-FSP for example, this is possible for each lane of a road.

In some distance, other real-world detector data are used to validate the simulation results. An external optimiser is then used to find a parameter set that fits best to the scenario. The next picture shows a preliminary result of the error development during the calibration process. A closer look will be given in another publication.

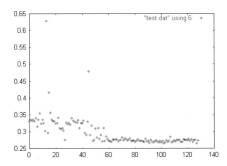

Fig. 5. The development of the error over simulation iterations.

4 Conclusion

We hoped to be able to show that some prominent traffic predicates may be simulated with a small effort using the open source simulation package SUMO. Beside simple demonstrations as the ones described herein, the software is capable to deal with large scenarios and is already in use for such cases within the IVF. As open source software highly lives from his users, we hope so to encourage some of the readers to try it and share their experiences or ideas with us.

References

1. E. Brockfeld , R. Kühne, A. Skabardonis, P. Wagner. 2002. „Towards a Benchmarking of microscopic Traffic Flow Models". TRB, in submission
2. E. W. Dijkstra. 1959. „A note on two problems in connection with graphs". In: Numerische Mathematik, 1:269-271.
3. Christian Gawron. 1998. „Simulation-Based Traffic Assignment". Inaugural Dissertation.
4. Stefan Janz. 1998. „Mikroskopische Minimalmodelle des Straßenverkehrs". Diploma Thesis.
5. "SUMO (Simulation of Urban MObility); An open-Source Traffic Simulation". D. Krajzewicz, G. Hertkorn, C. Rössel, P. Wagner. In: Proceedings of the 4th Middle East Symposium on Simulation and Modelling (MESM2002), Edited by: A.~Al-Akaidi, pp. 183 - 187, SCS European Publishing House, ISBN 90-77039-09-0.
6. D. Krajzewicz, G. Hertkorn, C. Rössel, P. Wagner; "An Example of Microscopic Car Models Validation using the Open Source Traffic Simulation SUMO"; In: Proceedings of Simulation in Industry, 14th European Simulation Symposium, 2002, Edited by: A. Verbraeck and W. Krug, pp. 318-322, SCS European Publishing House, ISBN 3-936150-21-4.
7. Stefan Krauß. 1998. „Microscopic Modelling of Traffic Flow: Investigation of Collision Free Vehicle Dynamics"; Hauptabteilung Mobilität und Systemtechnik des DLR Köln. ISSN 1434-845.
8. Stefan Krauß, Peter Wagner, Christian Gawron. 1997. "Metastable States in a Microscopic Model of Traffic Flow"; Physical Review E, volume 55, number 304, pages 55-97; May, 1997.

Approximating Traffic Flow by a Schrödinger Equation - Introduction of Non-Reflecting Boundary Conditions

R. Woesler, K.-U. Thiessenhusen, and R.D. Kühne

Institute of Transportation Research – German Aerospace Center
Rutherfordstr. 2, 12489 Berlin – Germany
richard.woesler@dlr.de

Summary. We show that some simple urban traffic flow equations can be approximated by equations which are equivalent to a Schrödinger equation. For a simulation of the Schrödinger equation as well as for analytical computations it is useful that waves of traffic which travel along a road are not reflected at the boundaries of the simulated region. We present the non-reflecting boundary condition for a corresponding one-dimensional Schrödinger equation, and show simulation results for a wave package of traffic moving towards such a boundary.
Keywords: traffic flow theory, macroscopic equations, Schrödinger equation, traffic simulation, boundary conditions

1 Introduction

Numerous ways to describe traffic flow are known [1,2]. A simple way to approximate urban traffic with one-dimensional lanes is to use the following macroscopic equations

$$\dot{\rho} + \nabla q = 0 \tag{1}$$

$$\dot{v} = -v\nabla v + \mu \nabla^2 v + \sum_j F_j(t) \tag{2}$$

The first equation is the equation of continuity, the second one includes external forces $\sum_j F_j(t)$ representing the effects of traffic light signals at intersections. A second order viscosity like term like $\mu \nabla^2 v$ often is introduced in order to smooth shock waves [3-6]. A relaxation term for a velocity adaptation as well as an anticipation term are neglected as a rough approximation.

2 Theory

Using the transformation $\Psi = \sqrt{\rho}\, e^{\frac{i}{\eta}\Phi}$ and $\nabla \Phi = v$, the equations (1,2) can be written in the following way

$$i\eta\, \dot{\Psi} = \left(-\frac{\eta^2}{2}\nabla^2 + U \right)\Psi \quad \text{with } \nabla U = -\sum_j F_j(t) \tag{3}$$

when modifying the viscosity term as $\dfrac{\eta^2}{2}\nabla\dfrac{\nabla^2\sqrt{\rho}}{\sqrt{\rho}}$. The interpretation of a fluid dynamical equation as a Schrödinger equation first was pointed out by Madelung [7].

To derive the non-reflecting boundary conditions consider a road for x from minus infinity to plus infinity. A right boundary of the system shall be established at the point $x = 0$. For $x < -b$, $b > 0$, there are assumed to be one or some positive potential barriers, and zero potential for $x \geq -b$. There exist eigenfunctions with positive energy eigenvalues E_m . Waves travelling from inside the system from $x < 0$ to the boundary $x = 0$ are assumed, and no waves in the other direction. No reflections of waves means for values x between $-b$ and 0 for zero mean waves:

$$\Psi(x,t) = \sum_{k_m>0} a_m e^{ik_m x - i\frac{E_m}{\eta}t} . \tag{4}$$

The positive values k_m are given by

$$E_m = \eta^2 k_m^2 / 2. \tag{5}$$

In order to simplify computations we restrict m to fulfil $m \in \{1,2,3...\}$, and we define $k_m < k_{m+1}$ for all m. In order to obtain the corresponding boundary conditions, the function

$$J(t) = \int_{-\infty}^{t} \Psi(0,t') \cdot (t - t')^{-1/2} dt' \tag{6}$$

is considered. Replacing $t - t'$ by T, and replacing the integral over T from close to zero to close to infinity by possibly residuals and the sum of three integrals in the complex plane $T = T' + i\,T''$ starting from close to zero with $T'' = 0$ along a circle to $T' = 0$ into the part of the plane with positive T'', then from this point to close to $T = +i\infty$, and from there along a circle back to $T'' = 0$ the integral finally arrives close to $T = T' = \infty$, as desired. Both integrals along the parts of the circles converge to zero when taking the limit of the radius of the former circle to zero, and that of the latter to infinity. Within the relevant region, i.e. for positive T' and positive T'', there are no singularities, i.e., no residuals. The integration path is illustrated in Fig. 1.

Using the Γ-function, $J(t)$ can be computed which yields

$$J(t) = e^{i\pi/4}\sqrt{\pi} \cdot \sum_{k_m>0} \dfrac{a_m}{\sqrt{E_m/\eta}} \cdot e^{-i\frac{E_m}{\eta}t} . \tag{7}$$

From here it follows that the function $\dfrac{d}{dt}J(t)$ is proportional to $\nabla_x\Psi(x,t)\big|_{x=0}$, i.e.

$$\dfrac{d}{dt}\int_{-\infty}^{t} \Psi(0,t') \cdot (t - t')^{-1/2} dt' = -\sqrt{\eta\,\pi/2} \cdot e^{i\pi/4} \cdot \nabla_x\Psi(x,t)\big|_{x=0} \tag{8}$$

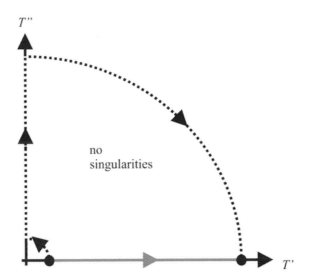

Fig. 1. The integration paths for computation of $J(t)$ (6,7) are shown. The variable $t-t'$ is replaced by $T = T' + i\ T''$, where T', T'' are real numbers. Further explanation see main text.

which are the desired non-reflecting boundary conditions.

Using (8) it is possible to calculate the wave function values $\Psi(0,t)$ at the boundary from the previous values of Ψ at the boundary and from the values of the wave function $\Psi(x<0,t)$ which are defined by (3).

3   Simulation Results

An initial traffic wave package for $t = 0$ is shown in Fig. 2 (dashed dotted line). The value $\rho(x,0)$ is constant from -4.8 to -3.2, i.e. constant traffic density. In the simulation, when computing the integral in (8) we neglect values for negative t. The result for $t = 5$ (Fig. 2 and Fig. 3) shows that no differences can be seen between the case with non-reflecting boundary conditions (thin line) and the case without boundary (dashed line), i.e., a good correspondence between theory and simulation.

Real and imaginary parts for $t = 7$ are depicted in Fig. 4, and Fig. 5 respectively. The results show a small deviation between simulation without boundary and simulation with non-reflecting boundary. The deviation can be seen in Fig. 5. The absolute values of $\mathrm{Im}\Psi$ at the local extrema for x between -1.3 and 0 with non-reflecting boundary are slightly larger than those without any boundary. This demonstrates that for larger t the discretisation has to be decreased in order to obtain optimal results.

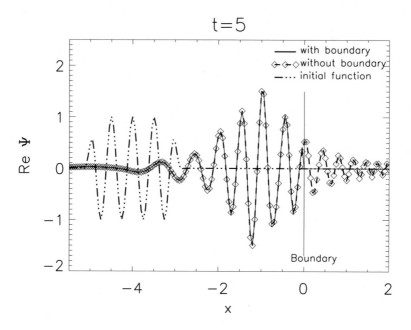

Fig. 2. Real part of $\Psi(x,t)$ for $t = 0$ (dashed dotted line), for $t = 5$ without boundary (dashed line), and for $t = 5$ with non-reflecting boundary (8) at $x = 0$ (thin line) according to the Schrödinger equation (3). The discretisation parameters are $dx = 0.1$, $dt = 0.001$, and $\eta = 0.1$ is chosen.

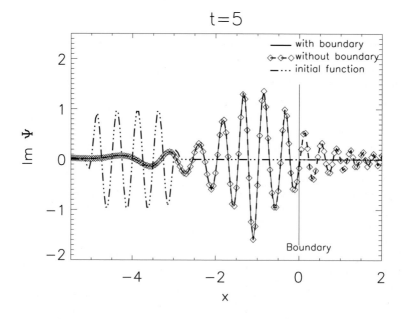

Fig. 3. Imaginary part of $\Psi(x,t)$ for $t = 0$ (dashed dotted line), for $t = 5$ without boundary (dashed line), and for $t = 5$ with non-reflecting boundary (8) at $x = 0$ (thin line) according to the Schrödinger equation (3). The discretisation parameters are $dx = 0.1$, $dt = 0.001$, and $\eta = 0.1$ is chosen.

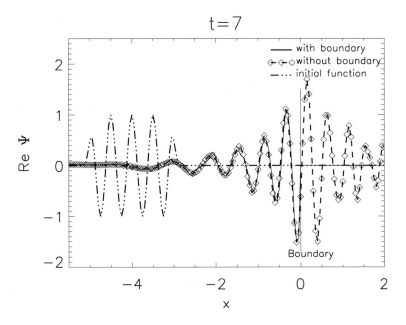

Fig. 4. Real part of $\Psi(x,t)$ for $t = 0$ (dashed dotted line), for $t = 7$ without boundary (dashed line), and for $t = 7$ with non-reflecting boundary (8) at $x = 0$ (thin line) according to the Schrödinger equation (3). The discretisation parameters are $dx = 0.1$, $dt = 0.001$, and $\eta = 0.1$ is chosen.

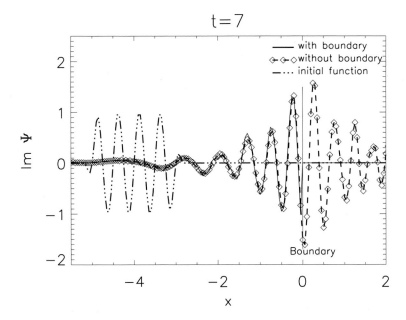

Fig. 5. Imaginary part of $\Psi(x,t)$ for $t = 0$ (dashed dotted line), for $t = 7$ without boundary (dashed line), and for $t = 7$ with non-reflecting boundary (8) at $x = 0$ (thin line) according to the Schrödinger equation (3). The discretisation parameters are $dx = 0.1$, $dt = 0.001$, and $\eta = 0.1$ is chosen.

4   Conclusions

Generally, the solution of the Schrödinger equation (3) leads for initially localized functions, which travel towards a usual boundary, to reflections back into the system which falsifies the simulation results. We introduced a non-reflecting boundary condition (8) where the characteristics of the function are maintained.

References

1. D. Helbing. 2001. Reviews of Modern Physics 73, 1067-1141.
2. K. Nagel, P. Wagner, and R. Woesler. Still flowing: Approaches to traffic flow and traffic jam modelling, 2003. Operations Research 51 (5), 681-710.
3. H.J. Payne. Models of freeway traffic and control. In G.A. Bekey, editor. Mathematical Models of Public Systems, Simulation Council, La Jolla, CA, 1971. number 28 in Simulation Council Proceedings, 51-61.
4. G.B. Whitham. Linear and nonlinear waves. New York, 1974. Wiley.
5. R. D. Kühne. Macroscopic freeway model for dense traffic stop-start waves and incident detection. In J. Volmuller and R. Hamerslag, editors, Proceedings of the 9th International Symposium on Transportation and Traffic Theory, Utrecht, Netherlands, 1984. VNU Science Press, 21.
6. B.S. Kerner, P. Konhäuser. Cluster effect in initially homogeneous traffic flow, 1993. Physical Review E 48, R2335.
7. E. Madelung. 1926. Zeitschrift für Physik 40, 322-326.

Modeling and Managing Traffic Flow Through Hyperbolic Conservation Laws

R.M. Colombo

Department of Mathematics – Brescia University
Via Branze, 38, 25123 Brescia – Italy
rinaldo@ing.unibs.it

Summary. Continuum models for traffic flows may serve as a tool for the optimal management of traffic. From the mathematical point of view, the development of a control theory for conservation laws is still at its beginning. From the engineering point of view, both the optimality criteria and the selection of the controllable parameters seem far from being standardized.

This note presents first some recently introduced continuum models, then some optimal management problem are discussed.
Keywords: Continuum models, Optimal Management of Traffic Flow.

1 Introduction

A standard distinction in physics separates the theory of systems composed by a few rigid bodies from that of continuum mechanics. In the former case, typically, one is led to systems of ordinary differential equations: based on the principles of dynamics they model the interactions among the various individual bodies. In the latter case, typically, the particles that constitutes the system are so many that the use of statistics is well justified and leads to systems of partial differential equations.

The study of vehicular traffic is somewhere in the middle between these two situations. Both *microscopic* and *macroscopic* models can be used as tools for the description of the evolution of traffic.

The former, typically, are based on rules describing the interaction among individuals and lead to ordinary differential (or difference) equations. The size of these systems is usually rather high and a considerable number of parameters needs to be specified. A careful use of computers allows the integration of these systems, leading to satisfactory descriptions of several phenomena.

The latter approach is not justified *a priori* by the number of individuals. The very introduction of continuum variables is questionable[1]. On the

---

[1] The usual analytical theory [1] considers rather irregular "continuum" variables

other hand, macroscopic models depend on the choice of very few parameters. Moreover, the well defined scope of these models makes them easily falsifiable.

In the present note we investigate a possible use of continuum models, providing an *a posteriori* justification to their introduction. Indeed, the analytical structure of continuum models allows to state, tackle and (hopefully) solve problems on the *optimal* management of traffic.

First, in Section 2, we describe some continuum models for the description of some specific features of vehicular traffic. Then Section 3 is devoted to the optimal management of traffic flow.

2 Continuum Models

The starting point for any continuum model in traffic flow is the Lighthill-Whitham [2] and Richards [3] (LWR) model. The success of this model stems from its simplicity. Its role was once more underlined in [4], where it is shown that several more recent traffic models may hardly compete with it. Only two assumptions completely define the LWR: **1.** the total number of cars is conserved, whenever entries and exits are absent; and **2.** the vehicles' speed is a function of their density. While the first assumption is obviously a must, accepting the second may be more difficult. Indeed, experimental fundamental diagrams show two rather distinct zones. At low density and high speed, a line may reasonably interpolate the measured points in the density–flow plane. As density increases and speed decreases, the area covered by the data is less accurately described by a line. Indeed, these points cover a two dimensional area. This fact, clearly evidenced for instance in [5], directly contradicts the functional dependence of speed from density.

We consider below models based on conservation laws, i.e. on first order systems of partial differential equations stating the *conservation* of the unknown variables. The more general situations in which these quantities are not conserved, for instance when entries or exits are present, is dealt with the techniques presented by Corli *et al.* at this meeting.

2.1 Hyperbolic Phase Transitions

In connection with vehicular traffic, the term *"Phase Transition"* was introduced by Kerner. In several papers (see [5] and the references therein), backed by a huge experimental work, he distinguishes between different *phases* typical of vehicular traffic. The first distinction is between *free* and *congested* flows. It is remarkable that these observations are coherent with a continuum model similar to those used to describe phase transitions in elastodynamics. Indeed, we juxtaposed the LWR model for the free phase to the model [6] particularly suited for congested traffic. The result is a system composed by a scalar conservation law modeling free phase coupled with a 2×2 system:

$$\rho \in \text{(free phase)} \qquad (\rho, q) \in \text{(congested phase)}$$

$$\partial_t \rho + \partial_x [\rho \cdot v] = 0 \qquad \begin{cases} \partial_t \rho + \partial_x [\rho \cdot v] = 0 \\ \partial_t q + \partial_x [(q - Q) \cdot v] = 0 \end{cases} \qquad (1)$$

$$v = V \cdot \left(1 - \frac{\rho}{R}\right) \qquad v = \left(1 - \frac{\rho}{R}\right) \cdot \frac{q}{\rho}$$

where the unknown variables are the traffic density ρ, the speed v and the weighted flow q. The parameters are the maximal density R, the maximal speed in the free phase V and Q.

The solutions to (1) are selected first by requiring that the total number of cars be conserved. Secondly, we impose that if the initial datum is assigned in a given phase, then the solution remains in the same phase for all times. A final regularity condition on the solution operator singles out a unique solution to all Riemann problems for (1).

As a result, the usual minimal requirements for traffic flow models are met: traffic density and speed are non negative and bounded from above. Moreover, information travels not faster than vehicles. Furthermore, the descriptions provided by (1) are coherent with the observations in [5]. In particular, this system is capable to model persistent phenomena see [7] for more details. An example of persistent phenomenon is provided by *wide jams*, see [5, §2.3].

2.2 An n - Populations Model

The absence of a functional dependence of v from ρ can be explained also through the deep differences among drivers and vehicles. This observation leads to generalize the LWR model to the case of n-populations, as in [8]. The n equations for the evolution of the density ρ_i of the i-th population are

$$\partial_t \rho_i + \partial_x [\rho_i \cdot V_i \cdot \psi(\rho_1 + \cdots + \rho_n)] = 0. \qquad (2)$$

The various populations differ in the their speed law: the i-th population drives at speed $V_i \cdot \psi\left(\sum_{i=1}^n \rho_i\right)$, the total traffic density being $\sum_{i=1}^n \rho_i$. Here, V_i is the maximal speed of the i-population. ψ is a function, the same for all populations, modeling how drivers behave at a given density. More precisely, $\psi(r)$ is the percentage of the maximal speed chosen by drivers when the total traffic density is r.

In [8] it is proved that, independently from the number of populations, in (2) information travels not faster than vehicles. However, this model does not fit in the nowadays known results on the global well posedness of system of hyperbolic conservation laws.

However, numerical integrations lead to believe that the usual minimal requirements for traffic flow models are met: the traffic density is non negative and bounded from above, as also the traffic speed. Furthermore, this system appears stable and well posed. In particular, from the traffic point of view, (2) also allows the description of overtakings, differently from the LWR model. We defer further details to [8].

3 Traffic Management

Various means allow to regulate traffic flow. In the case of a freeway, typical control parameters are the maximal speed and the inflow rate through entries. In specific situations, also the rightmost lane can be dynamically treated either as a standard lane, or as the emergency lane.

Less obvious is the aim with which these or other parameters should be controlled. Several optimality criteria can be devised: minimize the traveling time for all vehicles or only for some categories, keep the traffic flow below given thresholds along dangerous road segments, maximize the efficiency of various toll collecting systems.

Assume that a continuum model describes the most important features of a given specific traffic situation. Then, once an optimality criterion is chosen, the problem of finding those control parameters that yield the optimal situation leads to an optimal control problem. Although the control theory for conservation laws is only at its beginning, we consider in the next two paragraphs a necessary and a sufficient condition for optimality. In both cases, the starting point is the initial boundary value problem for a continuum system. For instance, in the simplest case of the LWR model, we are lead to consider

$$
\begin{cases}
\partial_t \rho + \partial_x \left[\rho v(\rho) \right] = 0 & (t, x) \in [0, +\infty[\times [0, +\infty[\\
\rho(0, x) = \bar{\rho}(x) & x \in [0, +\infty[\\
\left[\rho v(\rho) \right] (t, 0) = \tilde{f}(t) & t \in [0, +\infty[.
\end{cases} \tag{3}
$$

Here, the road under consideration is the half line $x \in [0, +\infty[$, the initial data $\bar{\rho}$ is a given initial density distribution and the boundary data $\tilde{f}$ is the inflow entering the road at $x = 0$. $\tilde{f}$ is the control parameter. Once it is chosen, it defines the traffic density $\rho = \rho(t, x)$ as the solution to (3), say $\rho = S(\tilde{f})$. Optimizing the traffic flow amounts to minimize a function J of the traffic density, $J = J(\rho)$. Finally, the optimality criterion is translated into a cost functional $J = J(\rho) = J(S(\tilde{f}))$. The optimization thus consists in finding the boundary data $\tilde{f}$ that minimizes J. Schematically:

$$
\begin{array}{ccccccc}
\tilde{f} & \longrightarrow & \rho & \longrightarrow & J & \\
\text{boundary} & \text{solution} & \text{traffic} & \text{cost} & & \text{cost} & (4) \\
\text{data} & \text{to (3)} & \text{density} & \text{evaluation} & &
\end{array}
$$

The structure above does *not* depend on the specific choice of the LWR model.

For the sake of completeness, we remark here that for the definition of *solution* to the initial boundary value problem for (3) we refer to that introduced in [9], see also the presentation by Lebacque in these proceedings.

3.1 Minimizing Stop and Go Waves

As a first example we consider the problem of minimizing the so called *stop & go* waves, see [5, §1.4] and the references therein. Reducing the stop &

go phenomenon amounts to reduce the variations of the speed, so that the traffic speed is as near to a constant as possible. From the analytical point of view, this means that the quantity that needs to be minimized is the total variation of the traffic speed $v = v(\rho)$. With reference to (3)–(4), this leads to $J = \int_0^T \text{TV}\big(v(S\tilde{f})(t, \cdot)\big)\, dt$.

More generally, it is reasonable to assume that the relevance of stop & go waves may differ along different road segments, leading to a sort of weighted total variation:

$$J(\tilde{f}) = \int_0^T \int_0^{+\infty} \varphi(t, x)\, d\left|\partial_x v\big((S\tilde{f})(t, \cdot)\big)\right|\, dt \qquad (5)$$

Here, φ is the weight and the inner integral above is computed with respect to the measure $\left|\partial_x v\big((S\tilde{f})(t, \cdot)\big)\right|$, the total variation of the Radon measure $\partial_x v\big((S\tilde{f})(t, \cdot)\big)$ which, in turn, is obtained as (weak) space derivative of the traffic speed $v\big((S\tilde{f})(t, \cdot)\big)$.

We thus obtained a rather unusual optimization problem: the integral functional J that needs to be minimized depends upon the control variable $\tilde{f}$ through the measure of integration. Nevertheless, an existence result for a minimizer is available, see [10].

3.2 Other Optimization Criteria

More standard optimization problems are obtained when the functional J in (4) is of integral type, i.e.

$$J(\rho) = \int_0^T \int_0^{+\infty} \varphi(t, x)\, \psi\big(\rho(t, x)\big)\, dx\, dt \qquad (6)$$

Various reasonable optimality criteria can be written in the form (6). An example is the mean traveling time between $x = a$ and $x = b$, obtained through suitable φ (non zero for $x \in\,]a, b[$) and ψ.

Existence results for minimizers of (6) are straightforward, provided φ and ψ are sufficiently regular. We are thus interested in necessary conditions for optimality. In other words, we seek an equation equivalent to the Euler–Lagrange equations of the classical calculus of variations.

In the search for the minima of the map $\tilde{f} \to J\big(S(\tilde{f})\big)$, a major analytical issue is the lack of differentiability of S. This difficulty is bypassed through the introduction of an *ad hoc* differential structure, see [11, 12], called *shift differentiability*. This analytical tool allows to differentiate J with respect to $\tilde{f}$ under mild regularity assumptions. An equation satisfied by the stationary points of J is hence derived. Its detailed analytical description is beyond the scopes of this note. However, here we underline its full explicit nature. In a toy problem motivated by traffic flow, for instance, this equation is explicitly written and solved in [12, 13].

Acknowledgments

The author thanks the GNAMPA project "Analysis of Hydrodynamic Models for Vehicular Traffic Flows" for having partially supported this work.

References

1. A. Bressan. *Hyperbolic systems of conservation laws*, volume 20 of *Oxford Lecture Series in Math. and Applications*. Oxford Univ. Press, Oxford, 2000.
2. M.J. Lighthill and G.B. Whitham. On kinematic waves. II. A theory of traffic flow on long crowded roads. *Proc.Roy.Soc London.Ser.A.*, 229:317–345, 1955.
3. P.I. Richards. Shock waves on the highway. *Operations Res.*, 4:42–51, 1956.
4. C.F. Daganzo. Requiem for high-order fluid approximations of traffic flow. *Trans. Res.*, 29B(4):277–287, August 1995.
5. B.S. Kerner. Phase transitions in traffic flow. In D. Helbing, H.J. Hermann, M. Schreckenberg, and D.E. Wolf, editors, *Traffic and Granular Flow '99*, pages 253–283. Springer Verlag, 2000.
6. R.M. Colombo. A 2×2 hyperbolic traffic flow model. *Math. Comput. Modelling*, 35(5-6):683–688, 2002. Traffic flow—modelling and simulation.
7. R.M. Colombo. Hyperbolic phase transitions in traffic flow. *SIAM J. Appl. Math.*, 63(2):708–721, 2002.
8. S. Benzoni Gavage and R.M. Colombo. An n-populations model for traffic flow. *Europ. J. Appl. Math.*, 14(5):587–612, 2003.
9. François Dubois and Philippe Lefloch. Boundary conditions for nonlinear hyperbolic systems of conservation laws. *J. Diff. Equations*, 71(1):93–122, 1988.
10. R.M. Colombo and A. Groli. Minimising stop & go waves to optimise traffic flow. *Appl. Math. Letters*, To appear.
11. A. Bressan and G. Guerra. Shift-differentiability of the flow generated by a conservation law. *Discrete Contin. Dynam. Systems*, 3(1):35–58, 1997.
12. R.M. Colombo and A. Groli. On the optimization of a conservation law. *Calc. Var. Partial Differential Equations*, 19(3):269–280, 2004.
13. R.M. Colombo and A. Groli. On the optimization of the initial boundary value problem for a conservation law. *J. Math. Analysis Appl.*, 291(1):82–99, 2004.

Conservation Versus Balance Laws in Traffic Flow

P. Bagnerini[1], R.M. Colombo[2], A. Corli[1], and S. Pedretti[2]

[1] Department of Mathematics – University of Ferrara
   Via Machiavelli 35, 44100 Ferrara – Italy
   crl@unife.it
[2] Department of Mathematics – University of Brescia
   Via Branze, 38, 25123 Brescia – Italy

Summary. This note discusses the role of source terms in the modeling of vehicular traffic through conservation laws. As is well known, a source term in the equation for the vehicular density may represent entries or exits. When a second conservation laws is present, suitable source terms may describe various inhomogeneities of the road, such as ascents, descents or fog banks. In each of these cases, we provide an analytical framework. Finally, numerical experiments show specific phenomena that may not be reproduced by purely conservative models.

Keywords: Hyperbolic balance laws, Temple systems, Invariant domains.

1 Introduction

Scalar continuum models for traffic flow are motivated by the conservation of the number of vehicles. The main example is the LWR [1, 2] model $\partial_t \rho + \partial_x (\rho \cdot v(\rho)) = 0$, ρ and v being the traffic density and speed. The introduction of a source term s leads to $\partial_t \rho + \partial_x (\rho \cdot v(\rho)) = s(t, x, \rho)$. The term s takes into account inflows (resp. outflows) at entries (resp. exits), for instance.

More recently, several 2×2 (or *"higher order"*) systems of conservation laws modeling traffic flow were proposed, see [3–5]. Besides the conservation of vehicles, these models postulate also that of another quantity related to traffic density and speed. In the case of fluid dynamics, the conservation of linear momentum justifies such a conservation principle. In the case of traffic flow, however, no such principle seems to be present. Source terms can therefore be added, not only because of the presence of entries and exits, but also to correct when necessary the unjustified conservation of this second quantity.

For example, in the case of [4], the model is

$$\begin{cases} \partial_t \rho + \partial_x (\rho \cdot v(\rho, q)) = 0 \\ \partial_t q + \partial_x ((q - q_*) \cdot v(\rho, q)) = 0 \end{cases} \qquad v(\rho, q) = \left(1 - \frac{\rho}{R}\right) \cdot \frac{q}{\rho}. \qquad (1)$$

Here, q is a sort of *weighted momentum* while R (the maximal car density) and q_* (related to wide jams [6]) are characteristics of the road. Notably, in this model as well as in those in [3, 5], the fundamental diagram admits positive upper and lower bounds on traffic density and speed. The conservation of q can be partially justified *a posteriori* by the reasonably good predictions provided by the model. Of course entries and exits interfere with the conservation of q. This is the case also when inhomogeneities of the road such as descent or ascents are considered. We focus here on the balance laws

$$\begin{cases} \partial_t \rho + \partial_x \left(\rho \cdot v(\rho, q) \right) = s_\rho(t, x, \rho, q) \\ \partial_t q + \partial_x \left((q - q_*) \cdot v(\rho, q) \right) = s_q(t, x, \rho, q) \end{cases} \qquad v(\rho, q) = \left(1 - \frac{\rho}{R} \right) \cdot \frac{q}{\rho} . \qquad (2)$$

First we state general geometric conditions on the source terms s_ρ and s_q ensuring that (2) keeps the same reasonable properties of (1). These conditions are then checked on real models. Finally, by means of numerical integrations, we show some possible effects of these source terms. The reader is referred to [7] many more details and examples and a wider bibliography.

2 Source Terms and Invariance

Consider first a general system of hyperbolic conservation laws, i.e.

$$\partial_t u + \partial_x f(u) = 0 \qquad (3)$$

with $f : \Omega \subseteq \mathbf{R}^n \mapsto \mathbf{R}^n$. System (3) is a *Temple system* if it is strictly hyperbolic and shock and rarefaction curves coincide, see [8]. If $f(u) = v(u)\, u$, with $u = (\rho, y)$ and v being a scalar function, we obtain

$$\begin{cases} \partial_t \rho + \partial_x \left(\rho \cdot v(\rho, y) \right) = 0 \\ \partial_t y + \partial_x \left(y \cdot v(\rho, y) \right) = 0 . \end{cases} \qquad (4)$$

This is a Temple system: indeed, it is strictly hyperbolic in any region Ω where

$$\rho\, \partial_\rho v + y\, \partial_y v \neq 0 . \qquad (5)$$

The eigenvalues are $\lambda_1 = v + \rho\, \partial_\rho v + y\, \partial_y v$ and $\lambda_2 = v$.

System (4) unifies the traffic models introduced in [3, 4]. Setting $y = q - q_*$ in (4), one obtains (1). Condition (5) is satisfied in the region $\{(\rho, q) \in (0, R] \times [0, +\infty);\ (\rho, q) \neq (R, 0)\}$. Moreover in this region $\lambda_1 < \lambda_2$, so the requirement that no information travels faster than the car speed is satisfied. Also the model presented in [3] can be written in the form (4), see [8]. On the contrary, the Payne model [9] with $p(\rho) \sim \rho^\gamma$ is not in the Temple class.

Several of the requirements that a traffic model should meet can be described by means of *invariant* sets. A set $\mathcal{U} \subset \Omega$ is invariant for (3) if for any initial data u_o with range in $\mathcal{U}$ also the solution u to (3) has range in $\mathcal{U}$; an analogous definition holds in the case of systems of balance laws.

We refer to [7] for a description of many invariant sets for (1). In these domains the states (ρ, q) in $\mathcal{U}$ have positive bounded density and speed; moreover condition (5) holds. Furthermore, for some domains the traffic speed vanishes if and only if the traffic density reaches the maximum R.

The addition of source terms in (1) may cause $\mathcal{U}$ to loose the invariance and, hence, the above properties. In [8] it is proven that this is not the case if the source terms satisfy suitable conditions. The main assumption there is that the set $\mathcal{U}$ must be invariant also for the differential system

$$\dot{u}(t) = s\left(t, x, u(t)\right) . \tag{6}$$

A sufficient condition for the invariance of (6) is

$$s(t, x, u) \cdot n(u) \geq 0 \tag{7}$$

for all $t \geq 0$, $x \in \mathbf{R}$ and $u \in \partial\mathcal{U}$, $n(u)$ being an inner normal to $\mathcal{U}$ at u.

If system (3) is of Temple type and $\mathcal{U}$ is invariant both for (3) and (6), then, for a wide class of functions f and g, weak solutions to $\partial_t u + \partial_x f(u) = s(t, x, u)$ exist and are stable, see [8].

3 Models

Consider now (2). We refer to [7] for invariance conditions, deduced from (7), under which the result in [8] applies. In particular if $s_q = 0$ we can take as s_ρ any positive function vanishing on the segment $\rho = R$, $q \in [Q_1, Q_2]$. If $s_\rho = 0$ then s_q may be any negative function vanishing on the segment $\frac{q - q_*}{Q_1 - q_*} = \frac{\rho}{R}$, $\rho \in [\rho_1, R]$. The choice $Q_2 = +\infty$ is also admissible.

3.1 Entries and exits

Denote $g_{\text{in}}(t)(1 - \rho/R)$, resp. $g_{\text{out}}(t)\rho/R$ the density per unit time of entering (exiting) cars, $g_{\text{in}} > 0$, $g_{\text{out}} > 0$. Let $\chi_{\text{in}}, \chi_{\text{out}}$ be two characteristic functions taking into account the width of the ingoing, resp. outgoing, ramp. Let $a_{\text{in}}(t, x) = g_{\text{in}}(t)\chi_{\text{in}}(x)$, and analogously $a_{\text{out}}(t, x)$. A simple model for entries/exits is

$$\begin{cases} \partial_t \rho + \partial_x (\rho v) = a_{\text{in}}(t, x) \left(1 - \dfrac{\rho}{R}\right) - a_{\text{out}}(t, x)\dfrac{\rho}{R} \\ \partial_t q + \partial_x \left((q - q_*)v\right) = -\left(\dfrac{a_{\text{in}}(t, x)}{R} + \dfrac{a_{\text{out}}(t, x)}{R}\right)(q - q_*). \end{cases} \tag{8}$$

Let now $0 < Q_1 < Q_2$. The following sets are invariant for (8):

(i) $\mathcal{U} = \{(\rho, q) \in [0, R] \times \mathbf{R}: q_* + (q - q_*)R/\rho \in [Q_1, Q_2]\}$;
(ii) $\mathcal{U} = \{(\rho, q) \in [0, R] \times \mathbf{R}: v \leq V,\ q_* + (q - q_*)R/\rho \in [0, Q_2]\}$ if $a_{\text{out}} = 0$ and $V \geq (\sqrt{5} - 2)q_*/R$;
(iii) $\mathcal{U} = \{(\rho, q) \in [0, R] \times \mathbf{R}: v \geq V,\ q_* + (q - q_*)R/\rho \in [Q_1, Q_2]\}$ if $a_{\text{in}} = 0$ and $V \geq 0$.

The above condition $V \geq (\sqrt{5} - 2)q_*/R$ is verified, for instance, in [6].

3.2 Changes in traffic speed

We focus on the case $s_\rho = 0$. Assume that a descent is present between $x = x_1$ and $x = x_2$. As source term for the second equation we propose

$$s_q(t, x, \rho, q) = \chi(x) \cdot \rho \cdot a(t, \rho, q) \tag{9}$$

where χ localizes the descent and $a \geq 0$ is the mean acceleration. Invariance holds if $a(t, \rho, q) = 0$ where $v(\rho, q) = V$ and $\frac{q - q_*}{Q_2 - q_*} = \frac{\rho}{R}$. The first requirement is quite natural: the mean acceleration is zero at the maximal speed. The second one is not essential: we still obtain an invariant domain if $Q_2 = +\infty$. In such a case the invariant domain becomes simply

$$\mathcal{U} = \left\{ (\rho, q) \in [0, R] \times \mathbf{R} : v \in [0, V], \; q_* + (q - q_*) \cdot R/\rho \geq Q_1 \right\}.$$

The case of an ascent is analogous; but now $a \leq 0$. Invariance holds if $a(t, \rho, q) = 0$ and $\frac{q - q_*}{Q_1 - q_*} = \frac{\rho}{R}$. Similar source terms may model different phenomena, for instance a fog bank or narrowings/widenings of the road.

The introduction of a source term as (9) with $a < 0$ leads to queue formation. Such behaviour cannot be obtained with a single scalar conservation law (as for instance in the LWR case), because of the maximum principle.

4 Numerical Examples

Example 1 (Exits). Consider model (2) with the following parameters: $R = 150 \frac{\text{vehicle}}{\text{Km}}$, $q_* = 2.000 \frac{\text{vehicle}}{\text{h}}$, on a stretch long $L = 30$ Km, with constant initial data $\rho(0, x) = 80 \frac{\text{vehicle}}{\text{Km}}$ and $q(0, x) = 15.000 \frac{\text{vehicle}}{\text{h}}$, so that $v(0, x) = 87.5 \frac{\text{Km}}{\text{h}}$. There is an exit in the interval $[15 \text{ Km}, 15.01 \text{ Km}]$.

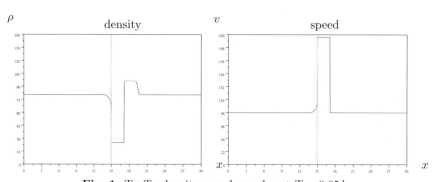

Fig. 1. Traffic density ρ and speed v at $T = 0.05$ h.

Case 1: Exit without slowdown. Let $s_\rho(t, x, \rho, q) = -0.25 \cdot \rho(t, x) \cdot v(t, x)$ if $x \in [15, 15.01]$ and 0 otherwise, while $s_q(t, x, \rho, q) = 0$. See Fig. 1.

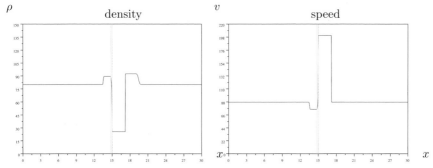

Fig. 2. Traffic density ρ and speed v at time $T = 0.05$ h.

Case 2: Exit with slowdown. Assume that 200 m before the exit, drivers slow down, i.e. $s_\rho(t, x, \rho, q) = -0.25 \cdot \rho(t, x) \cdot v(t, x)$ if $x \in [15, 15.01]$ and 0 otherwise, while $s_q(t, x, \rho, q) = -5.000\frac{\text{Km}}{\text{h}^2} \cdot \rho(t, x)$ if $x \in [14.8 \text{ Km}, 15 \text{ Km}]$ and 0 otherwise. See Fig. 2.

Note the qualitative difference before the entry. The presence of the source term in the second equation causes an increase in the density and a decrease in the speed *before* the exit.

Example 2 (Variations in traffic speed). Consider (2) with: $R = 150\frac{\text{vehicle}}{\text{Km}}$, $q_* = 2.000\frac{\text{vehicle}}{\text{h}}$ on a stretch long $L = 30$ Km with constant initial data $\rho(0, x) = 80\frac{\text{vehicle}}{\text{Km}}$ and $q(0, x) = 18.000\frac{\text{vehicle}}{\text{h}}$, so that $v(0, x) = 105\frac{\text{Km}}{\text{h}}$.

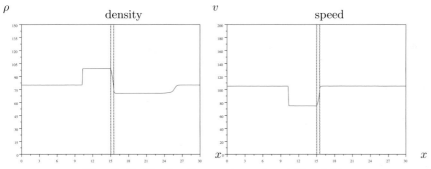

Fig. 3. Traffic density ρ and speed v at $T = 0.1$ h.

Case 1: Decrease of the speed. In the stretch $[15 \text{ Km}, 15.5 \text{ Km}]$ the traffic speed decreases. Let $s_\rho(t, x, \rho, q) = 0$ while $s_q(t, x, \rho, q) = -5.000\frac{\text{Km}}{\text{h}^2} \cdot \rho(t, x)$ in $x \in [15 \text{ Km}, 15.5 \text{ Km}]$ and 0 otherwise. See Fig. 3.

Deceleration causes a queue to propagate backwards. The increase of the speed along the stretch corresponds to the decrease of the density there. Examples of decreasing of the speed could be ascents, fog banks.

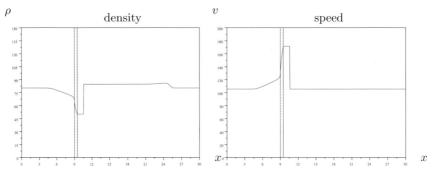

Fig. 4. Traffic density ρ and speed v at $T = 0.15\,\mathrm{h}$.

Case 2: Increase of the speed. The traffic speed increases in the stretch $[9\,\mathrm{Km}, 9.5\,\mathrm{Km}]$. Let $s_\rho(t, x, \rho, q) = 0$, $s_q(t, x, \rho, q) = 5.000\frac{\mathrm{Km}}{\mathrm{h}^2} \cdot \rho(t, x)$ if $x \in [9\,\mathrm{Km}, 9.5\,\mathrm{Km}]$ and 0 otherwise. See Fig. 4.

This acceleration causes a wave with higher speed and lower density to propagate forward. As in the previous case, the increase of the speed matches the descrease of the density. A descent is an example of this phenomenon.

Acknowledgements

P. Bagnerini acknowledges support from University of Ferrara. R.M. Colombo and A. Corli thank the GNAMPA project "Analysis of Hydrodynamic Models for Vehicular Traffic Flows" for having partially supported this work.

References

1. M.J. Lighthill and G.B. Whitham. On kinematic waves. II. A theory of traffic flow on long crowded roads. *Proc. Roy. Soc. London. Ser. A.*, 229:317–345, 1955.
2. P.I. Richards. Shock waves on the highway. *Operations Res.*, 4:42–51, 1956.
3. A. Aw and M. Rascle. Resurrection of "second order" models of traffic flow. *SIAM J. Appl. Math.*, 60(3):916–938 (electronic), 2000.
4. R.M. Colombo. A 2×2 hyperbolic traffic flow model. *Math. Comput. Modelling*, 35(5-6):683–688, 2002. Traffic flow—modelling and simulation.
5. R.M. Colombo. Hyperbolic phase transitions in traffic flow. *SIAM J. Appl. Math.*, 63(2):708–721, 2002.
6. B.S. Kerner. Phase transitions in traffic flow. In D. Helbing, H.J. Hermann, M. Schreckenberg, and Wolf D.E., editors, *Traffic and Granular Flow '99*, pages 253–283. Springer Verlag, 2000.
7. P. Bagnerini, R.M. Colombo, and A. Corli. On the role of source terms in continuum traffic flow models. Preprint, 2004.
8. R.M. Colombo and A. Corli. On a class of hyperbolic conservation laws with source terms. Preprint, 2004.
9. H. J. Payne. Models of freeway traffic and control. In G.A. Bekey, editor, *Mathematical Models of Public Systems, Vol I*, pages 51–61. Simulation Council, 1971.

Introducing the Effects of Slow Vehicles in a LWR Two-Flow Traffic Model

S. Chanut and L. Leclercq

Laboratoire ingénierie circulation transports LICIT (ENTPE/INRETS)
Rue Maurice Audin, 69518 Vaulx-en-Velin Cedex – France
stephane.chanut@entpe.fr, ludovic.leclercq@entpe.fr

Summary. The aim of this article is to study how the obstruction caused by slow vehicles (like trucks for example) on other ones can be modelled in a two-flow macroscopic model based on the Lighthill-Whitham-Richards (LWR) theory. This will be done by using the results of moving bottleneck models which describe the effect of a single slow vehicle on the rest of the flow.

Keywords: heterogeneous flow, slow vehicles, macroscopic traffic flow model.

1 Introduction

Taking account of the heterogeneousness of vehicles in macroscopic traffic flow models is a recurring subject of research works today. A first approach consists in modelling the variety of speeds and/or lengths (see [1-3] for example) in order to represent the interactions between several flows. Such models are able to reproduce differences in terms of spatial occupancy or kinematics capacity of each category of vehicles and their consequences on the behaviour of the global flow. Nevertheless these models do not explicitly consider the direct interactions between vehicles, like the perturbations due to overtaking.

Otherwise some models, like [4-7], try to represent the effect of a single slow vehicle, considered as punctual, on the surrounding flow. This slow vehicle is represented as a granularity moving among the traffic and restricts its capacity to flow. Thus, this second approach considers the slow vehicle as a moving bottleneck. Interactions between the slow vehicle and the rest of the flow are explicitly described but these models can only be applied for an isolated singular vehicle which is represented exogenously.

The aim of this paper is therefore to see how the results of the latter class of models can be used to improve the models of the former type by explicitly modelling the perturbations due to overtaking. This will be done by using two existing models both based on the LWR (Lighthill-Whitham-Richards) theory [8-9].

2 A Two-Flow Mixed Model with Different Speeds and Lengths

Several macroscopic models have been developed in the past in order to take account of heterogenous traffic [1-2]. The model we will use here has been presented in detail in [3]. It consists of an extension of the LWR model with two types of vehicles: trucks and passenger cars, where trucks are longer and slower than cars.

The conservation equation is applied for the two types of vehicles, leading to the following system (1=cars, 2=trucks):

$$\begin{cases} \dfrac{\partial K_1}{\partial t} + \dfrac{\partial Q_1}{\partial x} = 0 \\[2mm] \dfrac{\partial K_2}{\partial t} + \dfrac{\partial Q_2}{\partial x} = 0 \end{cases} \Leftrightarrow \quad \dfrac{\partial \boldsymbol{K}}{\partial t} + \boldsymbol{A}\,\dfrac{\partial \boldsymbol{K}}{\partial x} = 0$$

where $\boldsymbol{K}$ is the vector with K_1 and K_2 as coordinates, and $\boldsymbol{A}$ the following matrix:

$$\boldsymbol{A} = \begin{pmatrix} \dfrac{\partial Q_1}{\partial K_1} & \dfrac{\partial Q_1}{\partial K_2} \\[3mm] \dfrac{\partial Q_2}{\partial K_1} & \dfrac{\partial Q_2}{\partial K_2} \end{pmatrix} = \begin{pmatrix} V_1 + K_1 \dfrac{\partial V_1}{\partial K_1} & K_1 \dfrac{\partial V_1}{\partial K_2} \\[3mm] K_2 \dfrac{\partial V_2}{\partial K_1} & V_2 + K_2 \dfrac{\partial V_2}{\partial K_2} \end{pmatrix}$$

Two different fundamental diagrams have been built to highlight the bigger space occupied by trucks and their lower speeds. Fig.s 1 and 2 present the fundamental diagrams used in the model.

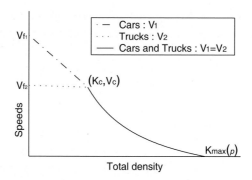

Fig. 1. Fundamental diagram for the two-flow model (example for a given p).

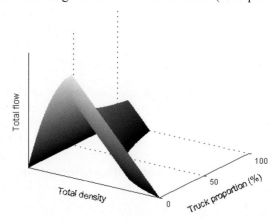

Fig. 2. Resulting curve $Q = f(K)$ for the two-flow model.

Thus the jam capacity depends on the truck rate $p = \dfrac{K_2}{K_1 + K_2}$:

$$K_{\max}(p) = \frac{N}{(1-p)L_1 + pL_2}$$

where L_1 and L_2 are the lengths of the vehicles, N the number of lanes.

Moreover we suppose that the vehicle speeds linearly decrease from the desired speeds (lesser for trucks) to the same critical speed which is reached for a critical density $K_c(p) = \alpha K_{\max}(p)$. So in congested state the two kinds of vehicles have the same speeds.

The resolution of the Riemann problems with two different vehicle densities is obtained by building an intermediate state linking the two initial states using the same analytical method as for higher order macroscopic models [10] (cf. Fig. 3). This intermediate state is linked to the upstream and downstream states by two discontinuities (shock or rarefaction or contact waves) whose type and propagation speed depend on the eigenvalues of the matrix A. In the case of our model, upstream and intermediate states are linked by a 1-field discontinuity (*ie* determined by the lower eigenvalue), which is rather a change in speeds, and intermediate and downstream states are linked by a 2-field discontinuity (*ie* determined by the bigger eigenvalue) which is rather a change in truck proportion.

An example of such a transition in the plane (K_1,K_2) is given in Fig. 3.

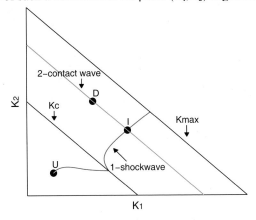

Fig. 3. Resolution of a two-flow Riemann problem (U, I, D: Upstream, Intermediate and Downstream states).

In this model where truck speed is endogeneously described in the model, the only interactions between cars and trucks are caused by the bigger space occupied by trucks, which increases the total density and therefore decreases the truck speeds but also the car speed. No perturbation linked to overtaking is taken into account.

3   Effect of a Singular Vehicle on a Homogeneous Traffic Flow

When a single vehicle moves at low speed among a traffic flow, it is perceived by the other vehicles as an obstruction which limits locally the capacity of the road. Several studies have been made to extend the LWR model by taking the effect of such a vehicle into account [4-6]. These studies have been grouped together in a unique frame by [7] where the influence of the obstruction is defined by the phenomenological flow/density relationship level with it (the bottleneck diagram). The aim of this paragraph is to present this frame which will then be used to improve the above two-flow mixed model.

If we suppose that the bottleneck diagram does not depend on the speed U_b of this discontinuity, it is possible to deduce the behaviour of the flow upstream (K_u, Q_u) and downstream (K_d, Q_d) of it. In fact, it is possible to demonstrate (cf. [5] and [7]) that the relative flow $q=Q_u-K_uU_b= Q_d-K_dU_b$ is conserved on both side of the moving bottleneck. Furthermore, when the moving bottleneck is active (that is the case where the traffic states are different on both side of the bottleneck due to its presence), the upstream and downstream traffic states are associated with an equilibrium state on the bottleneck diagram where the relative flow is maximal.

The two above properties can be translated graphically by using the fundamental diagrams. The conservation of the relative flow imposes that the traffic states (K_u, Q_u) upstream, (K_d, Q_d) downstream and (K_a, Q_a) level with the moving bottleneck are linked by a straight line whose slope is U_b, which is called the capacity line (cf. Fig. 4). As the relative flow is maximal when the bottleneck is active, the straight line is necessarily tangent to the bottleneck diagram. The tangential point describes the traffic state along the moving bottleneck. This point is the only point where the slope of the characteristics (locus where density or flow is constant) is equal to the moving bottleneck speed, that is to say that it is the only equilibrium state for which the flow is naturally constant through the bottleneck.

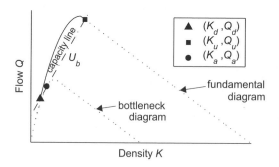

Fig. 4. The LWR moving bottleneck model.

To conclude this presentation of the moving bottleneck model, it is important to notice that the bottleneck is active only if the traffic flow is in an equilibrium state corresponding to the full line part of the fundamental diagram (cf. Fig. 4). Outside this area, the singular vehicle which is represented by the bottleneck has no effect on the other vehicles.

4   Introduction of the Overtaking Obstruction in the Two-Flow Mixed Model: a First Step

We are now going to study how to improve the two-flow mixed model described in part 2 by using the results of the above model to take perturbations due to overtaking into account.

Here we suppose that the bottleneck diagram corresponds to a *(N-1)*-lane road used by passenger cars only: trucks are supposed to stay on the right lane. Furthermore the fluid part of the bottleneck diagram is supposed to be the same as the fluid part of the diagram for *N* lanes with a truck proportion *p* (see Fig. 5). In this case the results of the bottleneck model enable us to say that pertubations due to overtaking only occur for states in the A to D part of the car fundamental diagram. Note that these states exist only if the car demand is strong enough, *ie* the capacity of the fundamental car diagram is higher than the capacity of the bottleneck diagram: it occurs if the truck rate is lower than $L_1/(L_1+(N-1)L_2)$. Note also that for the D to C part of the diagram the difference in speeds between cars and trucks is too low to cause overtaking problems (for the B to C part speeds are equal besides).

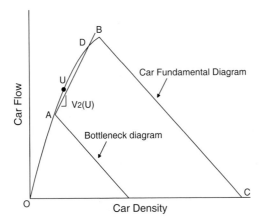

Fig. 5. Determining the impossible states (here U is an impossible state).

Then, for a truck rate *p*, we search which states U between A and B raise a problem due to overtaking in the two-flow model. To do that we just notice that each state U corresponds to a truck speed v_2: so, just like in the bottleneck model, we can draw the straight line coming through A and whose slope is v_2. If U stands between the points A and D, U is an impossible state.

Figure 6 gives the resulting area of the impossible states. We can note that this area appears when car demand is high enough, and disappears when the difference in speeds becomes too low (when we are next to the capacity limit). Therefore some intermediate states built in the two-flow model are impossible, and the way to build the analytical solution in this model has to be modified. An example of an impossible transition between two states is given in Fig. 6.

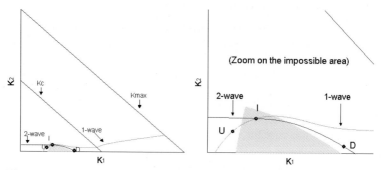

Fig. 6. Area of the impossible states (gray area) (here I, intermediate state between U and D, is an impossible state).

5   Conclusion

From the results of the moving bottleneck model we have defined impossible states of the two-flow model. Defining such states enables us to determine which transitions between couples of upstream and downstream states are impossible. Further research has to be led in order to determine in these cases how to build one or several intermediate states which can link these upstream and downstream states while avoiding the "forbidden" area.

References

1. Daganzo, C. F. A behavioral theory of multi-lane traffic flow. Part I: Long homogeneous freeway sections. In Transportation Research Part B: Methodological 36 (2), 2002, pp. 131-158.
2. Zhang, H. M. and W. L. Jin. A kinematic wave traffic flow model for mixed traffic. In Transportation Research Record 1802, TRB, 2002, pp. 197-204.
3. Chanut, S. and Ch. Buisson. A macroscopic model and its numerical solution for a two-flow mixed traffic with different speeds and lengths. To be published in Transportation Research Record, TRB, 2003.
4. Gazis, D.C. and R. Herman. The moving and 'phantom' bottlenecks. Transportation Science, 1992, Vol 26, p. 223-229.
5. Lebacque, J.P., J.B. Lesort and F. Giorgi. Introducing buses in first order traffic flow models. Transportation Research Record, 1998, Vol 1644, p. 70-79.
6. Munoz, J. and C.F. Daganzo. Moving Bottlenecks: A Theory Grounded on Experimental Observation. In: Taylor, M (Ed.), Proceedings of the 15th International Symposium on Transportation and Traffic Theory, 16-18th July, Adelaide (Australie). Amsterdam: Pergamon, 2002, p. 441-462.
7. Leclercq, L., S. Chanut and J.-B. Lesort. Moving bottlenecks in the LWR model: a unified theory. Submitted to the 83rd meeting of the Transportation Research Board, 2004.
8. Lighthill, M.J. and J.B. Whitham. On kinematic waves II. A theory of traffic flow on long crowded roads. Proceedings of the Royal Society, Vol A229, 1955, p. 317-345.
9. Richards, P.I. Shockwaves on the highway. Operations Research, Vol 4, 1956, p. 42-51.
10. Zhang, H.M. A finite difference approximation of a non-equilibrium traffic flow model. In Transportation Research Part B: Methodological 35 (4), 2001, pp. 337-365.

Bifurcation Analysis of Meta-Stability and Waves of the OV Model

P. Berg[1] and E. Wilson[2]

[1] Department of Mathematics – Simon Fraser University
8888 University Drive, Burnaby, V5A 1S6 – Canada
`pberg@sfu.ca`
[2] Department of Engineering Mathematics – University of Bristol
University Walk, BS8 1TR, Bristol – UK
`re.wilson@bristol.ac.uk`

Summary. This contribution re-visits the classical optimal-velocity model of Bando *et al* [1], describing vehicles on a loop, and explains the different wave types and meta-stable behaviour from the mathematical viewpoint of bifurcation theory. We apply the numerical continuation package AUTO for a rapid derivation of phase diagrams to investigate the relation between coexisting states, flow transitions and hysteresis in traffic flow. Some extensions which involve (i) multiple look-ahead effects and (ii) delay are also discussed.
Keywords: OV model, sub-critical, bifurcation, meta-stability, travelling wave, co-existence

1 The OV Model: Meta-stability and Travelling Waves

In what follows, we investigate the dynamics of the optimal-velocity (OV) car-following model as proposed in [1] for N vehicles on a loop governed by

$$\dot{x}_n = v_n, \tag{1}$$

$$\dot{v}_n = a \left[V(h_n) - v_n \right]. \tag{2}$$

Here, $h_n = x_{n-1} - x_n$ is the *headway*, the distance between two successive cars, a the *sensitivity*, and $V(h) = \tanh(h - 2) + \tanh(2)$ the OV function representing the desired speed of a driver encountering a headway h.

For uniform trivial flow, the linear stability criterion for waves of wave number $k = 2\pi m/N$ (where $m = 1, ..., N - 1$) reads [1]

$$\frac{V'(h^*)}{a} = \frac{1 - \cos k}{\sin^2 k}. \tag{3}$$

The dash denotes the derivative with respect to the headway taken at the average headway h^* on the loop. Note that we exclude $m = 0$ to preserve the

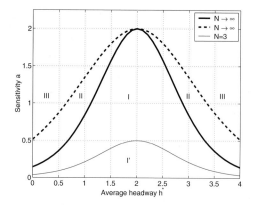

Fig. 1. Instability in the (a, h^*)-plane for $N \to \infty$: linear instability in region I and I', meta-stability in II, linear stability in III. $N = 3$: linear instability in I'.

length of the loop, and that the most unstable mode is at $m = 1$ [1] with the limit $k \to 0$ as $N \to \infty$.

For $m = 1$, Eq. (3) defines a phase diagram in the (a, h^*)-plane, whose shape varies with N, separating linear stable and unstable regions. However, there exist also meta-stable states which co-exist with linear stable states [2], represented by region II in Fig. 1. Here, the travelling wave structures which form on the loop as $t \to \infty$, range from kink-antikink (Fig. 2(a)) over saw tooth (d) to (dark) solitons (b-c). For given N, their shape depends on where the system is located in region II, determined by h^* and a. Generally, kink-antikink solutions are found for large N away from the boundary II-III, dark solitons near the boundary II-III for $h^* > 2$, solitons near the boundary II-III for $h^* < 2$, and saw tooth for very small N ($N < 10$).

The kink-antikink solutions ("stop-and-go" traffic) can be explained as two connecting heteroclinics by a travelling-wave-plane analysis of an analogous continuum model [3, 4]. The solitary waves do not occur as solutions of a KdV-equation derived from the analogous continuum model as for on-ramp simulations [5] (due to higher-order corrections), but as homoclinics in the corresponding $2N$-dimensional phase space. Correspondingly, one might regard the saw tooth solution as a limit cycle.

Meta-stable states can be found for $N > 2$ (see Fig. 2(d)), whereas for $N = 2$ we find global asymptotic stability of the trivial state $h_1 = h_2 = L/2$ and $v_1 = v_2 = V(L/2)$, i.e. this trivial state is stable towards any perturbations. The latter is proved by an energy argument derived from subtracting Eqs. (1-2) for $n = 2$ from $n = 1$, using $\ddot{h}_2 = \dot{v}_1 - \dot{v}_2$ and substituting $h_2 = L/2 + y$ to obtain an equation for a damped oscillator

$$\ddot{y} = -a\dot{y} + a\left[V(L/2 - y) - V(L/2 + y)\right] \tag{4}$$

with global stability $y = 0$, hence $h_2 = L/2\,(= h_1)$.

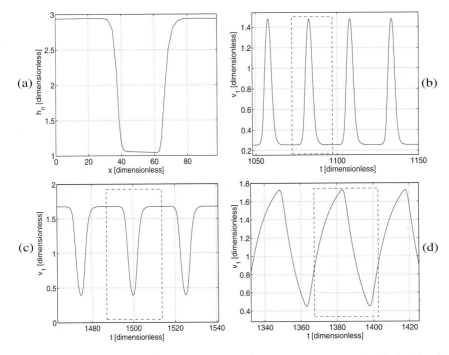

Fig. 2. Meta-stable states of the OV model (loop corresponds to the dashed box): (a) kink-antikink (shown on the whole loop): $N = 70$, $a = 1.5$, $h^* = 1.43$; (b) soliton: $N = 20$, $a = 1.5$, $h^* = 1.42$; (c) "dark" soliton: $N = 20$, $a = 1.5$, $h^* = 2.57$; (d) saw tooth: $N = 3$, $a = 0.1$, $h^* = 3.22$ (Note that we represent the discrete system by lines for illustration purposes.).

2 Bifurcation Analysis

Much attention has been drawn to meta-stability in continuum models and in the long wave length limit of car-following models [6]. The latter corresponds to a large number of vehicles on a loop in the above mentioned limit $N \to \infty$ with $m = 1$, hence $k \to 0$. It is then possible to analyse the supercritical bifurcation at the critical point $(a, h^*) = (2, 2)$ analytically [6], as a is varied from $a > 2$ through $a = 2$ to $a < 2$. Numerical efforts have been made to sketch the phase diagram (Fig. 1) [2] by perturbing trivial flow solutions. However, this method does not guarantee the correct determination of folds.

In contrast, our goal is to explain meta-stability, hysteresis and flow (phase) transitions in car-following models through bifurcation theory using the continuation package AUTO [7]. Ultimately, it can be used for rapid derivations of phase (bifurcation) diagrams for a variety of models and system parameters.

2.1 Derivation of Phase Diagrams Using Continuation Software Package AUTO

The basic question that we consider is whether we can derive the boundary (fold) of region II and III in Fig. 1 accurately and rapidly. We would like to differentiate between sub-critical and supercritical bifurcations when moving across the (a, h^*)-plane, and investigate the properties of the fold.

One way to do so is to use AUTO, a continuation package for dynamical systems with many (up to about 200) degrees of freedom. The advantage of AUTO is that it allows for one and two parameter continuation, hence, for moving through the (a, h^*)-plane. We can continue with the trivial solution or with periodic orbits until we hit a fold, the boundary between region II and III. We can thus investigate discrete systems with few (or many) degrees of freedom, equivalent to few or many cars on the loop. Moreover, even unstable periodic orbits are found (see Fig. 3).

We start by choosing the OV model for $N = 6$ vehicles, and investigate the bifurcations for one average headway h^* and continuation in the sensitivity parameter a. Figure 3 shows the bifurcation diagram for this particular case, exhibiting linearly stable and unstable trivial flow solutions, and a region where two linearly stable and one linearly unstable solution co-exist, the meta-stable regime. It contains the linearly stable trivial flow solution, and both an unstable and stable periodic solution. The latter is commonly referred to as the "jam" solution or "stop-and-go" traffic, in the case of many vehicles on the loop (Fig. 2a). The numerical value at which the trivial flow solution becomes unstable, $a_{crit} \approx 0.445$, approximates very well the solution of Eq. (3) for $h^* = 3.22$, $N = 6$, $m = 1$, given by $a = 0.4425$. The second branch, branching off the linearly unstable trivial solution at about $a = 0.16$, corresponds to the higher wave number perturbation $m = 2$, with the solution of Eq. (3) for $m = 2$ being $a = 0.1475$. This bifurcation analysis was carried out by AUTO on a time scale of seconds and promises to be a good candidate for rapid derivations of phase diagrams like Fig. 1, even for a much larger amount of vehicles.

We found meta-stability for as little as $N = 3$ vehicles on the loop (Fig. 2d). In this case, AUTO proves particularly useful as it allows the tracking of the fold (boundary of region II and III), even though region II is a very small band adjacent to region I. We cannot see any other way of determining the properties of such small meta-stable regions other than by continuation. The reason is the very small probability of "exciting" such periodic (meta-)stable solutions co-existing with the linearly stable trivial solutions for a given point (a, h^*). In other words, we have to start from the unstable manifold of the trivial solution in order to find periodic solutions, which proves to be difficult when every possible point of region II must be considered.

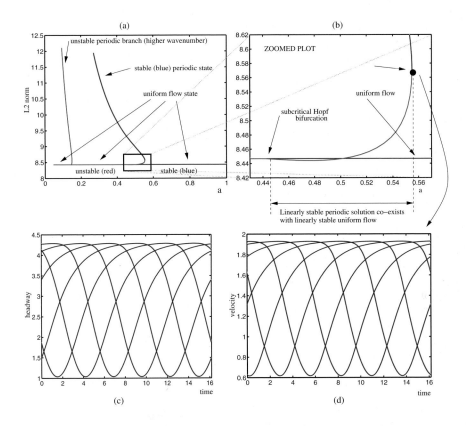

Fig. 3. Bifurcation diagram for $N = 6$ at $h^* = 3.22$: a) Sub-critical bifurcation along sensitivity parameter a; b) Zoom of a), coexistence of 3 states (1 unstable); c) Structure of periodic solution at fold point in terms of headways; d) Structure of periodic solution at fold point in terms of velocities. Note large deviations from mean headway ($h^* = 3.22$) and velocity.

2.2 Multiple Look-Ahead and Delay Models, Multi-Species Traffic and ACCS

The analysis presented in this paper can be applied to other models and systems, such as multiple look-ahead, delay [8] (biftool), autonomous cruise control systems (ACCS) and multi-species traffic. Here, the following points are of particular interest.

- The dependence of stability, meta-stability, hysteresis and flow transitions on the system parameters such as the number of vehicles on the loop, sensitivity, coupling to further vehicles ahead (number of vehicles, strength of coupling to each vehicle), OV function or other parameters specific to the underlying car-following model.

- Multi-species traffic: modelling small systems of different drivers to understand flow transitions and stabilisation, as well as safety issues such as the evolution towards harsh braking situations.
- ACCS and delay: impact of delay due to reaction time or vehicle control system response time on flow solutions and (meta-)stability; safety issues as mentioned above.

3 Conclusion

We have shown briefly how the continuation package AUTO can be used as a suitable tool for the traffic theory community. It exhibits how previously inconcise statements about meta-stability and co-existence of solutions can be made water-tight by using continuation software and exhibiting sub-critical bifurcations and folds. This analysis can also be applied to e.g. multiple look-ahead and delay car-following models, and is currently under investigation.

References

1. M. Bando, K. Hasebe, A. Nakayama, A. Shibata, and Y. Sugiyama. Dynamical model of traffic congestion and numerical simulation. *Phys. Rev. E*, 51:1035–1042, 1995.
2. Y. Sugiyama and H. Yamada. Aspects of OV model of traffic flow. In M. Schreckenberg and D.E. Wolf, editors, *Traffic and Granular Flow '97*. Springer, 1998.
3. A.D. Mason P. Berg and A.W. Woods. Continuum approach to car-following models. *Phys. Rev. E*, 61:1056–1066, 2000.
4. R.E. Wilson and P. Berg. Existence and classification of travelling wave solutions to 2nd-order highway traffic models. In M. Fukui, Y. Sugiyama, M. Schreckenberg, and D.E. Wolf, editors, *Traffic and Granular Flow '01*. Springer, 2003.
5. P. Berg and A.W. Woods. On-ramp simulations and solitary waves in a car-following model. *Phys. Rev. E*, 64:035602, 2001.
6. T.S. Komatsu and S. Sasa. Kink soliton characterizing traffic congestion. *Phys. Rev E*, 52:5574–5582, 1995.
7. E.J. Doedel, A.R. Champneys, T.F. Fairgrieve, Y.A. Kuznetsov, B. Sandstede, and X. Wang. AUTO 97. *ftp.cs.concordia.ca*.
8. M. Bando, K. Hasebe, K. Nakanishi, and A. Nakayama. Analysis of optimal velocity model with explicit delay. *Phys. Rev. E*, 58:5429–5435, 1998.

Pinch Effect in a Cellular Automaton (CA) Model for Traffic Flow

H.K. Lee[1], R. Barlović[2], M. Schreckenberg[2], and D. Kim[1]

[1] School of Physics – Seoul National University
Seoul 151-742, Seoul – Korea
`hklee@phya.snu.ac.kr`
[2] Theoretische Physik, Fakultät 4 – Universität Duisburg-Essen
47048 Duisburg – Germany
`barlovic,schreckenberg@traffic.uni-duisburg.de`

Summary. A traffic model based on a cellular automaton (CA) approach with the focus on limited braking capabilities is presented. It is shown that the known empirical features of traffic flow can be reproduced including even the so called *pinch effect* being the object of actual investigations.

Since the early 90's traffic flow phenomena have been a subject among physicists and the methods of statistical physics have been successfully performed. The investigation of empirical data led to an appropriate description of traffic flow. Three traffic phases (free, synchronized, and jammed) have been founded [1–7]. Parallel to the investigation of empirical data various traffic models have been put forward for the reproduction of *reality*. These approaches can be divided into microscopic [8,9] and macroscopic ones [10]. Note, the focus is on microscopic modelling in this paper. Actual traffic models mostly focus on the explanation of synchronized flow (considerable high flux compared to the velocity without clear density-flux relation) [20,21]. However, the results are often unsatisfactory since synchronized flow is usually established as a transient process due to the strong phase separation between free flow and the jammed state. This leads to unrealistic strong fluctuations. In this paper a CA model is presented which overcomes the problem of strong fluctuations in synchronized flow. As a further goal it is demonstrated that the presented approach is cabable to reproduce the pinch effect often occuring in the vicinity of on- off-ramps.

In the following a sketch of the investigated CA traffic model is given. For a more detailed describtion it is referred to [11]. All values are assumed to be integers in the following, also for the case of arithmetic results only the integer part is used. The dynamics of the model is bounded within a limited acceleration capability (a) and its deceleration-counterpart (D). These do not stand for the extremal abilities declared by motor companies but for the ones frequently adapted in a normal traffic situation. The acceleration part is just inherited from existing models while the deceleration part contains some new concepts. Note, that most of the CA approaches for traffic flow allow an unrealistic high (infinite) deceleration since this is the simplest method to establish collision free driving. Our aim here is not on collision freeness itself but on the collective behavior of vehicles equipped with ordinary braking maneuvers. In this sense, even the collision freeness of daily highway traffic can be viewed as a collective phenomenon emerging from moderate driving which should be distinguished from an emergency. Beside the two capabilities themselves, an asymmetry between them $(D > a)$ is also introduced in order to reflect the wider spectrum of braking action.

First of all an inequality is given in terms of a CA expression with the scope on successive braking maneuvers.

$$x_n^t + \Delta_n + \sum_{i=1}^{\tau_f(c_n^{t+1})} \left[c_n^{t+1} - Di\right] \le x_{n+1}^t + \sum_{i=1}^{\tau_l(v_{n+1}^t)} \left[v_{n+1}^t - Di\right] \tag{1}$$

x_n^t (v_n^t) is the location (velocity) of the n-th vehicle at time t and the increased index represents the vehicle in front. The reduction of the summation index by 1 in LHS of (1) can be related to the averaged response time of a driver. The remaining variables (τ, c, and Δ) are dynamical. The equation can be understood naively the way that a gap of Δ is guaranteed during the next τ time steps if a velocity c is assigned to the n-th vehicle at the next time step. Note, if Δ is chosen to be 1 (minimal gap) and $\tau(x) = x/D$, then the inequality corresponds to the case that a full break of the predeccesor is supposed for safe driving. In the next paragraph the driver's opinion on the traffic condition is divided into *optimistic* driving or not.

The typical driving strategy for the most traffic models is to accelerate until the maximum velocity is reached, assumed that there is enough space left in front of a vehicle. In detail this means that for every velocity a minimum "safety gap" exists and the drivers are not allowed to fall below. However, such a behavior seems contrary to the daily driving experience. Consequently, it should temporarily be allowed to take on a small time headway, i.e., to disobey strict safety conditions, if the situation is optimistic. In the following we assume that i) the driver's opinion is determined according to the locally detectable traffic situation and ii) this opinion governs the short term driving strategy. For this sake, a binary parameter γ is introduced which describes if a situation is *optimistic* 0 or not 1. A situation is assumed to be optimistic if the velocity of a vehicle is at least equal to $v_{max} - 1$ or if the velocities of the two preceding vehicles are successively larger. Thereby, γ governs the dynamical variables in (1) in the following way:

$$\tau_{\mathrm{f}}(v) = \gamma_n^t v/D + (1 - \gamma_n^t)\max\{0, \min\{v/D, t_{\mathrm{safe}}\} - 1\},$$
$$\tau_{\mathrm{l}}(v) = \gamma_n^t v/D + (1 - \gamma_n^t)\min\{v/D, t_{\mathrm{safe}}\},$$
$$\Delta_n^t = l + \gamma_n^t \max\{0, \min\{g_{\mathrm{add}}, v_n^t - g_{\mathrm{add}}\}\},$$

where t_{safe} is the upper bound of $\tau_{\mathrm{f}} - 1$ in the *optimistic state* and g_{add} is introduced as an additional security gap in the *defensive state*.

The update rules of the model can be written in the following form:

1. $p = \max\{p_{\mathrm{d}}, p_0 - v_n^t(p_0 - p_{\mathrm{d}})/v_{\mathrm{slow}}\}$,
2. $\tilde{c}_n^{t+1} = \max\{c_n^{t+1}| c_n^{t+1} \text{ satisfies (1) and (2)}\}$,
3. $\tilde{v}_n^{t+1} = \min\{v_{\mathrm{max}}, v_n^t + a, \max\{0, v_n^t - D, \tilde{c}_n^{t+1}\}\}$,
4. $v_n^{t+1} = \max\{0, v_n^t - D, \tilde{v}_n^{t+1} - \eta\}$
   where $\eta = 1$ if $\mathrm{rand}() < p$, or 0 otherwise,
5. $x_n^{t+1} = x_n^t + v_n^{t+1}$.

Herein, the stochastic parameter in step 1 is a generalization of the well known *slow to start* (sts) rule (see [17]). The sts rule is known to be an important ingredient in respect to the formation of congested traffic states. In [18] it is shown that a simple CA model with sts rule is capable to produce stripped jam patterns which can often be found in traffic flow. Note that the stochastic deceleration in step 4 is also limited by the braking capability D. The following model parameters, motivated by empirical facts, are used: $a = 1$, $D = 2$, $l = 5$, $t_{\mathrm{safe}} = 3$, $g_{\mathrm{add}} = 4$, $p_0 = 0.32$, $p_{\mathrm{d}} = 0.11$, $v_{\mathrm{slow}} = 5$, $v_{\mathrm{max}} = 20$. The length of one cell is chosen to be $\Delta x = 1.5$ meter and one time-step to $\Delta t = 1$ sec.

In Fig. 1 the traffic phases occurring under periodic boundary conditions are depicted. Three traffic states (free flow, synchronized traffic, jams) can be clearly identified in the fundamental diagram (FD) Fig. 1(a). The straight line with the positive slope corresponds to free flow. The synchronized states form a 2-dimensional region in the middle of the diagram while jammed vehicles produce the scattered points below. The flux and the velocity are averaged at a fixed position every minute while the density is obtained via the hydrodynamical relation $J = v\rho$. A typical spatiotemporal shape of each phase is shown in Fig. 1(b). Note that the so-called *universal* constants of traffic flow [4–6] as the jam velocity $v_{\mathrm{g}} \approx -15$ km/h and the flux out of a jam $q_{\mathrm{out}} \approx 1800$ veh./h are also reproduced by simply adjusting p_0 and p_{d}. For the issue of demonstrating the inner structure of synchronized flow a snapshot of a part of the road is presented in Fig. 1(c). As mentioned before the vehicles exhibit a feasible smooth profile.

The results for an open system with on-ramp are presented in Fig. 2. Thereby the focus is on the so-called *pinch effect* being the object of actual investigations [6, 7, 19]. The pinch effect describes the process of a local self-

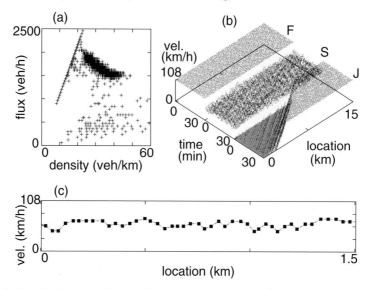

Fig. 1. Results for a road (periodic boundary conditions) consisting of 40.000 cells (60 km). A homogeneous distribution of standing vehicles is used as initial condition while the density is varied from $10 - 50$ veh./km.

compression in synchronized regions which leads to the formation of small narrow jams. These small jams evolve finally into a few wide jams through a merging process while moving upstream. This process can be seen in Fig. 2(a). Some small jams are first formed near the on-ramp out of synchronized flow. The small jams are combined into wide jams after several kilometers upstream. Empirically, it was found that this is the most frequent type of congested traffic near a bottleneck and thus named the *general pattern* (GP) [7]. This is captured more clearly in Fig. 2(b). The other known [7] types of congested traffic near bottlenecks are *localized synchronized flow patterns* (LSP), *widening synchronized flow patterns* (WSP), and *moving synchronized flow patterns* (MSP). It is stressed here that these states can also be reproduced in the presented model as shown in Figs 2(c,d,e). Furthermore, it should be noticed that the phase diagram in Fig. 2(f) is quite comparable to the one in [7, 19] and emphasizes that jams can not emerge directly out of free-flow.

In conclusion, a new CA traffic model focusing on mechanical restrictions realized by limited acceleration and braking capabilities is introduced. As a further element *human overreaction* is implemented in order to reflect the tendency for biased decisions in respect to the local traffic conditions. Moreover, it is shown that the model reproduces most empirical findings including the three known traffic phases, the so-called *pinch effect* and several types of congested traffic patterns as well as a small time-headway observed frequently

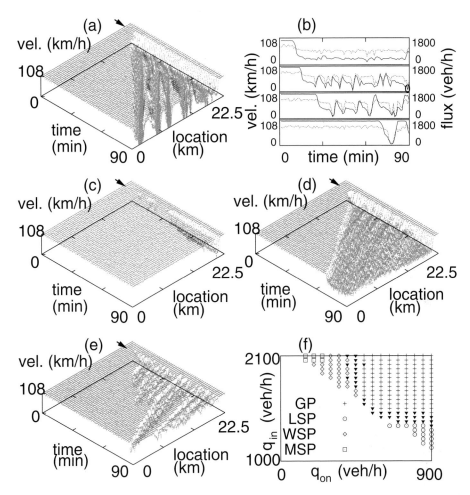

Fig. 2. The impact of an on-ramp at position 20.4 km (black arrow) consisting of a 0.6 km long merging road is demonstrated. The left (right) end of the road is positioned at −45 km (45 km). The influx at the left is kept constant during simulation. After q_{in} is supplied for 10 hours, the simulation time is set to 0 and then 8 minutes later the on-ramp injection q_{on} is turned on. Within the merging road the widest leading gap is selected and then a vehicle is inserted at the mid point. For its velocity 0.7 times that of the follower's is assigned. Below, a coupled number means (q_{in}, q_{on}). (a) Wide jam formation at $(1800, 700)$. (b) Pinch effect in (a): Each figure shows one minute averaged velocity (solid) and flux (dotted) at different location. Thus the starting velocity (flux) is v_{max} (1800). From the top, detectors are located at 20.4, 18.9, 15.9, and 5.4 km, respectively. (c) Localized synchronized flow at $(1300, 800)$. (d) Widening synchronized flow at $(2000, 250)$. (e) Moving synchronized flow at $(2000, 250)$. (f) Phase diagram of the congested traffic patterns.

especially in free flow. It should be remarked, since no artificial collision-free requirement is imposed on the model that collisions can occur near the on-ramp if the vehicle insertion is not treated carefully enough.

H.K.L. was supported by the Brain Korea 21 Program and thanks D. Kim for his critical reading. The authors are grateful to the BMBF within the project DAISY for financial support.

References

1. B. S. Kerner, in *Traffic and Granular Flow '01*, edited by M. Fukui, Y. Sugiyama, M. Schreckenberg, and D. Wolf, Springer, Heidelberg, 2003.
2. D. Helbing, Rev. Mod. Phys. **73**, 1067 (2001).
3. D. Chowdhury, L. Santen, and A. Schadschneider, Phys. Rep. **329**, 199 (2000).
4. I. Treiterer and J. A. Myers, in Proceedings of the 6th International Symposium on Transportation and Traffic Theory, edited by D. J. Buckley Elsevier, New York, 1974.
5. B. S. Kerner H. Rehborn, Phys. Rev. E **53**, R4275 (1996).
6. B. S. Kerner, Phys. Rev. Lett **81**, 3797 (1998).
7. B. S. Kerner, Phys. Rev. E **65**, 046138 (2002).
8. K. Nagel and M. Schreckenberg, J. Physique I **2**, 2221 (1992).
9. M. Bando, K. Hasebe, A. Nakayama, A. Shibata, and Y. Sugiyama, Phys. Rev. E **51**, 1035 (1995).
10. B. S. Kerner and P. Konhäuser, Phys. Rev. E **48**, R2335 (1993).
11. H. K. Lee, R. Barlović, M. Schreckenberg, and D. Kim, Phys. Rev. Lett **92**, 238702 (2004).
12. P. Berg and A. W. Woods, Phys. Rev. E **61**, 1056 (2000).
13. H. K. Lee, H.-W. Lee, and D. Kim, Phys. Rev. E **64**, 056126 (2001).
14. D. Helbing, A. Hennecke, V. Shvetsov, and M. Treiber, Math. Comput. Model. **35**, 517 (2002).
15. H. K. Lee, H.-W. Lee, and D. Kim, Phys. Rev. E **69**, 018116 (2004).
16. W. Knospe, L. Santen, A. Schadschneider, and M. Schreckenberg, J. Phys. A **33**, L477 (2000).
17. R. Barlović, L. Santen, A. Schadschneider, and M. Schreckenberg, Eur. Phys. J. B. **5**, 793, 1998.
18. R. Barlović, T. Huisinga, A. Schadschneider, and M. Schreckenberg, Phys. Rev. E **66**, 046113 (2002).
19. B. S. Kerner and S. L. Klenov, J. Phys. A: Math. Gen. **35**, L31 (2002); B. S. Kerner, S. L. Klenov, and D. E. Wolf, J. Phys. A: Math. Gen. **35**, 9971 (2002).
20. W. Knospe, L. Santen, A. Schadschneider, and M. Schreckenberg, Phys. Rev. E. **70**, 016115 (2004).
21. N. Eissfeldt and P. Wagner, Eur. Phys. J. B **33**, 121 (2003).

Traffic Networks and Driver's Behaviour

Intersection Modeling, Application to Macroscopic Network Traffic Flow Models and Traffic Management

J.P. Lebacque[1,2]

[1] INRETS – GRETIA
2, Avenue du Général Malleret-Joinville, F 94114 Arcueil – France
lebacque@inrets.fr
[2] ENPC – DR
6-8, Avenue Blaise Pascal, Cité Descartes, Champs-sur-Marne,
F 77455 Marnes-la-Vallée – France

Summary. The object of the paper is to analyze intersection modeling in the context of macroscopic traffic flow models. The paper begins with a brief review of classical boundary conditions of the Dubois-LeFloch and the Bardos-Nédélec-LeRoux type, and their relation to the concepts of local traffic supply and demand. It will be shown that the local traffic supply and demand concept extends and simplifies these classical approaches. The resulting constraints on phenomenological intersection models will be discussed. Several examples of intersection models are deduced. Some of these recapture earlier models; others are specifically designed for congested traffic conditions and take into account the bounds on car acceleration. The last part of the paper is devoted to network modeling and to applications to network traffic management.
Keywords: intersection modeling, boundary conditions, LWR model, ramp metering, speed control

1 Introduction

Macroscopic modeling of traffic flow on networks is a difficult problem. Many solutions have been proposed in the past, based on heuristic approaches, resulting in numerous discretized traffic flow models: FREFLOW [22], METANET [20], METACOR [6], NETCELL [4, 18] for instance.

From a theoretical point of view the problem is unsolved. At the core of the problem lies the difficulty of specifying proper boundary conditions on links for solutions of hyperbolic equations such as traffic flow equations. Further, link boundary conditions must be combined at intersections in order to yield intersection models [17, 1].

Mathematical textbooks on systems of conservation equations ususaly skip the subject of boundary conditions, with the exception of Kröner [9] who devotes a chapter to the topic of boundary conditions. Two pionneering works addressed the problem of boundary conditions for hyperbolic systems of conservation laws: Bardos-LeRoux-Nédélec [2] used the viscosity method and Dubois-LeFloch [5] the Riemann problem approach. The Bardos-LeRoux-Nédélec (BLN) and Dubois-LeFloch (DL) boundary conditions are known to be equivalent in the scalar case (Kröner [9]), which applies to the LWR (Lighthill-Whitham-Richards [19, 23]) model. Both yield existence and unicity of entropy solutions under mild hypotheses in the scalar 1-D case.

In the non-scalar case, which applies to 2nd order traffic flow models, the theory of boundary conditions is still tentative. In the field of transportation little research effort has been devoted to the topic of boundary conditions, mainly for the LWR model [12, 17].

Mathematicians have been starting to show some interest in the problem (papers by Holden-Risebro [7] Coclite-Piccoli [3] Klar and Herty [8] for instance. The resulting intersection models still lack realism.

In the field of traffic flow theory some intersection models have been proposed in the past ([11, 12, 4, 17, 10]). The difficulty of specifying boundary conditions has limited the scope of these earlier efforts.

The first part of the paper aims at proving the equivalence between the BLN (Bardos-LeRoux-Nédélec) boundary conditions and the supply-demand boundary condition introduced in [12]. The classical intersection models of Holden-Risebro and Coclite-Piccoli are reinterpreted in view of this equivalence. Reference here is [15].

Several models can be deduced from this first step: pointwise intersection models without inner state description, pointwise intersection models with inner state description, intersection models taking into account the physical extension of the intersection node.

The Holden-Risebro and Coclite-Piccoli models, already mentionned, fall into the category of pointwise intersections without inner state description. Let us also mention the merge model in [4] and the model [12]. At the opposite end of the spectrum, we find models with physical extension for discretized macroscopic traffic flow models (STRADA [1], SSMT [11], included as its urban part into METACOR [6]).

Finally, pointwise intersection models with inner state description, that is to say with a simple description of vehicle dynamics inside the intersection (total number of vehicles, capacity, movements) combine the simplicity of pointwise models and the precision of models with physical extension.

It is possible to add another behavioral element, in order to account for congestion downstream of intersections ([13, 14]): the bounded acceleration of traffic flow. Thus it becomes possible to account for phenomena such as histeresis, capacity drops. The last part of the paper will illustrate such effects.

2 Boundary Conditions

2.1 The LWR (Lighthill-Whitham-Richards) Model

This model [19, 23] is simple, robust, and describes correctly many traffic phenomena and situations. It admits analytical solutions for most traffic situations and is presently the only traffic flow model for which an exhaustive analysis of boundary conditions is possible. Thus most of the paper is centered on the use of this model.

Let us recall the prominent features of this model. On an axis, with position x and time t the LWR model is described by the following conservation equation:

$$\frac{\partial K}{\partial t} + \frac{\partial}{\partial x} Q_e(K, x) = 0 \tag{1}$$

with:

- Q : the flow
- K : the density
- V : the speed
- $Q_e(K, x)$: the equilibrium flow (fundamental diagram)
- $V_e(K, x)$: the equilibrium speed: $Q_e(K, x) \overset{def}{=} KV_e(K, x)$.

Equilibrium flow Q_e and speed V_e are described by Fig. 1: Note that the

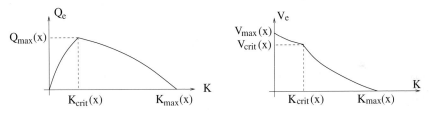

Fig. 1. Fundamental diagram: equilibrium flow and speed.

parameters of these relationships are: K_{max} the maximum density, K_{crit} the critical density, V_{max} the free speed, V_{crit} the critical speed, Q_{max} the maximum flow (capacity).

From the fundamental diagram two equilibrium function can be constructed, the *equilibrium supply* Σ_e and *demand* Δ_e functions. The construction is illustrated by Fig. 2. At a given point, the *local supply and demand* are defined as:

$$\Sigma(x, t) = \Sigma_e(K(x+, t), x+)$$
$$\Delta(x, t) = \Delta_e(K(x-, t), x-) \tag{2}$$

These quantities can be interpreted as the *greatest possible inflow* resp. *outflow* at any given location x. The symbols $+, -$ in equation (2) represent right-hand

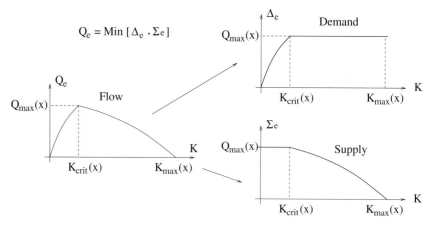

Fig. 2. Fundamental diagram: equilibrium supply and demand functions.

side (resp. left-hand side) limits. This precision is necessary: all functions are piecewise continuous only.

Flow must be less than both supply and demand. The *entropy* solution of (1) is locally flow maximizing [12]. Thus the entropy solution can be characterized by:

$$Q(x,t) = \text{Min} \left[\Sigma(x,t), \Delta(x,t) \right] \tag{3}$$

a formula which will be refered to as the **min formula** in the rest of the paper.

2.2 Boundary Conditions for (1)

BLN (Bardos-LeRoux-Nédélec) Boundary Conditions

The BLN boundary data is the prescription on the boundary of a *density-like* quantity A. More specifically, Bardos-LeRoux-Nédélec prove that (1) (and more generally scalar conservation equations) admits a unique solution on a link $\mathcal{D} \overset{def}{=} [a,b]$ given an initial density on the link $\mathcal{D}$ and such that the density at the boundary $\partial \mathcal{D} \overset{def}{=} \{a,b\}$ relates at all positive times to the boundary data A according to:

$$\{\text{sgn}\left[K(c,t) - \kappa\right] - \text{sgn}\left[A(c,t) - \kappa\right]\} \left[Q_e(K(c,t),c) - Q_e(\kappa)\right].n(c) \geq 0 \tag{4}$$
$$\forall c \in \partial \mathcal{D} = \{a,b\} \quad \text{and} \quad \forall \kappa \geq 0$$

The symbol sgn represents the sign function, $n(c)$ is the normal to the domain boundary at c, that is $n(a) = -1$ and $n(b) = 1$ since (1) is 1-D (Fig. 3). This boundary condition (4) is obtained by analyzing viscosity solutions of (1), and generalizing standard boundary conditions of Dirichlet type for parabolic equations. The reader is refered to [9] chapter 6.

$$D$$

A(a,t) .... A(b,t)

n(a) ⟵ • ——————————————— • ⟶ **n(b)**

a b

Fig. 3. BLN data for a link $\mathcal{D} = [a, b]$.

The Dubois-Lefloch boundary conditions

The more recent Dubois-Lefloch boundary condition specification, based on the Riemann problem approach [5], is known to be equivalent to the BLN approach in the 1-D scalar case, which makes the equivalence hold true for the LWR model. An upstream boundary data (for extremity a of link in Fig. 3) in the Dubois-Lefloch sense is the solution of a Riemann problem with *origin* a and right-hand side initial condition equal to the link density at point a. This idea is very close to the supply-demand method.

The Supply-Demand Boundary Conditions

The method has been introduced in [12]. Upstream boundary data is upstream

Fig. 4. BLN data for a link $\mathcal{D} = [a, b]$.

demand $\Delta_u(t)$, the downstream boundary data is the downstream supply $\Sigma_d(t)$. Given the link supply and demand:

$$\text{Link supply} \;:\; \Sigma(a, t) = \Sigma_e(K(a+, t), a)$$
$$\text{Link demand} : \Delta(b, t) = \Delta_e(K(b-, t), b)$$

we apply the *min formula* (3) to derive the link inflow $Q(a, t)$ and link outflow $Q(b, t)$:

$$Q(a, t) = Min\,[\Delta_u(t), \Sigma(a, t)]$$
$$Q(b, t) = Min\,[\Delta(b, t), \Sigma_d(t)] \tag{5}$$

The density at the boundaries is given by:

$$K(a, t+) = \begin{bmatrix} \Delta_e^{-1}(\Delta_u(t)) & \text{if } \Delta_u(t) < \Sigma(a, t) \\ K(a, t) & \text{if } \Delta_u(t) \geq \Sigma(a, t) \end{bmatrix}$$

$$K(b, t+) = \begin{bmatrix} \Sigma_e^{-1}(\Sigma_d(t)) & \text{if } \Sigma_d(t) < \Delta(b, t) \\ K(b, t) & \text{if } \Sigma_d(t) \geq \Delta(b, t) \end{bmatrix} \tag{6}$$

Thus upstream demand and downstream supply determine completely the traffic inside the link.

It is not possible to *prescribe* demand. Indeed, physically, if the link supply is insufficient, i.e.

$$\Delta_u(t) > \Sigma(a, t)$$

a queue forms upstream of the link at $t+$, implying that the upstream demand switches to maximum flow:

$$\Delta_u(t+) = Q_{max}(a)$$

Thus really only the *cumulative demand* can be imposed in a meaningful fashion. A symmetric observation can be made for downstream boundary conditions.

2.3 Equivalence of BLN and Supply/Demand Boundary Conditions.

The main result can be stated as follows: in the case of left-hand-side (upstream) boundary conditions, (4) is equivalent to an upstream demand boundary condition. More precisely *the BLN boundary data $A(a, t)$ is equivalent to a demand boundary data $\Delta_e(A(a, t), a)$.*

Symmetrically, in the case of right-hand-side (downstream) boundary conditions, *the BLN boundary data $A(b, t)$ is equivalent to a supply boundary data $\Sigma_e(A(b, t), b)$.*

Let us denote by A^* the *conjugate* of A with respect to Q_e: A^* must satisfy both $Q_e(A) = Q_e(A^*)$ and $A^* \neq A$. (4) is equivalent to (see [15] for a proof):

$$\begin{aligned}
A \leq K_{crit} &: K \in \{A\} \cup [A^*, K_{max}) \\
A \geq K_{crit} &: K \in [K_{crit}, K_{max})
\end{aligned} \tag{7}$$

Note that we skip the time dependency of variables: K means $K(a+, t)$ with a the upstream extremity of the link, and A stands for $A(a, t)$.

A cannot be prescribed: K is $\neq A$ in most cases. This fact is a consequence of information propagation in the LWR model, the velocity of which depends on the traffic state. Let us interpret the cases in which $K \neq A$.

- $A \leq K_{crit}$ and $K \in [A^*, K_{max})$:

$$\Sigma_e(K) = Q_e(K) \leq Q_e(A) = \Delta_e(A)$$

- $A \geq K_{crit}$ and $K \in [K_{crit}, K_{max})$:

$$\Sigma_e(K) = Q_e(K) \leq Q_{max} = \Delta_e(A)$$

All these cases are characterized by: $\Sigma_e(K) \leq \Delta_e(A)$ (*supply regime*). The case $A \leq K_{crit}$, $K = A$ corresponds to:

$$\Delta_e(A) = Q_e(A) \le Q_{max} = \Sigma_e(K)$$

which characterizes to a *demand regime*.

In other words, the boundary data A in the sense of BLN is equivalent to the boundary data $\Delta_e(A)$ in the supply-demand sense.

Let us note that in the case of downstream boundary conditions (i.e. $A = A(b,t)$, $K = K(b-,t)$), the following admissible values of K result:

$$A \le K_{crit} : K \in [0, K_{crit}]$$
$$A \ge K_{crit} : K \in \{A\} \cup [0, A^*]$$
(8)

which is equivalent to the boundary data $\Sigma_e(A)$ in the supply-demand sense.

3 Intersection Modeling.

3.1 Holden-Risebro and Coclite-Piccoli pointwise Intersection Models

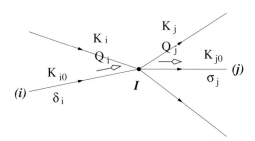

Fig. 5. Holden-Risebro and Coclite-Piccoli pointwise intersection.

Both aim at solving a generalized Riemann problem at the intersection. Initial data are the densities K_{i0}, K_{j0}, which are assumed uniform on semi-infinite upstream links $[i]$ and downstream links $[j]$. The main problem addressed by Holden and Risebro is: what are the densities K_i, K_j, and flows $Q_i = Q_e(K_i)$, $Q_j = Q_e(K_j)$, at the node.

In both approaches, the following constraints are applied to the K_is, K_js:

$$\left[\begin{array}{l} \left\{ \begin{array}{ll} K_i \in [K_{i,crit}, K_{max}] & \text{if } K_{i0} \ge K_{i,crit} \\ K_i \in \{K_{i0}\} \cup [K_{i0}^*, K_{i,crit}] & \text{if } K_{i0} \le K_{i,crit} \end{array} \right. \\ \\ \left\{ \begin{array}{ll} K_j \in [0, K_{j,crit}] & \text{if } K_{j0} \le K_{j,crit} \\ K_j \in \{K_{j0}\} \cup [0, K_{j0}^*] & \text{if } K_{j0} \ge K_{j,crit} \end{array} \right. \end{array} \right.$$
(9)

(9) expresses the BLN boundary conditions between K_{i0} (boundary data in the BLN sense) and K_i, and between K_{j0} (boundary data in the BLN sense)

and K_j respectively. The reader will observe the similitude with (7). In the supply-demand framework, upstream boundary data is demand $\Delta_e(K_{i0})$ and downstream boundary data is supply $\Sigma_e(K_{j0})$. Thus (9) is equivalent to the simpler condition:

$$
\left[
\begin{array}{l}
Q_i \leq \Delta_e(K_{i0}) \overset{def}{=} \delta_i \\[4pt]
\text{and} \quad \left\{ \begin{array}{ll} K_i = \Delta_e^{-1}(Q_i) & \text{if } Q_i < \delta_i \\ K_i = K_{i0} & \text{if } Q_i = \delta_i \end{array} \right. \\[16pt]
Q_j \leq \Sigma_e(K_{j0}) \overset{def}{=} \sigma_j \\[4pt]
\text{and} \quad \left\{ \begin{array}{ll} K_j = \Sigma_e^{-1}(Q_j) & \text{if } Q_j < \sigma_j \\ K_j = K_{j0} & \text{if } Q_j = \sigma_j \end{array} \right.
\end{array}
\right.
\tag{10}
$$

If states $(i) \overset{def}{=} (K_i, Q_i)$ and $(j) \overset{def}{=} (K_j, Q_j)$ satisfy the above conditions (9) or (10), it is clear that on each link $[i]$ or $[j]$, the $K_{i0} - K_i$ resp. $K_j - K_{j0}$ discontinuities propagate into the right direction, with < 0 resp. > 0 speed. Thus the basic variables of the generalized Riemann problem for a pointwise

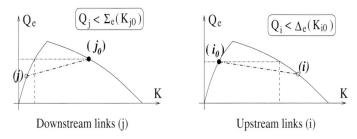

Fig. 6. Admissible node states, for downstream and upstream links

intersection are the node inflows Q_i and outflows Q_j. The constraints that apply to these variables are posyivity constraints, conservation constraints, and constraints resulting from (10):

$$
\left[
\begin{array}{ll}
0 \leq Q_i \leq \delta_i & \forall i \\
0 \leq Q_j \leq \sigma_j & \forall j \\
\sum_i Q_i = \sum_j Q_j
\end{array}
\right.
\tag{11}
$$

The Holden-Risebro and Coclite-Piccoli approaches differ only in how they resolve the underdeterminacy of (11).

Holden and Risebro propose to maximize a concave function of the Q_is, Q_js. A similar idea was developed from a different basis in [17]. For instance in the case of a merge, the following function can be maximized:

$$
\sum_i \left(Q_i - \frac{Q_i^2}{2\, p_i\, Q_{max}} \right)
\tag{12}
$$

with p_i some priority coefficients and Q_{max} the downstream capacity. The resulting model, discretized, can be shown [17] to be equivalent to Daganzo's merge model [4].

Coclite and Piccoli propose to maximize the total throughflow of the node, $\sum_i Q_i = \sum_j Q_j$, and add node outflow constraints that take into account proportions of turning movements in the intersection. The coefficients are α_{ij} the prportion of users of link $[i]$ that chose downstream link $[j]$. The resulting model can be stated as:

$$
\text{Max } \sum_i Q_i
$$

$$
\left|
\begin{array}{ll}
0 \leq Q_i \leq \delta_i & \forall i \\
0 \leq Q_j \leq \sigma_j & \forall j \\
\sum_i Q_i = \sum_j Q_j & \\
Q_j = \sum_i \alpha_{ij} Q_i & \forall j
\end{array}
\right.
\tag{13}
$$

3.2 Intersection Models with Physical Extension

Such models allow a detailed description of the dynamics of traffic in intersections. They have been designed for discretized macroscopic traffic flow models.

Exchange Zone Models

This model was introduced as part of the STRADA project [1]. The idea of an exchange zone is illustrated by the following figure: An exchange zone is

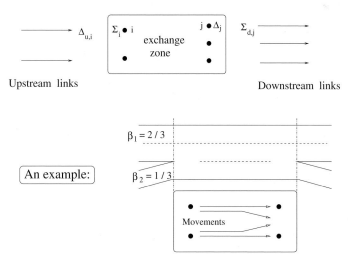

Fig. 7. Exchange zone: principle, example.

like a cell, but it is endowed with several entry points i, exit points j, and traffic inside the zone is disaggregated according to entry and exit points. More specifically, the following traffic state variables need be defined in an exchange zone:

- N: total number of vehicles
- NI_i: number of vehicles having entered the zone through entry point i
- NO_j: number of vehicles bound to exit the zone through exit point j
- N_{ij}: number of vehicles having entered the zone through entry point i and bound to exit it through exit point j

The following defintion identities hold:

$$NO_j = \sum_i N_{ij}, NI_i = \sum_j N_{ij}, N = \sum_j NO_j = \sum_i NI_i = \sum_{ij} N_{ij}$$

The dynamics of the zone result from the following items:

- A global zone fundamental diagram $Q_e(N)$ is given for N, yielding total zone supply $\Delta_e(N)$ and total zone demand $\Sigma_e(N)$ by the usual procedure (Fig. 2).
- Rules for split of total zone supply and demand among upstream and downstream links are:

$$\begin{cases} \Sigma_i = \beta_i \Sigma_e(N) \\ \Delta_j = \frac{NO_j}{N} \Delta_e(N) \end{cases}$$

The supply split model is a standard FIFO model (partial demand proportional to traffic composition). The linear supply split model tells us that partial supply should be proportional to the number of lanes available, as suggested by Fig. 7.

If we consider a downstream link $[i]$ (or a zone) with supply $\Sigma_{d,j}$, the zone outflow through exit point j during a time-step is given by:

$$Q_j = \text{Min}\,(\Delta_j, \Sigma_{d,j}) \tag{14}$$

Similarly the zone inflow through entry point i is given by

$$Q_i = \text{Min}\,(\Delta_{u,i}, \Sigma_i) \tag{15}$$

if the demand of link $[i]$ is $\Delta_{u,i}$.

Formulae (14) and (15) generalize the min formula. If applied with conservation equations for state variables N, NO_j, NI_i and N_{ij}, during each time-step, they yield the equation of traffic dynamics in the zone.

The SSMT Concept

The SSMT concept was introduced in [11] and resulted in a package for urban traffic. This package was included later on in the METACOR model as its urban submodel [6].

SSMT is a first order discretization of the LWR model, based on a Godunov-like discretization of the LWR model based on Osher's formula [21]. The idea of the intersection model in SSMT is that of overlapping discretization cells (Fig. 8). Thus vehicles belong to several cells. One of the difficulties of the

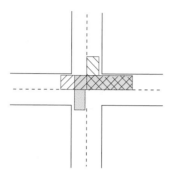

Fig. 8. The SSMT intersection model.

model is how to assign the right quantities of vehicles to each cell. Provision is made in the model to account for conflicts between opposing movements. The common supply of the three overlapping cells is determined by the geometry of the cells, i.e. their overlap ratio.

3.3 Pointwise Intersections with Internal State Dynamics

Basic Concepts

This intermediate model has been recently introduced in [15] and is obtained by neglecting the physical extension of an exchange zone and simplying its internal dynamics. The basic idea can be described as follows, on the simple case of a merge (a single downstream link).

- N is the number of vehicles contained in the intersection node.
- The dynamics of traffic in the node results from a fundamental diagram, resulting in a global node supply $\Sigma_e(N)$ and demand $\Delta_e(N)$.
- The global node supply $\Sigma_e(N)$ is split between upstream links [i] according to a *linear split model*:

$$\Sigma_i \stackrel{def}{=} \beta_i \Sigma_e(N) \tag{16}$$

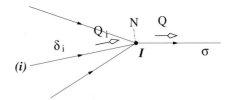

Fig. 9. Merge.

- Given the demands of upstream links $[i]$, δ_i and the supply of the downstream link, σ, the intersection dynamics equations follow (applying the Min formula (3) and the conservation of vehicles):

$$
\begin{cases}
Q_i = \text{Min } [\delta_i, \Sigma_i] \\
Q = \text{Min } [\Delta_e(N), \sigma] \\
\dot{N} = \sum_i Q_i - Q
\end{cases}
\tag{17}
$$

Congestion in Merges

It is a known experimental fact that congestion in merges occurs not only upstream of the intersection, but also in the intersection node and *downstream* of the intersection.

In order to account for this fact, we include two elements into the pointwise merge model (17).

1. The first concerns the supply split model. We assume that:

$$
\beta \overset{def}{=} \sum_i \beta_i > 1
\tag{18}
$$

a frequent situation in practice. Thus N can increase and eventually reach N_{max} (node storage capacity), since

$$
\sum_i Q_i = \sum_i \text{Min } (\beta_i \Sigma_e(N), \delta_i) > \Sigma_e(N)
$$

becomes possible. N cannot exceed N_{max}: $\sum_i Q_i < \beta \Sigma_e(N)$ and $\Sigma_e(N)$ vanishes as N reaches N_{max}.

2. The second concerns demand. We replace a regular demand function such as depicted on Fig. 2 by a modified demand function $\Delta_e^r(N)$. This modified demand function accounts, at al level of approximation of the order of 100 meters, for the fact that the acceleration of vehicles is bounded [13, 14]. For congested traffic conditions ($N > N_{crit}$):

$$
\Delta_e^r(N) < \Delta_e(N) = Q_{max}
$$

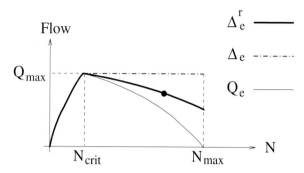

Fig. 10. Congested merge: demand function and modified demand function.

which expresses that vehicles accelerating out of the node do not reach maximum flow.

$\Delta_e^r(N)$ has no analytical form, but can be shown to be the solution in q of the following equation:

$$\left[\begin{array}{l} q - Q_e(N) = \xi(N)\left(\Delta_e^{-1}(q) - N\right) \\ \quad \text{with } \xi(N) = \left(Q_e(N) - Q_{max}\right)/N \end{array} \right.$$

4 Applications to Traffic Control

We consider applications of the pointwise merge model of subsection 3.3.

4.1 Node Equilibrium. Recovery Flow, Hysteresis

For a node such as depicted in Fig. 9, with dynamics (17), node equilibrium is given by $\dot{N} = 0$, i.e.:

$$\sum_i \text{Min} \left(\delta_i, \beta_i \Sigma_e(N)\right) = \text{Min} \left(\Delta_e^r(N), \sigma\right) \tag{19}$$

Assuming that $\sigma = Q_{max}$ and that the δ_is are large (congested merge), the following equation results:

$$\beta \Sigma_e(N) = \Delta_e^r(N) \tag{20}$$

(with $\beta > 1$). This equation admits a unique solution N_r, as shown by Fig. 11.

N_r is called the *recovery state* of the node, and $q_r \overset{def}{=} Q_e(N_r)$ its *recovery flow*. This equilibrium state is stable with respect to the node dynamics (20).

From now on, node capacity will be denoted q_{max}, and $Q_{max} \overset{def}{=} \beta q_{max}$ will denote the total capacity of upstream links (Fig. 11).

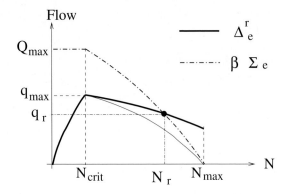

Fig. 11. Congested merge: demand function and modified demand function.

If the total node inflow exceeds at some time the node capacity q_{max}, N increases, reaches N_r and node capacity drops to q_r. The node can only recover its full capacity q_{max} if N drops below N_{crit}, that is to say the node inflow drops below q_r (i.e. $\sum_i \delta_i < q_r$).

The figure 12 illustrates this behaviour:

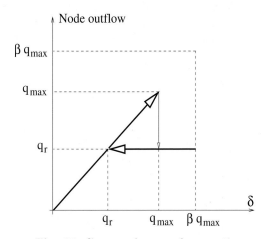

Fig. 12. Congested merge: hysteresis.

4.2 Speed Control

Let us consider a node with a single upstream link (Fig. 13). It represents the interface between a motorway (the upstream link, with capacity Q_{max}) and a urban area (the downstream link, with capacity q_{max}). Node dynamics are given by:

Motorway   Urban area

Fig. 13. Motorway - Urban area interface.

$$\dot{N} = \beta\Sigma_e(N) - \Delta_e^r(N)$$

with $\beta > 1$. The capacity of the node is q_{max} and its recovery flow is q_r. In order to prevent a capacity drop of the node from q_{max} to q_r, it is conceivable to apply speed control measures to the motorway. If VMSs (variable message signs) are spaced closely enough (every 500 meters or so), users can be assumed to perceive messages *instantaneously*. Thus changes in speed apply at *constant density* [16]. If current density at a point is K, in order to keep the flow below the q_{max} level it is sufficient to apply a speed limit u:

$$u = q_{max}/K \qquad (21)$$

The following figures illustrate the strategy, applied to a stretch of 15.5 km of motorway. The downstream half of the motorway is submitted to strategy (21). The motorway inflow presents a peak of demand during 15 minutes out of 1h15. The demand peak exceeds the node capacity. The simulation is based on the simplified two-phase model (LWR plus bounded acceleration) [14].

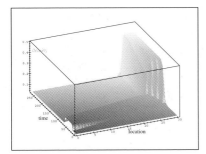
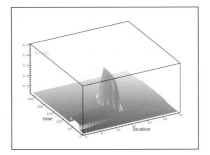

Fig. 14. density vs time x location, without and with speed control.

The following Fig. 15 shows that all users benefit from the speed control. The cumulative outflow with speed control is greater than the cumulative outflow without speed control.

4.3 Ramp Metering

The principle is the same but with two upstream links instead of one. The object of ramp metering is to keep total node inflow below its outflow capacity

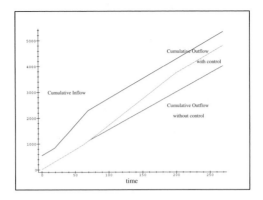

Fig. 15. Cumulative flows vs time: inflow, outflow with and without control.

q_{max}, in order to prevent its capacity drop tp q_r. Actually the well-known local controller ALINEA acts directly on the ration N/N_{max} by keeping it below critical occupancy N_{crit}/N_{max}. If we denote by [1] the onramp and [2] the motorway upstream of the merge, the onramp outflow q_1 must be bounded by $q_{max} - q_2$, with q_2 the flow of the motorway into the merge ($q_2 = \text{Min} (\delta_2, \beta_2 \Sigma_e(N))$). Not all users benefit from ramp metering. Some users on the on ramp may loose during queue formation (nevertheless they are liable to catch up the delay downstream of the node, [16]).

Figure 16 illustrates the response of the system to a simultaneous demand peak on both links (onramp and motorway). In Fig. 16, r_i denotes $\beta_i q_r$, and Q_i denotes the maximum of the peak demand, q_i the off-peak demand on each link [i].

5 Conclusion

The object of the paper was to show that:

- local supply and demand concepts provide the natural framework for analyzing boundary conditions and for expressing intersection models in the LWR model,
- intersection models are necessary for modeling traffic flow on networks and for designing control strategies, of which they constitute an essential component.

Further research includes specifying a greater variety of intersection models, more realistic movement and conflict description, disaggregation of traffic according to destination, user and/or vehicle type.

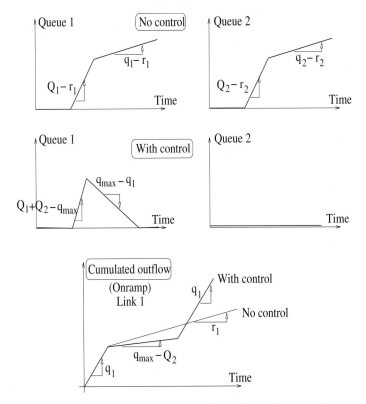

Fig. 16. Ramp meterig: queue lengths and cumulated outflows.

References

1. Buisson, C., Lebacque, J.P., Lesort, J.B. *1. Macroscopic modelling of traffic flow and assignment in mixed networks.* Proc. of the Berlin ICCCBE Conf. 1367-1374. (ed. Pahl, P.J., Werner, H.). 1995. *2. STRADA, a discretized macroscopic model of vehicular flow in complex networks based on the Godunov scheme.* Proc. of the CESA'96 IEEE Conference. 1996. *3. The STRADA model for dynamic assignment.* Proceedings of the 1996 ITS Conference. 1996.
2. Bardos, C., LeRoux, A., Nédélec, J.C. *First order quasilinear equations with boundary conditions.* Comm. partial Differ. Equations 4, 1017-1034, 1979.
3. Coclite, G.M., Piccoli, B. *Traffic flow on a road network.* Technical Report, SISSA, 2002.
4. Daganzo, C.F. *The cell transmission model 2: network traffic simulation.* Transportation Research 29B. 2: 79-93. 1995.
5. Dubois, F., LeFloch, P. *Boundary conditions for nonlinear hyperbolic systems of conservation laws.* J. Differential Equations 71, 93-122, 1988.
6. Elloumi, E., Haj-Salem, H., Papageorgiou, M. *METACOR, a macroscopic modelling tool for urban corridors.* TRISTAN II Int. Conf., Capri, 1994.
7. Holden, H., Risebro, N. *A mathematical model of traffic flow on a network of unidirectional roads.* SIAM J. Math. Anal. 4, 999-1017, 1995.

8. Klar, A., Herty, M. *Modeling of traffic flow networks.* Proceedings of the Autumn school: "Modélisation mathématique du trafic automobile" (Editors: J.P. Lebacque, J.P. Quadrat, M. Rascle). In press, 2003.

9. Kröner, D. *Numerical schemes for conservation laws.* Wiley-Teubner. 1997.

10. Jin, W.L., Zhang, H.M. *On the distribution schemes for determining flows through a merge.* In press, Transportation Research B, 2002.

11. Lebacque, J.P. *Semimacroscopic simulation of urban traffic.* Int. 84 Minneapolis Summer Conference. AMSE. 1984.

12. Lebacque, J.P. *The Godunov scheme and what it means for first order traffic flow models.* Transportation and traffic flow theory, Proceedings of the 13th ISTTT (J.B. Lesort Editor). Pergamon, 1996.

13. Lebacque, J.P. *A two-phase extension of the LWR model based on the boundedness of traffic acceleration.* Transportation and traffic flow theory in the 21st century, Proceedings of the 15th ISTTT (M.A.P. Taylor Editor). Pergamon, 2002.

14. Lebacque, J.P. *Two-phase extension of the LWR model: analytical solutins and applications.* TRB 2002 and TRR (in press). 2003.

15. Lebacque, J.P. *Problèmes de modélisation des réseaux par les modèles du premier ordre: conditions aux limites, intersections, affectation.* Proceedings of the Autumn school: "Modélisation mathématique du trafic automobile" (Editors: J.P. Lebacque, J.P. Quadrat, M. Rascle). In press, 2003.

16. Lebacque, J.P., Haj-Salem, H. *Speed limit control: a problem formulation and theoretical discussions.* Preprints of the TRISTAN IV Symposium, tome 2, 421-426, 2001.

17. Lebacque, J.P., Khoshyaran, M.M. *Macroscopic flow models.* Presented at the 6th Meeting of the EURO Working Group on Transportation. Published in "Transportation planning: the state of the art" (Editors: M. Patriksson, M. Labbé), 119-139. Kluwer Academic Press, 2002.

18. Lo, H.K. *A dynamic traffic assignment formulation that encapsulates the Cell-transmission model.* Transportation and traffic flow theory, Proceedings of the 14th ISTTT (A. Ceder Editor). Pergamon, 1999.

19. Lighthill, M.H., Whitham, G.B. *On kinematic waves II: A theory of traffic flow on long crowded roads.* Proc. Royal Soc. (Lond.) A 229: 317-345. 1955.

20. Messner, A., Papageorgiou, M. *METANET: a macroscopic modelling simulation for motorway networks.* Technische Universität München, 1990.

21. Osher, S. *Riemann solvers, the entropy condition, and difference approximations.* SIAM J. Num. Analysis 21: 217-235, 1984.

22. Payne, H.J. *Models of freeway traffic and control.* Simulation Council Proceedings 1: ch 6 , 1971.

23. Richards, P.I. *Shock-waves on the highway.* Opns. Res. 4: 42-51, 1956.

Granular Transport Policy
Is Cooperation Better than Optimization?

H.J. van Zuylen and H. Taale

Section Transportation and Planning – Delft University of Technology
P.O. Box 5048, 2600 GA Delft – The Netherlands
h.j.vanzuylen@ct.tudelft.nl, h.taale@citg.tudelft.nl

Summary. The present situation in transport policy is that many authorities have a responsibility for a part of the transport system and that each authority has its own goals and priorities. Central steering is only on main policy goals. Public transport and municipal, regional, provincial and national road authorities have their own policies, targets and instruments. The interaction between different policies themselves exists indirectly, but policies are 'granular'.

In this paper the granular transport policies are described as a game with many players. Integrated traffic control and traffic assignment problem are studied for a situation with two road authorities. The road authorities try to optimize their own objectives and the same is done by the travelers. This leads to a two-level three-player, multi-stage optimization problem with complete information. Game theory gives a suitable framework to analyze the problem and to find solutions for different situations, e.g. no cooperation, cooperation between the two authorities and a system optimum where all actors cooperate to minimize the total costs for all travellers.

In this paper two approached are used: an analytical one and an approach based on a simulation and assignment framework. Both approaches are described and used to study a simple example, for which the results are given and discussed.

The results show that separate or integrated anticipatory control gives better results than iterative reacting to the current situation. If one road authority takes the lead and anticipates the reactions of both the road users and the other road authority a sub-optimum is reached. The model calculations give evidence that cooperation of road authorities improves the utilization of the infrastructure and that a global optimization does not necessarily give a worse situation for one road authority.

Keywords: multi-level optimisation, game theory, urban ring roads, traffic control

1 Introduction

Urban accessibility is a big problem at present. A healthy city generates much transport demand. Allocating too much space for transport degrades the quality of the city, but on the other hand, a city without good accessibility may gradually loose its economic importance and becomes less attractive for living, recreation and shopping. The use of different infrastructure and transport services is necessary because a single mode or level of infrastructure is not sufficient to meet the challenges.

For road transport we can distinguish two levels of roads: local, urban roads and motorways, often organized as ring roads and / or diagonal connections. Further we can distinguish between private and public transport. For the sake of simplicity this paper is limited to passenger transport, with the assumption that trips are shifted between modes and routes, under the influence of travel costs.

Urban motorways, especially ring roads, have become a dominant element of the transport infrastructure of many cites. In most cases the municipal road authority tries to use the urban ring road as the main road infrastructure that has to attract as much traffic as possible [1]. This relieves the environmental stress on the lower order roads. Apart from the fact that the capacity of urban roads is often insufficient to handle all the traffic demand, the environmental capacity of most urban roads is rather low, since traffic noise and exhaust gasses make the quality of the environment around the roads below the standards that are applicable for residential areas. Furthermore, the urban intersections cannot handle large volumes of motorized traffic and offer a reasonable waiting time for crossing pedestrians at the same time. Problems also occur due to the conflicts between cars and vehicles of public transport using the same road space.

Rerouting of motorized traffic to ring roads is a suitable strategy to realize the goals of municipalities. The consequence is that the urban ring roads carry much intra-urban traffic [2]. Another consequence is that municipal authorities have a preference for motorways with as many entry and exit ramps as possible. Such road structure makes it possible to use the motorways as shortcuts for the urban trips. Furthermore, these connections between motorways and urban roads increase the value of the land and make it attractive for enterprises to allocate new settlements at such points. The consequence is that these new settlements attract and generate more traffic, which further increases the flow on the motorways.

The switch of private to public transport is slightly more complicated that the choice of routes. An improvement of public transport services often creates new demand and the transfer from car to public transport is less likely as the one from slow modes (bicycle, walking) to the bus. In this paper we shall still consider the transfer from travelers between routes and between modes as similar processes.

Both the municipal authority responsible for the urban roads and intersections and the motorway authority have their own objectives and will follow their own strategies to achieve these. Dynamic traffic management, e.g. traffic control, provides possible tools to realize the strategies. In the optimization of traffic control the authorities have to take into account how drivers react to the conditions created by the traffic control [3]. Drivers try to minimize the disutility of their trip, which might be expressed in the travel time. The most commonly used assumption is that drivers try to minimize their travel time and that equilibrium exists in which no driver can reduce his or her (perceived) travel time by choosing another route [4].

A third authority involved in urban accessibility is the public transport company. Their objective is to run a profitable enterprise, or at least, to have a limited loss, to be subsidized by the province, the municipality or the central government. Their objective is to have many passengers as possible; their instruments are to adjust the frequency of the service, the number of stops, the lines or to adapt the connection between different lines. Each change has consequences in terms of costs and number of passengers.

Several other authorities have influence on the urban accessibility. For instance, spatial planning has an influence on transport demand, while the availability of good access to a certain part of a city makes it attractive for settlements of enterprises that depend on transport. The policy of spatial urban development has, therefore, a considerable influence on the urban accessibility and vice versa. The analysis given in this paper will not give further attention to this relationship. It is assumed that the traffic demand is fixed, but the interaction between spatial planning policy and transport policy is obvious.

We see at least four different classes of actors involved in the optimization of urban accessibility: the motorway authority, the municipal road authority, the public transport company and the population of travelers. Each of them has an own objective and strategy, but each strategy has an impact on the conditions for the possibilities for the other actors to realize their objectives. If each actor tries to optimize his own objective without any form of cooperation, it is likely that a sub-optimal situation emerges.

Game theory [5, 6, 7] gives a suitable framework to analyze the problem and to find solutions for different situations, e.g. no cooperation, cooperation between the two authorities and a system optimum where all actors cooperate to maximize the weighted benefits for all actors.

In this paper we start with the generic approach to formulate the problems as a game. The approach is afterwards limited by omitting public transport. A two-level, three actor problem, formed by two road sub-networks with each its own control options and by the following actors: two road authorities trying to optimize their objective and the collective of road users travelling from an (urban) origin to and (urban) destination is studied in two ways: analytically and by using a simulation and assignment framework. Both approaches are described and used to study a simple example, for which the results are given and discussed. Finally, some conclusions are drawn.

2 Problem Statement

First of all, we have to understand that the problem described in the introduction is a rather complicated one. It is not the objective of this paper to solve real-life problems for realistic networks. The intention is to analyze the properties of the multi-actors optimization and to find out what the impact is of different strategies. This does not mean that the outcomes of the study have no meaning for practice: the structure of the simplified model is similar to the structure of existing urban situations and it is rather obvious that the results give evidence for the quality of certain policies of road administrators.

This paper analyzes the multi-actors problem in a simplified form, using several assumptions about the behavior of the actors, their objectives and strategies and about the characteristics of the road network and public transport. We will study a situation where the authorities optimize in a rational way the management of their infrastructure and services with simple objectives and that travelers are aware of and always choose the cheapest mode and route, in this case with lowest fare and shortest in travel time or perceived travel time. For this situation we will study the consequences of different optimization strategies for the performance of the traffic system.

The main question is, whether the independent optimization of objectives by the actors gives a worse performance than a cooperative optimization of the use of the infrastructure.

Let's assume that the urban road authority chooses the objective to minimize the total time spent on the urban roads. The road administrator of the ring road chooses to maximize the average speed on the road, which means that he minimizes the total time spent divided by the total distance traveled. The travelers are assumed to minimize their travel costs. Then four optimization problems can be formulated:

For public transport

$$\underset{S_p}{Max} \sum_{i \in R} (V_i f_i - C_i) \tag{1}$$

for the urban roads

$$\underset{S_1}{Min} \sum_{i \in U} T_i(V_i, S_i, S_j) \quad j \in M \tag{2}$$

for the motorways

$$\underset{S_2}{Min} \sum_{j \in M} T_i(V_i, S_i, S_j) \quad i \in U \tag{3}$$

and for the travelers

$$Min_V Z(\mathbf{V}, \mathbf{S}) \tag{4}$$

$$Z(\mathbf{V}, \mathbf{S}) = \sum_{i \in M \cup U \cup R} \int_0^{V_i} TC_i(S, x) dx \tag{5}$$

where T_i is the travel time on the road section i consisting of the free travel time, determined by distance and speed, and the delay at the signals, TC_i represents he total travel costs over I, i.e. the value of the time spent plus a possible fee. L_i is the length of link i, V_i is the volume or number of passengers, $T_i(S,x)$ is the travel time as a function of the volume x and traffic control S and the summation is over all links in the two sub networks: M the set of all links belonging to the motorway and U all links belonging to the urban roads. R is the set of all public transport lines and f_i is the fee for trip I while C_i represents the costs for the public transport operator for trip i.

Entry links to the motorway, which are controlled by the motorway authority, belong to M; exit links belong to the set U. It is clear that the decision variables for the municipal road manager are the signal settings $S_1 = \{S_i, i \in U\}$ and for the motorway authority $S_2 = \{S_j, j \in M\}$. The travellers have route choice and modal choice as decision variable, which can be represented as the volumes on all roads and number of passengers V. Expression (5) is a generalization of the expression derived by Beckman et al. [8] for the equilibrium condition where the traffic distribution satisfies Wardrop's principle that no traveler can improve his travel costs by choosing another mode or route.

For simplicity we omit here the constraints for the signal settings, with respect to minimum and maximum green times, conflicts, maximum allowed queue lengths etc. and on the volumes and OD-matrix. The complete mathematical framework for the optimization problem can be found e.g. in [3]. In the calculations of the next sections all necessary constraints are taken into account. If multimodal trips are to be included of the formulation of this problem, constraints have to be added for trips made partly by car, partly by public transport.

The travelers choosing their mode and routes will use public transport, urban roads and / or the highway, their travel choice depends on the travel costs for the different alternatives. The municipal and motorway road authorities take traffic control measures, which influence the travel times on different routes. The public transport operator chooses which routes to serve, and the frequency. The impact might be that travelers divert their trips. In this way traffic control can be used to influence the route choice and the utilization of the motorway and urban roads and fares and frequencies can be used to influence mode choice. The possibilities of the motorway authority to optimize the traffic situation for his roads depend on the traffic control chosen by the urban authority and vice versa. Also the choice options for the travelers depend on the signal settings, while the optimum signal settings depend on the flows, i.e. the route choice. This situation is characteristic for a four-actor optimization problem as studied by e.g. Fisk [5] and Chen and Wang [6] for two optimization processes (route choice and traffic control).

For simplicity and to demonstrate the properties of the problem, the analysis in the following sections is limited to one single mode, the car, and the two road administrators using traffic management measures as control instruments.

3 Game Theory Applied to Urban Traffic Management

The situation where N actors try to realize their own objectives, while they know that their decisions influence the possibilities of the other actors to realize their objectives, is analogous with a game for N persons. If two or more persons play a game with the intention to win, they choose actions that maximize their chances to win. Winning for one person does not necessarily mean a loss for the other players. All players try to maximize their own profit. The profit structure, the actions that the players can take and the consequences of the actions for the gains of each of the players, determine the character of the game. Since the action of one player has an influence on the chances of the other players to win, and vice versa, the players have to anticipate the possible actions of their partners.

If we assume that one player (the leader) knows how the other players (the follower) will respond to any decision he may make, we have a game that is known as a *Stackelberg game*. If, on the contrary, we assume that the players do not know each other's strategy, they will each optimize their own objective function assuming that the other player remains unchanged. The equilibrium that can be achieved in that situation is that no player can improve his objective function by changing his own decision without cooperation of the other player. This equilibrium is called the *Nash* equilibrium. The term 'Nash equilibrium' is used for systems with many non-cooperating players. If there are only two players, the term *Cournot* equilibrium is used.

In the case of the Nash equilibrium, when the road authorities have no a-priori knowledge of each other objectives, we can see the iterative equilibrium as follows:

1. the first road authority takes some traffic measure
2. the road users react and the effects in terms of flows on his infrastructure is observed by the second road authority
3. the second road authority takes measures to optimize the traffic conditions on his roads
4. road users react again and after a shift of routes a new user equilibrium emerges
5. the first road authority observes the shifted flows and adjusts his control measures etc.

The reality might be a little more complicated, because the control measures taken by a road authority are often based on the present traffic conditions. After the realization of the measures traffic flows may change in the whole network, which makes an adjustment of the control measures necessary on both levels of infrastructure. However, as shown in e.g. [3] it is possible to optimize control measures taking in advance already into account the reaction of the road users. It is not unlikely that, if control measures are taken without taking into account the reaction of the road users, that the state that emerges after the adaptation process is worse than the initial state before the optimization.

In the situation of an urban road infrastructure with two authorities with their own objectives and drivers that optimize their own routes, the following strategies can be distinguished:

1. The motorway and municipal authorities optimize their objectives independently, the drivers react and both authorities adjust their strategies if necessary, until equilibrium is reached. This is an example of the Nash equilibrium.
2. The motorway authority optimizes his objectives, while he anticipates the reaction of the drivers. The municipal road authority does the same.
3. The motorway authority knows the reaction of the drivers and the municipal road authority and optimizes his strategy, anticipating their response. This is a Stackelberg game with the motorway authority as leader.
4. Similar to 3, but now with the municipal road administrator as leader who optimizes his control anticipating the reaction of drivers and the motorway authority.
5. Both road administrators cooperate and define an objective for the whole road infrastructure, e.g. the minimization of the total delay. They optimize their traffic control while they anticipate the reaction of the drivers.
6. Based on the shared objective for the whole network, the drivers are forced to take the route that optimizes the network.

Of course, there are some alternatives in between, apart from these 6 alternatives. Most interesting is, however, the question whether cooperation brings much advantage compared to single actor optimization and whether the Stackelberg solution is much better than the Nash equilibrium. There is a lot of discussion in several countries about the need to have a single road authority responsible for the traffic in a region or that the existing road administrators can do their work individually, with the assumption that coordination emerges automatically due to the interaction between the different sub-networks.

4 The Analytical Approach

The optimization problem for a real life situation is rather complicated and the results might be difficult to interpret. Therefore a simple network is chosen for a first analysis. In Fig. 1 we see a representation of a motorway with a parallel urban road.

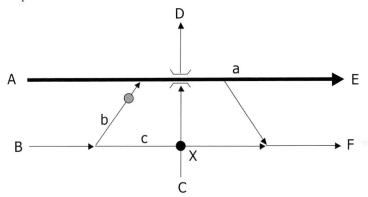

Fig. 1. Hypothetic network with urban roads and motorways.

Road a is the motorway, b the entry ramp. Road c is the parallel urban road with a controlled intersection X. Traffic from B to F can choose to follow the urban road c or to travel by the motorway, using on-ramp b and motorway a. The length of the two routes between B and F is supposed to be equal so that the travel speed and delays at the entry ramp and the intersection X determine the travel times. Thanks to this assumption it is also possible to convert the objective for the motorway by a minimization of time spent, just as the objective for the urban network. The travel speed S at the motorway is assumed to be

$$S_M = 80 - \frac{V_a}{100} \quad km/hr$$

(6)

with V_a the volume in vehicles per hour on road a. The delay at the entry ramp is determined by the ramp metering rate. We assume that this is given by

$$D_b = \frac{x_b^2}{2V_b(1-x_b)}$$

(7)

where V_b is the volume on link b and $x_b = V_b / A$ with A the number of vehicles per time unit that is admitted to the motorway. The delay at the on-ramp is the concern of the motorway authority. The delay at the controlled intersection is assumed to consist only of the random delay term of the Webster formula for fixed-time traffic control [9]

$$D_c = \frac{x_c^2}{2V_c(1-x_c)}$$

(8)

where V_c is the volume on road c and x_c is the degree of saturation on approach c of intersection X, given by

$$x_c = \frac{V_c C}{S_c t_c}$$

(9)

S_c is the saturation flow at the stop line at intersection X of approach c, t_c is the green time given to approach c and C is the cycle time. The delay functions (7) and (8) are only valid for values of the degrees of saturation x_b and x_c less than 1. In this analytical analysis we assume that no spill back occurs and the travel speeds and delay
s are only functions of the flow. In the next section we removed this limitation and include also time varying traffic demand and spill back.

The optimization problem has now two parameters: A for the motorway and t_c /C for the urban road. If t_c /C is replaced by λ, the decision space consists of a two dimensional space with axis A and λ. The feasible domain can be defined and for each point in the domain the value of the objective function can be calculated. We assume that the travel time over route c is a free travel time plus the delay at intersection X. Furthermore the total delay for the municipal network is assumed to be the delay for trips over c and the trips from C to D. For the delay for the trips from C to D the assumption is used that only the random delay term of Webster's delay formula is sufficient to describe the main effect of the traffic control at intersection X. The third optimization, the route choice, can be represented as a constraint for the optimization of the control by the two road authorities. The user equilibrium requires that for every feasible state the travel time over both routes are equal:

$$tt_c + D_c = \frac{L_a}{80 - \frac{V_a}{100}} + D_b$$

(10)

or

$$tt_c + \frac{V_c}{2(1-\frac{V_c}{S_c \lambda})(S_c \lambda)^2} = \frac{L_a}{80 - \frac{T_{AE}+V_b}{100}} + \frac{V_b}{2(1-\frac{V_b}{A})A^2}$$

(11)

with tt_c the free travel time over c, T_{AE} is the flow from A to E and L_a is the length of the motorway section a. With the constraint that $V_c + V_b = T_{BF}$, the flow between origin B to destination F, a solution $V_b(\lambda,A)$ and $V_c(\lambda,A)$ can be found for equation (11). For this case with delay functions (7) and (8) and speed function (6), these volumes can even be given in closed form so that the optimization problem is reduced to a two parameter, two level problem.

The objective function for the urban network is (omitting the terms that do not directly or indirectly depend on λ or V_c)

$$O_U = V_c tt_c + \frac{V_c^2}{2(1-\frac{V_c}{S_c\lambda})(S_c\lambda)^2} + \frac{V_{CD}^2}{2(1-\frac{V_{CD}}{S_{CD}\lambda'})(S_{CD}\lambda')}$$

(12)

where V_{CD} is the flow between C and D, S_{CD} the saturation flow of the approach used by traffic going from C to D and λ' is the fraction of the cycle time that green is given to the traffic from C to D. The constraint on λ and λ' is that $\lambda + \lambda' < 1$. We shall assume that the fraction λ_0 is allocated for the internal lost time and green times for pedestrians and $\lambda' = 1 - \lambda - \lambda_0$.

For the motorway the objective is to minimize the total time spent. This can be represented by the following objective function

$$O_M = \frac{(T_{AE} + V_b)L_a}{80 - \frac{T_{AE} + V_b}{100}} + \frac{V_b^2}{2(1-\frac{V_b}{A})A^2}$$

(13)

The analysis of the three level optimization problem can now be presented in the (λ, A) space. For each value of the metering parameter A we can calculate the optimum value for λ. This gives the set of control settings that are optimum for the urban network under the assumption that the motorway control parameter A is fixed. In the same way, for each value of λ, we can find the optimum for A. This gives another line in the (λ, A) plane. The Nash optimum can be found as the points where both curves intersect. The Stackelberg optimum can be found by searching for the optimum value for your own objective function along the optimum curve for the other authority. It is clear that such an optimum is equal to or better than the Nash optimum.

5 Results for the Analytical Approach

We assume that the (static) OD matrix has the following content: AE 2500 veh/hr, BF 1080 veh/hr and CD 700 veh/hr. The road network has the following characteristics: $S_c = 1800$ veh/h, $S_{CD} = 1800$ veh/h, $\lambda_0 = 0.1$, $L_a = 5000$ meter and $tt_c = 360$ seconds. The constraints for the volumes can be transformed to a third order equation for the volumes, which can be solved using standard mathematical techniques.

The result is obvious: each road administrator makes the green fraction for the travellers from B to F as small as possible without causing oversaturation. By doing so, they favour the other vehicles on their network. The boundary conditions determine for a great deal the possible optimum signal settings, which are all close to the boundaries of the feasible range.

In this model we also investigated what the distribution of the traffic would be if all actors would cooperate: the control on both networks would be optimized to minimize total travel time. This is given as 'system optimum' in Tab. 1 and Fig. 2.

The shape of the curves in Fig. 2 depends on the volumes and capacities. If there is sufficient capacity, the optimum settings follow the boundaries of the feasible domain. The system optimum becomes a range instead of a single value for the signal setting.

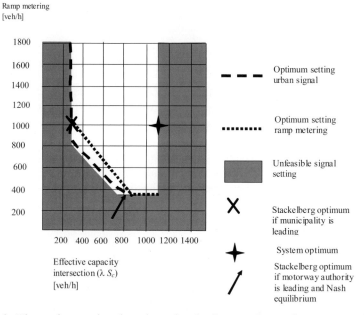

Fig. 2. The optimum signal settings for the intersection and ramp metering for the two road administrators, depending on the setting on the other network. The Stackelberg optimum coincides with the Nash equilibrium if the urban road administrator is leading and anticipates the reaction of both the motorway authority and the drivers.

Tab. 1. Assessment of optimum signal settings with different strategies.

	Stackelberg for urban leading	Nash equilibrium and Stackelberg for motorway leading	System optimum
urban road volume B to F	241 veh/h	775 veh/h	1002 veh/h
Motorway volume B to F	739 veh/h	305 veh/h	78 veh/h
Average speed motorway	40.9 km/h	46.0 km/h	49.4 km/h
Total time spent on motorway	496 veh.h/h	380 veh.h/h	341 veh.h/h
Total time spent on urban roads	31 veh.h/h	96 veh.h/h	109 veh.h/h
Total time spent	527 veh.h/h	476 veh.h/h	450 veh.h/h

6   A Simulation and Assignment Framework

For the network shown in Fig. 1 also the simulation and assignment framework described in [3] was used. With this framework it is possible to calculate Nash and Stackelberg equilibriums. First, an initial assignment is done to determine path flows. With these given path-flows green times are optimized. For the path flows and optimized green times the route travel times are calculated using simulation and the travel times are used to re-assign the traffic.

The optimization of the green times can be done using several local and network control strategies, e.g. Webster control and anticipatory control which leads to a Stackelberg equilibrium. For anticipatory control the optimization of green times includes a simulation and an assignment, given predicted path flows. Thus, the green times are optimized for future path flows. The optimization itself is done with genetic algorithms. The complete algorithm for anticipatory control is drawn in Fig. 3, where f represents the path flows and g the vector of green times. Note that to determine g_{new} a lot of vectors of green times are tried and evaluated in the loop on the right side.

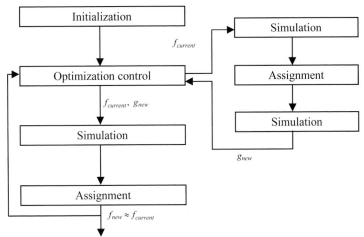

Fig. 3. Solution algorithm for anticipatory control.

For the assignment a path-based stochastic assignment is used. For every time period k the following objective function is minimized (note the extra term compared with equation (5)):

$$Z^k(h) = \frac{1}{\Theta} \sum_{(o,d)\in OD} \sum_{r\in R_{OD}} h^k_{odr} \log(h^k_{odr}) + \sum_{i\in M\cup U} \int_0^{V_i^k} T_i^k(S,x)dx \tag{14}$$

subject to

$$\sum_{r\in R_{OD}} h^k_{odr} = D^k_{od} \qquad\qquad \forall(o,d)\in OD$$

$$h^k_{odr} > 0 \qquad\qquad \forall r \in R_{OD} \quad \forall(o,d)\in OD$$

$$\sum_{(o,d)\in OD} \sum_{r\in R_{OD}} \delta_{odra} h^k_{odr} = V_i^k \quad \forall i \in M\cup U \tag{15}$$

where h denotes the vector of path flows, Θ a positive scaling factor to represent the variances in the perceived travel costs, (o,d) and OD-pair, R_{OD} the routes and D_{OD} the demand for the OD-pair and δ is the link-route incidence matrix, which relates the links to the routes. In the limit when Θ

approaches infinity, perfect knowledge is assumed and the deterministic user equilibrium solution is obtained. The optimization of equation (14) leads to route choice probabilities (see references [10] and [11]) defined by

$$P_{odr}^k = \frac{\exp(-\Theta c_{odr}^k)}{\sum\limits_{s \in R_{OD}} \exp(-\Theta c_{odr}^k)} \qquad \forall r \in R_{OD}, \forall (o,d) \in OD, \forall k$$

(16)

In this analysis the C-logit model, proposed by Cascetta *et al* [12], is used. This model takes into account overlap in routes with the so-called commonality factor, given, for route r of OD pair od per time period k, by

$$CF_{odr}^k = \beta \ln \left[\sum_{s \in R_{OD}} \left(\frac{L_{rs}}{\sqrt{L_r L_s}} \right)^\gamma \right] \qquad \forall r \in R_{OD}, \forall (o,d) \in OD, \forall k$$

(17)

where L_r and L_s are the 'lengths' of routes r and s belonging to OD pair od. L_{rs} is the 'length' of the common links shared by routes r and s and β and γ are positive parameters. 'Length' can be the physical length or the 'length' determined by travel costs. In our case travel times are used. With this commonality factor and the known travel costs, the probability to choose path r, for OD pair od and time period k, and flow h are given by

$$P_{odr}^k = \frac{\exp(-\Theta c_{odr}^k - CF_{odr}^k)}{\sum\limits_{s \in R_{OD}} \exp(-\Theta c_{odr}^k - CF_{ods}^k)} \qquad \forall r \in R_{OD}, \forall (o,d) \in OD, \forall k$$

$$h_{odr}^k = P_{odr}^k D_{od}^k$$

(18)

The traffic simulation model is a simple demand/capacity model that uses travel time functions to calculate link travel times and moves traffic through the network based on these travel times. For this purpose the travel time functions described by Akçelik [13] and functions from the HCM 2000 [14] are used. If demand exceeds capacity a queue builds up (all vehicles have a length of 4.6 meters) and blocking back is explicitly taken into account. The simulation time period is split into small time steps (e.g. 20 seconds), dependent on the length and free flow speed of the shortest link, and for these time steps all necessary variables (flow, speed and travel time) are calculated. From these results travel time per time period can be derived, which can be used to feed the assignment.

7 Results Using the Framework

The network configuration given in Fig. 1 is used. The free flow speed on the motorway is 80 km/hr and the urban roads 50 km/hr. In Tab. 2 the route length and flows for the four routes and five time periods are given after the initial assignment, which is based on the free flow speeds. There are three OD pairs and only OD pair BF has two routes: one via the motorway (route M) and one via the urban network (route U). From the table it is clear that the route via the motorway is preferred in all time periods, although it is longer.

Tab. 2. Route length and flows (veh/hr).

OD pair	Length (km)	Time periods				
		1	2	3	4	5
AE	5.0	3100	3600	3600	3000	2000
BF	5.2	271	406	352	271	163
U	5.8	729	1094	948	729	437
M						
CD	1.5	600	1100	900	500	400

All time periods have a length of 15 minutes. For the control strategies the green times of the intersection are allowed to vary between 7 and 40 seconds and the green times of the ramp metering between 2 and 12 seconds. The cycle time of the intersections is based on the green times and the intersection lost time is 6 seconds. For the ramp metering the cycle time is always 12 seconds. That means that with a saturation flow of 2000 veh/hr, the ramp metering allows between 367 and 2000 veh/hr to enter the motorway. Following Chen [15] the parameters for the stochastic assignment are chosen to be $\beta = 1.0$, $\gamma = 2.0$ and $\Theta = 1.0$. For anticipatory and system optimum control the genetic algorithm described in [16] was used. For anticipatory control the number of generations was set to 50, with a population size of 25. For the system optimum control the population size was set to 20 and 500 generations should lead to a near optimal solution.

It is well known that different initial assignments can lead to different equilibrium situations [17]. For this case the initial flows were based on the free flow travel times. The effects of other initial flows were not studied.

7.1   Control Alternatives

For network 1 several control alternatives were studied. The first alternative (situation 1) is the situation without metering and fixed-time control for the intersection. The fixed-time control is optimized for the busiest time period of the initial flows (see Tab. 1). Then the motorway authority introduces ramp metering, with a local metering strategy (situation 2). The amount of vehicles allowed to enter the motorway is based on the upstream flow and downstream capacity. Due to this measure traffic is now using the urban route. In response the urban authority locally optimizes the intersection with Webster [9] control (situation 3). This is the first situation described in the paragraph on game theory (Nash equilibrium). The next control alternative is when both authorities separately anticipate the reaction of the drivers with respect to route choice (situation 4).

If the motorway authority knows the reaction of the drivers and the control strategy of the municipal road authority and optimizes his strategy, anticipating their response, then we get a Stackelberg game with the motorway authority as leader (situation 5). The same can be obtained, but now with the municipal road administrator as leader (situation 6). The next situation is when both authorities cooperate and optimize their traffic control while they anticipate the reaction of the drivers as if the network is one (situation 7). Finally, a system optimum is calculated (situation 8). This is the situation when the control is optimal and drivers choose the routes that are best for the network as a whole.

7.2   Results

The results for all situations are shown in Fig. 4.

	Situation	Motorway	Urban	Total
No metering, fixed-time	1	141,28	21,45	162,72
Metering, fixed-time	2	33,00	31,18	64,18
Metering, Webster	3	34,64	27,35	61,98
Separate anticipatory	4	28,66	27,05	55,71
Motorway leader	5	21,51	39,78	61,30
Urban roads leader	6	34,38	23,20	57,58
Integrated anticipatory	7	34,07	22,37	56,44
System optimum	8	15,15	30,69	45,84

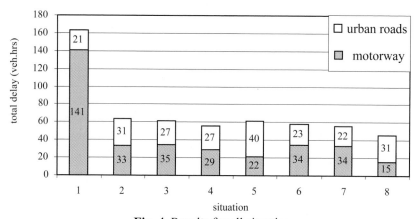

Fig. 4. Results for all situations.

The results are given in terms of the total delay of the equilibrium solution. The total delay is split in two parts: delay for the motorway and delay for the urban roads. From Fig. 4 it is clear that the introduction of ramp metering (situation 2) is beneficial for the motorway. It gives some extra delay on the urban network, but the total delay is much less then in the situation without ramp metering. The local optimization of the intersection (situation 3) improves the situation on the urban network. If both authorities anticipate the reaction of the drivers separately, only the situation on the motorway improves, compared with local optimization.

The Stackelberg equilibrium where the motorway is the leader (situation 5) improves the situation on the motorway, but this has negative consequences for the urban network and also the total delay is higher than for situation 4. The same is true for the Stackelberg equilibrium if the urban authority is the leader (situation 6): the urban delay is smaller, but the motorway delay is higher. It seems that separated or integrated anticipatory control is the best. If one of the authorities takes the lead and anticipates the reaction of the other, the situation for the other authority becomes worse and also in total the delay increases.

8 Conclusions

Road authorities have the freedom to optimize the control of their own part of the road infrastructure. Similarly, public transport operators can optimize their services to minimize their losses. By doing so, they influence the route and modal choice of the travellers which has an impact on the conditions on the part of the transport system that they do not control themselves. This situation can be considered as a social dilemma where self interest of one authority may lead to situations that are locally optimal, but globally, for the whole system, sub-optimal. One road authority may anticipate the reaction of the authorities of the roads that he does not control himself and optimize according to a Stackelberg strategy. This can improve the state of the system as controlled by the two agents. However, an integration of goals of the different road authorities and a joint optimization of the control of the network generally leads to a better optimum.

In terms of game theory the three options are:

- The Nash game, where each authority optimizes the control of his part of the network, given the flows that present; after a change of flows the control is adapted;
- The Stackelberg game, where one road authority optimizes his network, while he anticipates the reaction of both the road users and the authority that controls the rest of the network;
- The monopolist game, where the two authorities cooperate to realize an optimum for the whole network, according to a goal that they have agreed on.

The example discussed in this paper is very simple, but it shows the mechanism. In practice we often see that road authorities do not optimize traffic signal control and if they do, they often do not have the resources to adapt the control when the situation has changed. That means that even the worse strategy, according to the Nash game, already gives substantial improvements of the traffic conditions. The advantage of a monopolistic optimization especially becomes visible in congested situations, where small increases of the network capacity give substantial improvements of the performance.

The analysis explained in this paper can be easily extended to different classes of communication networks, such as public transport, data communication, freight distribution and, as been discussed above, road networks belonging to different authorities. Also the interaction between spatial development and traffic demand can be included. The demand for each network or sub-network is influenced by the conditions on all other networks and control of one network has impact on the other ones. Recently, Zhang et al. [18] gave an analysis of such a system of loosely coupled networks. They focus on the convergence towards equilibrium, but their approach and the one of this paper can be combined to show the value of a cooperative control of multiple communication networks. The advantages of having a transport authority with regional decision power over all transportation networks, instead of the fragmented decision making that is applied in nearly every country can be analyzed by this kind of models.

References

1. Chen, Y.-S., J. L. Hu, F. op de Beek, L. Sun and H.J. van Zuylen. An Integrated Approach for Solving Traffic Problems at the Shanghai Elevated Highway: the Case Study with Dynamic Simulation, Conference ITS Shanghai 2000.
2. Van Roon, J.J. Studie effecten openstelling ringweg Amsterdam. Dienst Verkeerskunde Rotterdam 1990.
3. Taale, H. and H.J. van Zuylen. The Effects of Anticipatory Control for Several Small Networks, Proceedings of the 82nd Annual Meeting of the Transportation Research Board, Washington D.C., January 2003.
4. Wardrop J.G.. Some theoretical aspects of road traffic research. Proceedings of the institution of Civil Engineers, 1952 Part.II, 1, pp. 325-378.
5. Fisk. C.S. Game theory and transportation system modeling. *Transportation Research*, Volume 18B, 1984, pp. 310 – 313.
6. Chen H.-K. and C.-Y. Wang. Dynamic Capacitated User-Optimal Route Choice Problem. Proceedings of the 78th Annual Meeting of the Transportation Research Board, Washington D.C., January 1999.
7. Fudenberg, D. and J. Tirole. *Game Theory,* The MIT Press, Cambridge, Massachusetts. 1991.
8. Beckman, M.J., C.B. McGuire and C.B. Winsten. *Studies in the Economics of Transportation.* Yale University Press, New Haven. 1956.
9. Webster, F.V. *Traffic Signal Settings.* Road Research Technical Paper No. 39, Road Research Laboratory, London, 1958.
10. Sheffy, Y. *Urban Transportation Networks: Equilibrium Analysis with Mathematical Programming Methods.* Prentice-Hall, Englewood Cliffs, NJ, USA.
11. Chen, H.-K. *Dynamic travel choice models: a variational inequality approach.* Springer, Heidelberg, Germany, 1999.
12. Cascetta, E., A. Nuzzolo, F. Russo and A. Vitetta. A modified logit route choice model overcoming path overlapping problems: specification and some calibration results for interurban networks. Proceedings of the 13th International Symposium on Transportation and Traffic Theory, Lyon, France, 1996, pp. 697-711.
13. Akçelik, R. Travel time functions for transport planning purposes: Davidson's function, its time-dependent form and an alternative travel time function. *Australian Road Research* 21 (3), 1991, pp. 49–59. (Minor revisions: December 2000).
14. Transportation Research Board. *Highway Capacity Manual.* Special Report 209, 4th Edition, National Research Council, 2000.
15. Chen, O.J. *Integration of Dynamic Traffic Control and Assignment.* PhD Thesis, Massachusetts Institute of Technology, 1998.
16. Houck, C.R., J.A. Joines and M.G. Kay. A Genetic Algorithm for Function Optimization: A Matlab Implementation. *Genetic Algorithm for Optimization Toolbox*, NCSU-IE TR 95-09, 1995.
17. Van Zuylen, H.J. and H. Taale. Traffic control and route choice: Conditions for Instability. In: *Reliability of Transport Networks*, (M.G.H. Bell and C. Cassir, ed), Research Studies Press Ltd, Baldock, United Kingdom, 2000, pp. 55-72.
18. Zhang, P., S. Peeta and T. Friesz. Dynamic game theoretic model of multi-layer infrastructure networks, 10th International Conference on Travel Behaviour Research, Luzern, August 2003.

Real-Time Motorway Network Traffic Surveillance Tool RENAISSANCE

Y. Wang[1], M. Papageorgiou[1], and A. Messmer[2]

[1] Dynamic Systems and Simulation Laboratory –
Technical University of Crete
73100 Chania – Greece
ywang@dssl.tuc.gr, markos@dssl.tuc.gr
[2] Groebenseeweg 2, 82402 Seeshaupt – Germany
Albert.Messmer@t-online.de

Summary. The paper presents a real-time motorway network traffic surveillance tool called RENAISSANCE that enables traffic state estimation and short-term prediction, travel time estimation and prediction as well as queue length estimation and prediction based on a limited amount of real-time measurements. Both simulation testing and real data testing were conducted to evaluate the RENAISSANCE performances.
Keywords: motorway networks, real-time traffic surveillance, RENAISSANCE

1 Introduction

Large-scale and medium-scale motorway networks are usually equipped with a number of measurement devices of various kinds (loops, video sensors, radar) that deliver real-time information about the current traffic conditions in corresponding locations. However, if the density of available sensors is not very high (e.g. lower than one sensor per 0.5 or 1 km), the delivered real time information may not be complete due to significant space inhomogeneities. This creates the need for a traffic state estimator that would deliver in real time the complete traffic state based on a more or less limited amount of traffic measurements. Moreover, a number of further real-time traffic surveillance tasks including traffic state prediction, travel time estimation and prediction, queue tail/head estimation and prediction are of interest to the network operators for various uses. The **RE**al-time motorway **N**etwork tr**A**ff**I**c State Surveill**ANCE (RENAISSANCE)** tool has been developed to address these traffic surveillance tasks in a unified macroscopic model-based approach. The RENAISSANCE software tool is applicable in real time to motorway networks of arbitrary size, topology, and characteristics based on any suitable traffic detector configuration. A particular feature of RENAISSANCE is its ability to handle in an efficient way real-time measurements collected via inductive loops, radar detectors, or video sensors, or via any arbitrary combination of them. Real-time information extended and enriched by RENAISSANCE may be exploited for further operations in motorway networks (e.g. traffic control, route guidance, etc.). The paper introduces briefly the macroscopic motorway network traffic flow model of RENAISSANCE, describes the addressed traffic surveillance tasks, and outlines the functional architecture of the tool. A simulation test was conducted via application of RENAISSANCE to a hypothetical motorway network, while the traffic state estimator of RENAISSANCE was also tested with 8-hour real data collected from a Bavarian motorway. Both testing results are presented in some detail. Final conclusions and future steps are outlined.

2 RENAISSANCE

2.1 Modelling of RENAISSANCE

A validated second-order macroscopic motorway traffic flow model is utilized in RENAISSANCE [1,2]. It describes the dynamic evolution of traffic flow in motorway networks (with any topology, size, and link characteristics) in terms of appropriate aggregated traffic flow variables (i.e. *traffic density*, *space mean speed*, and *traffic volume* defined for each motorway segment of about 500 m), while the distribution of traffic flow at bifurcations is modelled in terms of turning rates. The model can be expressed in a compact state-space form $\mathbf{x}(k+1) = \mathbf{f}[\mathbf{x}(k), \xi(k)]$, where $\mathbf{f}$ is a nonlinear differentiable vector function corresponding to a number of model equations [2], while the **traffic state** vector $\mathbf{x}(k)$ includes all segment speeds, segment densities, and all boundary variables (origin flows, origin speeds, destination densities, turning rates) of the motorway network model as well as the main model parameters (free speeds, critical densities, and capacities), while vector $\xi(k)$ includes modelling noise. Traffic measurements within a motorway network depend on the traffic state $\mathbf{x}$ via an output equation $\mathbf{y}(k) = \mathbf{g}[\mathbf{x}(k), \mathbf{\eta}(k)]$, where vector $\mathbf{y}$ consists of all available measurements of flow and mean speed; $\mathbf{g}$ is a nonlinear differentiable vector function; vector $\mathbf{\eta}$ is a function of modelling and measurement noise. Both equations compose the dynamic system model of RENAISSANCE.

2.2 Traffic Surveillance Tasks of RENAISSANCE

The traffic surveillance tasks of RENAISSANCE are interrelated, whereby the traffic state estimation and prediction are the most fundamental tasks because all other tasks are performed based on the traffic state estimation and/or prediction results.

Traffic State Estimation refers to estimating all segment speeds and densities of a motorway network at each current time instant, based on more or less limited real-time traffic measurements available up to the current time instant. The RENAISSANCE traffic state estimator is designed based on the extended Kalman filter. **Traffic State Prediction** refers to predicting at each current time instant all segment speeds and densities of a motorway network over a given future time horizon (e.g. 1 hours). The traffic state prediction is performed by use of the motorway network traffic flow model, based on the traffic state estimates at the current time instant and boundary value prediction over the prediction horizon. **Boundary value prediction** is indispensable to the traffic state prediction and can be performed based on historical data, or on the boundary variable estimates available up to the current time instant, or by an appropriate combination of both kinds of data. **Travel time estimation** refers to the estimation of the **instantaneous travel time** along user-specified routes within a motorway network and is based on the traffic speed estimation at each current time instant. On the other hand, **travel time prediction** predicts the **experienced travel time** along user-specified routes at each current time instant. When traffic speeds of some segments along a user specified route are below a user-selectable speed threshold value, one (or several) vehicular queue(s) are considered to be forming along the route. A vehicle queue has a queue head and a queue tail. Note that a queue tail is actually the shockwave of a traffic congestion. The **queue tail/head estimation (prediction)** aims at estimating (predicting) the location of any queue tail/head existing (to appear) at the current (future) time instant(s). With the identified locations of queue tails and queue heads, the queue lengths are readily estimated and predicted. For further details on the traffic surveillance tasks, the reader is referred to [2,3].

2.3   Functional Architecture of RENAISSANCE

Figure 1 displays a schematic diagram of the functional architecture of RENAISSANCE. The highlighted central block represents the main body of RENAISSANCE. The sub-blocks inside it represent its various functional modules, most of which correspond to the aforementioned traffic surveillance tasks. The directed lines between the functional modules represent the signal flows. Each module has both input and output. RENAISSANCE also includes an accessory visualization module. The inputs to RENAISSANCE include real time traffic measurements (flow and mean speed or occupancy) and, possibly, incident reports from the TCC operators, while the outputs of RENAISSANCE correspond to its various functionalities. In addition, RENAISSANCE provides the users with quite a few options in order for the traffic surveillance tasks to be performed according to the specific user needs. The readers may see [3] for further details.

3 Simulation Testing of RENAISSANCE

A preliminary simulation test was conducted for RENAISSANCE with regard to a hypothetical motorway network example [2].

3.1   Test Setup

Figure 2(a) depicts the example network, which consists of 13 nodes (N1 to N13) and 23 links (origins O1 to O4, destinations D1 to D5, and normal links L1 to L14). Each normal link is divided into several segments with respective lengths of 500 m. The digit in the parentheses next to each normal link name represents the corresponding number of segments. O1 and D1 have five lanes each, while the other origins and destinations have one lane each. Except L1 and L7 each with five lanes and L10 ~ L14 each with two lanes, all other normal links have three lanes each. Moreover, the capacity of each origin link or normal link is 2000 veh/lane/h. The values of the free speed, critical density, and exponent (which are the three parameters of the utilized fundamental diagram) are the same for all normal links. The utilized traffic demand of 6 hours at O1 is shown in Fig. 2(b), while Fig. 2(c) displays from the top to bottom the turning rates at N2, N10, N4, N6, N13, and N9. During the peak period, the total in-coming flow at N5 exceeds the maximum permissible flow of L5. Consequently, a congestion is built up initially at N5 and spills rapidly back in two directions via L4 and L11, respectively. The detector configuration is also displayed in Fig. 2(a), where each grey rectangle represents a detector that delivers flow and speed measurements.

Traffic dynamics in the motorway network (emulated reality) are simulated by use of the same macroscopic traffic flow model as employed in RENAISSANCE. It is assumed for the testing that RENAISSANCE does not have prior knowledge of the parameter values of the reality-emulating model, hence the RENAISSANCE model starts with some different parameter values that lead to an initial capacity bias of 50%. The utilized time steps for both models are 10 s. Totally 103 state variables are considered for this network example by RENAISSANCE that include 88 segment speeds and densities, 12 boundary variables, and three model parameters; however, only 20 among these 103 state variables are directly measured. It is noted that the traffic scenario is designed such that all boundaries of the network are congestion-free and thus traffic congestions only occur inside the network rather than mounting from the downstream boundaries.

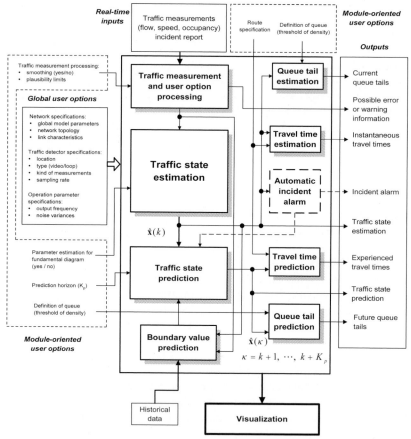

Fig. 1. Functional architecture of RENAISSANCE.

3.2   Testing Results

Traffic state estimation: During the peak period, two congestion queues built up at N5 and spread upstream along directions L4→L3→L2 and L11→L10, respectively. Eventually, the queue tails reach the second segment of L2 in the first direction and the second segment of L10 in the second direction, respectively, at the maximum lengths of the queues (with the speed threshold value for a queue set to 40 km/h). Fig. 3(a) shows that the speed in the last segment of L4 is around 100 km/h under free-flow conditions, but the traffic flow is getting dense in the fourth hour, leading to a speed drop down to about 20 km/h at the end of the fourth hour, which persists within the fifth hour. Meanwhile, the segment density rises up to 65 veh/km/lane from less than 10 veh/km/lane. Obviously, the RENAISSANCE traffic state estimator is capable of tracking the speed and density variation in this segment properly under various traffic conditions. Similarly satisfactory traffic state estimates are also obtained for any other segment. The model parameters are estimated simultaneously with all segment variables. The capacity estimation result is presented in Fig. 3(b), while the estimation results of the free speed, critical density, and exponent are even better. It should be highlighted that without on-line model parameter estimation (i.e. the RENAISSANCE model parameters are kept constant at their initial values over the whole simulation horizon), a significant estimation bias is observed for each segment variable. The reader may see [4] for the significance of on-line model parameter estimation.

Boundary value prediction and traffic state prediction: The boundary variables are predicted solely based on the boundary variable estimates (via extrapolation) available up to each current time instant, while the segment variables are predicted based on the RENAISSANCE model. The inflow prediction at O1, and the speed and density prediction in the second segment of L4 are presented in Figs. 3(c) and 3(d) as two examples; more specifically, the real trajectory and a number of selected prediction trajectories are displayed for each variable. The prediction horizon is 20 minutes. The boundary variables are generally well predicted. Since no historical data is employed for the boundary value prediction, some acceptable prediction biases occur but mostly at local extreme points (valleys and peaks). On the other hand, satisfactory prediction results are obtained for all segment variables under various traffic conditions.

Travel time estimation and prediction: In this investigation, one route is specified for travel time estimation and prediction, which is "L1, L2, L3, L4, L5, L6, L7". Fig. 3(e) shows that the instantaneous travel time along this route is estimated well; merely during congestion, the travel time estimation is lower than the instantaneous travel time; this is due to the high real-speed fluctuation and the nonlinear (inverse) character of the travel-time-versus-speed relationship. Note, however, that the peaks of the instantaneous travel time are not very realistic when compared to the real experienced travel time (Fig. 3(f)). In other words, the moderate estimated travel time proves to be more realistic. Fig. 3(f) illustrates that the experienced travel time is predicted quite accurately.

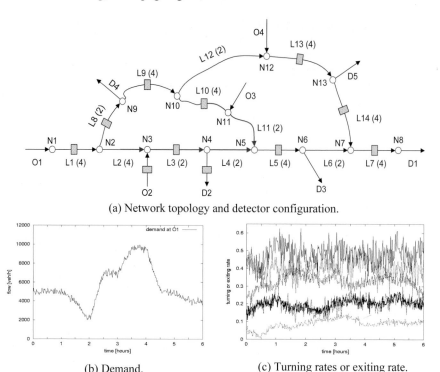

(a) Network topology and detector configuration.

(b) Demand.   (c) Turning rates or exiting rate.

Fig. 2. Test motorway network example.

Queue tail/head estimation and prediction: Both queue tail and head locations on this route are estimated and predicted well since all segment
speeds are estimated and predicted accurately. Then, proper queue length
estimations and predictions on the route are obtained; see Figs. 3(g) and 3(h).

4   Real Data Testing of the RENAISSANCE Traffic State Estimator

The traffic state estimator of RENAISSANCE was also tested with respect to
a real motorway stretch example based on real measurements.

4.1   Test Setup

The test example is a 2-lane westbound stretch of the A92 motorway, close to
Munich (Germany), which has a length of 4.1 km and covers A92's most
congested part between the Munich airport and the junction AK Neufahrn. A
schematic diagram of the test stretch is presented in Fig. 4(a). Except for an
on-ramp at the upper boundary of the stretch, the test stretch is almost homogeneous, without any lane-drop or slope but with a slight curvature. Fig. 4(a)
shows that four loop detectors are installed along the stretch (i.e. loops L0,
L1, L2, and L3 at the start, in the middle, and at the end of the stretch). These
loop detectors count passing vehicles and measure individual car speeds. The
resulting single-car data are converted into aggregated traffic measurements
of flow and space mean speed for every minute.

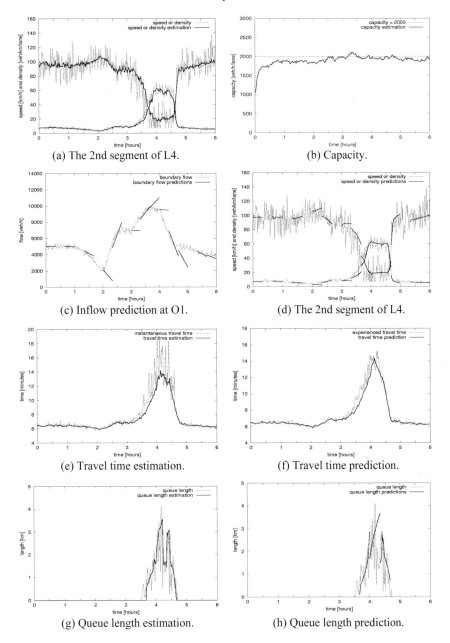

(a) The 2nd segment of L4.

(b) Capacity.

(c) Inflow prediction at O1.

(d) The 2nd segment of L4.

(e) Travel time estimation.

(f) Travel time prediction.

(g) Queue length estimation.

(h) Queue length prediction.

Fig. 3. The simulation testing of RENAISSANCE for the motorway network example.

Flow and speed measurements of 8 hours from L1, L2, and L3 are displayed in Figs. 4(b) and 4(c), respectively. It is noticed that two distinguished speed drops occurred at each mainstream measurement location within those 8 hours, during which the flow trajectories descended noticeably. This means that two different traffic congestions spilled back into the test motorway stretch from the downstream. Moreover, the flow measurements keep reduc-

ing on average after 6:00 PM due to the decreasing demand. Before the
measurement data was utilized, a plausibility checking was conducted [2,5].
The flow measurement data was found to be plausible, while the speed measurement data seems to be plausible as well, except at the middle measurement location where the speed trajectory recovers first at the end of the second congestion. In addition, under the free-flow conditions, the speed trajectory at the downstream location is always lower than those at the middle and
upstream locations by around 10% on average (Fig. 4c), for which there is no
ready explanation as the stretch is virtually homogeneous.

Figure 4(a) illustrates that the whole test stretch is subdivided into 8 segments numbered from 1 (upstream) to 8 (downstream), each with approximate segment length of 500 m such that L1 resp. L3 are located at the upstream resp. downstream boundaries of the stretch, while L2 is located at the
boundary between segments 4 and 5. The measurement data from L1 and L3
are used to feed RENAISSANCE, while the measurements from L2 are only
used to compare with the traffic state estimates for segment 4 so as to evaluate the performance of the RENAISSANCE estimator. The aggregated measurement data is updated every minute while the time step of the
RENAISSANCE model is 10 seconds. The free speed, critical density, and
capacity are assumed to be the same for whole stretch (since the stretch is
homogeneous), and their initial values are set equal to 90 km/h, 40
veh/km/lane and 2180 veh/h/lane, respectively.

4.2   Testing Results

A number of tests have been conducted for the traffic state estimator of
RENAISSANCE with on-line model parameter estimation. Fig.s 4(d) resp.
4(e) compare the flow resp. speed estimate in segment 4 with the flow resp.
speed measurements from L2. It may be seen that the flow and speed estimates match the corresponding measurements very well. It is interesting to
see that the speed estimate can track the real speed trajectory in a very accurate manner, i.e. even the oscillatory stop-and-go behaviour of the speed during the congestion is tracked reasonably well. It is again emphasized that the
displayed measurements from L2 are not used by RENAISSANCE, but only
used for the comparison purpose. Since traffic measurements from both upstream and downstream measurement locations are used by RENAISSANCE,
it is not surprising that the resulting estimates at both locations match the
corresponding measurements satisfactorily (figures omitted).

The on-line estimates of the free speed and capacity are displayed in Figs.
4(f) and 4(g). It should be pointed out that this particular real-data-based testing was actually conducted over a quadruple-time horizon, with the traffic
scenario of the first time horizon (12:00 AM ~ 8:00 PM) duplicated to the
second, third, and fourth time horizon in order to check the stability of the
estimator thoroughly. It is displayed in Figs. 4(f) and 4(g) that identical
model parameter estimation results are repeated over the last three time hori-
zons, which indicates that (a) the RENAISSANCE estimator is stable, (b) the
first time horizon can be taken as a warm-up period. As a matter of fact, except for the first time horizon, the same segment estimation results are delivered for each time horizon. Considering that the traffic state estimator is designed for its application in the field over a time horizon that is much longer
than 8 hours, the estimator performance over the warm-up period of the first

8 hours can be neglected. In fact, the segment estimation results presented in Figs. 4(d) and 4(e) were produced over the third time horizon, but even for the first time horizon the segment variable estimates are already quite acceptable.

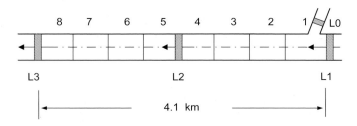

(a) Test motorway stretch and detector configuration.

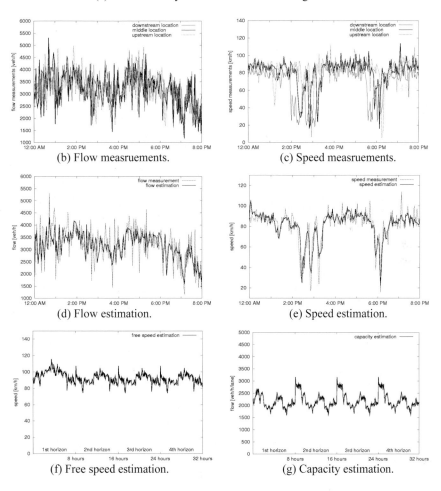

(b) Flow measruements.

(c) Speed measruements.

(d) Flow estimation.

(e) Speed estimation.

(f) Free speed estimation.

(g) Capacity estimation.

Fig. 4. The real data testing of the traffic state estimator of RENAISSANCE for the motorway stretch example.

For the same test example, the RENAISSANCE estimator was also tested with respect to the necessity of the on-line model parameter estimation, its sensitivity to initial model parameter values and to the utilized standard deviation values, as well as its capability of dealing with measurement biases. The interested reader is referred to [5].

5 Conclusion

A motorway network traffic surveillance tool RENAISSANCE has been developed. The RENAISSANCE model is highly nonlinear and hence the RENAISSANCE traffic state estimator, designed based on the extended Kalman filter, is only a sub-optimal estimator. Therefore, there are quite a few potential uncertainties in the RENAISSANCE performance before it is tested. For this reason, a thorough simulation-based testing was first conducted in order to evaluate RENAISSANCE under various traffic conditions. It is noted that in the simulation testing the RENAISSANCE model is virtually the same as the reality-emulating model, albeit with strongly different initial model parameter values. In view of this, the RENAISSANCE estimator was also tested with real measurement data, whereby the RENAISSANCE model only has limited knowledge of the test stretch. Both testing results are very encouraging for the subsequent field evaluation of RENAISSANCE.

Acknowledgement

This research was partly supported by the European Commission's IST (Information Society Technologies) Program under the project RHYTHM (IST-2000-29427). All statements of the paper are under the sole responsibility of the authors and do not necessarily reflect the European Commission's policies or views.

References

1. M. Papageorgiou, J.-M. Blosseville, and H. Hadj-Salem. Modelling and real-time control of traffic flow on the southern part of Boulevard Périphérique in Paris - Part I: Modelling. Transportation Research, Vol. 24A, pp. 345 – 359, 1990.
2. Y. Wang, M. Papageorgiou and A. Messmer. Algorithms and preliminary testing for traffic surveillance. Deliverable D3.2 of the RHYTHM Project IST-2000-29427, Brussels, Belgium, 2003.
3. Y. Wang, E. Smaragdis, M. Papageorgiou, and A. Messmer. State-of-the-art and functional specifications of traffic surveillance and isolated ramp metering algorithms. Deliverable D3.1 of the RHYTHM Project IST-2000-29427, Brussels, Belgium, 2002.
4. Y. Wang, and M. Papageorgiou. Real-time freeway traffic state estimation based on extended Kalman filter: a general approach. Submitted to Transportation Research B, 2003.
5. Y. Wang, M. Papageorgiou and A. Messmer. Real-time freeway traffic state estimation: a case study. Submitted to Transportation Science, 2003.

Minority Game - Experiments and Simulations

T. Chmura[1], T. Pitz[1], and M. Schreckenberg[2]

[1] Laboratory of Experimental Economics – University of Bonn
Adenauerallee 24-42, 53113 Bonn – Germany
chmura@uni-bonn.de
[2] Physics of Transport and Traffic – University Duisburg-Essen
47048 Duisburg – Germany

Summary: The paper reports laboratory experiments on a *minority game* with two routes. Subjects had to choose between a road *A* and a road *B*. Nine subjects participated in each session. Subjects played 100 rounds and had to choose between one of both roads. The road which the minority of players chose got positive payoffs. Two treatments with 6 sessions each were run at the Laboratory of Experimental Economics at Bonn University. Feedback was given in treatment I only about own travel time and in treatment II on travel time for road *A* and road *B*.

Keywords: minority game, travel behaviour research, information in intelligent transportation systems, laboratory experiments, payoff sum model

1 Introduction and Experimental Set-Up

The setup of the first minority Game, the El Farol Bar Problem (EFPB), introduced by Arthur [1] is the following: a number of agents *n* have to choose in several periods whether to go in room A or B. Those agents who have chosen the less crowded room win, the others lose.

Later on, the EFBP was put in a mathematical framework by Challet and Zhang [2], the so-called Minority Game (MG). An odd number *n* of players has to choose between two alternatives (e.g., yes or no, A or B, or simply 0 or 1). In the Literature are many examples, where the MG is discussed [2,3,4].
In this paper we transferred the minority problem into a route choice context. We did minority game experiments at the Laboratory of Experimental Economics (University of Bonn).

In these Experiments subjects are told that in each of 100 periods they have to make a choice between a road *A* and road *B* for travelling from X to Y.

Fig. 1. Participants had to choose between a road [A] and a road [B].

The number of subjects in each session was 9. They were told the time t_A and t_B depends on the numbers n_A and n_B of participants choosing *A* and *B*, respectively:

$$t_A = 1,\ t_B = 0 \iff n_A < n_B$$
$$t_B = 1,\ t_A = 0 \iff n_A > n_B.$$

The period payoff was t_A if A was chosen and t_B if B was chosen.

The total payoff of a subject was the sum of all 100 period payoffs converted to money payoffs in Euro [€] with a fixed exchange rate of 0.2 € for each experimental money unit (Taler). Additionally every participant received a show-up fee of 3 €. One session took roughly one hour. There are no pure equilibria in this game.

The pareto-optimum can be reached by 4 players on one road and 5 players on the other road.

Two treatments have been investigated. In treatment I the subjects received information about:

- whether own last choice was in the minority or majority
- last chosen route
- payoff of the last period in Taler
- cumulated payoff in Taler
- number of the actual period

In treatment II additional feedback was provided about distribution on both-routes in the last period. Six sessions were run with treatment I and six with treatment II. No further information was given to the subjects.

2   Observed Behaviour

2.1   Number of Players on the Road A

Figure 2 shows the number of participants on the road A as a function of time for a typical session of treatment I. Fig. 3 shows the number of participants on the road A as a function of time for a typical session of treatment II.

There are substantial fluctuations until the end of the session. The same is true for all sessions of both treatments.

The mean number of players on the road A is 4.5 in treatment I and 4.43 in treatment II. This was expected, because of the experimental setup, there is no preference for one road.

The fluctuations can be measured by the standard deviation of the number of participants choosing A per period. This standard deviation is between 0,67 and 1,5. In view of these numbers one can speak of substantial fluctuations in each of the 12 sessions.

The fluctuations are obvious larger under treatment I than under treatment II. The effect is significant. The null-hypothesis is rejected by a Wilcoxon-Mann-Whitney-Test on the significance level of 1 % (one sided).

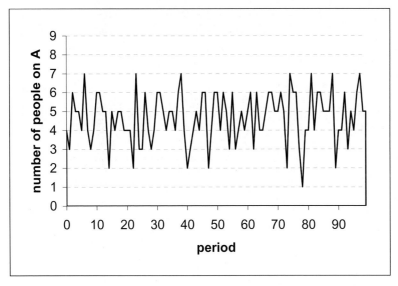

Fig. 2. Number of participants on *A* [a typical session of treatment I].

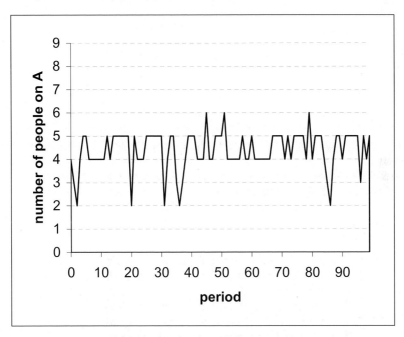

Fig. 3. Number of participants on *A* [a typical session of treatment II].

The non existence of pure strategy equilibria poses a coordination problem which may be one of the reasons for non-convergence and the persistence of fluctuations. Feedback on both travel times vs. feedback on only own travel time has a beneficial effect by the reduction of fluctuations. This effect is remarkable.

		number of players on A	
		mean	std. dev.
Treatment I	session I 01	4,33	1,36
	session I 02	4,74	1,50
	session I 03	4,41	1,50
	session I 04	4,40	1,31
	session I 05	4,65	1,33
	session I 06	4,44	1,28
	treatment I	**4,50**	**1,38**
Treatment II	session II 01	4,19	1,35
	session II 02	4,62	1,19
	session II 03	4,36	1,05
	session II 04	4,34	0,97
	session II 05	4,62	0,84
	session II 06	4,50	0,67
	treatment II	**4,44**	**1,01**

Tab. 1. Mean and standard deviation of participants on A.

2.2 Road Changes

Figure 4 shows an example of the number of road changes as a function of time for a typical session of treatment I. There was a negative trend in each session of treatment II. By comparison in treatment I there were four sessions with a positive and two with a negative trend. The fluctuations are connected to the total number of road changes within one session.

The median number of road changes is significantly higher in treatment I. The null-hypothesis is rejected by the Wilcoxon-Mann-Whitney-Test on a level of 1% (one sided). The mean number of road changes under treatment I is also higher than under treatment II. A Wilcoxon-Mann-Whitney-Test rejects the null-hypothesis only on a significance level of 1% (one sided).

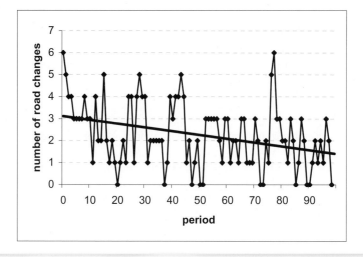

Fig. 4. Number of road changes [a typical session of treatment I].

		number road changes
		Mean
Treatment I	session I 01	0,430
	session I 02	0,466
	session I 03	0,434
	session I 04	0,442
	session I 05	0,374
	session I 06	0,464
	treatment I	0,44
Treatment II	session II 01	0,357
	session II 02	0,222
	session II 03	0,250
	session II 04	0,099
	session II 05	0,146
	session II 06	0,053
	treatment II	0,19

Tab. 2. Mean number of road changes.

Under treatment I subjects who mainly choose only one of the roads feel the need to travel on the other road from time to time in order to get information on both roads. Under treatment II there is no necessity for such information gathering. This seems to be the reason for the greater number of changes and maybe also for the stronger fluctuations under treatment I.

		number of road changes	
		mean	std. dev.
Treatment I	session I 01	5,08	2,298
	session I 02	3,87	1,865
	session I 03	5,16	1,934
	session I 04	5,19	1,931
	session I 05	5,28	2,391
	session I 06	4,35	2,083
	treatment I	4,82	2,084
Treatment II	session II 01	3,99	2,001
	session II 02	3,68	2,039
	session II 03	3,67	2,091
	session II 04	5,19	2,32
	session II 05	4,67	2,48
	session II 06	4,44	2,044
	treatment II	4,27	2,163

Tab. 3. Mean and standard deviation number of road changes.

2.3   Payoffs and Road Changes

In all sessions except one session of treatment I the number of road changes of a subject is negatively correlated with the subject's payoff. Tab. 4 shows that the negative correlation between the payoff and number of road changes in treatment II is higher than in treatment I.

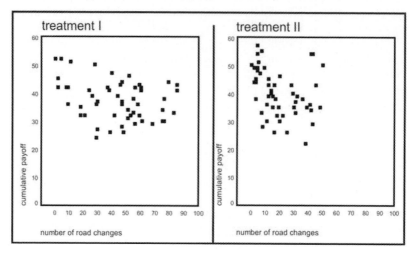

Fig. 5. Scatter diagram cumulative payoff/number of road changes for treatment I and II.

		Spearman rank correlation between cumulative payoffs and number of road changes
Treatment I	session I 01	-0,48
	session I 02	0,34
	session I 03	-0,44
	session I 04	-0,70
	session I 05	-0,18
	session I 06	-0,18
Treatment II	session II 01	-0,51
	session II 02	-0,54
	session II 03	-0,30
	session II 04	-0,82
	session II 05	-0,27
	session II 06	-0,78

Tab. 4. Spearman rank correlation between cumulative payoffs and number of road changes for treatment I and II.

In every eleven observations of both treatments the Spearman rank correlations between cumulative payoffs and the number of road changes is negative. The Spearman-correlation-coefficients in treatment II are lower than in treatment I. This effect is small but significant. A Wilcoxon-Mann-Whitney-Test rejects the null-hypothesis on a significance level of 10% (one sided).

Even if subjects change roads in order to get higher payoffs, they do not succeed in doing this on the average. This suggests that it is difficult to use the information provided by the feedback to one's advantage.

3   Response Mode

A participant who had no payoff on the road chosen may change his road in the next period in order to travel where it is less crowded. We call this the *direct* response mode.

The *direct* response mode is the prevailing one but there is also a *contrarian* response mode. The contrarian participant expects that a positive payoff will attract many others and that therefore the road chosen will be crowded in the next period.

For each subject let c_- (c_+) be the number of times in which a subject changes the roads when there was a payoff $p=0$ ($p=1$) in the period before. And for each subject let s_- (s_+) be the number of times in which a subject stays on the road when there was a payoff $s=0$ ($s=1$) in the period before.

	change	stay
p=0	c_-	s_-
p=1	c_+	s_+

Tab. 5. 2x2 table for the computation of Yule coefficients.

For each subject such a 2x2 table has been determined and a Yule coefficient Q has been computed as follows.

$$Q = \frac{c_- \cdot s_+ - c_+ \cdot s_-}{c_- \cdot s_+ + c_+ \cdot s_-}$$

The Yule coefficient has a range from -1 to $+1$. In our case a high Yule coefficient reflects a tendency towards direct responses and a low one a tende- The mean and the standard deviation of the Yule coefficients are shown in Tab. 6.

The mean Yule coefficients are significantly higher in treatment II. The null-hypothesis for both treatments is rejected by a Wilcoxon-Mann-Whitney-Test on the significance level of 1% (one sided). That means there are less contrarian response modes in treatment II.

		Yule coefficiants Q	
		mean	std. dev.
Treatment I	session I 01	0,14	0,62
	session I 02	0,15	0,43
	session I 03	0,27	0,76
	session I 04	0,01	0,47
	session I 05	0,11	0,75
	session I 06	-0,01	0,54
	treatment I	0,11	0,60
Treatment II	session II 01	0,21	0,75
	session II 02	0,42	0,39
	session II 03	0,48	0,61
	session II 04	0,72	0,40
	session II 05	0,68	0,66
	session II 06	0,87	0,33
	treatment II	0,56	0,52

Tab. 6. Mean and standard deviation of the Yule coefficients in both treatments.

4   Simulations

In order to get more insight into this theoretical significance of our result, we have run simulations based on a version of a well known reinforcement learning model, the payoff-sum model.

This model already described by Harley (1981) [5] and later by Arthur (1991) [6] has been used extensively by Ereth and Roth [7,8] in the experimental economics literature. Here we used an extended payoff-sum model, which is already published in Selten et al. [9].

Table 7 explains the version underlying our simulations. We are looking at player i who has to choose among n strategies $1,...,n$ over a number of periods t, $t=1..T$. The probabilities with which each strategy i is chosen is proportional to its "propensity" $x_{i,j}^t$. In period 1 these propensities are exogenously determined parameters. Whenever the strategy j is used in period t, the resulting payoff a_i^t is added to the propensity if this payoff is positive. If all payoffs are positive, then the propensity is the sum of all previous payoffs for this strategy plus its initial propensity. Therefore one can think of a propensity as a payoff sum.

In our simulations we chose the same conditions as in the experiments. For 100 periods 9 players (agents) interact with each other. Each player has four strategies:

1. road A: This strategy simply consists in taking the decision for the road A.

2. road B: This strategy consists in taking the road B.

3. *direct*:   This strategy corresponds to the direct response mode. The payoff of a player is 1, then the player stays on the road last chosen. If his payoff is 0 the players changes the road.

4. *contrarian*:   This strategy corresponds to the contrarian response mode. The payoff of a player is 1, then the player changes the road. If his payoff is 0 the players will stay on the road.

Initialisation: For each player i let $[x_{i,1}^1, ..., x_{i,n}^1]$ the initial propensity, where n is the number of strategies, which are used in the simulations.

1. period: Each player i chooses strategy j with probability $\dfrac{x_{i,j}^1}{\sum\limits_j x_{i,j}^1}$.

t+1. period: For each player i, let a_i^t the payoff of player i in period t, j the number of the chosen strategy in period t.

IF $a_i^t \geq 0$:
$$x_{i,j}^{t+1} := x_{i,j}^t + a_i^t$$
$$x_{i,k}^{t+1} := x_{i,k}^t, \ k \neq j$$

ELSE
$$x_{i,j}^{t+1} := x_{i,j}^t$$
$$x_{i,k}^{t+1} := x_{i,k}^t - a_i^t, \ k \neq j$$

Each player i chooses strategy j with the probability

$$\frac{x_{i,j}^{t+1}}{\sum\limits_j x_{i,j}^{t+1}}.$$

Tab. 7. The extended payoff-sum model.

In the first period only strategy one and two were available to the simulated subjects since strategy three and four cannot be applied because there is no previous payoff.

In the simulations we did not want to build in prejudices based on theoretical values. Our simulated players base their behaviour on initial propensities and observations only. Of course, it is assumed that as in the experiments the players get feedback about their own payoffs immediately after their choices. In the experimental treatment II additional feedback about the payoff on the route not chosen was given. The payoff sum model makes use of a player's own payoff only and therefore ignores the additional feedback of treatment II.

The differences between treatment I and treatment II cannot be explained by the payoff sum model since it does not process the additional feedback information given in treatment II. For the purposes of comparing our simulation data with the experimental data we ignore the differences between treatment I and II which are not big anyhow.

The difficulty arises that the initial propensities must be estimated from the data. We did this by varying the initial propensities for the strategies *road A* and *road B* over all integer values from 1 to 10 and the initial propensities for the strategies *direct* and *contrarian* over all integer values from 0 to 10. We compared the simulation results with the six variables listed in Tab. 8. We aimed at simulation results which were between the minimum and maximum experimental results over all twelve sessions of treatment I and II. For each of the 12100 parameter combinations we have run 1000 simulations. There were three parameter combinations which satisfied the requirement of yielding means for the six variables between the minimal and maximal experimentally observed values. This was the parameter combination *(1,1,2,1)* and *(2,2,1,1)* and *(3,3,4,2)*. The numbers refer to *road A*, *road B*, *direct* and *contrarian* in this order. The parameter combination is a reasonable vector of initial propensities. There is no difference between road *A* and road *B*, so it is reasonable to have the same propensities for both roads. In two of the three vectors the propensity of the direct mode is greater than the others propensities. There were especially in treatment II as you see on the yule coefficients more direct response modes. This could be an explanation for the direct propensities in the simulations.

	Treatment I & Treatment II				
		Simulations			
	Minimum	{1,1,2,1}	{2,2,1,1}	{3,3,4,2}	Maximum
Player on road A [mean]	4,19	4,48	4,50	4,54	4,74
Player on Road A [standard deviation]	0,67	1,45	1,48	1,50	1,50
Road Changes [mean]	0,59	4,32	4,18	4,51	5,17
Last Roadchange [mean]	54,44	96,11	97,67	97,44	98,11
Yule Coefficiant [mean]	-0,01	0,10	0,04	0,14	0,87
Yule Coefficiant [mean for every player]	0,33	0,50	0,40	0,35	0,76

Tab. 8. Experiments and simulations with 9 players.

It is surprising that a very simple reinforcement model reproduces the experimental data as well as shown by Tab. 8. Even the mean Yule coefficient is in the experimentally observed range. In spite of the fact that at the beginning of the simulation the behaviour of all simulated players is exactly the same. It is not assumed that there are different types of players.

5   Conclusion

Fluctuations persist until the end of the sessions in both treatments.

Feedback on both road times significantly reduces fluctuations in treatment II compared to treatment I. This effect is strong. There is a significant rank correlation between the total number of road changes and the size of fluctuations. In treatment I road changes may serve the purpose of information gathering. This motivation has no basis in treatment II. However, road changes may also be attempt to improve payoffs. The finding of a negative correlation between a subject's payoff and number of road changes suggests that on the average such attempts are not successful.

Two response modes can be found in the data, a *direct* one in which road changes follow bad payoffs and a *contrarian* one in which road changes follow good payoffs. One can understand these response modes as due to different views of the causal structure of the situation. If one expects that the road which is crowded today is likely to be crowded tomorrow one will be in the direct response mode but if one thinks that many people will change to the other road because it was crowded today one has reason to be in the contrarian response mode. We have presented statistical evidence for the importance of the two response modes.

We have also run simulations based on a simple payoff sum reinforcement model. Simulated mean values of six variables have been compared with the experimentally observed minimal and maximal of these variables. The simulated means were always in this range. Only four parameters of the simulation model, the initial propensities, were estimated from the data. In view of the simplicity of the model it is surprising that one obtains a quite close fit to the experimental data. The response modes *direct* and *contrarian* also appear in the simulations as the result of an endogenous learning behaviour by which initially homogeneous subjects become differentiated over time.

References

1. W.B. Arthur. Inductive reasoning and bounded rationality. Am. Eco. Rev. **84**, 406 (1994).
2. D. Challet and Y.C. Zhang. Emergence of cooperation and organization in an evolutionary game. Physica A 246, 407-418 (1997).
3. D. Challet and Y.C. Zhang. *On the minority game: Analytical and numerical studies*. Physica A 256, 514-532 (1998).
4. N.F. Johnson, S. Jarvis, R. Jonson, P. Cheung, Y.R. Kwong, and P.M. Hui. *Volatility and agent adaptability in a self-organizing market*. Physica A 258, 230-236 (1998).
5. C.B. Harley. Learning in Evolutionary Stable Strategy. J. Teoret. Biol. 89, 611-633 (1981).
6. W.B. Arthur. Designing economic agents that act like human agents: A behavioural approach to bounded rationality. Amer. Econ. Rev. Papers Proc. 81, May, 353-359 (1991).
8. A.E. Roth and I. Erev. Learning in extensive form games: Experimental data and simple dynamic models in the intermediate term, Games and economic Behavior 8, 164-212 (1995).
9. A.E. Roth and I. Erev. Predicting how people play games: Reinforcement learning in games with unique mixed strategy equilibrium, American economic review 88, 848-881 (1998).
10. R. Selten, M. Schreckenberg, T. Pitz, T. Chmura, and J. Wahle. Experiments on Route Choice Behaviour. In Interface and Transport Dynamics, Lecture Notes in Computational Science and Engineering, 317-321 (Springer, Heidelberg, 2003).

Experimental Investigation of a Two Route Scenario with Construction Areas

R. Selten[1], M. Schreckenberg[2], T. Chmura[1], and T. Pitz[1]

[1] Laboratory of Experimental Economics – University of Bonn
   Adenauerallee 24-42, 53113 Bonn – Germany
   chmura@uni-bonn.de
[2] Physics of Transport and Traffic – University Duisburg-Essen
   47048 Duisburg – Germany

Summary: The paper reports laboratory experiments on a *day-to-day route choice game* with two routes and alternating construction areas on both routes. Subjects had to choose between a main road *M* and a side road *S*. The capacity was in every period greater for the main road. 18 subjects participated in each session. In periods without construction areas the equilibrium is 12 players on *M* and 6 players on *S*. In periods with construction areas on M the equilibrium is 10 players on *M* and 8 players on *S*. In periods with construction areas on S the equilibrium is 15 players on *M* and 3 players on *S*. Two treatments with 6 sessions each were run at the Laboratory of Experimental Economics at Bonn University. Feedback was given in treatment I only about own travel time and in treatment II on travel time for *M* and *S*. Money payoffs increase with decreasing time. Subjects are told that in each of 200 periods they have to make a choice between the two routes.

Keywords: travel behaviour research, information in intelligent transportation systems, day-to-day route choice, laboratory experiments, construction area

1   Introduction

As reported in Selten et al. [1] understanding individual travel behaviour is essential for the design of Advanced Traveller Information Systems (ATIS), which provide real-time travel information, like link travel times [2,3]. However, the response of road users to information is still an open question [4,5,6]. It is not clear whether more information is beneficial [4]. Drivers confronted with too much information may become oversaturated in the sense that information processing becomes to difficult and users develop simple heuristics to solve the problem [7].

Drivers may also overreact to information and thereby cause additional fluctuations. Thus, the behaviour of the drivers has to be incorporated in the forecast [5,7,8]. ATIS can reduce fluctuations only if behavioural effects are correctly taken into account. The Literature reports a number of experiments on route choice behaviour [1,5,6].

In Selten et al. (2003) [1] we reported about laboratory experiments on a *day-to-day route choice game* with two routes. Subjects had to choose between a main road *M* and a side road *S*. The capacity was in every period greater for the main road. 18 subjects participated in each session. In equilibrium the number of subjects is 12 on *M* and 6 on *S*.

Two treatments with 6 sessions each were run at the Laboratory of Experimental Economics at Bonn University. Feedback was given in treatment I only about own travel time and in treatment II on travel time for M and S. Money payoffs increase with decreasing time. This paper reports about an additional setup with periodical alternating construction areas on both routes. Also in this paper our aim is to present experiments with a large number of periods and with sufficiently many independent observations for meaningful applications of non-parametric significance tests.

2   Experimental Setup

Subjects are told that in each of 200 periods they have to make a choice between a main road M and a side road S for traveling from A to B.

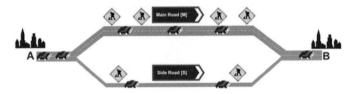

Fig. 1. Participants had to choose between a side road [S] and a main road M].

They were told that M is faster if M and S are chosen by the same number of people and that each of both roads is the faster, the fewer participants use it. The number of subjects in each session was 18. Subjects were informed that on one of both routes construction areas could appear. Before every period with a construction area a sign appears on the computer screen. Construction areas appeared in the following schema:

Ten periods without, 5 periods with a construction area on S, ten periods without, 5 periods on M and so on.

The time t_M and t_S depends on the numbers n_M and n_S of participants choosing M and S, respectively:

1. without construction areas

$$t_M = 6 + 2n_M$$
$$t_S = 12 + 3n_S$$

2. construction area on M

$$t_M = 2(n_M + 5) + 6$$
$$t_S = 12 + 3n_S$$

3. construction area on S

$$t_M = 6 + 2n_M$$
$$t_S = 3(n_S + 5) + 12 .$$

The period payoff was $42 - t$ with $t = t_M$ if M was chosen and $t = t_S$ if S was chosen. However the subjects were not informed about the quantitative relationships between t_M and n_M and between t_S and n_S.

The total payoff of a subject was the sum of all 200 period payoffs converted to money payoffs in DM with a fixed exchange rate of 0.015 DM for each experimental money unit (Taler). Additionally, every participant received a lump sum payment of 200 Taler and a show-up fee of 10 DM. One session took roughly one and a half hours.

All pure equilibria of the game are characterized by

1. without construction areas
$$n_M = 12 \text{ and } n_S = 6,$$

2. construction areas on S
$$n_M = 15 \text{ and } n_S = 3,$$

3. construction areas on M
$$n_M = 10 \text{ and } n_S = 8.$$

The equilibrium payoff is 12 Taler per player in periods without a construction area and 6 Taler per player in periods with a construction area on M or S.

Two treatments have been investigated. In treatment I the subjects received:

- travel time of the last chosen route
- last chosen route
- payoff of the last period in Taler
- cumulated payoff in Taler
- number of the actual period

In treatment II additional feedback was provided about the travel time on the non-chosen-route in the last period. Six sessions were run with treatment I and six with treatment II. No further information was given to the subjects.

3   Equilibrium Predictions and Observed Behaviour

3.1   Number of Players on the Side Road S

Figure 2 shows the number of participants on the side road S as a function of time for a typical session of treatment I.

There are substantial fluctuations until the end of the session. The same is true for all sessions of both treatments. The overall average in periods without construction areas of numbers of participants on S is very near to the equilibrium prediction.

The average number of participants on S changes in the right direction in periods with construction areas on S. In all sessions these averages are lower then the ones for periods without construction areas. This is significant on the 5 % level (two sided) for treatments I and treatment II separately according to the binomial test. However in every session the average number for periods with construction areas on S is above the equilibrium 3 for such periods. These numbers are 5.32 in treatment I and 5.26 in treatment II, which is well above the equilibrium value of 3. An adjustment to construction areas on S takes place but it is far from complete.

The case that there are construction areas on M shows a similar picture. In all sessions the averages of players on S are higher for periods with construction areas on M, then for periods without construction areas. We obtain the same significances for an adjustment in the right direction. The mean number of players on S is 6.72 in treatment I and 6.6 in treatment II. These numbers are far below the equilibrium number 8. Also here an adjustment takes place which, however, remains incomplete.

The fluctuations can be measured by the standard deviation of the number of participants choosing S per period.

The fluctuations are larger under treatment I than under treatment II. The effect is significant. The null-hypothesis in periods without construction areas is rejected by a Wilcoxon-Mann-Whitney-Test on the significance level of 5 % (one sided). In periods with construction areas on S and in periods with construction areas on M the significance level is 1 % (one sided).

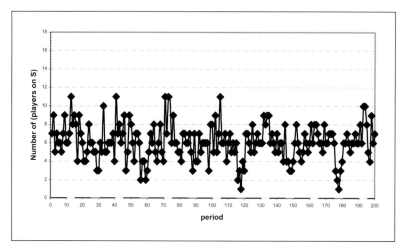

Fig. 2. Number of participants on S [a typical session of treatment I].

		number of players on S [without construction areas]		number of players on S [with construction areas on S]		number of players on S [with construction areas on M]	
		mean	sdt. dev.	mean	sdt. dev.	mean	sdt. dev.
	session I 01	6,07	1,99	5,27	2,85	6,83	3,41
	session I 02	5,73	1,88	5,17	2,68	6,69	3,03
	session I 03	5,96	2,04	5,5	2,67	6,43	2,96
T I	session I 04	6,12	1,89	5,37	2,63	6,89	2,95
	session I 05	6,09	1,96	5,2	2,78	6,91	3,2
	session I 06	5,99	1,88	5,43	2,74	6,6	3,28
	treatment I	**5,99**	**1,94**	**5,32**	**2,72**	**6,72**	**3,14**
	session II 01	6,1	1,34	5,17	2,49	6,71	2,92
	session II 02	6,07	1,53	5,1	2,22	6,89	2,52
	session II 03	5,8	1,9	5,33	2,23	6,29	2,49
T II	session II 04	5,89	1,9	5,43	2,65	6,34	2,42
	session II 05	6,12	1,87	5,47	2,47	7	2,93
	session II 06	5,83	1,5	5,07	2,55	6,34	2,72
	treatment II	**5,97**	**1,67**	**5,26**	**2,44**	**6,6**	**2,67**

Tab. 1. Mean and standard deviation of the number of players on S.

The players had more problems to reach the equilibrium during construction areas. The reason could be, that the construction time is shorter, than periods without constructions.

3.2   Road Changes

Figure 3 shows an example of the number of road changes as a function of time for a typical session of treatment II. This surprises, because of many construction areas on both routes.

There was a negative trend in each session of treatment II. In both treatments of most sessions are negative trends.

We splitted each session into 4 conditions. Condition I are the periods without construction areas, Condition II are the periods with construction areas on *M*. Condition III are periods with construction areas on *S*. Condition IV are periods during transitions to and away from construction areas.

The mean number of road changes is significantly higher in treatment I in every condition. The null-hypothesis is rejected by the Wilcoxon-Mann-Whitney-Test on a level of 5 % (one sided) in condition II, III and IV. In condition I the significance level is by 10% (one sided). The mean number of road changes under treatment I is also higher than under treatment II.

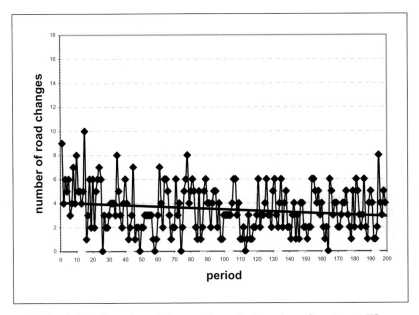

Fig. 3. Number of road changes [a typical session of treatment II].

In Selten et al. (2003) we show that under treatment I subjects who mainly choose only one of the roads feel the need to travel on the other road from time to time in order to get the information on both roads. Under treatment II there is no necessity for such information gathering. This seems to be the reason for the greater number of changes and maybe also for the stronger fluctuations under treatment I.

mean number of road changes								
	Condition I		Condition II		Condition III		Condition IV	
	T I	T II	T I	T II	T I	T II	T I	T II
1	5,53	3,42	4,68	2,96	5,7	3,86	7,42	5
2	4,64	4,54	4,8	4,42	5,66	5,68	6,62	5,42
2	5,84	5,21	5,26	4,71	6,49	5,54	6,81	7
2	6,68	5,5	5,05	4,21	6,84	4,68	7,42	6,62
2	5,6	5,95	4,55	5,13	5,41	6,43	7,04	7,12
6	5,21	3,64	5,34	1,88	6,16	3,25	6,96	4,54
T	**5,58**	**4,71**	**4,95**	**3,88**	**6,05**	**4,9**	**7,04**	**5,95**

Tab. 2. Number of Road Changes in 4 Conditions.

4 Response Mode

As we did in our paper Experiments and simulations on day to day route choice behaviour we classified for this experiments response modes of players. Participant who had a bad payoff on the road chosen may change his road in order to travel where it is less crowded. We call this the *direct* response mode. A road change is the more probable the worse the payoff was. The *direct* response mode is the prevailing one but there is also a *contrarian* response mode. Under the *contrarian* response mode a road change is more likely the better the payoff was. The contrarian participant expects that a high payoff will attract many others and that therefore the road chosen will be crowded in the next period.

The equilibrium payoff is E=7 or E=10. Payoffs perceived as bad tend to be below E and payoffs perceived as good tend to be above E. Accordingly we classified the response of a subject as direct if the road is changed after a payoff smaller than E or not changed after a payoff greater than E. An opposite response is classified as contrarian. Tab. 3 shows the numbers of times in which a subject changes roads (c_- for a payoff below E and c_+ for a payoff above E), or stays at the same road (s_- for a payoff below E and s_+ for a payoff above E).

	change	stay
payoff < E	c_-	s_-
payoff > E	c_+	s_+

Tab. 3. 2x2 table for the computation of Yule coefficients.

For each subject such a 2x2 table has been determined and a Yule coefficient Q has been computed as follows.

$$Q = \frac{c_- \cdot s_+ - c_+ \cdot s_-}{c_- \cdot s_+ + c_+ \cdot s_-}$$

The Yule coefficient has a range from −1 to +1. In our case a high Yule coefficient reflects a tendency towards direct responses and a low one a tendency towards contrarian responses.

In Tab. 4 the mean yule coefficients of the former experiments without construction areas and of the reported experiments with construction areas are shown for every treatment.

The mean yule coefficients are significantly higher in the experiments with construction areas. The null-hypothesis for both treatments is rejected by a Wilcoxon-Mann-Whitney-Test on the significance level of 5 % (one sided).

That means, that there are less contrarian response modes in experiments with construction areas. An explanation could be that in a more difficult situation, here the construction areas, people do react more directly.

The yule coefficients are not only connected to the players, there is also a connection to the situation.

| | | Yule coefficients Q | |
		Experiment without construction areas	Experiments with construction areas
Treatment I	session I 01	0,214	0,334
	session I 02	0,373	0,337
	session I 03	0,277	0,324
	session I 04	0,191	0,361
	session I 05	0,313	0,316
	session I 06	0,332	0,435
	treatment I	0,283	0,351
Treatment II	session II 01	0,365	0,362
	session II 02	0,374	0,349
	session II 03	0,308	0,378
	session II 04	0,271	0,363
	session II 05	0,246	0,438
	session II 06	0,122	0,392
	treatment II	0,281	0,380

Tab. 4. Mean of the Yule coefficients in both treatments.

5   Conclusion

The study has shown, if there is no construction area, that the mean numbers on both roads tend to be very near to the equilibrium. During periods with construction areas the mean number of players differ from the equilibrium. Nevertheless, fluctuations persist until the end of the sessions in both treatments.

Feedback on both road times significantly reduces fluctuations in treatment II compared to treatment I. As in the Experiments without construction areas two response modes can be found in the data, a *direct* one in which road changes follow bad payoffs and a *contrarian* one in which road changes follow good payoffs. One can understand these response modes as due to different views of the causal structure of the situation.

It is remarkable, that yule coefficients are significantly higher in experiments with construction areas.

References

1   R. Selten, M. Schreckenberg, T. Pitz, T. Chmura, and J. Wahle. Experiments on Route Choice Behaviour. In Interface and Transport Dynamics, Lecture Notes in Computational Science and Engineering, 317-321 (Springer, Heidelberg, 2003).

2   J. Adler and V. Blue. Toward the design of intelligent traveler information systems, Transpn. Res. C 6, 157 (1998).

3   W. Barfield and T.A. Dingus (eds.). Human Factors in Intelligent Transportation Systems. (Lawrence Erlbaum Associates, Publishers London, 1998).

4   M. Ben-Akiva, A. de Palma, and I. Kaysi. Dynamic network models and driver information systems, Transpn. Res. A 25, 251 (1991).

5   P. Bonsall. The influence of route guidance advice on route choice in urban networks, Transportation 19, 1-23 (1992).

6   H.S. Mahmassani and Y.H. Liu. Dynamics of commuting decision behaviour under advanced traveller information systems. Transp. Res. C 7, 91–107 (1997).

7   G. Gigerenzer, P.M. Todd, and ABC Research Group (eds.). Simple heuristics that make us smart. (Oxford, University Press, 1999).

8   J. Wahle, A. Bazzan, F. Klügl, and M. Schreckenberg. Decision dynamics in a traffic scenario, Physica A 287, 669-681 (2000).

Phase Diagrams of an Internet Model with Multi-Allocation of Sites

T. Huisinga[1], R. Barlović[1], A. Schadschneider[2], and M. Schreckenberg[1]

[1] Theoretische Physik Fakultät 4 – Universität Duisburg-Essen
47048 Duisburg – Germany
e-mail: huisinga, barlovic, schreckenberg@uni-duisburg.de
[2] Institut für Theoretische Physik – Universität zu Köln
50937 Köln – Germany
e-mail: as@thp.uni-koeln.de

Summary. A recently introduced cellular automaton model for Internet traffic is investigated in the context of boundary induced phase transitions and the appropriate phase diagrams. Since the model allows multi allocation of sites there are some deviations in the internal dynamics as well as in global properties compared to known driven lattice gas models. As a consequence the yielding phase diagram derived by numerical simulations reveals some interesting new features like a capacity shift in dependence to the allocation number.
Keywords: Cellular Automata Models, Internet, Driven Lattice Gas Models

In the field of statistical physics cellular automata models have become a very useful tool for investigating complex many-body systems. Their flexibility and in particular their simple applicability predestine these models for investigating topics of inter-disciplinary fields like biochemical systems, vehicular traffic or especially in this case Internet data transport. In these topics the main focus of interest lies on the survey of real data combined with the development of effective models describing the main features to understand the basic phenomena occurring in reality. In this paper, a recently introduced one-dimensional cellular automaton model with open boundaries for data packet transport in the Internet [1] is investigated in detail. This model can be classified as a driven lattice gas (DLG) model where particles move according to a driving force from one reservoir at one end to another reservoir at the other end of the system. In contrast to periodic systems with a conserved number of particles, these systems reside in a state far from equilibrium and therefor can not be described in the context of statistical equilibrium mechanics. A well understood DLG-model describing transportation processes is the Asymmetric Simple Exclusion Process (ASEP) a lattice gas automaton [3] with slightly changes first introduced to simulate biological systems or later vehicular traffic [4]. In fact, the investigated model is another derivate of the

ASEP. In contrast thereto, in this model each site can store more than one particle in a buffer, and moreover these particles are able to overtake each other. This feature is essential for describing the statistical features of Internet Ping time-series [1]. Similar to the ASEP, this model shows a rich behavior of non equilibrium boundary induced phase transitions which in the case of the ASEP are exactly known from analytical investigations [5]. In particular for such systems with open boundary conditions and a specific shape of the periodic fundamental diagram (FD) a theory was developed in [6] which is capable to depict the phase diagram of the open system by the so called extremal principle:

$$J = \begin{cases} \max_{\rho \in [\rho_R, \rho_L]} & \text{for} \quad \rho_L > \rho_R \\ \min_{\rho \in [\rho_L, \rho_R]} & \text{for} \quad \rho_L < \rho_R \end{cases} \tag{1}$$

which still holds for systems with meta-stable states [7]. Here ρ_R respectively ρ_L correspond to the densities at the right respectively left boundary, so that the current J in the system can be describe by the relations between these two densities.

In this paper, especially the shape of the phase-diagram is of interest. Therefore the FD of the periodic version of the model is investigated. Comparing the FDs of the ASEP and the Internet model there are some apparent differences. A striking feature of the Internet model is hereby a shift of the maximal flow to higher densities (the shape of the FD becomes asymmetric). This can also be found in other fields of transportation problems, e.g:, in pedestrian or cluster movement. Moreover, a strong increase of the flow for larger buffer sizes compared to the FD of the ASEP with buffer size $B = 1$ occurs. Typical FDs of the periodic version of the Internet model and the ASEP showing this features are depicted in the left part of fig. 2.

Definition of the Model

The exact definition of the investigated model with open boundaries (see fig. 1) is as follows:

Consider a one-dimensional lattice of $l = \{1, 2, 3, ..., L\}$ sites with open boundaries, whereby each site can be empty or store a maximum amount of B particles in a buffer. At the left side of the system a reservoir is introduced from which particles are inserted into the first buffer as far as the buffer is not completely occupied and move throughout the complete system to the right boundary. Here the particles are leaving the system (last buffer) and will deleted at once to return back into the reservoir.

At each time step t particles corresponding to the inflow rate α are inserted at the left boundary (site $l = 1$) as far as there is space left in the buffer of size B. Particles corresponding to the outflow parameter β are removed at the right end of the system. In fact the inflow respectively outflow is realized at each update step by choosing an identically independent distributed (i.i.d) variable for each site in the buffers of the first and the last router which determines if particles have to be inserted or removed. This means that for $\alpha = 1.0$, each position in the buffer of the first site is occupied by a particle from the reservoir as long as enough space is left there. An inflow rate of $\alpha = 0.5$ means that at each site of the first router a particle is inserted with probability 0.5 as far as this side is not already occupied. Contrary an outflow rate $\beta = 1.0$ denotes, that all particles in the last router will be erased in the next time step while an outflow rate of $\beta = 0.5$ is realized by erasing each particle with probability 0.5 from the last site N.

The parallel update of the system $t \to t+1$ for all particles in all buffers of site $l = 1 \to L-1$ is then as follows:

At time t the $n_l(t)$ particles in the buffers of each site $l = 1, \ldots, L-1$ are picked up respectively to their time of arrival and move with probability p to the next neighboring site $l+1$ as far as the buffer of this site is not completely occupied $(n_{l+1}(t) < B)$. In case of a complete occupation of the succeeding site $n+1$ the remaining data packets reside at their current position.

Note, that, particles are updated only once at each time step. Moreover, due to the stochastic character of the update in the case of the multi occupation $(B > 1)$, particles can overtake each other. For $B = 1$ this model is identically to the ASEP [2, 3] with open boundary conditions.

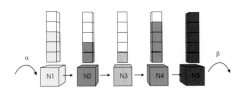

Fig. 1. Open system consisting of $N = 5$ routers with buffer size $B = 5$. Particles are inserted at the left boundary and move to the right end of the system.

Numerical Results

The presented model is capable to provide significant higher flows than the mentioned ASEP derivates, because of the fact that particles in a buffer which are not chosen to move because of the internal dynamics (hopping probability p) can be passed by other particles in the same buffer.

Considering an open system, it is clearly shown in the right part of fig. 2 that the complete FD of the periodic system (full line) can be reproduced by the open system for an appropriate choice of the parameters α and β. This means, that the extremal principle (1) can be utilized and allows to give a phenomenological description of the shape of the phase diagram for the open system as well as the nature of the phase transitions.

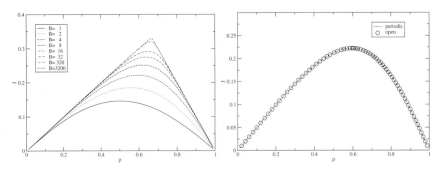

Fig. 2. Left: FD of periodic systems for buffer size $B = 1, 2, 4, 8, 16, 32, 320, 3200$ and $p = 0.5$. The number of cells $(N \cdot B)$ is kept constant. $B = 1$ is identical to the symmetric FD of the ASEP. **Right:** FD of a system with $N = 1600$ routers, buffer size $B = 4$ and $p = 0.5$. The dotted line corresponds to the simulation results of a periodic system while the symbols represent the simulation results of an open system for a suitable choice of parameters.

As mentioned before, buffer size $B = 1$ corresponds to the ASEP where the shape of the phase diagram is exactly known [5]. As well as in the ASEP, in the presence of noise three phases can be distinguished (see space time plots in fig. 3). First of all there is a low density phase where particles move nearly undisturbed from one reservoir to the other one. Small density fluctuations will dissolve quickly and no jams exist. The flow is determined by the inflow rate α at the left boundary while p and β do not have any restricting influence on the system flow. The bulk density corresponds to the density at the left border ρ_L. On the other hand it exists a high density phase where backwards moving jams are mostly induced at the right boundary, i.e., the outflow rate β out of the system determines the maximal flow in the system. Here, the bulk density is given by the density at the right border ρ_R. And finally there is a high flow phase or maximum current phase. In this phase the bulk flow is determined by the internal hopping probability p. The inflow rate α is larger than the capacity of the first router and so the system is overfed. This leads to small high density region at the left boundary while on the other hand the right boundary does not have any restrictive effect on the system flow, i.e., the bulk flow is less than the capacity of the right boundary and corresponds to the maximal flow a router is capable to provide for the given hopping probability p.

To elucidate the special characteristics of the models, the influence of larger buffer sizes onto the shape of the phase diagram is investigated. Thereto the bulk flow in dependence to the inflow respectively the outflow is analyzed.

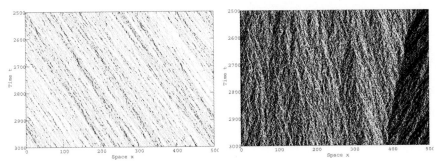

Fig. 3. Space-Time plots of $N = 500$ routers, buffer size $B = 5$ and hopping probability $p = 0.5$. **Left:** System in low density state. Particles are inserted at the left boundary with $\alpha = 0.1$ and move mostly undisturbed to the right end where they are removed with $\beta = 1.0$. **Right:** System in high density regime. Here $\alpha = 1.0$ which means that the system is overfed and the left part of the system resides in the maximum current phase. Because of the restriction of the outflow to $\beta = 0.3$ which is less than $p = 0.5$ the right part of the system is situated in the high density phase which is growing to the left boundary.

Obviously, the maximal current principle holds and can be used to predict the phase diagrams since the FD of the periodic system exhibits one single maximum and moreover can be reproduced by the open version of the model. This leads to a qualitatively similar shape of the phase diagram compared to the one of the ASEP [8].

In order to reveal the complete shape of the phase diagram in dependence of different buffer sizes, monte-carlo simulations were made to determine the point of maximal bulk flow in correspondence to the inflow rate α respectively the outflow rate β. As a result, the phase diagrams derived by monte-carlo simulations for deterministic and stochastic movement are in totally agreement with the results of the phenomenological approach introduced by Kolomeisky et al. (see fig. 4). The deterministic case with $p = 1.0$ reflects the expected results, namely two phases, a low density and a high density phase, which conform to earlier investigations of phase transitions in DLG-models [5, 7] are separated by a first order phase transition. Clearly, the influence of varying buffer sizes becomes apparent (see left part of fig. 4). For medium inflow rates the phase interface for larger buffer sizes is shifted to larger outflow rates which means that the free flow phase is strongly reduced. Considering the provided flow it has to be mentioned that the maximal flow is highly enlarged compared to ASEP. In case of the stochastic movement another phase occurs, namely a maximum current phase. This phase is entered for maximum inflow rates while the system flow is limited by the maximal system capacity. Explicitly, as well as in the deterministic case the diverse buffer sizes affect the shape of the phase diagram. Similar to the phase transitions found in the ASEP there is a first order phase transition between free flow and high density regime and there are second order phase transitions between low density and maximum current respectively high density and maximum current regime. The only difference thereto are the inflow respectively outflow rates where the phase interfaces are located, which as a result of the multi allocation, shifts the interfaces to larger α resp. β (see right part of fig. 4). Consequently as shown in the deterministic case the jammed phase is strongly increased while the free-flow phase is slightly enlarged both at the expense on the maximum current phase. Even here it is important to say that the provided flows in the free flow, jammed and maximum current phase are highly increased for larger buffer sizes compared to the ASEP.

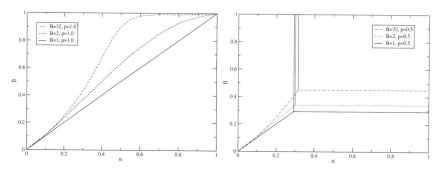

Fig. 4. Phase diagrams for the deterministic (left) and stochastic version (right) of the investigated model for buffer sizes $B = 1, B = 2$ and $B = 32$. In the deterministic version, the shift of the phase interface between high density and low density regime to larger outflow rates for larger buffer sizes can clearly be identified. The stochastic model exhibits the same characteristics for the phase interface between low and high density phase. In case of the transition from low density to the maximum current phase respectively from high density regime to the maximum current phase the transition is situated at larger outflows respectively inflows.

References

1. T. Huisinga, R. Barlović, W. Knospe, A. Schadschneider, and M. Schreckenberg. *Physica A*, **294**, 249-256 (2001).
2. J. Krug. *Phys. Rev. Lett.*, **67**, 1882-1885 (1991).
3. D. Chowdhury, L. Santen, and A. Schadschneider. *Physics Reports*, **329**, 199-329 (2000).
4. K. Nagel and M. Schreckenberg. *J. Physique I*, **2**, 2221-2229 (1992).
5. B. Derrida, M.R. Evans, V. Hakim, and V. Pasquier. *J. Phys. A*, **26**, 1493 (1993).
6. A.B. Kolomeisky, G.M. Schütz, E.B. Kolomeisky, and J.P. Straley. *J. Phys. A*, **31**, 6911-6919 (1998).
7. R. Barlović, T. Huisinga, A. Schadschneider, and M. Schreckenberg. *Phys. Rev. E*, **66**, 046113 (2002).
8. M. R. Evans, N. Rajewski, and E.R. Speer. *J. Stat. Phys.*, **95**, 45-96 (1999).

Adaptive Traffic Light Control in the ChSch Model for City Traffic

R. Barlović[1], T. Huisinga[1], A. Schadschneider[2], and M. Schreckenberg[1]

[1] Theoretische Physik, Fakultät 4 – Universität Duisburg-Essen
    47048 Duisburg – Germany
    `barlovic, huisinga, schreckenberg@uni-duisburg.de`
[2] Institut für Theoretische Physik – Universität zu Köln
    50937 Köln – Germany
    `as@thp.uni-koeln.de`

Summary. The impact of adaptive traffic light control is studied in the Chowdhury Schadschneider (ChSch) cellular automaton (CA) model for city traffic. Therefore, three adaptive strategies are presented being able to react flexible to the traffic conditions. It is shown that the adaptive control is capable to direct the system to its *system optimum*. Furthermore, the impact inhomogeneous densities is investigated with the aim to demonstrate the difference between a global traffic control and an adaptive one if varying traffic conditions are considered.
Keywords: Cellular Automata Models, City Traffic, Traffic Light Control

1 Introduction

In the recent years there were strong attempts to develop a theoretical framework of traffic science among the physics community [1–3]. Consequentially, a nearly completed description of highway traffic, e.g., the "Three Phase Traffic" theory, was developed [4]. This describes the different traffic states occurring on highways as well as the transitions among them. Also the concepts for modeling vehicular traffic are well developed. Most of the models introduced in the recent years are formulated using the language of cellular automata (CA) [5]. A newsworthy CA approach by Lee et al. [6] (see also this proceedings) is even capable to reproduce the "Three Phase Theory" in all details. Unfortunately, no comparable framework for the description of traffic states in city networks is present. In contrast to the highway networks, where individual highway segments can be treated separated, the structure elements of city networks exert an immense influence onto the traffic dynamics. The aim here is to investigate the impact of the most frequent structure element, i.e., traffic light, onto the dynamics of an idealized city network. For this purpose different traffic light strategies are analyzed with the help of the Chowdhury-Schadschneider (ChSch) [7] CA model for city traffic.

As one can see from Fig. 1 the network of streets builds a $N \times N$ square lattice. In addition, all intersections are assumed to be equal, i.e., there are no main roads in the network where the traffic lights have a higher priority. The streets parallel to the x-axis allow only single-lane east-bound traffic while the ones parallel to the y-axis manage the north-bound traffic. The separation between any two successive intersections consists of $D - 1$ cells so that the total length of a single street is $L = N \times D$. The dynamics of vehicles on the streets is given by the update rules of the well known NaSch model [5] while the vehicles break in front of a red light according to the rules of the ChSch [7] model.

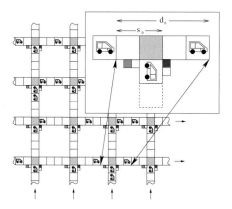

Fig. 1. Snapshot of the model topology. Vehicles move from west to east on the horizontal streets or from south to north on the vertical ones. The magnification shows a segment of a west-east street.

2 Global Strategies

One big advantage of the ChSch model is the fact that the "System Optimum" can be given by phenomenological arguments. This can act as an indicator for the performance of investigated cycle time strategies. The "System Optimum" is given for low densities by the free flow of the underlying NaSch model (no interaction with crossings) and for high densities also by the high density flow of the NaSch model (interaction among vehicles is dominant). In the intermediate region a plateau is formed corresponding to half of the maximum flow of the NaSch model since the network has two equitable directions.

In this section the results for three global strategies, these act as a base of comparison for the adaptive strategies, are presented (see [8] for a more detailed description). The considered global strategies are a "synchronized strategy", a "green wave strategy" and a "random strategy". In the case of

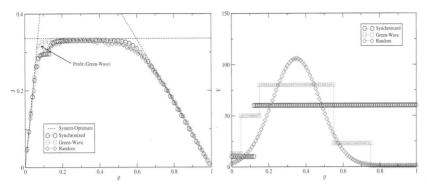

Fig. 2. For the global strategies the global flow (left) can nearly reach the "System Optimum" (dashed line). Note, that the optimal curves are obtained by systematically scanning all cycle times. This can be seen on the cycle time curves (right).

the synchronized strategy all traffic lights switch synchronously to green (red) for the east (north) bound vehicles and vice versa. The fundamental diagrams (FD) given in Fig. 2 were obtained by systematically scanning the model parameters. More precisely, the curve was obtained by picking the optimal cycle times corresponding to the maximal flow out of plots depicting mean flow vs. density.

As can be seen in the FD in Fig. 2(left) an optimal adjusted synchronized strategy nearly matches the "System Optimum" despite of densities in the area of $\rho = 0.12$. Below this density the optimum cycle time for the synchronized strategy is given by the free flow travel time between two intersections and above by the travel time of a jam. This can be seen in the left part of Fig. 2.

In the case of a green wave strategy, i.e., adjacent traffic lights switch with an defined offset, the optimal FD is obtained if a step function like cycle time dependence is used (see Fig. 2 (right)). Additionally, an appropriate offset has to be determined for the green wave strategy, this is equal to the free flow travel time for the depicted case. In Fig. 2 (left) it can be seen that the green wave is capable to match the "System Optimum" for all densities.

A further investigated candidate among the global strategies is the random strategy, i.e., adjacent traffic lights switch with a random offset. Also here a quite good results is achieved, except the region among $\rho = 0.12$ which is only matched by the green wave strategy. The optimal cycle time curve (see Fig. 2 (right)) shows a Gaussian shape this adapts the step function like shape of the optimal green wave strategy.

3 Adaptive Strategies

In the following three different adaptive strategies are presented for the ChSch model. These are kept simple in the sense that only one control parameter

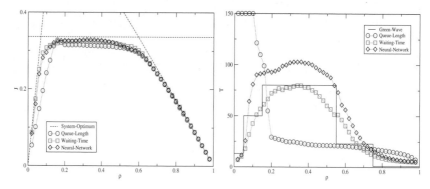

Fig. 3. The achieved flows for the adaptive strategies match the optimal curve relatively good, especially if considering that no information about the global traffic state is needed. Interestingly the shape of the cycle time curves adjust itself to the step function like curve of the optimal green wave strategy for the switching based on the waiting time and the switching in analogy to a neural network.

is used which can be easily related to a measurable quantity of real traffic. Furthermore one parameter should serve for all densities, i.e., the traffic lights should be able to react flexible to a varying traffic demand. As a consequence, the extensive procedure of identifying the optimal cycle time in dependence of the density as in the case of the global strategies is not needed anymore.

The first investigated adaptive strategy is the "switching based on the queue length". Here a traffic signal switches if the length of a vehicle queue in front of a red light trespasses a certain value. As one can see in Fig. 3 the best results are obtained if the control parameter (queue length) is set to $n = 16$. Then the FD matches the "System Optimum" relatively good, despite for low densities.

Further investigated adaptive strategies are the "switching based on waiting time" and the "switching in analogy to a neural network". In the first case a traffic light switches to red if the green phase is not used by a vehicle for a certain time. Here the best result is obtained if the control parameter (waiting time) is set to $n = 4$. The corresponding FD shown in Fig. 3 (left) matches the "System Optimum" very well even for low densities. An interesting feature of this strategy is that the cycle time curve adjusts itself to the step-function like shape of the green wave strategy, this is identical to the "System Optimum". In the case of switching in analogy to a neural network the traffic lights act like integrate-and-fire neurons. More precisely the number of passed vehicles is integrated and determines the cycle time (potential) of a traffic light. After the switching process (fire) the potential is reset to zero again. The switching in analogy to a neural network strategy leads to similar results like the switching based on the waiting time. Also the cycle time curve shows an Gaussian profile which adjusts around the optimal green wave curve (step-function).

4 Impact of Inhomogeneous Densities

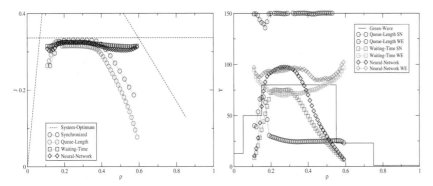

Fig. 4. The impact of inhomogeneous densities along the two directions of the network is illustrated. In the case of the global strategies the reachable flow falls clearly below the "System Optimum" while for the adaptive strategies an excellent agreement is achieved.

In the following it is shown how the adaptive strategies presented above compare under more realistic traffic conditions. Therefore the impact of inhomogeneous densities along the two directions of the network is presented exemplarily. For a detailed investigation about the impact of realistic traffic conditions, e.g., turning, to the choice of a proper traffic signal strategy it is referred to [9, 10]. In order to investigate inhomogeneous densities in the network, the density along the south-north direction is fixed ($\rho = 0.2$ in Fig. 4) and the density along the west-east direction is varied. This situation is somehow comparable to real city networks which are mostly build out of main streets or directions with a higher density, and minor ones with a lower density.

In the case of global strategies, the two directions of the network can be assumed to be completely decoupled. Consequently, the result for the whole network is given by the superposition of the results from the two directions. In order to illustrate this point the FD for the synchronized strategy is depicted in Fig. 4 (left). It can clearly be seen that the flow falls below the "System Optimum" since the direction of the network with the low density can not contribute enough to the overall system flow.

Turning to the adaptive strategies, one can see (Fig. 4 (left)) that the switching based on the waiting time as well as the switching based on a neural network produce excellent results. These strategies are capable to react flexible to the inhomogeneous traffic demand. The global flow lies completely on a plateau very close to the maximum possible system flow although the reachable density range is strongly limited. In Fig. 4 (right) it is shown that completely different cycle times are adapted for the two directions of the network. Furthermore it can be seen in Fig. 4 (left) that the switching based on the queue length strategy is unable to manage the inhomogeneous density distribution.

5 Summary

Recapitulating, in this paper a brief overview on adaptive traffic light strategies within the framework of the ChSch model is given. An important point is that the "System Optimum" can be given in the ChSch model this can be used as the base of comparison for the investigated strategies. Among the global strategies the green wave strategy revealed to match the "System Optimum". The main part of the investigation in this paper deals with adaptive signal strategies. Therefore, also here three different strategies are investigated. It is shown that these strategies lead to quite good results. Finally, it is demonstrated that the adaptive strategies are capable to react flexible to more realistic traffic conditions (inhomogeneous densities) while the global strategies fail.

References

1. D. Chowdhury, L. Santen, and A. Schadschneider, Phys. Rep. 329, 199 (2000).
2. D. Helbing, in *Verkehrsdynamik*, Springer, Berlin, 1997.
3. *Traffic and Granular Flow, '97, '99, '01*, Springer, (1996,1998,2000,2003).
4. B.S. Kerner, Phys. Rev. E **65**, 046138 (2002).
5. K. Nagel and M. Schreckenberg, J. Physique I **2**, 2221 (1992).
6. H.K. Lee, R. Barlović, M. Schreckenberg, and D. Kim, Phys. Rev. Lett. **92**, 238702 (2004).
7. D. Chowdhury and A. Schadschneider , Phys. Rev. E **59**, R1311 (1998).
8. E. Brockfeld, R. Barlović, A. Schadschneider, and M. Schreckenberg, Phys. Rev. E**64**, 056132 (2001).
9. R. Barlović, Phd thesis, Universität Duisburg-Essen, Germany (2004).
10. R. Barlović, A. Schadschneider, and M. Schreckenberg, in preparation.

Vehicle Flow and Phase Transitions in Traffic Networks

S. Jain

Information Engineering, School of Engineering and Applied Science –
Aston University
Birmingham B4 7ET – UK
S.Jain@aston.ac.uk

Summery. A computer simulation of a simple dynamical model of vehicles moving across a two-dimensional traffic network is performed. Traffic management is enforced through the roundabouts: providing the conditions are appropriate, vehicles can move through the roundabouts. When the density of the travelling traffic is low, the vehicles can move freely.
Increasing the vehicle density leads to structured patterns which initially enable a high density of vehicles to travel throughout the system but eventually leads to gridlock. A phase transition separates a congestion free phase from a congested one as the vehicle density increases.
Keywords: computer simulations, dynamical model, two-dimensional traffic, phase transitions

1 Introduction

The flow of traffic and the formation of traffic jams have been studied extensively in recent years [1]. The traffic flow is often characterised by the relation between the density of vehicles on the road and the traffic throughput. This quantity has been studied in a large variety of models [1, 2].

For a low density of vehicles, the traffic is free-flowing and grows linearly with the density. At high density, congestion occurs and traffic jams are formed. This can lead to a complete cessation of the movement of traffic [3].

Here we discuss the flow of two directional vehicles over a system of interconnected roundabouts.

2 Vehicle Flow as a Three-State System

We consider a simple two-dimensional square grid of roundabouts (Fig. 1). Traffic can flow left to right ("right traffic") or from top to bottom ("down" traffic).

Each roundabout can be in one of three states: it can contain vehicles (i) moving right (ii) moving down (iii) or no vehicles.

Vehicles are permitted to move to the next roundabout only if it's empty. If the roundabout is occupied, the vehicle remains where it is until the roundabout becomes free. Furthermore, only one move is allowed at each time step. This means that the roundabouts are effectively assumed to be one move apart and that they can accommodate only one vehicle at a time. Clearly, one can consider both serial and parallel updating of the entire system. In this work only parallel updating was considered with preference given to right traffic in cases of conflicts. Fig. 2 depicts a typical scenario where both right (R) and down (D) traffic could move to the empty round-about. We avoid such collisions by simply considering right moving traffic first.

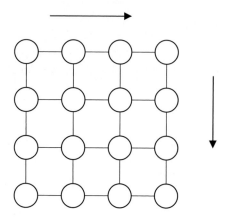

Fig. 1. Two-dimensional network of roundabouts.

Fig. 2. Parallel updating at roundabouts.

We also assume periodic boundary conditions. This means that vehicles leaving the system on the right/bottom re-enter from the left/top. As a result, total vehicle flow is conserved in the system.

3   Simulations

At the beginning of each simulation the two-dimensional set of roundabouts is initialised in one of three different states: empty, containing right traffic or down traffic.

A grid containing 2500 (= 50 x 50) roundabouts with a range of vehicle densities from 10 % - 90 % was used in the simulations and the maximum number of iterations employed was 10000.

As explained earlier, the right traffic is considered first and updated in par-allel, followed by the left traffic. The pattern of occupied and empty round-abouts is allowed to evolve and is monitored throughout the simulations.

4   Results

For vehicle densities less than 30%, we have free flowing traffic throughout the grid. For densities around 30%, we have the emergence of diagonal struc-tures from the bottom left to the top right which enable the traffic to continue flowing. However, for higher densities, between 40% - 60%, the right and down traffic merges into diagonal structures from the top left to the bottom right and we have gridlock. At even higher densities (above 70%), although the structures which emerged at lower densities breakdown, we still have complete gridlock.

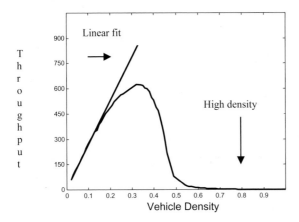

Fig. 3. Average vehicle throughput against density.

In Fig. 3 we plot the average vehicle throughput for right traffic against density. Note: we define the throughput as the total number of vehicles travelling right into a roundabout over an entire run, averaged over all the roundabouts. We can clearly see that at very high densities ($\geq 70\%$) we have complete gridlock. Furthermore, at low densities ($\leq 30\%$) the traffic is free flowing and grows linearly with the density.

5   Conclusions

We have used computer simulations to study a simple model of traffic flow on a square grid of roundabouts. We find that when the density of the travelling traffic is low, the vehicles can move freely and grows linearly with the density.

Increasing the vehicle density leads to structured patterns which initially enable a high density of vehicles to travel throughout the system but eventually leads to gridlock.

A phase transition separates a congestion free phase from a congested one as the vehicle density increases.

References

1. D. Chowdhury, L. Santen, and A. Schadschneider, Phys. Rep. 329, 199 (2000).
2. F.L. Hall, B.L. Allen, and M.A. Gunter, Trans. Res. A: Gen 20, 197 (1986).
3. W. Brilon and M. Ponzlet in Workshop on Traffic and Granular Flow, edited by D.E. Wolf, M. Schreckenberg, and A Bachem, 23 (World Scientific, Singapore, 1996).

Pedestrian Dynamics

Anomalous Fundamental Diagrams in Traffic on Ant Trails

A. Schadschneider[1], D. Chowdhury[2], A. John[1], and K. Nishinari[3]

[1] Institut für Theoretische Physik – Universität zu Köln
   50937 Köln – Germany
   [as,aj]@thp.uni-koeln.de
[2] Department of Physics – Indian Institute of Technology
   Kanpur 208016 – India
   debch@iitk.ac.in
[3] Department of Applied Mathematics and Informatics – Ryukoku University
   Shiga 520-2194 – Japan
   knishi@rins.ryukoku.ac.jp

Summary. Many insects like ants communicate chemically via chemotaxis. This allows them to build large trail systems which in many respects are similar to human-build highway networks. Using a stochastic cellular automaton model we discuss the basic properties of the traffic flow on existing trails. Surprisingly it is found that in certain regimes the average speed of the ants can vary non-monotonically with their density. This is in sharp contrast to highway traffic. The observations can be understood by the formation of loose clusters, i.e. space regions of enhanced, but not maximal, density. We also discuss the effect of counterflow on the trails.

1 Introduction

The occurance of organized traffic [1–4] is not only restricted to human societies. Prominent examples of self-organized motion in biology are herding, flocking and swarm formation [5–8]. In addition, especially ants build trail systems that have many similarities with highway networks [9].

Ants communicate with each other by dropping a chemical (generically called *pheromone*) on the substrate as they move forward [10–13]. Although we cannot smell it, the trail pheromone sticks to the substrate long enough for the other following sniffing ants to pick up its smell and follow the trail. This process is called *chemotaxis* [8, 11]. Ant trails may serve different purposes (trunk trails, migratory routes) and may also be used in a different way by different species. Therefore one-way trails are observed as well as trails with counterflow of ants.

In the following we will not discuss the process of trail formation itself. This has been studied in the past (see e.g. [14, 15] and references therein). Instead we assume the existence of a trail network that is constantly reinforced by the ants. Focussing on one particular trail it is natural to assume that the motion of the ants is one-dimensional.

In traffic flow interactions between the vehicles typically lead to a reduction of the average velocity $\bar{v}$. This is obvious for braking maneouvers to avoid crashes. Therefore $\bar{v}(\rho)$ decreases with increasing density ρ.[4] In the following we will show that this can be different for ant-trails. Even though due to the similar velocity of the ants overtaking is very rare, which implies that a description in terms of an exclusion process is possible, the presence of the pheromone effectively leads to an enhancement of the velocity.

We just briefly mention that similar ideas can be applied to the description of pedestrian dynamics. In [17, 18] we have developed a pedestrian model based on virtual chemotaxis. The basic principles are very similar to the ant-trail model introduced below. However, the motion in pedestrian dynamics is essentially two-dimensional.

2 Definition of the Model

In [19–21] we have developed a particle-hopping model, formulated in terms of a stochastic cellular automaton (CA), which may be interpreted as a model of uni-directional flow in an ant-trail. As mentioned in Sec. 1 we do not want to address the question of the emergence of the ant-trail [22], but focus on the traffic of ants on a trail which has already been formed. The model generalizes the asymmetric simple exclusion process (ASEP) [23] with parallel dynamics by taking into account the effect of the pheromone.

The ASEP is one of the simplest examples of a system driven far from equilibrium. Space is discretized into cells that can be occupied by at most one particle. In the totally asymmetric case (TASEP) particles are allowed to move in one direction only, e.g. to the right. If the right neighbour site is empty a particle hops there with probability q. For parallel (synchronous) dynamics this is identical to the limit $v_{\max} = 1$ of the Nagel-Schreckenberg (NaSch) model [24] with braking probability $p = 1 - q$.

In our model of uni-directional ant-traffic the ants move according to a rule which is essentially an extension of the TASEP dynamics. In addition a second field is introduced which models the presence or absence of pheromones (see Fig. 1). The hopping probability of the ants is now modified by the presence of pheromones. It is larger if a pheromone is present at the destination site. Furthermore the dynamics of the pheromones has to be specified. They are created by ants and free pheromones evaporate with probability f. Assuming

---

[4] A possible exception is a synchronized phase where the correlation between current and density is very small [16].

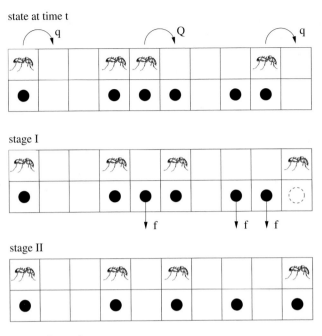

state at time t

stage I

stage II

state at time t+1

Fig. 1. Schematic representation of typical configurations of the uni-directional ant-traffic model. The symbols • indicate the presence of pheromone. This figure also illustrates the update procedure. Top: Configuration at time t, i.e. *before stage I* of the update. The non-vanishing probabilities of forward movement of the ants are also shown explicitly. Middle: Configuration *after* one possible realisation of *stage I*. Two ants have moved compared to the top part of the figure. The open circle with dashed boundary indicates the location where pheromone will be dropped by the corresponding ant at *stage II* of the update scheme. Also indicated are the existing pheromones that may evaporate in *stage II* of the updating, together with the average rate of evaporation. Bottom: Configuration *after* one possible realization of *stage II*. Two drops of pheromones have evaporated and pheromones have been dropped/reinforced at the current locations of the ants.

periodic boundary conditions, the state of the system is updated at each time step in two stages (see Fig. 1). In stage I ants are allowed to move while in stage II the pheromones are allowed to evaporate. In each stage the *stochastic* dynamical rules are applied in parallel to all ants and pheromones, respectively.

Stage I: Motion of ants

An ant in a cell cannot move if the cell immediately in front of it is also occupied by another ant. However, when this cell is not occupied by any other ant, the probability of its forward movement to the ant-free cell is Q or q, depending on whether or not the target cell contains pheromone. Thus, q (or Q) would be the average speed of a *free* ant in the absence (or presence) of pheromone. To be consistent with real ant-trails, we assume $q < Q$, as presence of pheromone increases the average speed.

Stage II: Evaporation of pheromones

Trail pheromone is volatile. So, pheromone secreted by an ant will gradually decay unless reinforced by the following ants. In order to capture this process, we assume that each cell occupied by an ant at the end of stage I also contains pheromone. On the other hand, pheromone in any 'ant-free' cell is allowed to evaporate; this evaporation is also assumed to be a random process that takes place at an average rate of f per unit time.

The total amount of pheromone on the trail can fluctuate although the total number N of the ants is constant because of the periodic boundary conditions. In the two special cases $f = 0$ and $f = 1$ the stationary state of the model becomes identical to that of the TASEP with hopping probability Q and q, respectively.

For a theoretical description of the process we associate two binary variables S_i and σ_i with each site i. S_i is the occupation number of ants and takes the value 0 or 1 depending on whether the cell i is empty or occupied by an ant. Similarly, σ_i is the occupation number of the pheromones, i.e. $\sigma_i = 1$ if the cell i contains pheromone; otherwise, $\sigma_i = 0$. Thus, we have two subsets of dynamical variables in this model, namely, $\{S(t)\} \equiv \{S_1(t), S_2(t), ..., S_i(t), ..., S_L(t)\}$ and $\{\sigma(t)\} \equiv \{\sigma_1(t), \sigma_2(t), ..., \sigma_i(t), ..., \sigma_L(t)\}$. In stage I the subset $\{S(t+1)\}$ at the time step $t+1$ is obtained using the full information $(\{S(t)\}, \{\sigma(t)\})$ at time t. In stage II only the subset $\{\sigma(t)\}$ is updated so that at the end of stage II the new configuration $(\{S(t+1)\}, \{\sigma(t+1)\})$ at time $t+1$ is obtained.

The rules can be written in a compact form as the coupled equations

$$S_j(t + 1) = S_j(t) + \min\{\eta_{j-1}(t), S_{j-1}(t), 1 - S_j(t)\}$$
$$- \min\{\eta_j(t), S_j(t), 1 - S_{j+1}(t)\}, \tag{1}$$
$$\sigma_j(t + 1) = \max\{S_j(t + 1), \min\{\sigma_j(t), \xi_j(t)\}\}, \tag{2}$$

where ξ and η are stochastic variables defined by $\xi_j(t) = 0$ with the probability f and $\xi_j(t) = 1$ with $1 - f$, and $\eta_j(t) = 1$ with the probability $p = q + (Q - q)\sigma_{j+1}(t)$ and $\eta_j(t) = 0$ with $1 - p$. This representation is useful for the development of approximation schemes.

3 Fundamental Diagram for Uni-Directional Motion

In vehicular traffic, usually, the inter-vehicle interactions tend to hinder each other's motion so that the average speed of the vehicles decreases *monotonically* with increasing density. This can be seen in Fig. 2 where also the

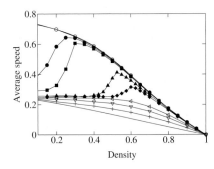
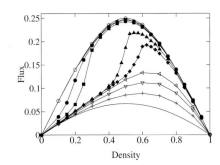

Fig. 2. The average speed (left) and the flux (right) of the ants, in the *uni-directional* ant-traffic model, are plotted against their densities for the parameters $Q = 0.75, q = 0.25$. The curves correspond to $f = 0.0001(\circ)$, $0.0005(\bullet)$, $0.001(\blacksquare)$, $0.005(\blacktriangle)$ $0.01(\blacklozenge)$, $0.05(\triangleleft)$, $0.10(\triangledown)$, $0.25(+)$. In both graphs, the cases $f = 0$ and $f = 1$ are also displayed by the uppermost and lowermost curves (without points); these are exact results corresponding to the TASEP or NaSch model with $v_{\max} = 1$ and hopping probability Q and q, respectively. Curves plotted with filled symbols have unusual shapes.

fundamental diagram of the NaSch model with $v_{\max} = 1$ (or the TASEP with parallel updating) is shown for different hopping probabilities. Note also the particle-hole symmetry of the flow that satisfies $F_{NS}(\rho) = F_{NS}(1 - \rho)$. Explicitly it is given by [25]

$$F_{NS}(\rho) = \frac{1}{2}\left[1 - \sqrt{1 - 4\, q_{NS}\, \rho(1 - \rho)}\right] \tag{3}$$

where $q_{NS} = 1 - p$ with p being the braking probability. Here $\rho = N/L$ is the density of vehicles and L the number of cells.

In contrast, in our model of uni-directional ant-traffic the average speed of the ants varies *non-monotonically* with their density over a wide range of small values of f (see Fig. 2) because of the coupling of their dynamics with that of the pheromone. This uncommon variation of the average speed gives rise to the unusual dependence of the flux on the density of the ants in the uni-directional ant-traffic model (Fig. 2). Furthermore the flux is no longer particle-hole symmetric.

3.1 "Loose" Cluster Approximation (LCA)

How can the unusual density-dependence of the average velocity be understood? Using mean-field-type theories that implicitly assume a homogeneous stationary state does not allow to reproduce the simulation results in a satisfactory way [20]. This indicates that the stationary state is characterized by some sort of clustering. This is confirmed by considering the probabilities

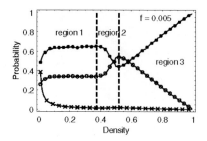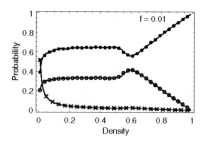

Fig. 3. Numerical results for the probabilities of finding an ant ($\bullet$), pheromone but no ant ($\circ$) and nothing ($\times$) in front of an ant are plotted against density of the ants. The parameters are $f = 0.005$ (left) and $f = 0.01$ (right).

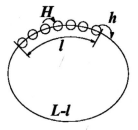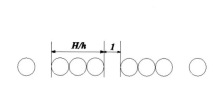

Fig. 4. (left) Schematic explanation of the loose cluster. H is the hopping probability of ants inside the loose cluster and h is that of the leading ant. (right) The stationary loose cluster. The average gap between ants becomes h/H, which is irrelevant to the density of ants.

of finding an ant (P_a), pheromone (P_p) and nothing (P_0) in front of a cell occupied by an ant. Typical results from computer simulations are shown in Fig. 3.

One can distinguish three different regions. "Region 1" is characterized by the flat part of the curves in Fig. 3 in the low density regime. Here, in spite of low density of the ants, the probability of finding an ant in front of another is quite high. This implies the fact that ants tend to form a cluster. On the other hand, cluster-size distributions obtained from computer simulations show that the probability of finding isolated ants are always higher than that of finding a cluster of ants occupying nearest-neighbor sites [20]. These two apparently contradictory observations can be reconciled by assuming that the ants form "loose" clusters in the region 1. The term "loose" means that there are small gaps in between successive ants in the cluster, and the cluster looks like an usual compact cluster if it is seen from a distance (Fig. 4). In other words, a loose cluster is just a loose assembly of isolated ants that corresponds to a space region with density larger than the average density ρ, but smaller than the maximal density ($\rho = 1$) of a compact cluster.

Let us assume that the loose cluster becomes stationary after sufficient time has passed. Then the hopping probability of all the ants, except the leading one, is assumed to be H, while that of the leading one is h (see Fig. 4). The values of H and h have to be determined self-consistently. If f is small enough, then H will be close to Q because the gap between ants is quite small. On the other hand, if the density of ants is low enough, then h will be very close to q because the pheromone dropped by the leading ant would evaporate when the following ant arrives there.

The typical size of the gap between successive ants in the cluster can be estimated [20] by considering a simple time evolution beginning with a usual compact cluster (with local density $\rho = 1$). The leading ant will move forward by one site over the time interval $1/h$. This hopping occurs repeatedly and in the interval of the successive hopping, the number of the following ants which will move one step is H/h. Thus, in the stationary state, strings (compact clusters) of length H/h, separated from each other by one vacant site, will produced repeatedly by the ants (see Fig. 4). Then the average gap between ants is $\frac{(H/h-1)\cdot 0 + 1\cdot 1}{H/h} = \frac{h}{H}$, independent of their density ρ. Interestingly, the density-independent average gap in the LCA is consistent with the flat part (i.e., region 1) observed in computer simulations (Fig. 3). In other words, region 1 is dominated by loose clusters.

Beyond region 1, the effect of pheromone of the last ant becomes dominant. Then the hopping probability of leading ants becomes large and the gap becomes wider, which will increase the flow. We call this region as region 2, in which "looser" clusters are formed in the stationary state. It can be characterized by a negative gradient of the density dependence of the probability to find an ant in front of a cell occupied by an ant (see Fig. 3).

Considering these facts, we obtain the following equations for h and H:

$$\left(\frac{h-q}{Q-q}\right)^h = (1-f)^{L-l}, \qquad \left(\frac{H-q}{Q-q}\right)^H = (1-f)^{\frac{h}{H}}, \qquad (4)$$

where l is the length of the cluster given by $l = \rho L + (\rho L - 1)\frac{h}{H}$. These equations can be applied to the region 1 and 2.

The total flux in this system is then calculated as follows. The effective density ρ_{eff} in the loose cluster is given by $\rho_{\text{eff}} = \frac{1}{1+h/H}$. Therefore, considering the fact that there are no ants in the part of the length $L - l$, the total flux F is

$$F = \frac{l}{L} f(H, \rho_{\text{eff}}), \qquad (5)$$

where $f(H, \rho_{\text{eff}})$ is given by

$$f(H, \rho_{\text{eff}}) = \frac{1}{2}\left(1 - \sqrt{1 - 4H\rho_{\text{eff}}(1 - \rho_{\text{eff}})}\right). \qquad (6)$$

Above the density $\rho = 1/2$, ants are assumed to be uniformly distributed such that a mean-field approximation works well [20]. We call this region as

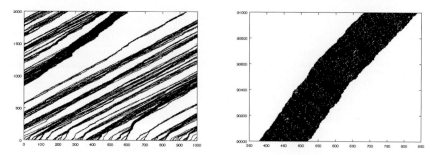

Fig. 5. Space-time plots showing the coarsening process of the loose clusters. The left part shows the state of a system of $L = 1000$ cells and $N = 100$ ants at early times. The right part corresponds to the stationary state. The evaporation probability of the pheromones is $f = 0.0005$.

region 3. Thus we have three typical regions in this model. In region 3, the relation $H = h$ holds because all the gaps have the same length, i.e. the state is homogeneous. Thus h is determined by

$$\left(\frac{h-q}{Q-q}\right)^h = (1-f)^{\frac{1}{\rho}-1}, \tag{7}$$

which is the same as our previous paper, and flux is given by $f(h, \rho)$. It is noted that if we put $\rho = 1/2$ and $H = h$, then (4) coincides with (7).

In region 1 we can simplify the analysis by assuming $h = q$ in (4). Then the flux-density relation becomes linear. Numerical results for regions 1 and 2 can be obtained by solving (4) using the Newton method for densities $\rho \leq 1/2$. Above this value of density, equation (7) can be used. Together these approximations can reproduce the results of the simulations rather well [20].

Another interesting phenomenon observed in the simulations is coarsening. At intermediate time usually several loose clusters are formed (Fig. 5). However, the velocity of a cluster depends on the distance to the next cluster ahead. Obviously, the probability that the pheromone created by the last ant of the previous cluster survives decreases with increasing distance. Therefore clusters with a small headway move faster than those with a large headway. This induces a coarsening process such that after long times only one loose cluster survives (Fig. 5). A similar behaviour has been observed in the bus-route model [26–28].

4 Bidirectional Motion

We develope a model of bi-directional ant-traffic [21] by extending the model of uni-directional ant-traffic described in Sec. 2. In the models of bi-directional ant-traffic the trail consists of *two* lanes of cells (see Fig. 6). These two lanes

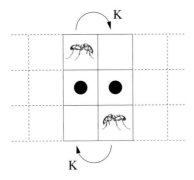

Fig. 6. A typical head-on encounter of two oppositely moving ants in the model of *bi-directional* ant-traffic. This is a totally new process which does not have any analog in the model of uni-directional ant-traffic.

need not be physically separate rigid lanes in real space; these are, however, convenient for describing the movements of ants in two opposite directions. In the initial configuration, a randomly selected subset of the ants move in the clockwise direction in one lane while the others move counterclockwise in the other lane. However, ants are allowed neither to take U-turn[5] nor to change lane. Thus, the ratio of the populations of clockwise-moving and anti-clockwise moving ants remains unchanged as the system evolves with time. All the numerical data presented here are for the *symmetric* case where equal number of ants move in the two directions. Therefore, the *average* flux of outbound and nestbound ants are identical. In all the graphs we plot only the flux of the nestbound ants.

The rules governing the dropping and evaporation of pheromone in the model of bi-directional ant-traffic are identical to those in the model of uni-directional traffic. The *common* pheromone trail is created and reinforced by both the outbound and nestbound ants. The probabilities of forward movement of the ants in the model of bi-directional ant-traffic are also natural extensions of the similar situations in the uni-directional traffic. When an ant (in either of the two lanes) *does not* face any other ant approaching it from the opposite direction the likelihood of its forward movement onto the ant-free cell immediately in front of it is Q or q, respectively, depending on whether or not it finds pheromone ahead. Finally, if an ant finds another oncoming ant just in front of it, as shown in Fig. 6, it moves forward onto the next cell with probability K.

Since ants do not segregate in perfectly well defined lanes, head-on encounters of oppositely moving individuals occur quite often although the frequency of such encounters and the lane discipline varies from one species of ants to another. In reality, two ants approaching each other feel the hindrance, turn by

---

[5] U-turns of so-called followers on pre-existing trails are very rare [29].

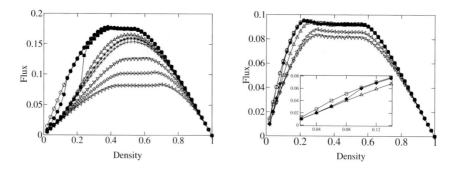

Fig. 7. Fundamental diagrams of the model for bi-directional traffic for the cases $q < K < Q$ (left) and $K < q < Q$ (right) for several different values of the pheromone evaporation probability f. The parameters in the left graph are $Q = 0.75, q = 0.25$ and $K = 0.5$. The symbols ○, ●, ■, △, *, +, ▽, ◇ and ◁ correspond, respectively, to $f = 0, 0.0005, 0.005, 0.05, 0.075, 0.10, 0.25, 0.5$ and 1. The parameters in the right graph are $Q = 0.75, q = 0.50$ and $K = 0.25$. The symbols ○, □, ♦, △, ◁ and ▽ correspond, respectively, to $f = 0, 0.0005, 0.005, 0.05, 0.5$ and 1. The inset in the right graph is a magnified re-plot of the same data, over a narrow range of density, to emphasize the fact that the unusual trend of variation of flux with density in this case is similar to that observed in the case $q < K < Q$ (left). The lines are merely guides to the eye. In all cases curves plotted with filled symbols exhibit non-monotonic behaviour in the speed-density relation.

a small angle to avoid head-on collision [15] and, eventually, pass each other. At first sight, it may appear that the ants in our model follow perfect lane discipline and, hence, unrealistic. However, that is not true. The violation of lane discipline and head-on encounters of oppositely moving ants is captured, effectively, in an indirect manner by assuming $K < Q$. But, a left-moving (right-moving) ant *cannot* overtake another left-moving (right-moving) ant immediately in front of it in the same lane. It is worth mentioning that even in the limit $K = Q$ the traffic dynamics on the two lanes would remain coupled because the pheromone dropped by the outbound ants also influence the nestbound ants and vice versa.

Fig. 7 shows fundamental diagrams for the two relevant cases $q < K < Q$ and $K < q < Q$ and different values of the evaporation probability f. In both cases the same unusual behaviour related to a non-monotonic variation of the average speed with density as in the uni-directional model can be observed [21].

An additional feature of the fundamental diagram in the bi-directional ant-traffic model is the occurrence of a plateau region. This plateau formation is more pronounced in the case $K < q < Q$ than for $q < K < Q$ since they appear for all values of f. Similar plateaus have been observed earlier [30–32] in models related to vehicular traffic where randomly placed bottlenecks slow down the traffic in certain locations along the route. Note that for $q < K < Q$ (Fig. 7(left)) the plateaus appear only in the two limits $f \to 0$ and $f \to 1$, but not for an intermediate range of values of f. In the limit $f \to 0$, most often the likelihood of the forward movement of the ants is Q whereas they are forced to move with a smaller probability K at those locations where they face another ant immediately in front approaching from the opposite direction (like the situations depicted in Fig. 6). Thus, such encounters of oppositely moving ants have the same effect on ant-traffic as bottlenecks on vehicular traffic.

But why do the plateaus re-appear for $q < K < Q$ also in the limit $f \to 1$? At sufficiently high densities, oppositely moving ants facing each other move with probability K rather than q. In this case, locations where the ants have to move with the lower probability q will be, effectively bottlenecks and hence the re-appearance of the plateau. As f approaches unity there will be larger number of such locations and, hence, the wider will be the plateau. This is consistent with our observation in Fig. 7(left).

5 Conclusions

We have introduced a stochastic cellular automaton model of an ant trail [19] characterized by two coupled dynamical variables, representing the ants and the pheromone. With periodic boundary conditions, the total number of ants is conserved whereas the total number of pheromones is not conserved. The coupling leads to surprising results, especially an anomalous fundamental diagram. This anomalous shape of the fundamental diagram is a consequence of the non-monotonic variation of the average speed of the ants with their density in an intermediate range of the rate of pheromone evaporation.

As the reason for this unusual behaviour we have identified the special spatio-temporal organization of the ants and pheromone in the stationary state. Three different regimes of density can be distinguished by studying appropriate correlation functions. At low densities (region 1) the behaviour is dominated by the existence of loose clusters which are formed through the interplay between the dynamics of ants and pheromone. These loose clusters are regions with a density that is larger than the average density ρ, but not maximal. In region 2, occuring at intermediate densities, the enhancement of the hopping probability due to pheromone is dominant. Finally, in region 3, at large densities the mutual hindrance against the movements of the ants dominates the flow behaviour leading to a homogeneous state similar to that of the NaSch model.

Furthermore we have introduced a model of *bi-directional* ant-traffic. The two main theoretical predictions of this model are as follows:
(i) The average speed of the ants varies *non-monotonically* with their density over a wide range of rates of pheromone evaporation. This unusual variation gives rise to the uncommon shape of the flux-versus-density diagrams.

(ii) Over some regions of parameter space, the flux exhibits plateaus when plotted against density.

In principle, it should be possible to test these theoretical predictions experimentally. Interestingly, various aspects of locomotion of *individual* ants have been studied in quite great detail [33–35]. However, traffic is a *collective* phenomenon involving a large number of interacting ants. Surprisingly, to our knowledge, the results published by Burd et al. [36] on the leaf-cutting ant *Atta Cephalotes* are the only set of experimental data available on the fundamental diagram. Unfortunately, the fluctuations in the data are too high to make any direct comparison with our theoretical predictions. Nevertheless it should be possible to observe the predicted non-monotonicity in future experiments. A simple estimate shows that the typical magnitudes of f, for which the non-monotonic variation of the average speed with density is predicted, correspond to pheromone lifetimes in the range from few minutes to tens of minutes. This is of the same order of magnitude as the measured lifetimes of real ant pheromones!

Acknowledgements

The work of DC is supported, in part, by DFG through a joint Indo-German research grant.

References

1. D. Chowdhury, L. Santen, and A. Schadschneider. Statistical physics of vehicular traffic and some related systems. *Phys. Rep.*, 329:199, 2000.
2. A. Schadschneider. Traffic flow: a statistical physics point of view. *Physica A*, 313:153, 2002.
3. D. Helbing. Traffic and related self-driven many-particle systems. *Rev. Mod. Phys.*, 73:1067, 2001.
4. T. Nagatani. The physics of traffic jams. *Rep. Prog. Phys.*, 65:1331, 2002.
5. J.K. Parrish and L. Edelstein-Keshet. Complexity, pattern, and evolutionary trade-offs in animal aggregation. *Science*, 284:99, 1999.
6. Y. Tu. Phases and phase transitions in flocking systems. *Physica A*, 281:30, 2000.
7. E.M. Rauch, M.M. Millonas, and D.R. Chialvo. Pattern formation and functionality in swarm models. *Phys. Lett. A*, 207:185, 1995.
8. E. Ben-Jacob. From snowflake formation to growth of bacterial colonies. part ii. cooperative formation of complex colonial patterns. *Contemp. Phys.*, 38:205, 1997.
9. N.R. Franks. Evolution of mass transit systems in ants: a tale of two societies. In I.P. Woiwod, D.R. Reynolds, and C.D. Thomas, editors, *Insect Movement: Mechanisms and Consequences*. CAB International, 2001.
10. E.O. Wilson. *The insect societies*. Belknap, Cambridge, USA, 1971.
11. B. Hölldobler and E.O. Wilson. *The ants*. Belknap, Cambridge, USA, 1990.

12. S. Camazine, J.L. Deneubourg, N. R. Franks, J. Sneyd, G. Theraulaz, and E. Bonabeau. *Self-organization in Biological Systems.* Princeton University Press, 2001.
13. A.S. Mikhailov and V. Calenbuhr. *From Cells to Societies.* Springer, 2002.
14. F. Schweitzer. *Brownian Agents and Active Particles.* Springer Series in Synergetics, 2003.
15. I.D. Couzin and N.R. Franks. Self-organized lane formation and optimized traffic flow in army ants. *Proc. Roy. Soc. London B*, 270:139, 2003.
16. B.S. Kerner. Complexity of synchronized flow and related problems for basic assumptions of traffic flow theories. *Netw. Spatial Econ.*, 1:35, 2001.
17. C. Burstedde, K. Klauck, A. Schadschneider, and J. Zittartz. Simulation of pedestrian dynamics using a 2-dimensional cellular automaton. *Physica A*, 295:507, 2001.
18. A. Kirchner, K. Nishinari, and A. Schadschneider. Friction effects and clogging in a cellular automaton model for pedestrian dynamics. *Phys. Rev. E*, 67:056122, 2003.
19. D. Chowdhury, V. Guttal, K. Nishinari, and A. Schadschneider. A cellular-automata model of flow in ant-trails: Non-monotonic variation of speed with density. *J. Phys. A: Math. Gen.*, 35:L573, 2002.
20. K. Nishinari, D. Chowdhury, and A. Schadschneider. Cluster formation and anomalous fundamental diagram in an ant trail model. *Phys. Rev. E*, 67:036120, 2003.
21. A. John, A. Schadschneider, D. Chowdhury, and K. Nishinari. Collective effects in traffic on bi-directional ant-trails. *J. Theor. Biol.*, 231:279, 2004.
22. D. Helbing, F. Schweitzer, J. Keltsch, and P. Molnar. Active walker model for the formation of human and animal trail systems. *Phys. Rev. E*, 56:2527, 1997.
23. B. Derrida. An exactly soluble non-equilibrium system: The asymmetric simple exclusion process. *Phys. Rep.*, 301:65, 1998.
24. K. Nagel and M. Schreckenberg. A cellular automaton model for freeway traffic. *J. Phys. I*, 2:2221, 1992.
25. M. Schreckenberg, A. Schadschneider, K. Nagel, and N. Ito. Discrete stochastic models for traffic flow. *Phys. Rev. E*, 51:2939, 1995.
26. O.J. O'Loan, M.R. Evans, and M.E. Cates. Jamming transition in a homogeneous one-dimensional system: the bus route model. *Phys. Rev. E*, 58:1404, 1998.
27. O.J. O'Loan, M.R. Evans, and M.E. Cates. Spontaneous jamming in one-dimensional systems. *Europhys. Lett.*, 42:137, 1998.
28. D. Chowdhury and R.C. Desai. Steady-states and kinetics of ordering in bus-route models: Connection with the Nagel-Schreckenberg model. *Eur. Phys. J. B*, 15:375, 2000.
29. R. Beckers, J.L. Deneubourg, and S. Goss. Trails and u-turns in the selection of a path by the ant *lasius niger*. *J. Theor. Biol.*, 159:397, 1992.
30. S.A. Janowski and J.L. Lebowitz. Finite-size effects and shock fluctuations in the asymmetric simple-exclusion process. *Phys. Rev. A*, 45:618, 1992.
31. S.A. Janowski and J.L. Lebowitz. Exact results for the asymmetric simple exclusion process with a blockage. *J. Stat. Phys.*, 77:35, 1994.
32. G. Tripathy and M. Barma. Steady state and dynamics of driven diffusive systems with quenched disorder. *Phys. Rev. Lett.*, 78:3039, 1997.

33. J.R.B. Lighton, G.A. Bartholomew, and D.H. Feener. Energetics of locomotion and load carriage and a model of the energy cost of foraging in the leaf-cutting ant *atta colombica*. *Physiol. Zoology*, 60:524, 1987.
34. C.P.E. Zollikofer. Stepping patterns in ants. *J. Experimental Biology*, 192:95, 1994.
35. J.A. Weier, D.H. Feener, and J.R.B. Lighton. Inter-individual variation in energy cost of running and loading in the seed-harvester ant *pogonomyrmex maricopa*. *J. Insect Physiology*, 41:321, 1995.
36. M. Burd, D. Archer, N. Aranwela, and D. J. Stradling. Traffic dynamics of the leaf-cutting ant, atta cephalotes. *American Natur.*, 159:283, 2002.

Models for Crowd Movement and Egress Simulation

H. Klüpfel[1], M. Schreckenberg[2], and T. Meyer-König[3]

[1] TraffGo GmbH
   Falkstraße 73–77, 47057 Duisburg – Germany
   kluepfel@traffgo.com
[2] Physik von Transport and Verkehr – Universität Duisburg-Essen
   47048 Duisburg – Germany
   schreckenberg@uni-duisburg.de
[3] TraffGo Flensburg
   Johannisstraße 42, 24937 Flensburg – Germany
   m-k@traffgo.com

Summary. This paper discusses basic findings on crowd movement and their application to simulation models. This includes empirical and experimental results concerning group behavior, pedestrian motion, and emergency egress. Next to a literature review, we will present own empirical investigations on walking speed distribution and the dependency of walking speed on group size.

The second part of the paper relates these findings to modelling and simulation of crowd movement. This comprises the representation of behavior, calibration and verification and the connection to many particle systems.

Finally, we will present an extension of the current microscopic theory (basically underlying all the "individual" models used for real world applications). This includes route-choice behavior and links microscopic and macroscopic behavior.

Keywords: crowd movement, simulation, egress, evacuation, cellular automata, pedestrian dynamics

1 Introduction: What is it Good for?

This paper describes fundamental properties of models for crowd movement. The real-world systems investigated can – as a crude but justified approximation – be mapped onto many particle systems far from equilibrium. The evacuation from, e.g. a buiding, is then equivalent to a transportation process. Of course, this mapping can only be done by simplifying the influences present in social systems and neglecting most of them, especially the psychological ones. An example for such a simplifications is the description of space as a grid of (usually quadratic cells), as it is done in the software package *PedGo*. The specific details of the PedGo model are described in another article in these proceedings [21].

In general, the aforementioned model assumptions lead to several questions. As a motivation, we will start with the most fundamental ones:

1. Why simulate crowd movement and evacuations?

2. And: How?

The answer to the first question is illustrated in fig. 1. We assume that the necessity for doing research on (empirically and theoretically) crowd movement and emergency egress is unquestioned. Then, the experimental investigation of crowd movement and especially emergency egress are limited by practical, ethical, financial, and logical constraints.

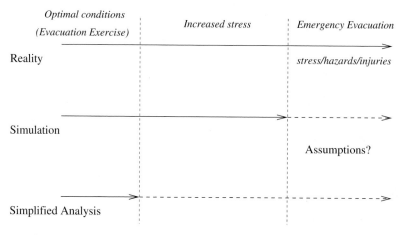

Fig. 1. A simplified analysis can basically cover the same range as an evacuation exercise. Aggravating circumstances like hazards can be included via extrapolation of the results obtained. A simulation, however, allows to include those influences via the adaption of the parameters, i.e., the extrapolation is made on the input and leads to an output different from the optimal case.

Furthermore, the scenarios we have in mind usually comprise large populations and structures, e.g. subway stations, football stadiums, passenger ships, etc.

At this point it is also worthwhile to clarify the meaning of the term crowd. Figure 2 sheds light especially on the distinction between gatherings and mobs. [4] When the term crowd or mass is used in this paper (as defined in the figure), only the former are addressed.

This paper focusses mainly on the theoretical aspects of crowd dynamics modelling. We will refer to our other paper in this book [21] when it comes to

---

[4] We would also like to note that the terms "mob" and "panic" are at least controversial [12, 33].

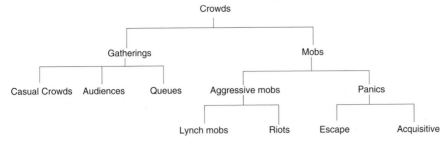

Fig. 2. Classification of crowds: [5]. Crowds are large groups that occupy a single location and share a common focus. One should probably also add passengers to the left branch.

specific model details of the cellular automaton model. First, however, let's have a look at empirical findings.

2 Investigation of Crowd Movement

Researchers from many disciplines – mostly traffic anc civil engineers – have investigated pedestrian and crowd movement for planning and design purposes. [3, 4, 6, 7, 29, 30, 37, 39]

The following figure 3 shows a rough classification of empirical and experimental situations.

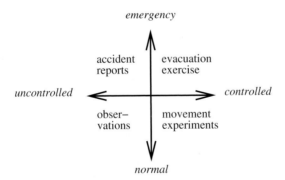

Fig. 3. Empirical data (including experiments) can be roughly classified according to controlled/uncontrolled and emergency/normal situations. Of course, there are other important criteria, like validity, reliability and objectivity. This becomes especially important for the uncontrolled situations, where those criteria can usually be questioned due to the lack of an operational definition of the situation. Whether an evacuation exercise can be called controlled depends largely on the number of repetitions.

Empirical in this context corresponds to observations (uncontrolled) and experimental to controlled situations.

2.1 Empirical Investigations

The following table 1 summarizes some of the major sources for data on pedestrian (single persons) and crowd (many persons) movement. In [14] the distribution for the directed walking speed (x–component) is Maxwellian and for its absolute value (velocity) – according to Maxwell's theory for kinetic gases – Gaussian.

This summary is restricted to quantitative results and therefore mainly comprises data on walking speed, interpersonal distances and flow-density-relations.

Figure 4 shows the curve of a flow-density-relation [38]. It was obtained by averaging over several results for walk- and passageways. Such a curve is especially useful for model calibration [21]: The same curve can be obtained by simulating a long and narrow hallway for all accessible densities (0 to ρ_{max}, where ρ_{max} is the maximum density the model allows) [18, 23]. !

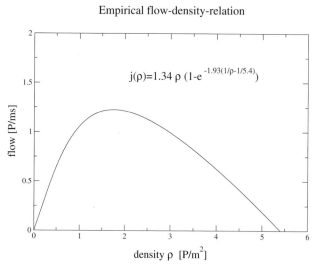

Empirical flow-density-relation

$$j(\rho)=1.34\,\rho\,(1-e^{-1.93(1/\rho-1/5.4)})$$

Fig. 4. Flow density relation for pedestrian movement. The analytical expression is shown at the top of the figure. The curve is a fit to empirical data [38].

For the case of stairs, the flow depends on the slope, of course. However, it does not seem to decrease significantly compared to flat terrain for the scenario depicted in fig. 5. This is probably due to the fact that stairs are equipped with handrails. However, the walking speed might well decrease (note that $j = \rho \cdot \langle v \rangle$) [2].

Table 1. Summary of the empirical data found in the literature. The results are described in more detail in the text. FWHM is short for Full width at half maximum (θ). If no explicit formula or type of distribution is given θ is used to characterize the width of the distribution (for the probability density it holds $f(\mu \pm \theta) = 1/2\, f(\mu)$). Additional reviews for the walking speed on stairs can be found in [6] and for the flow on stairs and and surface level in [9].

Environment	Type of data	Main result	
Walkways	frequency distr.	$\mu = 1.34\,\text{m/s}$, $\sigma = 0.26\,\text{m/s}$	[38]
Urban	frequency distr.	$\mu = 1.19\,\text{m/s}$, FWHM=$0.21\,m/s$	[7]
Campus	frequency distr.	$\mu = 1.53\,\text{m/s}$	[14]
Zebra crossing	frequency distr.	$\mu = 1.44\,\text{m/s}$	[14]
Walkways	flow vs. density	$\rho_{\max} = 5.4\,\text{P/m}^2$	[38]
Urban	speed vs. density	$j(\rho)$, $\rho \sim 1\,\text{P/m}^2$	[7]
Commuters	walking speed	$v(\text{age})$	[2]
School yard	walking speed	sexual differences	[15]
Aircraft mockup	egress time	critical exit width	[27]
Ships	behavior	panic is very rare	[12]
Stair Mockup	upstairs/downstairs	$d_{\text{gap}} \geq 0.25\,\text{m}$, $v_\uparrow$, $v_\downarrow$	[6]
Overviews			
Walkways/Urban			[7, 37, 38]
Buildings			[29, 30]

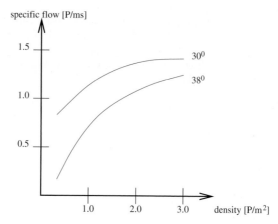

Fig. 5. Flow density relation for pedestrian movement on stairs. The dependence on the slope is most prominent for low densities, where steeper stairs perform worst. The experiments have been carried out in a Dutch football stadium. Additionally, the influence of the motivational level on the flow has been investigated [9].

We have also carried out studies on person flow at the world exhibition (EXPO) 2000 in Hannover, Germany. The frequency distribution for the walking speed on a (flat) pedestrian bridge is shown in fig. 6.

A normal distribution has been fit to the data using the mean and standard deviation of the empirical distribution as well as adapted values.[5] The data and the fitted curves are shown in fig. 6. The mean value obtained was $\mu = \langle v_x \rangle = 1.30\,\text{m/s}$ and the standard deviation $\sigma = 0.21\,\text{m/s}$. The third moment of the distribution $E(x^3 - \langle x^3 \rangle)$ is $0.41\,(\text{m/s})^3$. This shows that the distribution is not symmetric but slightly skewed towards the origin. The parameters for the second fit-curve shown in fig. 6 are $\mu = 1.28\,\text{m/s}$ and $\sigma = 0.2\,\text{m/s}$.

A second aspect of this investigation is the dependence of the walking speed on the group size. Several persons were identified as a group if the distance between at least two of them was not larger than about $1\,\text{m}$, they walked at the same speed, and in the same 'formation', i.e., they actually formed a social group.

Table 2 shows the decrease of the walking speed with increasing group size. It is interesting to note that groups larger than 6 persons were not observed. Of course the statistics for the larger groups are less reliable since they rarely occurred.

A second aspect of this investigation is the dependence of the walking speed on the group size. Several persons were identified as a group if the distance between at least two of them was not larger than about $1\,\text{m}$, they walked at the same speed, and in the same 'formation', i.e., they actually formed a social group.

---

[5] This differs from [14], where a Maxwell-Boltzmann distribution was used.

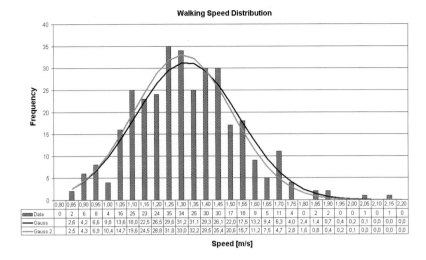

Walking Speed Distribution

	0,80	0,85	0,90	0,95	1,00	1,05	1,10	1,15	1,20	1,25	1,30	1,35	1,40	1,45	1,50	1,55	1,60	1,65	1,70	1,75	1,80	1,85	1,90	1,95	2,00	2,05	2,10	2,15	2,20
Data	0	2	6	8	4	16	25	23	24	35	34	25	30	17	18	9	5	11	4	0	2	2	0	0	1	0	1	0	
Gauss	2,6	4,2	6,6	9,8	13,6	18,0	22,5	26,5	29,6	31,2	31,1	29,3	26,1	22,0	17,5	13,2	9,4	6,3	4,0	2,4	1,4	0,7	0,4	0,2	0,1	0,0	0,0	0,0	
Gauss 2	2,5	4,3	6,9	10,4	14,7	19,6	24,5	28,8	31,8	33,0	32,2	29,5	25,4	20,6	15,7	11,2	7,5	4,7	2,8	1,6	0,8	0,4	0,2	0,1	0,0	0,0	0,0	0,0	

Speed [m/s]

Fig. 6. Shown is the walking speed distribution for the pedestrian bridge at the World Exhibition (Expo) 2000 in Hannover. The length that was walked by the pedestrians was 15.5 m. A Gaussian distribution is fitted to the results with μ and σ obtained either from the sample or using adapted values where the medians of the fitted and empirical curves are closer to each other. The data is shown in the table below the horizontal axis. Groups are represented by one data point (cf. table 2).

Table 2. Walking speed vs. group size for the pedestrian bridge. The speed is obtained by dividing the distance of 7 m by the travel time, i.e., $\langle v_x \rangle$, if the 'direction' of the bridge is denoted x.

Group size	Number of groups	Mean Velocity
1	95	1.38
2	149	1.28
3	59	1.24
4	17	1.24
5	10	1.22
6	2	1.10
	700	1.30

Table 2 shows the decrease of the walking speed with increasing group size. It is interesting to note that groups larger than 6 persons were not observed. Of course the statistics for the larger groups are less reliable since they rarely occurred.

Nevertheless, this information could be useful when integrating the influence of group size into a theory for crowd movmeent. At the moment there is – to our knowledge – no model that incorporates the influence of groups on the dynamics of crowd movement. (i.e., walking speed of groups vs. group size). This could – as a first approximation – be done by reducing the walking speed according to the group size. Of course, this would also require knowledge about the division of the population into groups and the distribution of group sizes.

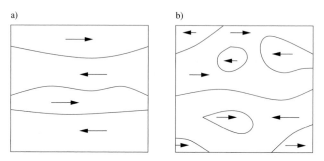

Fig. 7. A typical phenomenon in pedestrian movement is the formation of lanes (a) and clusters (b) with uniform movement direction. [41] has introduced a band index, which is basically the ratio of pedestrians in lanes to the overall number. For a) this would be nearly 1. High band indices have only been observed for large numbers of pedestrians (100 and above).

Finally, we would like to mention two phenomena that can be observed in moving crowds: lane formation [41] (cf. fig. 7 and clogging [28]. For a more detailed overview and discussion, please refer to [20].

2.2 Experiments

Hitherto,we have described empirical results (uncontrolled situations, cf. fig. 3. Another important source of knowledge are ewxperiments, of course. However, they are often expensive and complicated to carry out. A full scale trial for the evacuation of a Ro-Ro passenger ferry, e.g., cost about GBP 30,000 in 1999 [40].

An experiment on the evacuation of passenger aircraft is described in [27]. One of the major results, namely the existence of a minimal critical exit width, could quantitatively be reproduced in simulations! [19]. The empirical data are shown in fig. 8.

2.3 Accident Reports

Finally, data about behavior in a real emergency can usually only be obtained via accident reports. Sometimes, there are also CCTV (closed circuit television) recordings available.

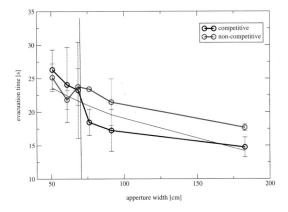

Fig. 8. The evacuation time increases with decreasing aperture width. For non-competitive situations, this decrease is rather smooth. However, if there is competition, at a certain aperture width ($w_c \approx 70$ cm) the increase in the egress time is quite drastic and the performance is worse than in the case of non-competition [27].

The awareness and response (pre-movement) times for complex buildings were investigated in [31]. The collapse of the World Trade Center in 2001 is subject of an intensive investigation by NIST [10].

Figure 9 shows the basic strategies of occupants in case of an emergency in complex buildings.

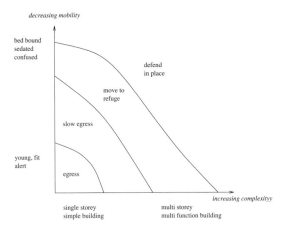

Fig. 9. Egress vs refuge in building evacuation: The more complex a building is and the less the mobility the more difficult is the egress from a building. This leads to a distinction between four different strategies: egress, slow egress, move to refuge, and defend in place [1].

3 Simulation of Crowd Movement

3.1 Basics

The method of choice to extent the results to scenarios not accessible (due to the aforementioned constraints) via observations and experiments are simulations.

On the other hand, looking at fig. 10 it is clear, that one cannot model crowd movement based on firtst principles (like it is possible for, e.g., Molecular Dynamics simulations).

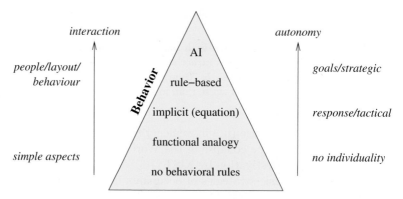

Fig. 10. The representation of behavior can range from its neglecting to artificial intelligence (AI), which aims at even modeling the decision making process [17]. However, the more complex approaches are not completely different from the basic ones but include additional features. Therefore, the autonomy and interaction increase first the top but are not completely absent and completely present in one case or the other. These qualities are rather a question of interpretation than of direct representation in a set of rules.

Once it is clear, that a model for crowd movement is phenomenological anyway, there seems to be no reason why not to chose the most simple approach one can come up with. However: As simple as possible but no simpler. This means – looking at fig. 10 – starting at the bottom and then adding complexity when necessary. The corresponding scenarios range from a single room or hallway to measure flow-density-relations (corresponding to "no individuality" in fig. 10), via structures with one floor and few exits (plattforms, small buildings), to very complex scenarios (large arenas, cruise ships, airports, etc.) corresponding to "strategic" in fig. 10.

We will first use grid-based models as an example to specify some of the thoughts previously outlined.

3.2 Model Details – Grid Based Models

The scenarios considered in an investigation of pedestrian dynamics can be classified according to several criteria. These criteria are summarized in fig. 11.

specific $\longleftrightarrow$ general

estimation $\longleftrightarrow$ first principles

discrete $\longleftrightarrow$ continuous

numerical $\longleftrightarrow$ analytical

stochastic $\longleftrightarrow$ deterministic

quantitative $\longleftrightarrow$ qualitative

macroscopic $\longleftrightarrow$ microscopic

Fig. 11. Modeling criteria that can be used for classifying different theories and models [8]. The major choices for models and simulations of crowd movement are discrete vs. continuous and stochastic vs. deterministic.

According to these modelling criteria, a grid-based model (or cellular automaton, cf. table. 3 is microscopic, general, quantitative, and usually stochastic.

Table 3. Definition of a cellular automaton. The assumption of a regular lattice and a uniform neighborhood is in accordance with complex geometries, since the set of states S also contains information about whether a cell is accessible or not (i.e., a wall cell, w in table 4).

Definition	Description
$\mathcal{L}$	consists of a regular discrete lattice of cells
$t \to t+1$	evolution takes place in discrete time steps
S	set of 'finite' states
$f : S^n \to S$	each cell evolves according to the same rule (transition function) which depends only on the state of the cell, and a finite number of neighboring cells
$\mathcal{N} : \forall c \in \mathcal{N}, \forall r \in \mathcal{L} : r + c \in \mathcal{L}$ the neighborhood relation is local and uniform	

For the ease of the reader, table 4 summarizes the characteristics of the *PedGo* model [21, 24–26, 36]. Of course, the criteria shown in table 3 apply also.

Table 4. Assumptions for the model and the empirical correlate. The symbols are explained in the text. The parameters can vary and the values given are typical ones.

empirical	model
orientation at exit signs	$V(r) \sim d(r, r_{\text{exit}})$
$\rho_{\text{max}} = 6.25 \text{P}/\text{m}^2$	$a = 0.4 \text{ m}$
$v_{\text{max}}^{\text{emp}} \approx 2 \text{ m/s} \wedge \Delta t \approx 1 \text{ s}$	$v_{\text{max}}^{\text{mod}} = 5$
stopping due to orientation	$p_{\text{dec}} = 0 \ldots 0.1$
deviations from the optimal direction	$p_{\text{sway}} = 0 \ldots 0.03$
walls are 'black' cells	$w = 1$
hard core exclusion	$o \in \{0, 1\}$

3.3 Other Models

We have already mentioned (in the introduction) that crowd movement can be mapped onto transportation processes. In this context, the connection between transportation processes and spin models might well become of interest in the future also for such phenomena like lane formation, shock waves, or jamming. Especially, the connection between person flows and granular materials research could be fruitful. This connection is briefly outlined in fig. 12.

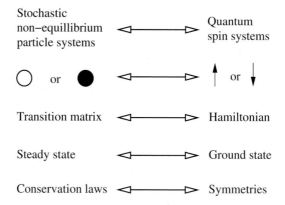

Fig. 12. Correspondence between stochastic non-equilibrium particle systems and quantum many body systems. For single-species exclusion processes the mapping can be to spin systems, as indicated in the second row [35].

3.4 Calibration

Concerning the calibration of the basic properties of crowd movement, there are three major issues:

- Maximum density

- Walking speed

- Time scale (decision making or reaction time) [6]

For the case of grid-based models, theses issues have been investigated by varying cell size and walking speed in [18].

3.5 Full Scale Trials – Verification

Finally, real world scenarios comprise many influences, which usually cannot be separated into isolated factors. Therefore, it is also necessary to reproduce th results of full scale trials in a simulation, where these factors influence each other and might lead to phenomena not present in less complex situations. An illustrative example is counterflow, which will – due to the lack of space – not be elaborated at this point, however. The interested reader might refer to [16, 32].

3.6 The Micro-Macro Link

Microscopic models might not be able to cover all aspects of crowd movement, even when neglecting most of the psychological and social influences (cf. section 1). The most prominent example in this context is presumably route-choice, which is a strategic (cf. fig. 10) and non-local (cf. fig. 3, bottom and fig. 13) task. Figure 13 (right) exemplifies an elegant solution of the route-choice problem within the microscopic framework: A graph superstructure is defined to represent the topology of the floorplan. Then, standard graph theoretical algorithms (e.g., Dijkstra's) can be applied to find the shortest path using the appropriate weigths. These weigths are set to represent the scenario under consideration (e.g., distance, familiarity, availability, crew influence, etc.).

For more and general information on network-flow approaches to the evacuation problem, please refer to [11].

---

[6] Please note that we used the terms awareness and response time in section 2.3 to quantify the pre-movement phase, which is in accordance with the generally accepted terminology [17].

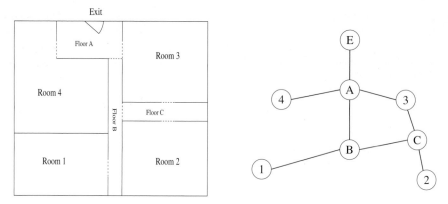

Fig. 13. A simple geometry and its graphical representation.

4 Summary and Conclusions

In this paper, the basic properties of a specific grid based model for simulating the dynamics of crowd movement was described. The focus was mainly on egress simulation. However, the model is applicable also to non-emergency situations.

When comparing the discrete space (and time) model to other similar and especially continuous models, it becomes clear that the differences are subtle. It depends mainly on the application one has in mind, what the preferable model or model variant is. However, I am aware of the fact, that this statement is controversial and other researchers argue, that models for pedestrian dynamics can be universal in the sense that there is no adaption of the parameters necessary for different scenarios.

Acknowledgements

We would like to thank Michael Schreckenberg and the members of his research team (*Physics of Transport and Traffic, Duisburg, Germany*) for many fruitful discussions and their extensive support in general. My (HK's) special thanks goe to Andreas Schadschneider and Ansgar Kirchner (formerly) at *Cologne University, Germany* and Katsuhiro Nishinari *Ryukoku University, Shiga, Japan*).

References

1. J. Abrahams. Fire escape in difficult circumstances. In P. Stollard and L. Johnson, editors, *Design against fire*, London, New York, 1994. SFPE.

2. K. Ando, H. Ota, and T. Oki. Forecasting the flow of people. *Railway Research Review*, 45:8–14, 1988. (in Japanese).

3. D. Canter, editor. *Fires and Human Behaviour*. David Fulton Publishers, London, 2nd edition, 1990.

4. P. DiNenno, editor. *SFPE Handbook of Fire Protection Engineering*. National Fire Protection Association, Washington, 2nd edition, 1995.

5. D. R. Forsyth. *Group Dynamics*. Wadsworth Publishing, Belmont, CA, 3 edition, 1999.

6. H. Frantzich. Study of movement on stairs during evacuation using video analysing techniques. Technical report, Department of Fire Safety Engineering, Lund Institute of Technology, Lund University, 1996.

7. J. Fruin. *Pedestrian Planning and Design*. Metropolitan Association of Urban Designers and Environmental Planners, New York, 1971.

8. N. Gershenfeld. *The Nature of Mathematical Modelling*. Cambridge University Press, Cambridge, 1999.

9. E. Graat, C. Midden, and P. Bockholts. Complex evacuation; effects of motivation level and slope of stairs on emergency egress time in a sports stadium. *Safety Science*, 31:127–141, 1999.

10. W. Grosshandler, S. Sunder, and J. Snell. Building and fire safety investigation of the world trade center disaster. In E. Galea, editor, *Pedestrian and Evacuation Dynamics 2003*, pages 279–281, London, 2003. University of Greenwich, CMS press. `http://wtc.nist.gov`.

11. H. Hamacher and S. Tjandra. Mathematical modelling of evacuation problems – a state of the art. In Schreckenberg and Sharma [32], pages 227–266.

12. J. Harbst and F. Madsen. The behaviour of passengers in a critical situation on board a passenger vessel or ferry. Technical report, Danish Investment Foundation, Kopenhagen, 1996.

13. D. Helbing, L. Buszna, and T. Werner. Self-organized pedestrian crowd dynamics and design solutions. `http://www.helbing.org`, Dec 2003.

14. L. Henderson. The statistics of crowd fluids. *Nature*, 229:381–383, 1971.

15. L. Henderson. Sexual differences in human crowd motion. *Nature*, 240:353–355, 1972.

16. IMO, London. *Interim Guidelines for Evacuation Analyses for New and Existing Passenger Ships*, 2002. MSC/Circ. 1033.

17. ISO. Fire safety engineering. Technical Recommendation TR 13387, International Organization for Standardization, Geneva, 1999.

18. A. Kirchner, H. Klüpfel, K. Nishinari, A. Schadschneider, and M. Schreckenberg. Discretization effects and influence of walking speed in cellular automata models for pedestrian dynamics. *In preparation for Journal of Statistical Mechanics: Theory and Experiment (JSTAT)*, 2004.

19. A. Kirchner, K. Nishinari, and A. Schadschneider. Friction effects and clogging in a cellular automaton model for pedestrian dynamics. *Phys. Rev. E*, 67:056122, 2003. cond-mat/0209383.

20. H. Klüpfel. *A Cellular Automaton Model for Crowd Movement and Egress Simulation*. PhD thesis, University Duisburg–Essen, 2003. `http://www.ub.uni-duisburg.de/ETD-db/theses/available/duett-08012003-092540`.

21. H. Klüpfel and T. Meyer-König. Simulation of the evacuation of a football stadium. In Schreckenberg and Sharma [32].

22. H. Klüpfel, T. Meyer-König, and M. Schreckenberg. Microscopic modelling of pedestrian motion – comparison of simulation results with an evacuation exercise in a primary school. In M. Fukui, Y. Sugiyama, M. Schreckenberg, and D. Wolf, editors, *TGF '03*, Berlin, 2003. Springer.

23. H. Klüpfel, T. Meyer-König, J. Wahle, and M. Schreckenberg. Microscopic simulation of evacuation processes on passenger ships. In S. Bandini and T. Worsch, editors, *ACRI 2000*, pages 63–71, London, 2000. Springer.

24. T. Meyer-König. Mikroskopische Simulation von Evakuierungsprozessen. *VDI Technische Überwachung*, 43(9):10–13, September 2002. available for downloag at www.traffgo.com.

25. T. Meyer-König, H. Klüpfel, A. Keßel, and M. Schreckenberg. Simulating mustering and evacuation processes onboard passenger vessels: Model and applications. In *The 2nd International Symposium on Human Factors On Board (ISHFOB)*, 2001. CD-Rom.

26. T. Meyer-König, H. Klüpfel, and M. Schreckenberg. Assessment and analysis of evacuation processes on passenger ships by microscopic simulation. In Schreckenberg and Sharma [32], pages 297–302.

27. H. Muir. Effects of motivation and cabin configuration on emergency aircraft evacuation behavior and rates of egress. *Intern. J. Aviat. Psych.*, 6(1):57–77, 1996.

28. K. Müller. Die Evakuierung von Personen aus Gebäuden – nach wie vor ein nationales und internationales Problem. *vfdb-Zeitschrift*, 3:131, 1999.

29. J. Pauls. Movement of people. In DiNenno [4], pages 3–263—3–285.

30. W. Predtetschenski and A. Milinski. *Personenströme in Gebäuden – Berechnungsmethoden für die Modellierung*. Müller, Köln-Braunsfeld, 1971.

31. G. Proulx. Evacuation time and movement in apartment buildings. *Fire Safety Journal*, 24:229–246, 1995.

32. M. Schreckenberg and S. Sharma, editors. *Pedestrian and Evacuation Dynamics*, Berlin, 2002. Springer.

33. J. Sime. The concept of panic. In Canter [3], chapter 5, pages 63–82.

34. R. Smith and J. Dickie, editors. *Engineering for Crowd Safety*. Elsevier, Amsterdam, 1993.

35. R. Stinchcombe. Stochastic non-equilibrium systems. *Advances in Physics*, 50(5):431–496, 2001.

36. TraffGo GmbH, Duisburg. *PedGo Users' Manual*, 2004. www.traffgo.de.

37. Transportation Research Board, Washington, D.C. *Highway Capacity Manual*, 1994.

38. U. Weidmann. Transporttechnik der Fußgänger. Schriftenreihe des IVT 90, ETH Zürich, 1992. (in German).

39. U. Weidmann. Grundlagen zur Berechnung der Fahrgastwechselzeit. Schriftenreihe des IVT 106, ETH Zürich, Juni 1995.

40. A. Wood. *Validating Ferry Evacuation Standards*, 1997. Available from MCA, 105 Commercial Road, Southhampton.

41. K. Yamori. Going with the flow: Micro-macro dynamics in the macrobehavioral patterns of pedestrian crowds. *Psychological Review*, 105(3):530–557, 2001.

Self-Organization in Pedestrian Flow

S. Hoogendoorn and W. Daamen

Faculty of Civil Engineering and Geosciences –
Delft University of Technology
Stevinweg 1, 2628 CN, Delft – The Netherlands
s.hoogendoorn@citg.tudelft.nl, w.daamen@citg.tudelft.nl

Summary. Microscopic simulation models predict different forms of self-organization in pedestrian flows, such as the dynamic formation of lanes in bi-directional pedestrian flows. The experimental research presented in this paper provides more insight into these dynamic phenomena as well as exposing other forms of self-organization, i.e. in case of over-saturated bottlenecks or crossing pedestrian flows. The resulting structures resemble states occurring in granular matter and solids, including their imperfections (so-called vacancies). Groups of pedestrians that are homogeneous in terms of desired walking speeds and direction appear to form structures consisting of overlapping layers. This basic pattern forms the basis of other more complex patterns emerging in multi-directional pedestrian flow: in a bi-directional pedestrian flow, dynamic lanes are formed which can be described by the layer structure. Diagonal patterns can be identified in crossing pedestrian flows. This paper both describes these structures and the conditions under which they emerge, as well as the implications for theory and modeling of pedestrian flows.

Keywords: pedestrian flow, self-organization, experimental research

1 Introduction

In the early seventies, a substantial body of research was established on pedestrian behavior and the interaction of pedestrians with their environment and other pedestrians. These studies focused primarily on the social and psychological aspects of walking, such as the way pedestrians observe the walking infrastructure via subliminal scanning [1], interact with other pedestrians via subconscious communication and cooperation [2-5]. Also, attention was paid to the way social and cultural differences affect these different processes [2,3,6].

For the design of walking infrastructure, a working knowledge of the characteristics of pedestrian flows is required in order to design the infrastructure as well as to assess its efficiency and safety. In particular, a good understanding of the emergent patterns is required to predict how the flow will behave under different circumstances. This knowledge is generally based on results of microscopic simulation studies [7-9] rather than detailed empirical or experimental research on how individual pedestrians act under a variety of traffic flow or environmental conditions. These models predict the formation of a variety of structures in pedestrian flows, such as dynamic lanes in bi-directional flows, or strips in crossing flows. Empirical evidence indeed points towards the existence of these self-organizing phenomena [10,11], but so far no experimental or empirical studies have provided definite results regarding quantitative characteristics of these self-organized structures.

2   Behavioral Theory of Self-Organization

Before presenting the results of the walking experiments and the observed self-organization therein, we propose a theory of self-organization in pedestrian flow. The theory is based on the assumption that each pedestrian aims to minimize his or hers *predicted disutility of walking* (i.e. the *pedestrian economicus* [9]): for all available options (e.g. accelerating, decelerating, changing direction, do nothing), a pedestrian tries to choose the option that will yield the smallest predicted disutility. In psychology, this behavioral assumption is referred to as the *principle of least effort*, i.e. an individual will try to adapt to his or her environment (or will try to change the environment to suit its needs, whichever is easier); see [15]. In predicting the disutility of walking, a pedestrian values the different attributes characterizing the available options (e.g. risk to collide with another pedestrian, straying from the intended walking path, physical contact with other pedestrians, etc.) differently.

Under specific conditions, we may expect that given this assumption, the pedestrian flow will evolve to an *user-equilibrium state* (which is either stable or meta-stable) in which no pedestrian can improve his or hers condition by unilaterally undertaking an action. In fact, under special conditions, differential game theory predicts the occurrence of a Nash equilibrium solution to the *n*-person non-cooperative differential game [16]. The hypothesis is that the self-formation of homogeneous patterns (e.g. dynamic lane formation, or the formation of diagonal strips) is equivalent to the evolution of the aforementioned user-equilibrium state.

In illustration, consider a situation in which all pedestrians have the same physical characteristics (e.g. desired walking speed, size, etc.). Clearly, in case of a unidirectional flow, the theory predicts a homogeneous distribution of pedestrians over the walking: each pedestrian has the same speed and there is no use of overtaking other pedestrians. This results a homogeneous pattern in which pedestrians walk behind each other in layers that overlap partially. Note that this staggered pattern is optimal from the viewpoint of efficient use of available walking space.

When the flow is heterogeneous (e.g. regarding the walking speeds), changing speed or direction may improve the conditions of an individual, since slower pedestrians with hold back faster ones. Faster pedestrians will overtake only if overtaking will result in a substantially improved situation. When such benefits are not sufficient, that is, when the switching costs are higher than the disutility of staying in the current layer, the pedestrian aiming to walk fast will stay behind the slow pedestrian: the disutility of not being able to walk at the desired velocity is in this case compensated by the reduced probability of colliding with other pedestrians and the switching cost from the current layer to another (caused by e.g. needing to cross through another layer). Depending on the pedestrian population, for inhomogeneous unidirectional flows, the theory thus predicts the formation of homogeneous structures (pedestrian have equals speeds) with in between individual pedestrians (or small groups) walking at a higher speed.

For bi-directional flows, user-optimal states are states in which lanes of uniform walking directions. In this case, the system optimal situation would be two regions of pedestrians moving in opposite directions, since this situation yields the least friction between the regions, and thus the smallest collective disutility. This ideal state may not occur due to initial conditions, heterogeneous composition of the flows (with respect to the desired walking speed) combined with overtaking opportunities, etc. This also implies that the number of lanes that are formed will also depend on the density, since the latter determines overtaking opportunities.

Applying the theory to crossing pedestrian flows will also lead to the formation of homogeneous structures, i.e. to the formation of diagonal strips. The shape of these diagonal strips yields a situation in which the system has the least collective disutility when the two flows cross.

Clearly, which self-organized structures will appear depends on a number of things amongst which are the initial and boundary conditions. The theory does also not preclude the emergence of non-optimal meta-stable states (e.g. the formation of multiple lanes in a bi-directional flow), which can only change in an optimal state when pedestrians 'invest' by (temporarily) accepting high disutilities (e.g. by crossing a flow moving in the opposite direction).

3   Walker Experiments

Motivated by the lack of experimental knowledge of self-organization in pedestrian flow, Delft University of Technology conducted a large-scale walking experiment to gain a better understanding of emergent structures in pedestrian flow. In total, 10 different walking experiments were performed, 5 of which are relevant for the results presented in this paper. In each of these experiments, approximately 60-90 individuals were involved. With a high-positioned digital camera, an area of approximately 14 m by 12 m was observed. Trajectories of all pedestrians were determined from the video data using dedicated tracking software. The experiments that are relevant within the scope of this contribution are shown in Tab.1..

Nr.	Experiment type	Size of walking area	Description
1.	←	10 x 4	Homogeneous walking conditions.
2.	←	10 x 4	Unidirectional flow through wide bottleneck (2 meter width)
3.	←	10 x 4	Unidirectional flow through narrow bottleneck (1 meter width)
4.	⇆	10 x 4	50% from East to West; 50% from West to East
5.	⊕	8 x 8	50% from East to West; 50% from North to South

Tab. 1. Overview of relevant walker experiments.

In all considered experiments, participants were asked to walk as they normally would. The composition of the populations was heterogeneous, both

both with respect to age, gender, etc. The pedestrians were asked to enter the walking area, walk across, leave, and walk around to the entry of the walking areas. Upon entering, care was taken to distribute pedestrians across the entire width of the area. For a detailed description of the experimental set-up, we refer to [12].

4   Observed Spatial Patterns

This section discusses the different spatial patterns that have been observed in the pedestrian experiment, and how these observations support our theory on self-organization in pedestrian flow.

4.1   Standard Pattern

The standard patterns occur in case of unidirectional pedestrian flows or sub flows (i.e. homogeneous clusters in multi-directional flows) when there is insufficient space to overtake slower pedestrians walking in the same direction (experiments 1-3). The standard pattern consists of a number of partly overlapping *layers*. We emphasize that overlapping does not imply *physical overlapping*. In [13] it is shown how pedestrians make a swerving motion while walking, thus requiring more space than just their width of their shoulders. For walking speeds higher than 0.3 *m/s*, the swerving amplitude $a(v)$ (in *m*) equals approximately [13]:

$$a(v) = 0.068 - 0.017v \tag{1}$$

where v is the pedestrian speed in *m/s*.

The photographs shown in Fig. 1 depict the four types of homogenous basic patterns that emerge, in this case in the narrow bottleneck experiment. Note that these patterns are in line with the prediction of the formation of staggered layers based on our theory presented in section 2. The arrows in the figure all have the same length and are used to show that the distance headways between pedestrians in a layer are approximately constant (except for pattern 4). Pattern 1 shows the case that the distances between pedestrians in different layers are also approximately constant, i.e. the lead gap and the lag gap are almost equal. Patterns 2 and 3 are similar, and reflect cases in which the lead gap for the pedestrians in one layer is relatively small, while the lag gap is relatively large (or vice-versa).

Pattern 4 is a special pattern, the properties of which can be described using analogies with defects in solids. This particular defect is referred to as a *vacancy*[1] in solid-state physics. In a pedestrian flow, this defect is caused by inefficient merging behavior at the bottleneck entry, e.g. due to overly polite or aggressive behavior. The latter may be more proficient in case of panic situations, causing more defects in the pattern and a less efficient usage of the bottleneck. Despite the fact that quite a number of defects were observed, it is noted that patterns 1 to 3 are more common. The average distances between pedestrians in longitudinal and lateral sense depends primarily on the average speed of the pedestrian flow. For saturation flows (i.e. speeds in the bottleneck after congestion has occurred), this speed is around 1 *m/s*; the average longitudinal distance between pedestrians (within a layer) is 1.3 *m*; the centers of the overlapping layers are on average 45 cm apart.

---

[1] A vacancy is defined by a lattice position that is vacant because an atom is missing.

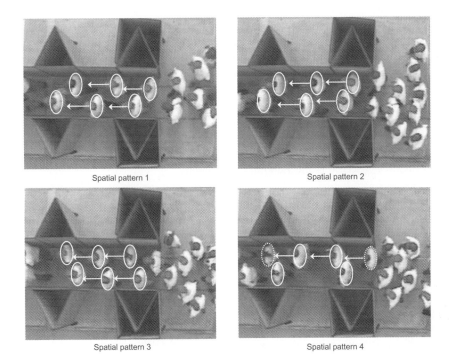

Fig. 1. Emergent standard patterns, in this case for the narrow bottleneck experiment.

Experi-ment	Direc-tion	Following pedes-trian fraction	Empty zone distribution	
			Mean (s)	Standard devia-tion (s)
1	←	28%	1.43	0.49
2	←	61%	1.36	0.45
3	←	72%	1.37	0.43
4	←	36%	1.42	0.50
	→	52%	1.42	0.54
5	←	16%	1.43	0.64
	↓	19%	1.37	0.52

Tab. 2. Overview of mean and standard deviation of empty zone distribution for different experiments.

By estimating a so-called *composite headway distribution model*, we have established that the minimal time headway needed for safe and comfortable following[2] (the so-called *empty zone*[3]) of pedestrians is on average 1.4 s for all considered experiments; the standard deviation of the empty zone distribution is between 0.4 s and 0.5 s. This empty zone describes the time headway

---

[2] The time headway is defined by the difference between the time instants of consecutive pedestrians in the same layer passing a cross-section x

[3] The empty zone is defined by the minimum time headway of a following pedestrian with respect to the leading pedestrian in the same layer.

that a pedestrians requires to safely and comfortably take a step, and thus depends on the step size, stepping frequency and walking speed. As is shown in the following section, this also holds for the self-organized structures that appear in bi-directional or crossing flows. For details, we refer to [13].

4.2   Self-Organization of Standard Patterns

One of the typical phenomena occurring in pedestrian flow is self-organization. Several types of self-organization have been described in literature [7,8,10,11]. This section provides an overview of the self-organized structures observed in the walking experiments.

4.2.1   Lane formation in bi-directional pedestrian flows

For the bi-directional flow experiment (experiment 4), it was observed that standard patterns are formed in both walking directions. An example is shown in Fig. 2, where two of these patterns are formed. Note that due to the fact that relatively much space is available, overtaking opportunities do exist and the patterns are relatively short-lived. Nevertheless, analyzing the formed structures shows that their characteristics are similar for the bi-directional and the unidirectional flow experiments. Using the composite headway model estimation method of [13], it was determined that the mean empty zone of pedestrians in the same layer is approximately 1.4 s (see Tab.2).

Fig. 2. Example of formation of standard pattern in bi-directional flow.

This type of self-organization in pedestrian flow has been observed by other researchers and appears to be characteristic of bi-directional pedestrian flows. In [14], it has been proposed that the number of formed lanes depends on the density in the walking area. In this paper, we have studies this quantitatively by application of cluster analysis. A cluster is in this case defined using the locations $\vec{r}_i(t)$ and velocities $\vec{v}_i(t)$, according to the following criteria:

1.   The distance $\| \vec{r}_i(t) - \vec{r}_j(t) \|$ between i and j is less than threshold c_1, and

2.   The velocity difference $\| \vec{v}_i(t) - \vec{v}_j(t) \|$ is less than some value c_2.

Using standard cluster analysis, clusters of pedestrians can be identified. Please not that the clusters and dynamic lanes are not be definition the same. In general, a dynamic lane can be made up from several clusters (see Fig. 3).

A cluster can however not belong to more than one lane. This distinction is due to the distance criterion 1 used to define a cluster.

Figure 3 shows results from application of cluster analysis for four different time-slices. The figure depicts different kinds of lane formation. Note that upon entering the walking area, pedestrians were instructed to spread uniformly over the width of the area.

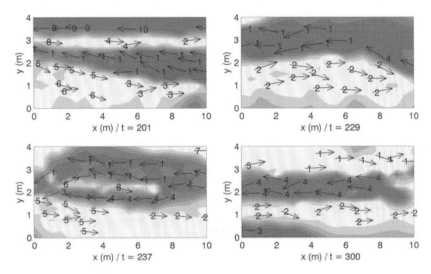

Fig. 3. Example results of cluster analysis for four time slices; numbers indicate pedestrian clusters. The color indicates the average direction at a particular location. Note that a lane can consist of multiple clusters.

As predicted by the theory proposed in section 2, it turns out that in the experiment often two lanes are formed that are largely right-hand-sided (e.g. Fig. 3, $t = 229\ s$). This is typical for countries in which traffic regulations stipulate walking on the right side of the street. However, quite frequently different patterns emerge: Fig. 3, $t = 300\ s$ shows a fairly common situation in which three dynamic lanes can be identified, and in which pedestrians obviously do not walk on the right side of the walking area. These situations are caused by the random conditions at the boundaries of the walking area and the fact that the *transition* from this sub-optimal three-lane state to the optimal two-lane situation will inflict a very high disutility (pedestrians having to cross a lane flowing into the opposite direction). Note that also situations where four lanes are formed have been recorded (e.g. Fig. 3, $t = 201\ s$).

Figure 4 shows the average number of clusters that are identified during the experiment each $0.5\ s$ as a function of density. The figure shows how the number of clusters first increases, reaching its maximum value of 12 near a density of $0.6\ P/m^2$. Please note that the number of formed clusters depends the cluster definition. Also note that a cluster may consist of a single pedestrian, and that a lane may be split up in several clusters.

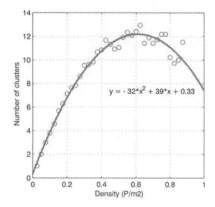

Fig. 4. Number of formed clusters as a function of density. Note that the number of pedestrians per cluster increases with density.

4.2.2   Strip formation in crossing flows

Self-organization is not restricted to bi-directional flows: also in crossing pedestrian flows self-organization will occur: clusters of pedestrians having the same direction and speed will form. The homogeneous patterns that emerge have the form of diagonal strips (see [9-11,14]). The characteristics of the formed structures are similar to the structures formed in the unidirectional and bi-directional experiments: for instance, Tab. 2 shows that the mean empty zone is approximately 1.4 s; the standard deviation of the empty zone is however much larger than for the other experiments. An explanation for this can be found by examining the way in which the diagonal strips are formed, as is explained below.

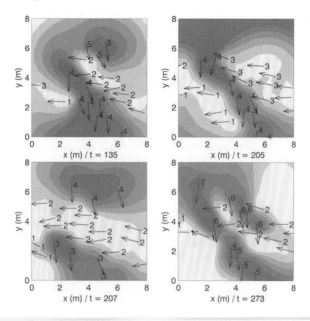

Fig. 5. Formation of diagonal strips in crossing pedestrian flow experiment. The colors indicate the average direction at a particular location.

From analyzing the video data, the following observation was made: consider a situation where pedestrian i meets pedestrian j coming from the side. Suppose that the conflict is resolved in such a way that i will undertake some action to prevent colliding with pedestrian j. It turns out that rather than changing direction, i will *generally reduce speed until j passes*. When j has passed, i will accelerate and continue along its intentional path. Apparently, pedestrian i predicts that the disutility of reducing speed is less than the disutility of changing directions and walking around j. As a result, pedestrians will accelerate and decelerate frequently, explaining the larger variability in the empty zone distribution.

Figure 5 shows the results of applying the cluster analysis to the crossing flow experiment 5. The figure clearly shows the self-formation of diagonal basic structures, as predicted by the theory proposed.

5   Implications for Pedestrian Theory and Modeling

Self-organization is an intrinsic property of pedestrian flow, which can be explained by considering a pedestrian as a utility optimizer, minimizing the predicted disutility of walking. Under this assumption, self-organized structures can be considered as user-optimal equilibrium states, in which none of the pedestrians can improve his or her predicted utility by unilaterally changing his or her situation.

As a result, predicting traffic flow operations accurately requires theory and models that somehow capture this property, such as the models described in [7-9,14]. For one, this is important to correctly capture the highly efficient flow operations in case of bi-direction or crossing flows. Furthermore, practical applications often require a correct reproduction of the different spatial structures that will result in order to correctly capture primary and secondary congestion (spill-back) effects. Failing to capture self-organization phenomena may thus result in the incorrect prediction of these spatial patterns, and thus yield faulty predictions of pedestrian flow in sport-stadiums, transfer stations, etc.

On top of this, correct description of the formation of diagonal strips in crossing pedestrian flows requires correct predicted of pedestrians actions in case of crossing another pedestrian. That is, it is essential that the model captures the fact that pedestrians will generally not change their directions but will decelerate and wait until the crossing pedestrian passes. As an example, the NOMAD model [9], which is based on the concept of utility maximising pedestrians, captures this by penalizing lateral acceleration more severely than longitudinal acceleration. As a result, the model predicts the formation of diagonal strips correctly.

6   Summary and Conclusions

This paper presents both theory and experimental results of self-organization in pedestrian flows. We identified a basic homogeneous pattern consisting of pedestrians having the same speed and direction. The basic pattern consist of overlapping layers, i.e. pedestrians are aligned in a staggered fashion. Furthermore, it was argued that self-organized structures, for instance in bi-directional or crossing flows consist of these basic patterns.

The self-organized phenomena (i.e. dynamic lane formation in bi-directional flow and formation of diagonal strips in crossing flows) that are predicted by the theory of the utility maximizing pedestrian are all identified in the different walking experiments. Furthermore, composite headway model estimation results show how the characteristics inside the formed patterns are similar for the different experiments, thereby providing quantitative evidence for the hypothesis that the formed patterns are indeed the identified basic patterns.

References

1. Goffman, E. Relations in Public: Microstudies in the Public Order. New York. Basic Books, (1971).
2. Sobel, R.S., and N. Lillith. Determinant of Nonstationary Personal Space Invasion. Journal of Social Psychology **97**, 39 - 45, (1975).
3. Dabbs, J.M., and N.A. Stokes. Beauty is Power: the Use of Space on the Sidewalk. Sociometry **38** (4), 551 - 557, (1975).
4. Wolff, M. Notes on the Behaviour of Pedestrians. In: Peoples in Places: the Sociology of the Familiar, 35 - 48, New York, Praeger, (1973).
5. Stilitz, I.B. Pedestrian Congestions. Architectural Phsychology (Canter, D., editor). London Royal Institute of British Architects, 61 - 72, (1970).
6. Willis et al. Stepping aside: correlates of Displacements in Pedestrians. Journal of Communication. **29** (4), 34 – 39, (1979).
7. Blue, V. & Adler, J.L. Cellular Automata Microsimulation for Modeling Bidirectional Pedestrian Walkways. Transportation Research B **35**, 293 - 312, (2001).
8. Helbing, D. & Molnar, P. Self-Organisation Phenomena in Pedestrian Crowds. Self-Organisation of Complex Structure: From Individual to Collective Dynamics (Schweitzer, F., editor), Amsterdam. Gordon and Breach Science Publisher (1997).
9. Hoogendoorn, S.P. & Bovy , P.H.L. Normative Pedestrian Behavior Theory and Modeling". Transportation and Traffic Theory in the 21st Theory – proceedings of the 15th International Symposium on Transportation and Traffic Theory (Taylor, A.P., editor)), 219 - 246, (2002).
10. Toshiyuki, A., Prediction Systems of Passenger Flow. Engineering For Crowd Safety (Smith, R.A. & Dickie, J.F., editors), Elsevier Amsterdam, 249 - 258, (1993).
11. Weidmann, U. Transporttechnik der Fussgänger. ETH Zürich, Schriftenreihe IVT-Berichte 90, Zürich (In German), (1993).
12. Daamen, W. and S.P. Hoogendoorn, Experimental Research of Pedestrian Walking Behavior. *Transportation Research Board Annual Meeting Pre-print CD-Rom*, Washington D.C., (2003).
13. Hoogendoorn, S.P., and W. Daamen. *Pedestrian Behavior at Bottlenecks*. Accepted for publication in Transportation Science, (2003).
14. Helbing, D., *Traffic Dynamics: New Physical Modeling Concepts* (In German). Springer-Verlag, (1997)
15. Zipf, G. K. (1949) *Human Behavior and the principle of least effort*, Cambridge, MA: Addison-Wesley Press.
16. Bressan, A., and W. Shen (2003), *Small BV Solutions of Hyperbolic Non-cooperative Differential Games*, Norwegian University of Science and Technology, Department of Mathematical Sciences. Preprint 2003-021 available via http://www.math.ntnu.no/conservation/.

A Comparison of Video and Infrared Based Tracking of Pedestrian Movements

J. Kerridge, R. Kukla, A. Willis, A. Armitage, D. Binnie, and L. Lei

Napier University
Edinburgh Scotland – UK
{initial.surname}@napier.ac.uk

Summery. We provide a comparison of the relative merits of video and infrared based methods for collecting pedestrian movements from the real world and also from experimental environments. We describe the underlying technological basis of both methods and the tools we have developed to help in collection and analysis of the data. The desire to collect such data is driven by the need of modellers and simulation packages to use base data that is founded in valid empirical evidence, rather than some form of inspired supposition, as is the case with many of the current systems. In addition to the collection of speeds we are also interested in understanding and quantifying the ranges of distances people deviate from a straight-on path when confronted by some obstruction in front of them.

1 Introduction

There are many groups who are building pedestrian simulators for a variety of applications, notably for example Exodus, Aseri, Evi, plus others described at PED 2003 [2]. In the main these groups are using little or no quantitative data concerning the movement patterns exhibited by pedestrians in their simulations. Some of them even make suppositions about the normal walking speed of pedestrians. The PEDFLOW project is perhaps unique in that it attempted to build a model of pedestrian movement based upon data about pedestrian movements that had been obtained from observation of the real world [7]. This data was extracted from CCTV video footage from various UK cities using a method that is described in the next section. The PERMEATE project [1] investigated the use of low-cost infrared detectors to capture pedestrian movement data. This technology provides the opportunity to collect pedestrian movement data in real-time. This is described in section 3. In section 4 we compare the two technologies from a number of different points of view.

2 Video Based Data Capture

Video analysis in the widest sense has been used to study the behaviour of pedestrians for a long time, as the video itself provides a permanent record that can be subjected to repeated, possibly different, kinds of analyses. However, before the advances in computer technology that allowed digital image processing, it was a laborious task observing pedestrians in reality and hence limited to counting people in an area, or crossing a datum line or measuring the time it takes for an individual to cover the distance between two datum lines.

In PEDFLOW we developed a methodology that uses digitised video footage and semiautomatic image tracking to greatly improve the accuracy and variety of data that can be extracted from video. It allows precise instant measures of position and time (and hence speed), but most importantly a way to identify points of interest (such as the start of a deviation) as a change in an instant measurement. By having well-defined criteria that characterise such a change, the extracted data becomes independent of the human factor and therefore repeatable. This data combined with information about the scenario where it was collected and data about the observed pedestrians such as, age and gender, can then be used as input to a pedestrian model.

2.1   Filming and Digitising

The first step is to obtain raw video data. Ideally this footage should be recorded with the camera pointing down vertically (90 degrees) as described, for example, by Hoogendoorn [3]. However, since this is difficult to achieve, an angle up to 45 degrees is still acceptable. Depending on the selected survey location, video footage of pedestrians was collected using either a standard hand-held digital camcorder (miniDV) mounted on a tripod, or by CCTV cameras operated by collaborators within local City Councils. We also contracted private survey companies to film areas that cannot be accessed using either of these methods. In all cases it is imperative that the filmed pedestrians are unaware of the fact that they are under observation, so their behaviour is natural.

If a digital camcorder was used the data can be transferred into the computer directly by means of a firewire connection. It is then stripped of its audio component and recompressed using an MPEG4 codec to save space. Also it is converted to an avi file as required by the tracking software. The method results in very good quality footage of 720x576 pixel resolution (CIF) and 25 fps frame rate (PAL standard). The conditions are not as good for the (more easily obtainable) analogue video material. In order to reliably capture it without dropped frames (which would result in unreliable time measurements) and taking into account the generally poorer image quality, the footage was digitised with a quarter PAL resolution of 352x288 pixels (QCIF) at 25fps.

The large and unwieldy files are then split into short clips according to predefined, objective sampling criteria that characterise a scenario of interest. If, for example, we were interested in investigating the deviation distance associated with an item of street furniture (such as a lamp post), all instances in which a pedestrian walks within a certain distance of the object would be clipped for inclusion in the sample. The 'clipping' reduces the amount of footage that will need to be tracked and it also simplifies the workflow. The clips are time stamped so that the global time distribution is available.

2.2   Tracking

Tracking is the semi-automatic extraction of pedestrian co-ordinates from every individual video frame over a period of time (the length of the clip). We use a commercial package from Mikromak [10] that was originally developed for use in the medical environment but proved to be flexible enough for our purpose. It has the advantage that it can transform the tracked pixel position to real-world co-ordinates directly, provided it is calibrated correctly.

Calibration of the software requires the exact position of 20 objects/points that are visible in the frame. The corners of paving slabs are suitable, but we also used road markings, street furniture and even chalk markers depending on the situation. We developed a method to efficiently record these positions by measuring the distance of each point to two fixed co-ordinates on a (virtual) base line. Using a spreadsheet, these distances are then converted into co-ordinates using trigonometric functions. In the software package the same 20 points are identified in the video frame and associated with their real-world co-ordinates. Technical data of the recording equipment (such as CCD size and resolution) also needs to be specified. The program then uses an interactive, iterative process to build a transformation model between the two co-ordinate systems.

Image matching does the actual tracking: an area marked in Frame A is matched to a similar area in Frame B within a specified area surrounding the original position of that patch. The new position and pattern is recorded and used as the basis for the matching in frame C and so on. Selecting the head and upper body was found to be most effective – the position on the ground is obtained by adding an offset. This works so well that human intervention is only required if lighting conditions change (shadow) or the person is occluded. For movements towards or away from the camera the offset needs to be corrected as well. The software can track several people simultaneously.

The precision is so great that the software also records the sideways movement of the body caused by the shifting of weight between the two feet while making a step, which results in a wavy line superimposed on the underlying trajectory (see Fig. 1 where the pedestrian is walking in the y-direction to emphasize the sideways movement). This higher frequency can be filtered out using the 'smooth' function of the tracking software, which averages between neighbouring points. This has proved to be more effective than the polygon approximation, which is also available.

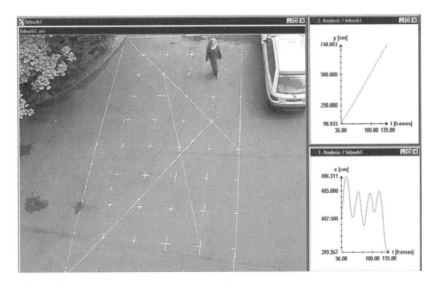

Fig. 1.

The result of the tracking process is a comma-separated file for every tracked person. A file of the same structure is also generated to capture the static mark up of the scenario such as building edges and street furniture as polygons

2.3   Data Exploration and Extraction

Having the raw numerical data is only the starting point of the analysis. Relevant configurations need to be identified with regards to time (frame) and place (co-ordinates) and the information recorded.

To help with this task, a software tool has been developed (Fig. 2). It will load all the data files related to one clip and draw the trajectories in an x-y co-ordinate system. A slider at the bottom of the window is used to adjust the time. The program displays the position of the pedestrian(s) at the selected time as a circle. By moving the slider, one can explore the data dynamically and find areas of interest such as two pedestrians who are about to avoid each other.

The program supports the following additional features to aid the exploration; a) online speed visualization b) online direction visualization and c) visualization of local maxima in direction change. The speed distribution over the time period is displayed at the bottom of the tool with a line indicating the current frame/time. It can be used to quickly identify maxima and minima of walking speed. The current walking direction is drawn as a line from the centre of the circle for every pedestrian in the tool. The black dots help to identify the points at which the person changed direction. Since this is usually not instantaneous, the tool defines one point of maximum change as being the representative. These points of greatest change are calculated by taking the second derivative of the direction with parameters that can be adjusted by further sliders.

Having found a configuration of interest, the program greatly simplifies the collection of associated measures. It uses two markers, which are set by mouse clicking on the area. The software immediately updates frame, x co-ordinate and y co-ordinate. It also calculates and displays the distance between the markers and (if they are on different frames) the time-difference between them. If the markers are associated with the same person, average speed is presented as well. At the press of a button this information, together with the filename, a label and a comment is stored in a file. After analysing several clips, the collected data can easily be imported into a spreadsheet or statistical package for further

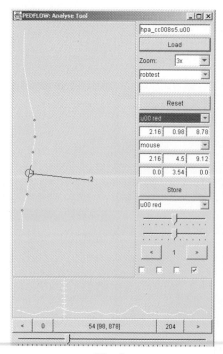

Fig. 2.

analysis [8], which is outside the scope of this paper.

3   Infrared Based Data Capture

The Irisys detector [5,6,9] is primarily designed for counting people moving back and forth across a datum line specified by the user. In this mode a single detector can be used, for instance, to count movements along a corridor or through a doorway. Multiple detectors can be aligned to create a larger counting line. In this mode detectors have been used successfully: to count pedestrian movements in and out of supermarkets, in station concourses, and over the Millennium footbridge in London during trials after the bridge was modified. We have been able to extract more information from the detectors, in particular to record pedestrian trajectories [4].

The detector uses a 16 x 16 array of pyroelectric ceramic detectors to measure changes in temperature. The detectors we have been using only measure temperature differences, and rely on the pedestrian being at a different temperature from the background. This has the advantage that the background disappears from the image, leaving pedestrians as clear targets. In Figure 3 the white areas indicate a person, with the darker areas indicating the background.

The normal mode of using the detector is to mount it at a height of three to four metres vertically above the region of interest. At this height, the detector covers a ground area three to four metres square. The detector has a processor on-board that undertakes image pre-processing. Likely targets have an ellipse fitted to them, and the center of the ellipse is then calculated and communicated down a communications link to the data-gathering computer. The X and Y positions are recorded approximately three times per second. These values are presented as a floating-point number in the range 0 – 16 for each axis. One unit, being the width of a pixel, corresponds approximately to 20 centimeters on the ground, when mounted 3 meters above the ground and 25 centimeters when mounted at 4 meters. However the centering process effectively averages data from a number of pixels, the position is actually recorded with sub-pixel accuracy; hence we can locate pedestrians to an

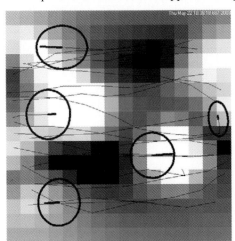

Fig. 3.

accuracy of 2.5 cm or better. Fig. 3 shows the ellipses that have been matched and the trajectories of these targets and other previous targets. The line within the ellipse shows the direction of travel. Due to the time it takes to process the image the fitted ellipse always lags behind the actual image. Fig. 3 also demonstrates that it is impossible to identify an individual from the image.

3.1   Multiple Detector Systems

The main problem with multiple detector systems is tracking the trajectory from one field of view to the next. We cannot guarantee the relative alignment of the detectors and thus the data analysis systems have to be able to overcome such a situation and cannot rely on a perfect alignment of the fields of view of adjacent detectors to a predefined specification. For each detector a set of data is captured in the central computer of the form shown in Fig. 4. This represents a single person moving through the field of view of a detector. A line starting with C gives a time stamp for the data reading with an accuracy of milliseconds, which is generated by the data gathering computer. A line starting with T gives data about the ellipse that has been fitted to a person by the detector's internal processor. If there were more than one person in the field of view then there would be as many T-type records for each C-type record. If no targets are detected a C-type record will be immediately followed by another. Thus the person in Fig. 4 starts from a position of [0.416, 4.789] at time Thu May 22 10:36:41 531 2003 and leaves the field of view at location [13.064, 3.179] at time Thu May 22 10:36:43 000 2003, if we ignore data points with S = 1. A simple calculation, assuming each pixel is 0.2m and we ignore movement in the y direction, which is just over 1 pixel yields a walking speed of 1.72 m/sec, from (13.064 - 0.416)*0.2)/(43.000 – 41.531)

Timestamp	Target Data				
C Thu May 22 10:36:40.906 2003	ID	S	X	Y	Mode
C Thu May 22 10:36:41.218 2003	T 2	1	0.304	5.089	257
C Thu May 22 10:36:41.531 2003	T 2	0	0.416	4.789	261
C Thu May 22 10:36:41.859 2003	T 2	0	0.776	3.303	261
C Thu May 22 10:36:42.171 2003	T 2	0	4.256	2.860	261
C Thu May 22 10:36:42.375 2003	T 2	0	7.636	3.020	261
C Thu May 22 10:36:42.687 2003	T 2	0	10.831	2.749	261
C Thu May 22 10:36:43.000 2003	T 2	0	13.064	3.179	261.
where					
ID - Target Identifier	S – Status; 1 – invalid; 0 – valid				
X – x co-ordinate	Y – y co-ordinate				
Mode – Further data when S = 1					

Fig. 4.

The detector's internal processor operates as follows. The identification number is unique to a detector. The sequence of identification numbers is reused from the initial value once the detector has been unable to detect any targets in the field of view for some time. This means that the lower identification numbers are reused many times during an observation, hence we have to undertake additional processing between the streams of data if we are to track people across multiple detectors.

3.2   Algorithmic Approach

For the purposes of explanation we shall consider the scenario shown in Fig. 5. Consider a corridor surveyed by three detectors such that people can enter through the left-hand edge of Detector 0 and then exit from the right-hand edge of Detector 2 and vice-versa. We have built a system in Java in which each detector is managed by its own process. Thus as a person is detected as having left the field of view, of say, Detector 0 their trajectory details are passed to Detector 1. Once the person has been detected as having arrived in Detector 1 then the process dealing with Detector 1 can search incoming trajectories from Detector 0 to find the best match. In this manner we can build up the complete trajectory of a person as they move from an entry to an exit edge. The system is capable of processing the incoming data and consequent saving of trajectory information for each person in real time.

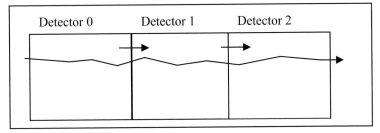

Fig. 5.

Figure 6 shows part of the printed output from the system for one person as they moved from Detector 0 to Detector 2. Each person is given an identifier derived from the edge of the detector by which they entered the observed area together with a unique sequence number. The Entry and Exit points with corresponding times are stored so as to make it easier to calculate average speed through the observed area. The remainder of the person record comprises the sequence of points through which the person moved in passing through the observed area. The location data comprises; [Detector Number, Identifier Tag, X-value, Y-value, Status followed by the time the person was at that point]. By observation it can be seen that in Detector 0 (Locations 0 to 6) the person was given the identifier tag 12, then in Detector 1 (Locations 7 to 11) the tag 2 and finally in Detector 2 (Locations 12 to 19) the tag 10. This highlights the difficulty in forming the complete trajectory.

A relatively simple, though not perfect, heuristic has been used to determine a match between an incoming partial trajectory and one that is being generated in an adjacent detector. It deals with the cases where detector fields of view both overlap and are disjoint. The amount of under and overlap is always kept as small as possible when setting up the detectors. Consider a person moving from Detector 0 to Detector 1 of Fig. 5; then the exit time from Detector 0 must be within 1.5 seconds of the entry time to Detector 1. However depending on whether there is underlap or overlap we may have started to build the trajectory in Detector 1 before the data from Detector 0 arrives. This time differences also takes account of the fact that edge effects in detector image processing may mean there is some time before a valid target is returned in Detector 1. If we assume that the base edges of the detectors are more or less in line, that is the edges of the detectors abut almost perfectly in the y-direction, then we also require that the y-pixel value at the point of exit from Detector 0 is within 2 pixels of that observed on entry to Detector 1. This is probably reasonable for people walking down a corridor as in this case but is less likely to be true if we are observing a larger rectangular area where people may move with more freedom.

```
[PersonId, 0, 16, 11]
 Entry point PointData - [ Detector 0, x 0.833, y 9.925]
 Exit  point PointData - [ Detector 2, x 14.901, y 11.065]
 Entry time [Time, 10:37:28.220 ]
 Exit  time [Time, 10:37:35.480 ]
 Location 0, [Location, 0, 12, 0.833, 9.925, 0, [Time, 10:37:28.220 ]]
 Location 1, [Location, 0, 12, 2.58, 10.458, 0, [Time, 10:37:28.550 ]]
 Location 2, [Location, 0, 12, 4.768, 10.734, 0, [Time, 10:37:28.880 ]]
 Location 3, [Location, 0, 12, 7.824, 11.004, 0, [Time, 10:37:29.210 ]]
 Location 4, [Location, 0, 12, 10.87, 11.069, 0, [Time, 10:37:29.540 ]]
 Location 5, [Location, 0, 12, 13.151, 10.798, 0, [Time, 10:37:29.870 ]]
 Location 6, [Location, 0, 12, 15.335, 10.472, 0, [Time, 10:37:30.200 ]]
 Location 7, [Location, 1, 2, 4.058, 11.696, 0, [Time, 10:37:31.190 ]]
 Location 8, [Location, 1, 2, 7.195, 11.673, 0, [Time, 10:37:31.520 ]]
 Location 9, [Location, 1, 2, 10.152, 11.55, 0, [Time, 10:37:31.850 ]]
 Location 10, [Location, 1, 2, 12.481, 11.332, 0, [Time, 10:37:32.180 ]]
 Location 11, [Location, 1, 2, 14.212, 10.979, 0, [Time, 10:37:32.510 ]]
 Location 12, [Location, 2, 10, 0.8, 10.74, 0, [Time, 10:37:33.170 ]]
 Location 13, [Location, 2, 10, 2.011, 11.596, 0, [Time, 10:37:33.500 ]]
 Location 14, [Location, 2, 10, 4.287, 11.955, 0, [Time, 10:37:33.830 ]]
 Location 15, [Location, 2, 10, 7.096, 12.015, 0, [Time, 10:37:34.160 ]]
 Location 16, [Location, 2, 10, 9.964, 11.902, 0, [Time, 10:37:34.490 ]]
 Location 17, [Location, 2, 10, 11.896, 11.682, 0, [Time, 10:37:34.820 ]]
 Location 18, [Location, 2, 10, 13.472, 11.337, 0, [Time, 10:37:35.150 ]]
 Location 19, [Location, 2, 10, 14.901, 11.065, 0, [Time, 10:37:35.480 ]]
```

Fig. 6.

Once the pixel locations have been transformed into actual offsets in the observed area we have data from that shown in Fig. 6 in the correct format for analysis by the tool developed for the video based data capture method. Hence we can process the data to determine deviation points and distances.

4 Comparison of Methods

Video based measurement has the obvious advantage that a human can immediately interpret the image, extracting information about the scene such as the layout, type of any obstruction, the weather and possibly the gender and any mobility impairment a person may have. It utilizes consumer-based products and thus is cheap to buy. Its main disadvantages are that processing the data is expensive either in terms of human time required to process the images (typically, 1 hour of video requires 1 week to analyze) or automatic processing requires very expensive image processing equipment. It may also cause data protection and privacy issues.

Conversely, the infrared detection equipment uses non-consumer technology and thus is more expensive but does allow the capture of trajectory information in real-time due to the incorporation of an internal processor in the detector. It is not possible to obtain mobility, gender and other aspects directly but the method does not have the data protection and privacy pitfalls of video images.

The best compromise is probably a combination of both techniques whereby trajectories are captured using infrared technology and the scene is observed by a web-cam, which allows an observer to mark points of interest in the data stream as it is collected.

Acknowledgements

The research has been supported by a grant from the UK Department of Transport in the LINK programme Future Integrated Transport with the project PERMEATE (GR/N33706). It has also been supported by a research grant from the UK Engineering and Physical Science Research Council for the project PEDFLOW (GR/M59792) for the video based data collection.

References

1. Armitage, A., Binnie, T.D., Kerridge, J.M., & Lei,L., (2003), "Measuring Pedestrian Trajectories with Low Cost Infrared Detectors: Preliminary Results", Pedestrian Evacuation and Dynamics – 2003, ER Galea (ed) University of Greenwich, London, UK.
2. Galea, E.R. (ed), "Pedestrian Evacuation and Dynamics 2003", University of Greenwich, London UK.
3. Hoogendoorn, S.P., Daamen, W. and Bovy, P.H.L., "Extracting Microscopic Pedestrian Characteristics from Video data", Transportation Research Board 2003 Annual Meeting, CD-ROM, Paper No 477, National Academies, Washington USA, 2003.
4. Kerridge, J.M., et al, Monitoring the Movement of Pedestrians Using Low-cost Infrared Detectors: Initial Findings, to appear in Transportation Research Board 2004 Annual Meeting, Washington USA, 2004.
5. Mansi, M. V., Porter, S. G. Galloway, J. L. &. Sumpter N., "Very low cost infrared array based detection and imaging systems" (SPIE) Aerosense 2001, Orlando, Florida USA, 17-19 April 2001.
6. Stogdale, N., Hollock, S., Johnson, N. & Sumpter, N. (2003), "Array based infrared detection: an enabling technology for people counting, sensing, tracking and intelligent detection", SPIE, USE 3, 5071-94.
7. Willis, A., Kukla, R., Kerridge, J. & Hine, J. (2001), "Laying the foundations: The use of video footage to explore pedestrian dynamics in PEDFLOW", Proceedings of Pedestrian and Evacuation Dynamics, Schreckenberg, M. and Sharma, S.D. (eds), Springer, Berlin, pp 181-186.
8. Willis, A., Gjersoe, N., Harvard, C. Kerridge, J.M., and Kukla, R., "Human Movement Behaviour in Urban Spaces", submitted for publication Environment and Planning B.
9. www.irisys.co.uk (accessed 9/7/2003).
10. http://www.mikromak.com/ (accessed 27/10/2003).

Statistical Analysis of the Panic in the Pedestrian Flow

T. Iizuka

Department of Physics, Faculty of Sciences – Ehime University
Matsuyama, 790-8577 – Japan

Summary. We consider a numerical model of pedestrian movement in a counter flow footway using a cellar automaton model. The enter rate of a pedestrian into the footway is the control parameter in our study. The probability of the panic occurrence is calculated statistically. If the width of footway is large, we find critical behavior of the panic probability which is similar to the phase transition obtained from statistical mechanics. We also consider the asymmetric case where the number of pedestrian moving one direction is different form that of the couter direction.
Keywords: pedestrian flow, cellar automaton model, panic probability

1 Introduction

Pedestrians flow is one of the interesting parts in the field of the traffic engineering. Actually pedestrian behavior in public place such as architectural space and street crossing has been analyzed in many situations. For designing public facilities, quantitative analysis of pedestrian flow is important[2].

Pedestrian counter flow[3] is remarkable situation as compared with the car traffic flow. In particular, panic is one of interesting phenomenon of the pedestrian flow dynamics. Shocking accident on July, 21, 2001 in the footbridge, Akashi, Japan is nothing but panic of the pedestrians. To avoid such accident, analysis of two-directional counter flow in facilities such as footway seems very important problem.

The traffic flow is modeled in some kinds manners. The celler automaton model is one of popular method to investigate traffic motion, which is applicable to the two-dimensional case[1].

This paper also employs a simple celler automaton model which will be explained in the next section. In §3, panic occurrence are analyzed statistically. Section §4 is devoted to the case that the number of pedestrians are different from that of couter-walking pedestrians.

2 Cellar Automaton Model

Here we introduce a model which simulates the two-way pedestrian flow. The footway is assumed to be a two-dimensional square lattice.

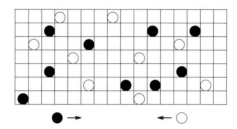

Fig. 1. Cellar automaton model to the two-way pedestrian move. The walkway is a rectangle and each pedestrian is located in a square lattice site.

Each lattice site is occupied by one person at most. There are two groups of pedestrian, one of which tends to move one direction and the other opposite direction. Figure 1 shows a configuration of the footway, where solid and blank circles indicate the pedestrians. A solid circle tends to move rightward (on paper) and a blank circle leftward. We apply the following two celler automaton rules for the motion of the pedestrians. One of them (rule A) impose that if a pedestrian is likely to head-on collide with another pedestrian, he tend to avoid right side. In the other rule (rule A), he chose his direction randomly. Detail of the rules are followings.

Rule A (right-hand)

1) If the front site is vacant, a pedestrian steps in it.
2) If the front site and left-front site are occupied and right-front site is vacant, the pedestrian steps in it.
3) If the front site and right-front site are occupied and left-front site is vacant, the pedestrian steps in it.
4) If front site is occupied and both left-front and right-front sites are vacant, the pedestrian steps in right-front site.
5) Front, left-front and right-front sites are occupied the pedestrian cannot move.

Rule B (random)

1)-3) Same as rule A.
4)   If front site is occupied and both left-front and right-front sites are vacant, the pedestrian *chooses a site randomly* for the next step.
5)   Same as rule A.

The above rules are for a motion of a pedestrians. In our model, all pedestrians do not move simultaneously to avoid long range correlation in the automaton rule. Therefore, we must fix the turn of pedestrians in applying the rules.

Ordering the pedestrians

6)   In one time step, every pedestrians applied once.
7)   Solid circle pedestrians (see fig.1)have priority to be applied.
8)   Hindward pedestrians (left on the paper) have priority to be applied.
9)   Order of the solid circles in the same column are fixed randomly.
10) Next, the rules are applied to blank circles . The ordering of them are determined in the same method as solid ones.

Rules 1)-5) and application order 6)-10), consist of the cellar automaton model for one time step. Rule 7) is introduced for convenience. Actually solid circle need not have the initial priority. Since the simulation are done many time steps, initial "difference" should not affect the long time behavior of the pedestrians.

On the other hand, rule 8) is important. As a simple case, consider single directional(rightward) flow and a congestion cluster in it. As seen from fig.2, if the hindward pedestrians have the priority to be applied the motion, only the front pedestrian can move in one time step. In case of the forward priority, all pedestrians can move simultaneously in one time step. In real footpath, the pedestrians tend to keep interval when walking. Therefore, the hindward priority is better choice in our simulation.

Hideward priority   Forward priority

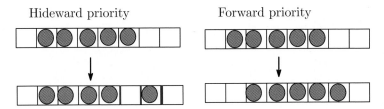

Fig. 2. Difference between hindward and forward priority. The former case seems realistic.

In the exit or entrance of the walkway we treat the pedestrians in the following manner.

On the exit or entrance

11) Pedestrians in the exit of the footway can always escape next time.
12) After 1-step movement, new pedestrians step into the footway. If one site in the entrance is vacant, new one can occupy it with probability $RATE$, which plays a role as a control parameter.

We note that the side wall is considered to be occupied by a pedestrian. In our analysis all sites are assumed to be vacant as the initial condition. $RATE$ and width and length of the footway ($Length, \quad Width \in \mathbf{N}$) are the parameters of our simulation.

3 Pedestrian Flow

Here, $RATE$ is treated as a control parameter and $Flow$ is defined by,

$$Flow = \frac{\text{\# of pedestrians escaped from the exits}}{2}.$$

In this paper total time step for one simulation is set 1000. In low $RATE$ area, the motion of pedestrians very smooth, while it getting larger, pedestrian often stop (see rule 5)). If $RATE$ is large enough, collisions between counter directional pedestrians occur frequently and we observe a freeze of the flow. It is nothing but the panic.

Our main aim is statistical behavior of the pedestrian flow. We take the ensemble average of the number of pedestrians who path thorough the footway ($Flow$).

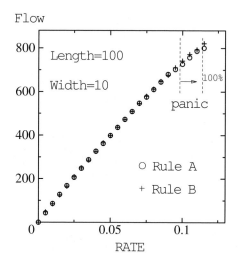

Fig. 3. Average of total number of pedestrians who path through the footway for rule A and B. Clear difference is not seen. If $RATE > 0.098$, panic can occur, and the panic probability is almost 100% at $RATE = 0.12$. The ensemble are taken only from non-panic case.

The probability that panic occurs seems interesting quantity for the footway designing. In the next section, we comprehensively treat the 'panic probability'.

4 Panic Probability

As seen before, if $Length = 100$, $Width = 10$, and $0.098 < RATE < 0.12$, the panic *possibly* occurs. Ensembles of simulations gives the probability of the panic occurrence. In this section we pay attention on the panic probability applying rule A. Ensemble number is set 200.

Figure 4 show the panic probability for a fixed length (=200). We change the $Width$ from 10 to 640. As $Width$ becomes larger, the transitions of panic probability in the figure becomes steeper.

We expect phase transition like phenomenon for a large width. Remark that the saturation $RATE$ on which the panic probability is 100% are common (~ 0.12).

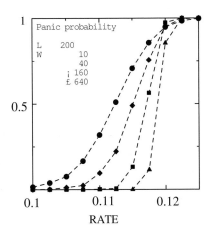

Fig. 4. Panic probability for a fixed length (=200). If width is large enough, the diagram exhibits phase transition like behavior.

This results represented by fig. 4, suggest that if the size of footpath is large enough, transition of the panic probability is steep. This resembles the phase transition of macroscopic matter. According to the statistical mechanics, if the system size becomes larger, a statistical quantity (order parameter) changes critically at a point of the control parameter. Here we can regard as the panic probability as the the order parameter.

5 Asymmetric Case

So far the enter rate ($RATE$) of the two groups is assumed common. In real situation, the flow usually asymmetric. Therefore, we need consider such case in which $RATE$ of the solid circles and blank circles are different.

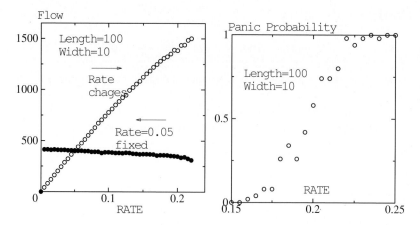

Fig. 5. *Flow* − *RATE* diagram in asymmetric case.(left) *RATE* of one group is fixed 0.05. Panic probability in asymmetric case.(right) The ensemble number(= 50) is not large enough. Then the curve seems ugly.

We fix the *RATE* of the pedestrian moving leftward (solid circles), and change that of rightward. The *Flow*−*RATE* diagram is shown in fig.5 for *Width* = 10 and *Length* = 100. The *Flow* of the fixed group is almost flat, though it gradually decay with *RATE* . This is because the collisions occures frequently as the *RATE* increases.

Similarly to the 'symmetric' case the panic occures above *RATE* ∼ 0.15 Panic probability is also shown in fig.5.

6 Summery and Discussions

We have considered pedestrian couter flow in a footway. If the rate of pedestrians entrance (*RATE*) is large enough, panic phenomena occur. Employing *RATE* as the control parameter, we have statistically calculated panic probability which is regardes as a order parameter. If the size of the footpath becomes larger, the transition of the panic probability becomes steeper, which resembles the behavior of an order parameter in statistical mechanics.

References

1. O. Biham, A.A. Middleton, and D. Levine, Phys. Rev. A **46**, R6124 (1992).
2. F. Feurtey, Simulating the Collison Avoidance Behavior of Pedestrians. Master thesis -the University of Tokyo, dep. Electronic Eng. (2000).
3. M. Muramatsu and T. Nagatani, Physica A**267**, 487-498 (1999).

Group Formation of Organisms in 2-Dimensional OV Model

A. Nakayama[1] and Y. Sugiyama[2]

[1] Gifu Keizai University
 Ohgaki, Gifu 503-8550 – Japan
 g44153g@cc.nagoya-u.ac.jp
[2] Graduate School of Information Science – Nagoya University
 Nagoya, 464-8601 – Japan
 genbey@eken.phys.nagoya-u.ac.jp

Summary. We propose a dynamical model for collective biological motion, which is based on 2-dimensional optimal velocity model. The property of the model is investigated by numerical simulation. Various patterns of group emerge by changing parameters of the model and density of organisms.
Keywords: simulation; biological motion, nonlinear dynamical systems

1 Introduction

There are interesting phenomena in traffic flow, pedestrian flow and other systems. Many physicists have proposed various models to clarify the properties of such systems [1–5]. The optimal velocity (OV) model is a dynamical model for 1-dimensional traffic flow [6] and can explain the behavior of traffic flow, especially the formation of congestion based on a simple idea for phase transition. We expect the same idea can be applied to other systems. However, pedestrian flow [4] and some biological systems [7–10], which show interesting collective motion, are essentially 2-dimensional systems. For the study of these systems, we have extended the OV model to 2-dimensional model [11, 12]. There are two types of the extended OV model: one is the model with repulsive interaction, which is suitable for pedestrian flow. The other is the model with attractive interaction, which is suitable for biological motion.

As for collective biological motion, many models have been proposed from various viewpoints: famous "Boids" [13], models for schooling of fish [14–17], and models from the physical viewpoint [7–10, 18]. In this paper, we introduce 2-dimensional OV model with attractive interaction and modify it to incorporate both attractive and repulsive interactions for comprehensive study. All interactions, however, are unified into single form which is usually called the OV function and the equation of motion remains simple. We investigate the model as a dynamical model for collective biological motion by use of numerical simulation. In order to clarify the global structure of the model, simulations are carried out in various settings and in a wide range of parameters.

2 Model

We consider 2-dimensional OV model [19]

$$\frac{d^2}{dt^2}\mathbf{x}_n(t) = a\left\{\sum_k \mathbf{F}(\varDelta\mathbf{x}_{kn}) - \frac{d}{dt}\mathbf{x}_n(t)\right\}, \tag{1}$$

$$\mathbf{F}(\varDelta\mathbf{x}_{kn}) = \mathbf{n}_{kn}\left(\frac{1+\cos\theta}{2}\right)\left(\frac{\tanh 4(\varDelta x_{kn}-1)+c}{2}\right), \tag{2}$$

where $\varDelta\mathbf{x}_{kn} = \mathbf{x}_k - \mathbf{x}_n$, $\mathbf{n}_{kn} = \varDelta\mathbf{x}_{kn}/\varDelta x_{kn}$ and $\varDelta x_{kn} = |\varDelta\mathbf{x}_{kn}|$. Bold symbols are 2-dimensional vectors. θ is the angle between $\varDelta\mathbf{x}_{kn}$ and $\mathbf{v}_n$ which is the velocity of nth organism $(d\mathbf{x}_n/dt)$. $\sum_k \mathbf{F}(\varDelta\mathbf{x}_{kn})$ corresponds to the OV function in the original OV model. The summation is taken over organisms within a certain range around nth one. The parameter a is usually called sensitivity, which represents the magnitude of reaction.

The numerical parameters in tanh-function are chosen for simplicity, and the values are not significant. The interaction of the model can be varied by changing the parameter c. For $c = 1$, only attractive interaction exists. For $-1 < c < 1$, there exist both attractive and repulsive interactions. It is well-known that organisms cannot move if there is no self-driving force. Then we add an additional term, which is proportional to $\mathbf{v}_n$, to the OV function. In this work, we set the amplitude at 0.75 as an example, that is, $\sum_k \mathbf{F} \rightarrow \sum_k \mathbf{F} + 0.75\mathbf{v}_n/|\mathbf{v}_n|$.

3 Simulation

Numerical simulations are carried out for various value of parameters a and c. The system size is fixed to 26×15, but we confirmed that the size dependence is small. For each simulation, we observe the number of groups after sufficiently large relaxation time. Then the average size of groups, which is the average number of members, is calculated from the number of groups. Here a group is defined as follows: Any member of a group must exist within the interaction range of at least one other member which moves in the same direction. It is considered that they move in the same direction, if the inner product of their velocities is positive.

Case 1
First, we consider the case that the interaction affects all the organisms within a range $\varDelta x_{kn} < 2.0$. Figure 1 shows the average size of groups. At the points

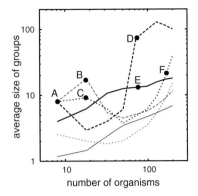

Fig. 1. Average size of groups: The size is defined by the number of members of each group. Bold solid line is $c = 1.0$, $a = 3.0$ and thin solid line is $c = 1.0$, $a = 1.0$. Bold dashed line is $c = 0.0$, $a = 3.0$ and thin dashed line is $c = 0.0$, $a = 1.0$. Bold dotted line is $c = -0.5$, $a = 3.0$ and thin dotted line is $c = -0.5$, $a = 1.0$.

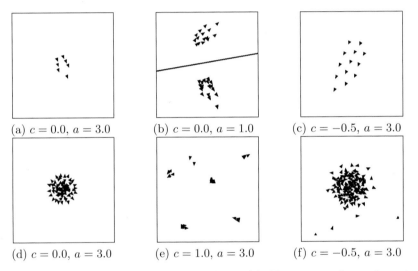

Fig. 2. Typical shapes of groups: Each figure (a)–(f) corresponds to the points (A)–(F) in Fig.1.

(A), (B) and (D), single group is formed finally but the shapes of them are different. Typical shapes at the points (A)–(F) are shown in Figs.2(a)–(f), respectively. Groups in Figs.2(a), (b) and (e) take a similar form. Only in the case $c = -0.5$, triangular structure appears (see Fig.2(c)). The stability of these groups are weak and they are easily destroyed by another organism. On the other hand, groups in cases (d) and (f) are completely different. The groups do not move and are more stable. Especially in the case (d), any organism cannot escape from the group. The shape of groups transits from one to another continuously as the change of parameters. It is unclear whether there exists phase transition or not.

Case 2

Next, we consider the case that the interaction is modified to act only on the

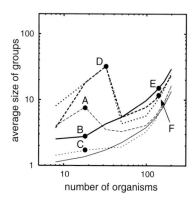

Fig. 3. Average size of groups: Bold solid line is $c = 1.0$, $a = 3.0$ and thin solid line is $c = 1.0$, $a = 1.0$. Bold dashed line is $c = 0.0$, $a = 3.0$ and thin dashed line is $c = 0.0$, $a = 1.0$. Bold dotted line is $c = -0.5$, $a = 3.0$, and thin dotted line is $c = -0.5$, $a = 1.0$.

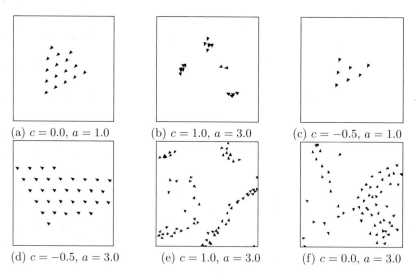

(a) $c = 0.0$, $a = 1.0$ (b) $c = 1.0$, $a = 3.0$ (c) $c = -0.5$, $a = 1.0$

(d) $c = -0.5$, $a = 3.0$ (e) $c = 1.0$, $a = 3.0$ (f) $c = 0.0$, $a = 3.0$

Fig. 4. Typical shapes of groups:

nearest neighbors, those are defined as follows. We divide the interaction range into six sectors, and find the nearest organism in each sector. The number of sectors is not so important. The same results are obtained also in the case where the range is divided into eight sectors, for example.

Figure 3 shows the average size of groups and typical shapes at the points (A)–(F) are shown in Figs.4(a)–(f), respectively. In the case of high sensitivity, single group is formed at low density (D). The group is rather stable but is broken by the collision with another group. Therefore the size of group does not increase even in high density.

The shapes of group are classified into two classes. One is the cases (A), (C), (D) and (F), where repulsive interaction exists. The other is the cases (B) and (E), where no repulsive interaction exists. In former class, each group shows triangular structure (see Figs.4(a), (c), (d), (f)). The distance between two organisms can be simply estimated by $\mathbf{F}(\Delta \mathbf{x}_{kn}) = 0$. The stability of the group can be understood in a similar way to the 1-dimensional OV model or the 2-dimensional OV model for pedestrians. The stability condition is decided by the sensitivity and the gradient of the OV function. The group is stable for large sensitivity and is unstable for small sensitivity, if we fix the OV function. In the latter class (the cases (b) and (e)), the group takes a compact form, which is similar to that in Fig.2(e). The transition between these forms of group is also continuous, and therefore it seems that definite phase transition does not exist.

4 Summary and Discussion

In this paper, we investigated the property of the 2-dimensional OV model. The model incorporate both attractive and repulsive interaction in a unified form. By use of numerical simulation, we observed various types of group formation but could not find any clear transition between these forms of groups. It is an open problem whether they exist in different phases or not. Generally, the group in this model is less stable than that in the 2-dimensional model for pedestrian. The analysis of stability is a future problem.

There remain some ambiguities in our model. One is the interaction range. As shown in section 3, the rough behavior of organisms does not changes but the shape of groups changes, when the interaction range is altered. The other is the magnitude of self-driving force. In this work, we chose an appropriate value. It should be investigated how the behavior changes if the magnitude is altered.

Acknowledgements

This work is partly supported by a Grant-in-Aid for Scientific Research (C) (No.15560051) of the Japanese Ministry of Education, Science, Sports and Culture.

References

1. M. Schreckenberg and D.E. Wolf, editors. *Workshop on Traffic and Granular Flow '97*. Springer, Singapore, 1998.
2. D. Helbing, H.J. Herrmann, M. Schreckenberg, and D.E. Wolf, editors. *Traffic and Granular Flow '99*. Springer, Berlin, 2000.
3. D. Chowdhury, L. Santen, and A. Schadschneider. Statistical physics of vehicular traffic and some related systems. *Physics Reports*, 329:199–329, 2000.
4. D. Helbing. Traffic and related self-driven many-particle systems. *Rev. Mod. Phys.*, 73:1067–1141, 2001.
5. M. Fukui, Y. Sugiyama, M. Schreckenberg, and D.E. Wolf, editors. *Traffic and Granular Flow '01*. Springer, Berlin, 2003.
6. M. Bando, K. Hasebe, A. Nakayama, A. Shibata, and Y. Sugiyama. Dynamical model of traffic congestion and numerical simulation. *Phys. Rev. E*, 51:1035–1042, 1995.
7. T. Vicsek, A. Czirók, E. Ben-Jacob, I. Cohen, and O. Shochet. Novel type of phase transition in a system of self-driven particles. *Phys. Rev. Lett.*, 75:1226–1229, 1995.
8. A. Czirók, E. Ben-Jacob, I. Cohen, and T. Vicsek. Formation of complex bacterial colonies via self-generated vortices. *Phys. Rev. E*, 54:1791–1801, 1996.
9. A. Czirók and T. Vicsek. Phase transition in the collective motion of organisms. In M. Schreckenberg and D.E. Wolf, editors, *Traffic and Granular Flow '97*, pages 3–19. Springer, Singapore, 1998.
10. T. Vicsek, A. Czirók, and D. Helbing. Collective motion and optimal self-organisation in self-driven systems. In D. Helbing, H.J. Herrmann, M. Schreckenberg, and D.E. Wolf, editors, *Traffic and Granular Flow '99*, pages 147–159. Springer, Berlin, Heidelberg, 2000.
11. A. Nakayama, K. Hasebe, and Y. Sugiyama. Optimal velocity moldel and its applications. In M. Fukui, Y. Sugiyama, M. Schreckenberg, and D.E. Wolf, editors, *Traffic and Granular Flow '01*, pages 127–140. Springer, Berlin, 2003.
12. Y. Sugiyama, A. Nakayama, and K. Hasebe. Modeling pedestrians and granular flow in 2-dimensional optimal velocity moldels. In M. Fukui, Y. Sugiyama, M. Schreckenberg, and D.E. Wolf, editors, *Traffic and Granular Flow '01*, pages 537–542. Springer, Berlin, 2003.
13. C.W. Reynolds. *Computer Graphics*, 21:25, 1987.
14. N. Sannomiya and K. Matsuda. A mathematical model of fish behavior in a water tank. *IEEE Trans. on Systems Man Cybernetics*, SMC-14:157, 1984.
15. H.-S. Niwa. Self-organizing dynamic model of fish schooling. *J. theor. Biol.*, 171:123–136, 1994.
16. H.-S. Niwa. Newtonian dynamical approach to fish schooling. *J. theor. Biol.*, 181:47–63, 1996.
17. N. Sannomiya and H. Nakamine. A simulation study on cooperative behavior in a fish school. In *Proc. of 3rd Int. Symp. on Artificial Life and Robotics*, volume 1, page 17. 1998.
18. N. Shimoyama, K. Sugawara, T. Mizuguchi, Y. Hayakawa, and M. Sano. Collective motion in a system of motile elements. *Phys. Rev. Lett.*, 76:3870–3873, 1996.
19. A. Nakayama and Y. Sugiyama. 2 dimensional optimal velocity moldel for pedestrians and biological motion. In P.L. Garrido and J. Marro, editors, *MODELING OF COMPLEX SYSTEMS:Seventh Granada Lectures*, pages 107–110. American Institute of Physics, 2003.

Simulations of Evacuation by an Extended Floor Field CA Model

K. Nishinari[1], A. Kirchner[2], A. Namazi[2], and A. Schadschneider[2]

[1] Department of Applied Mathematics and Informatics – Ryukoku University
   Shiga 520-2194 – Japan
   knishi@rins.ryukoku.ac.jp
[2] Institute for Theoretical Physics – University of Cologne
   50923 Köln – Germany
   [aki,an,as]@thp.uni-koeln.de

Summary. The floor field CA model for studying evacuation dynamics is extended in this paper. A method for calculating the static floor field, which describes the shortest distance to an exit door, in an arbitrary geometry of rooms is presented. The wall potential and contraction effect at a wide exit are also proposed in order to obtain realistic behavior near corners and bottlenecks.

1 Introduction

Recently pedestrian dynamics and evacuation behaviour attact much attention of physicists as well as engineers [1]. There are different types of models for pedestrians proposed so far, such as the social force model [2] and the floor field model [3, 4]. The former model is based on a system of coupled differential equations, and pedestrian interactions are modelled via long-ranged repulsive forces. The latter is a CA model, where two kinds of floor fields, i.e., a static and a dynamic one, are introduced to translate a long-ranged spatial interaction into an attractive local interaction, but with memory, similar to the phenomenon of chemotaxis in biology [5]. It is interesting that, even though these two models employ different rules for pedestrian dynamics, they share many properties including lane formation, oscillations of the direction at bottlenecks [3], and the so-called faster-is-slower effect [2]. Although these are important basics for pedestrian modelling, there are still many things to be done in order to apply the models to more practical situations such as evacuation from a building with complex geometry.

In this paper, we will propose a method to construct the static floor field for complex rooms of *arbitrary* geometry. Moreover, the effect of walls and contraction at a wide exit will be taken into account which enables us to obtain realistic behavior in evacuation simulations even for the case of panic situations.

2 An Extended Floor Field Model

In this section we will summarize the update rules of an extended floor field model for modelling panic behavior of people evacuating from a room. The space is discretized into cells of size $40\,\text{cm} \times 40\,\text{cm}$ which can either be empty or occupied by one pedestrian. Each pedestrian can move to one of the unoccupied next-neighbor cells (i, j) or stay at the present cell at each time step $t \rightarrow t + 1$ according to certain transition probabilities p_{ij} as explained below. For the case of evacuation processes, the *static floor field* S describes the shortest distance to an exit door. The field strength S_{ij} is set inversely proportional to the distance from the door. The *dynamic floor field* D is a *virtual trace* left by the pedestrians similar to the pheromone in chemotaxis [5]. It has its own dynamics, namely diffusion and decay, which leads to broadening, dilution and finally vanishing of the trace.

The update rules of our CA have the following structure:

1. The dynamic floor field D is modified according to its diffusion and decay rules, controlled by the parameters α and δ.
2. For each pedestrian, the transition probabilities p_{ij} for a move to an unoccupied neighbor cell (i, j) are determined by the two floor fields. The values of the fields D (dynamic) and S (static) are weighted with two sensitivity parameters k_D and k_S:

$$p_{ij} = N \exp\left(k_D D_{ij}\right) \exp\left(k_S S_{ij}\right) p_I(i, j) p_W(i, j), \tag{1}$$

with the normalization N. Here p_I represents the inertia effect [3] given by $p_I(i, j) = \exp\left(k_I\right)$ for the direction of one's motion in the previous time step, and $p_I(i, j) = 1$ for other cells, where k_I is the sensitivity parameter. p_W is the wall potential which is explained below. In (1) we do not take into account the obstacle cells as well as occupied cells.
3. Each pedestrian chooses randomly a target cell based on the transition probabilities p_{ij} determined by (1).
4. Whenever two or more pedestrians attempt to move to the same target cell, the movement of *all* involved particles is denied with probability $\mu \in [0, 1]$, i.e. all pedestrians remain at their site. Which one is allowed to move is decided using a probabilistic method [3, 6].
5. The pedestrians who are allowed to move perform their motion to the target cell chosen in step 3. D at the origin cell (i, j) of each *moving* particle is increased by one: $D_{ij} \rightarrow D_{ij} + 1$.

The above rules are applied to all pedestrians at the same time (parallel update). Some important details are explained in the following subsections.

2.1 Effect of Walls

People tend to avoid walking close to walls and obstacles. This can be taken into account by using "wall potentials". We introduce a repulsive potential

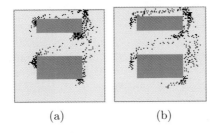

(a)   (b)

Fig. 1. Snapshot of evacuation (a) without ($k_W = 0$) and (b) with ($k_W = 0.5$) wall potential. We can clearly see the artifact of jamming at every corner without the wall potential. Parameters are $D_{\max} = 10, k_S = 2.0, k_D = 1.0, k_I = 0.2, \mu = 0.2$ and the initial density is $\rho = 0.03$.

inversely proportional to the distance from the walls. The effect of the static floor field is then modified by a factor (see eq. (1)):

$$p_W(i, j) = \exp(k_W \min(D_{\max}, d_{ij})), \qquad (2)$$

where d_{ij} is the minimum distance from all the walls, and k_W is a sensitivity parameter. The range of the wall effect is restricted up to the distance $D_{\max}$ from the walls.

Fig. 1 shows an example for an evacuation from a room with obstacles using the wall potential. Without wall potentials ($k_W = 0$), jamming areas near every corner can be observed, because everybody tries to evacuate along the same path of minimum length. For $k_W \neq 0$ these areas are clearly suppressed. Thus the introduction of this additional potential improves the realism of the model.

2.2 Calculation of the Static Field in Arbitrary Geometries

In the following we propose a combination of the visibility graph and Dijkstra's algorithm to calculate the static floor field. These methods enable us to determine the minimum Euclidian (L^2) distance of any cell to a door with arbitrary obstacles between them.

Let us explain the main idea of this method by using the configuration given in Fig. 2(a) where there is an obstacle in the middle of the room. We will calculate the minimum distance between a cell P and the door O by avoiding the obstacle. If the line PO does not cross the obstacle $A - H$, then the length of the line, of course, gives the minimum. If, however, as in the example given in Fig. 2(a), the line PO crosses the obstacle, one has to make a detour around it. Then we obtain two candidates for the minimum distance, i.e., lines $PBAO$ and $PCDHO$. The shorter one finally gives the minimum distance between P and O. If there are more than one obstacle in the room, then we apply the same procedure to each of them repeatedly. Here it is

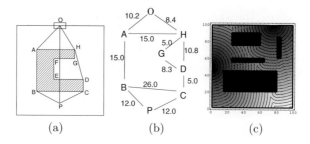

Fig. 2. Example for the calculation of the static floor field using the Dijkstra method. (a) A room with one obstacle. The door is at O and the obstacle is represented by lines $A - H$. (b) The visibility graph for this room. The real number on each bond represents the distance between them as an illustration. (c) A contour plot of the static floor field. The darkness of shading is inversely proportional to the distance from the nearest door.

important to note that all the lines pass only the obstacle's edges with an acute angle. It is apparent that the obtuse edges like E and F can never be passed by the minimum lines.

To incorporate this idea into the computer program, we first need the concept of the *visibility graph* in which only the nodes that are visible to each other are bonded [9] ("visible" means here that there are no obstacles between them). The set of nodes consists of a cell point P, a door O and all the acute edges in the room. In the case of Fig. 2(a), the node set is $\{P, O, A, B, C, D, G, H\}$ and the bonds are connected between $A - B$, $A - H$, and so on (Fig. 2(b)). Each bond has its own weight which corresponds to the Euclidian distance between them.

Once we have the visibility graph, we can calculate the distance between P and O by tracing and adding the weight of the bonds between them. There are several possible paths between P and O, and the one with minimum total weight represents the shortest route between them. The optimization task is easily performed by using the Dijkstra method [9] which enables us to obtain the minimum path on a weighted graph.

Performing this procedure for each cell in the room, the method allows us to determine the static floor field for arbitrary geometries. We will call this metric *Dijkstra metric* in the following. Results for a complex static floor field obtained by this method are shown in Fig. 2(c). There are two doors in this room, thus we calculate the minimum distance for each door from each cell in the room and take the shorter one as value of the static floor field.

2.3 Model Parameters and Their Physical Relevance

There are several parameters in our model, and the most important ones are listed below with their physical meaning which is helpful in understanding the collective behavior in the simulations.

1. $k_S \in [0, \infty) \cdots$ The coupling to the static field characterizes the knowledge of the shortest path to the doors, or the tendency to minimize the costs due to deviation from a planned route [7].

2. $k_D \in [0, \infty) \cdots$ The coupling to the dynamic field characterizes the tendency to follow other people in motion (*herding behavior*). The ratio k_D/k_S may be interpreted as the degree of panic. It is known that people try to follow others particulary in panic situations [2].

3. $k_I \in [0, \infty) \cdots$ This parameter determines the strength of inertia which suppresses quick changes of the direction of motion. It also reflects the tendency to minimize the costs due to deviation from one's desired route and acceleration [7].

4. $\mu \in [0, 1] \cdots$ The friction parameter controls the resolution of conflicts in clogging situations. Both cooperative and competitive behavior at a bottleneck are well described by adjusting μ [8].

5. $\alpha, \delta \in [0, 1] \cdots$ These constants control diffusion and decay of the dynamic floor field. It reflects the randomness of people's movement and the visible range of a person, respectively. If the room is full of smoke, then δ takes large value due to the reduced visibility.

6. $k_W, D_{\max} \cdots$ These parameters specify the wall potential. Pedestrians tend to avoid walking close to walls and obstacles. $D_{\max}$ is the maximum distance at which people feel the walls. It reflects one's range of sight or so-called personal space.

2.4 Contraction at a Wide Exit

If the width of an exit becomes large, a more careful treatment is needed in the calculation of the static floor field. People tend to rush to the center of the exit to avoid the walls. Thus one should introduce an effective width of the exit by neglecting certain cells from its each end. We call this effect *contraction* in this paper, due to its similarity with the contraction effect in hydrodynamics where fluid runs through a orifice with a smaller diameter than that of the orifice immediately after the fluid goes out of it. Introducing the contraction makes the evacuation behavior more realistic [10].

3 Simulations

We focus on measuring the total evacuation time by changing the parameter k_I. We put $D_{\max} = 10$, $\alpha = 0.2$ and $\delta = 0.2$. The size of the room is set to 100×100 cells. Pedestrians try to keep their preferred velocity and direction as long as possible. This is taken into account by adjusting the parameter k_I. In Fig. 3, total evacuation times from a room without any obstacles are shown as function of k_D in the cases $k_I = 0$ and $k_I = 3$. We see that it is monotonously increasing in the case $k_I = 0$, because any perturbation from other people becomes large if k_D increases, which causes the deviation

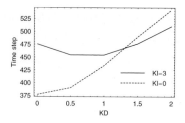

Fig. 3. Total evacuation time versus coupling k_D to the dynamic floor field in the dependence of k_I. The room is a simple square without obstacles and 50 simulations are averaged for each data point. Parameters are $\rho = 0.04, k_S = 2, k_W = 0.3$ and $\mu = 0$.

from the minimum route. Introduction of inertia effects, however, changes this property qualitatively, and the *minimum* time appears around $k_D = 1$ in the case $k_I = 3$. This is well explained by taking into account the physical meanings of k_I and k_D. If k_I becomes large, people become less flexible and all of them try to keep their own minimum route to the exit according to the static floor field regardless of congestion. By increasing k_D, one begins to feel the disturbance from other people through the dynamic floor field. This perturbation makes one flexible and hence contributes to avoid congestion. Large k_D again works as strong perturbation as in the case of $k_I = 0$, which diverts people from the shortest route largely.

References

1. M. Schreckenberg and S.D. Sharma (Eds.), "Pedestrian and Evacuation Dynamics," Springer-Verlag, Berlin, 2001.
2. D. Helbing, I. Farkas, and T. Vicsek, Nature, vol. 407, pp. 487–490, 2000.
3. C. Burstedde, K. Klauck, A. Schadschneider, and J. Zittartz, Physica A, vol. 295, pp. 507–525, 2001.
4. A. Kirchner and A. Schadschneider, Physica A, vol. 312, pp. 260–276, 2002.
5. E. Ben-Jacob, Contemp. Phys. vol. 38, pp. 205–241, 1997.
6. A. Kirchner, K. Nishinari, and A. Schadschneider, Phys. Rev. E, vol. 67, p. 056122 (2003).
7. S. P. Hoogendoorn, "Walker Behavior Modelling by Differential Games," Computational Physics of Transport and Interface dynamics, Springer, 2003.
8. A. Kirchner, H. Klüpfel, K. Nishinari, A. Schadschneider, and M. Schreckenberg, Physica A, vol. 324, p. 689 (2003).
9. M. de Berg, M. van Kreveld, M. Overmars, and O. Schwarzkopf, "Computational geometry," Springer-Verlag, Berlin, 1997.
10. K. Nishinari, A. Kirchner, A. Namazi, and A. Schadschneider, IEICE, Trans. Inf. & Syst. E87-D, 726 (2004).

Pedestrian Pulse Dispersion in an Underground Station

H. Moldovan, D.R. Parisi, and B.M. Gilman

Urbix
(ci) FADU, Subsuelo Pabellón III – Ciudad Universitaria
(1428) Buenos Aires – Argentina
hmoldovan@urbix.com.ar

Summary. Based on flow rate curves collected by two synchronized and separated cameras, the propagation and dispersion of pedestrian pulses were investigated. Microscopic simulations were performed to reproduce the real data. By means of a simple macroscopic analysis, it is possible to find microscopic parameters of the social force model. We found, for the case study that the ranges of these parameters are: A = 900-1500 N and B = 0.8-1.2 m.
Keywords: microscopic parameter determination, pedestrian flow simulation

1   Introduction

Recently, we have performed a pedestrian flow simulation in one of the most important underground stations of Buenos Aires. This station receives pulsated flow of passengers from the upward train station. These pulses have typically a thousand passengers and last about 2 to 3 minutes.

The basic flow model implements a microscopic algorithm considering that flow is made of discrete self-driven particles. The model takes into account forces (contact, desired and social) as described by Dirk Helbing [1-2]. The desired force is characterized by a list of targets where passengers want to go.

Data from the real system was collected in several ways. Particularly, for the present work, we focused on the data obtained by videotape recording. The video images allow measuring pedestrian flow rates counting the number of passengers as a function of time.

Two synchronized cameras were placed in two separated locations of the station. One near the main access and the other over the turnstiles line, approximately seventy meters far. The shape and the shift between both curves (number of pax vs. time) provide information on how the pulse is propagated through the system: the mean velocity of the pulse and its dispersion.

In order to extract microscopic information from the data, comparison with the simulation was made. The curve measured by the first camera was used as the input of the simulated system. The output can be compare to the data obtained by the second camera. By doing this, it is possible to adjust the social repulsive parameters.

2   The Real System

The layout of the system is shown in Fig. 1.

As can be observed, passengers arrived at entrance E1 going toward the turnstiles lines through the stairway S1.

Synchronized cameras were placed at E1 and S1. These two locations were approximately seventy meters apart.

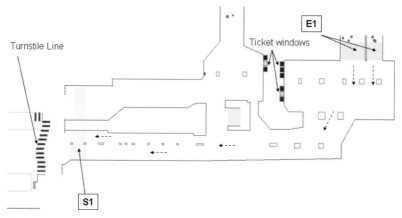

Fig. 1. Location of synchronized cameras at entrance E1 and at stairway S1 in the transfer station.

The video images allow counting the number of passengers passing through delimited areas as a function of time. The time interval chosen was 10 seconds. Typical curves collected from both cameras are shown in Fig. 2.

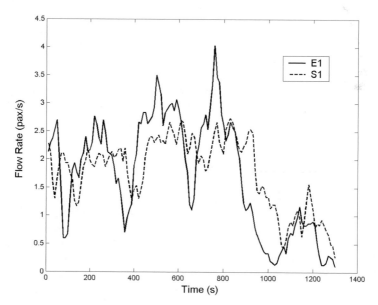

Fig. 2. Example of simultaneous data collected by cameras at locations E1 and S1 when consecutive pulses of pax go across the system.

Curve S1 is the transformation of curve E1 after going across the system. The shape and the shift between curves provide information about how the pulse is propagated through the system: the mean velocity of the pulse and its dispersion.

3   Data Analysis

Studying the cross correlation of time series recorded at E1 and S1 the lag that maximizes this correlation can be found. This lag is an indicator of the velocity of the pulse.

After analyzing nine pairs of experimental curves (as those shown in Fig. 2) the lag found was $t = 60 \pm 10$ s. At this lag the correlation is almost 0.8.

Considering that the distance between cameras S1 and E1 are about 70 ± 5 m, the mean velocity of pulses are computed as $v_{pulse} = 70$m / 60 s $= 1.2 \pm 0.3$ m/s.

This velocity is slightly slower than the mean velocity of individual pax moving freely ($v_{desire} = 1.50 \pm 0.25$ m/s) this is a natural result because passengers in the pulse are perturbed by congestions, obstacles, etc.

In order to measure the dispersion of the pulse, we define its width as follows. First the incoming signal at E1 is considered. We say that a pulse is the curve between to local minimums (less than 1 pax/min). For example, in Fig. 2, four pulses can be individualized. The integral of one pulse represents the total number of passengers in the pulse.

Beginning at the center of the pulse (t_c), we choose growing intervals ([t_c-k10s, t_c+k10s] with k=1,2,3,...), and for each interval the integral is calculated. The width of the pulse at E1 is found when the integral of the kth interval is equal to half the integral of the whole pulse.

To find the width of the propagated pulse (at S1), we considered that the center of the S1 pulse is t_c+ 60 s. Then, the same procedure as above is used.

By doing this, is possible to compare the width of the pulse at E1 and at S1.

Calculation from the available data shows that dispersed pulse at S1 is 26 ± 7 % wider than the original pulse at entrance E1.

The methodology used above allows characterizing the transformation suffered by a pulse moving across the system. It also provides a method for comparing the results of simulation with the data from the real system, as it will be seen in the next section.

4   Microscopic Simulation of the System

The discrete flow was simulated by implementing the social force model postulated by D. Helbing. This model considered basically three kinds of forces acting on each self driven particulate. They are: desire force, granular (or contact) force and social force. The social force represents that each person likes being far from other unknown perThe social force parameters A and B will be tuned in order to adjust simulation results to real data.

A model considering the geometry of the transfer station was built. The inputs of simulations were equal to the data from the real incoming pulse at E1.

The distribution of the desired velocity used was measured directly from the system under lsons. Therefore it is described by a smooth repulsive force given by $F_s=A$ exp $(-x/B)$ where x is the distance between two persons.

ow density situation and it value is $v_{desire} = 1.50 \pm 0.25$ m/s.

After performing each simulation the resulting pulse at S1 can be obtained. The lag and dispersion of the simulated pulse (at S1) can be found with respect to the input pulse at E1 (simulated exactly equal to the experimental one) in the same form as described in section 3.

In order to quantify the error of a simulation (with given parameters A and B) the lag and dispersion can be compared to those obtain from experimental data.

The relative error is defined by equation 1.

$$E_{(A,B)} = \left[\frac{\left| t_{real} - t_{sim(A,B)} \right|}{t_{real}} + \frac{\left| \left(w_{real}^{S1} \middle/ w_{real}^{E1} \right) - \left(w_{sim(A,B)}^{S1} \middle/ w_{real}^{E1} \right) \right|}{\left(w_{real}^{S1} \middle/ w_{real}^{E1} \right)} \right] \times 100\% \qquad (1)$$

where t is the lag between pulse at E1 and S1 and w is the width of the pulses, calculated as described in section 3. Note that w_{real}^{E1} is equal to w_{sim}^{E1} beacause the experimental curve at E1 was used as input in the simulations. Changing parameters A and B in simulations, the error surface $E_{(A,B)}$ can be found. Fig. 3 show this error surface.

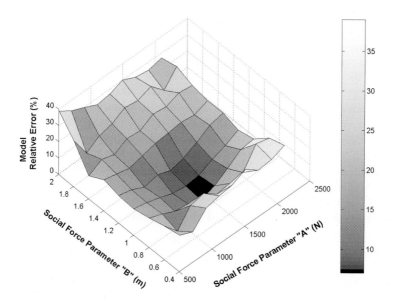

Fig. 3. Relative error surface as a function of the social force parameters.

As can be seen, the minimum is near $A = 900$-1500 N and $B = 0.8$-1.2 m.
It must be noted that the method proposed allows exploring microscopic parameters using macroscopic data.

5   Conclusion

In the present work, the propagation and dispersion of pedestrian pulses were investigated.

The velocity of the pulse can be estimated by making cross correlation between the flow rate curves of pulses measured at separated locations.

The found pulse's velocity was about 1.2m/s which is slower than the mean velocity (1.5 ± 0.25 m/s) of individual pedestrians moving freely in the station.

Also the definition of pulse's width presented in this work allows measuring the dispersion of the pulse going across the system.

Using these macroscopic quantities it is possible to compare experimental and simulated data and therefore an error can be defined. By minimizing this error microscopic parameters can be adjusted.

The social force model of D. Helbing was used to simulate the problem. The social repulsive parameters found were A = 900-1500 N and B = 0.8-1.2 m.

Summarizing, a simple macroscopic analysis useful for finding microscopic parameters of the model was presented in this work. We also showed the application of this analysis to a real case: an underground station.

6   Acknowledgment

Authors whish to acknowledge to Metrovias S.A. (Argentina) for their collaboration and contributed data.

References

1. D. Helbing and P. Molnár. Social force model for pedestrian dynamics. Physical Review E, 51, (5). pag. 4282-4286. May 1995.
2. D. Helbing, I. Farkas, and T. Vicsek. Simulating dynamical features of escape panic. Nature, 407. pag. 487-490. Sep. 2000.

Population and Distance Criteria for Pedestrian Decisions

D.R. Parisi, H. Moldovan, and B.M. Gilman

Urbix – Universidad de Buenos Aires
(ci) FADU, Subsuelo Pabellón III – Ciudad Universitaria
(1428) Buenos Aires – Argentina
dparisi@urbix.com.ar

Summary. Based on data collected by a turnstile line of an important subway station, the problem of decisions making was studied. Microscopic simulations were performed to reproduce the real behavior. A simple selection model was proposed to simulate how persons choose between different possibilities. The functionality of the criteria consider how far and how populated are each turnstile. The model was able to reproduce the utilization curve of a given layout and to predict the curve for a different geometry.

Keywords: decisional dynamics, pedestrian flow simulation

1 Introduction

Recently, we have performed the simulation of a pedestrian flow in one of the most important underground station of Buenos Aires. This station receives pulsated flow of passengers from the upward train station. These pulses have typically a thousand passengers and last about 2 to 3 minutes.

The main access is composed of two corridors which flow into a twelve turnstile line. Due to restriction on the accessibility, the utilization of turnstiles is not uniform. The utilization curve changes with the total number of incident passengers. This phenomenon may have an important influence in the accumulation of passengers at the turnstiles zone.

The basic flow model implements a microscopic algorithm considering that flow is made of discrete self-driven particles. The model takes into account forces (contact, desired and social) as described by Dirk Helbing [1-2]. The desired force is characterized by a target to where the person wants to go.

The turnstile line was model by putting a target in front of each turnstile. All the twelve targets are in parallel. This means that a person must choose one and only one turnstile among all of them. Once the person reaches the target, he/she takes about 3.5 seconds for finishing the ticket control transaction.

The central task of this work is to investigate the selection mechanism applied by persons when selecting a given turnstile of the line.

2 The Real System

The layout of the system is shown in Fig. 1. As can be observed, passengers arrived by corridors **C1** and **C2**. The mean velocity of the passengers was estimated around 1.5 m/s. They must cross the turnstiles in order to access the subway platform.

Each turnstile records the number of passengers per minute. Analyzing the data of 10 days, it was possible to build the utilization curves that should be reproduced by the model.

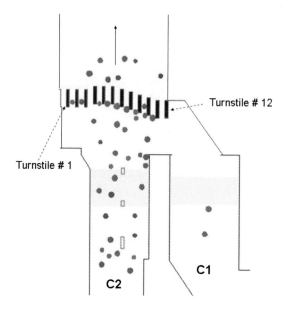

Fig. 1. Turnstile line and its access geometry.

3   Proposed Decisional Model

In the real system, when an individual faces the turnstile line, he/she has to make a decision in order to choose the turnstile that minimizes his/her subjective waiting time.

We postulate, as a first approximation of the problem, that the decision is made considering both, the distance to the turnstile and its "occupancy".

As stated before, the microscopic model assigns one target to each turnstile. Each passenger facing them must choose one among these parallel targets.

The virtual passengers take decisions every integration step (0.015 second).

The probability for a pedestrian j choosing turnstile i is $P(d_{ji}, n_i)$, where d_{ji} is the distance between the pedestrian j and the turnstile i, and n_i is the number of waiting passenger at that turnstile.

The functional form of $P(d_{ji}, n_i)$ is stated in equation (1),

$$P(d_{ji}, n_i) = P_d(d_{ji}, \sigma_d^j) \times P_n(n_i, \sigma_n^j) \tag{1}$$

We proposed that the probability functions P_d and P_n are Gaussian functions.

Trial and error in conjunction with a big number of simulations allowed tuning of standard deviation parameters σ_d and σ_n in order to reproduce de data collected by the turnstiles. Each passenger has its own parameters.

Figure 2 shows the results of simulations performed with $\sigma_d = 2\text{-}5$ and $\sigma_n = 1\text{-}2$.

It can be observed that the tendencies shown by real data was reproduced satisfactorily by the simulations, in particular with the distance population criteria proposed in equation (1). It can be also compared the utilization curve for different flow rates.

The general tendency was reproduced satisfactorily. The central turnstiles are more utilized, but for high flow rates these differences decrease.

The experimental data was fitted by polynomial functions in order to make them smoother. The noisy behavior of simulated points could be reduced by performing longer simulations.

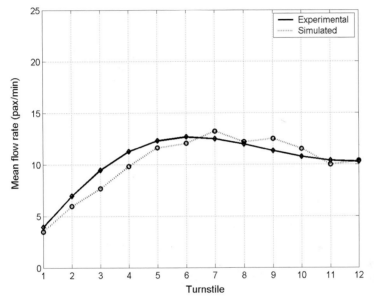

Fig. 2. Comparison of the utilization profile of the turnstile line, averaged from 50 to 200 (pax/min) flow rates.

4   Prediction Under Geometry Variation

Using the previously found parameters values, the model was able to predict the utilization curves of the turnstile line under different conditions, for instance when the geometry of the real system was modified.

Due to a redefinition of the layout, a new corridor (**C3**) was built and the central corridor (**C2**) was temporarily closed (see Fig. 4).

This situation allows studying the passengers' behavior for a different access condition to the same turnstile line (which remained unmodified).

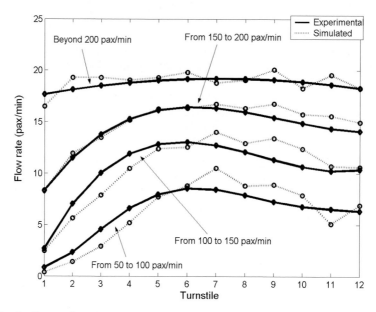

Fig. 3. Comparison between real and simulated utilization profile of the turnstile line for different incident flow rates.

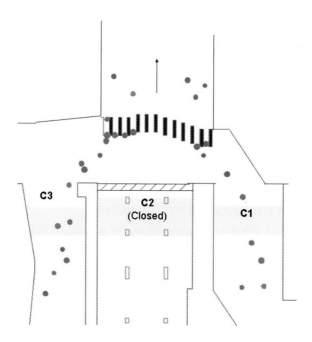

Fig. 4. A different access geometry towards the same turnstile line.

Also in this case, the mean utilization curve of the turnstiles is reproduced in agreement with the data as it can be observed in Fig. 5.

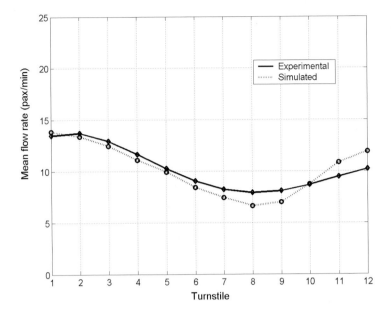

Fig. 5. Prediction of the averaged utilization profile of the turnstile line for the new geometry (Fig. 4).

5   Conclusion

Good quality data was obtained from a real system (a 12 turnstile line of a subway station). Each turnstile is able to record the number of passengers per minute. With this information it was possible to build utilization curves of the turnstile line.

These curves depend on the layout and on the selection criteria used by each passenger. We postulate a simple functional form as decisional law. This function depends on how far and how populated is a given turnstile.

The utilization curve of the twelve turnstiles, for different incident flows, was reproduced in good agreement with the data.

The two parameters, found for the studied layouts, show that the model is robust under geometry variations.

The universality of the proposed law and of the found parameters should be further investigated using data from radically different geometries.

6   Acknowledgment

Authors whish to acknowledge to Metrovias S.A. (Argentina) for their collaboration and contributed data.

References

1. D. Helbing and P. Molnár. Social force model for pedestrian dynamics. Physical Review E, 51, (5). pag. 4282-4286. May 1995.
2. D. Helbing, I. Farkas, and T. Vicsek. Simulating dynamical features of escape panic. Nature, 407. pag. 487-490. Sep. 2000.

Simulation of the Evacuation of a Football Stadium Using the CA Model PedGo

H. Klüpfel[1] and T. Meyer-König[2]

[1] TraffGo GmbH
Falkstr. 73-77, 47057 Duisburg – Germany
`kluepfel@traffgo.com`
[2] TraffGo GmbH
Johannisstr. 42, 24937 Flensburg – Germany
`m-k@traffgo.com`

Summary. Computer simulations have become an important tool for analysing egress processes and assessing evacuation concepts. Especially so called microscopic models can by now be considered state of the art. In this paper we will describe the software PedGo which is based on a 2D cellular automaton and its application to the simulation of evacuations form large and complex structures. The focus is on the practical application to full-scale scenarios. As an example, we show results for the egress from a football stadium.
Keywords: evacuation, egress, simulation, 2D cellular automaton, pedestrian

1 Description of the CA-Model

PedGo is based on a stochastic Multi-Agent Model where the space is divided into quadratic cells of size $(40cm)^2$ (cf. Fig. 1). Consequently, the time is measured also in discrete steps (identified as 1s in reality). Each cell can be either empty, occupied by at most one person, or be non-accessible (e.g. wall or furniture). Since the laws for human behaviour and interaction are not known in detail, PedGo is not based on a first principles approach for human behaviour modelling but the model parameters are rather an estimation of the effective (phenomenological) behaviour of and interaction between individuals.

The orientation of the persons when leaving a building or vessel is usually based on the signage. The routes are represented in the model by a piece of information about the distance to the next exit, respectively destination put into each cell (cf. Fig. 1, right).

For complex geometries, several potentials can be defined. Then, the population is divided into groups and each group is assigned to a different potential.

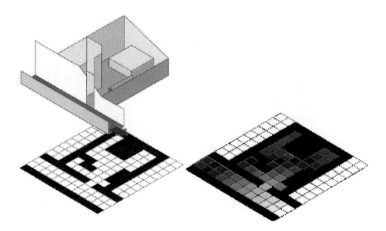

Fig. 1. Representation of the geometry as a grid of cells. The orientation potential is indicated by the shading. The darker the shade is the larger is the distance to the exit.

The transition probabilities are then given by:

$$p_i = e^{-\frac{(P_i - P_0)}{S}}$$ (1)

$$p_{i,\,\text{current direction}} \rightarrow p_{i,\,\text{current direction}} \cdot \Theta$$ (2)

where P_i is the potential value of cell i, P_0 the potential value for the current cell, and p_i is the probability for transition to cell i (without the necessary normalization constant). The so called inertia Θ favours the movement into the current direction, i.e. every pedestrian has a direction, even if she stands still. The computation of the potential values P_i is based on an extended Manhattan metric (with eight neighbouring cells) and an additional smoothing algorithm. Basically, for a hallway with the exit at the left, the potential value is 10 times the distance in cells: If the cell index along the hallway is i, then $P_i = 10 \cdot i$. This must of course be in accordance with the choice of the parameter S in the formulas above (see also Tab. 1 and the following section).

2   Validation of the Model and Calibration of the Paramters

There is a set of six parameters: v_{max}, $t_{patience}$, t_{react}, S, p_{orient}, and Θ. S and Θ are explained by equations (1) and (2). v_{max} is the free walking speed, after a pedestrian has been waiting in a queue ($v=0$) for the time $t_{patience}$ he changes his strategy and follows another route (routes are represented by potentials as explained above). The reaction time t_{react} is the individual off-set before movement starts and finally p_{orient} is the probability for stopping ($v=0$), which is applied in every time-step.

The fundamental diagram of the model (for a long hallway) is shown in Fig. 2. The parameters used in the simulation of the evacuation of the sports stadium can be found in Tab. 1. For the calculation of the fundamental diagram, the walking speed was set to 2 to 5 cells per time-step and the orientation frequency to 0 to 0.3. All the other parameters are as in Tab. 1.

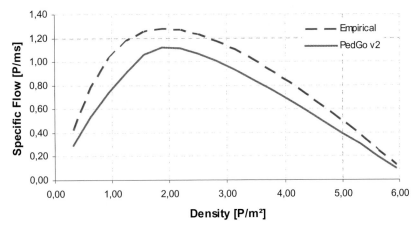

Fig. 2. Fundamental diagram for a hallway with uni-directional flow. Empirical curve adapted from [2].

3   Simulation of the Evacuation of a Football Stadium

The structure investigated in this example is a football stadium which consists of seven levels. Of course, these levels can not really be considered floors in the same sense as for, e.g. an office building. However, in order to simulate the process correctly, seven levels had to be defined, since all the stairs are projected on the floor below in the model. Accordingly, the horizontal free walking speed is divided by two on stairs, which is a combination of the reduced walking speed along the incline and the geometrical factor (cos α, where α is the angle of the stairs).

The parameters used in the simulation are shown in Tab. 1. The model defined in the previous section is sufficient to simulate also very complex scenarios like the football stadium without additional enhancements. Figure 3 shows simulation results for the stadium.

	Minimum	Maximum	Mean	Standard Deviation	Unit
Free walking speed v_{max}	4	4	-	-	cells/s
Patience $t_{patience}$	5000	5000	-	-	S
Sway S	1	5	3	2	0.1a (=4cm)
Reaction time t_{react}	0	0	-	-	S
Orientation frequency p_{orient}	0	0	-	-	%
Inertia Θ	1	5	-	-	%

Tab. 1. The parameters are distributed according to a Gaussian with the tails cut off.

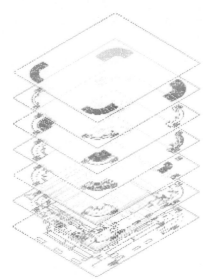

Fig. 3. Sketch of the stadium.

An important information is where congestion occurred. Congestion is easily defined by the local density. We use the following measure: Local congestion is defined by a density that is higher than 4 Persons per square meter. If this density is exceeded for longer than 10% of the overall egress time, the congestion is called significant. Fig. 1 shows the areas of significant congestion.

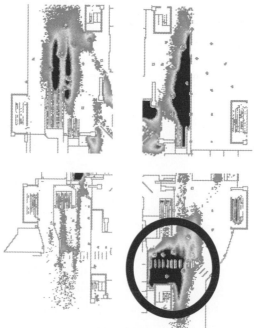

Fig. 4. Analysis of the densities occurred during the simulation. The red areas indicate significant congestion as defined in the text.

We have investigated three different scenarios concerning the population characteristics:

3.1   Case 1

	Min	Max	Mean	StdDev..	Units
Walking speed.	4	4	-	-	cells/s
Patience	5000	5000	-	-	S
Sway	1	5	3	2	0.1a (=4cm)
Reaction time	0	0	-	-	S
Orientation	0	0	-	-	%
Inertia	1	5	-	-	%

3.2   Case 2

	Min	Max	Mean	StdDev..	Units
Walking speed.	2	5	3	1	cells/s

3.3   Case 3

	Min	Max	Mean	StdDev..	Units
Walking speed.	2	5	3	1	cells/s
Reaction time	0	30	15	5	S
Orientation	0	30	15	5	%

The results for the different cases are compared in Fig. 4.

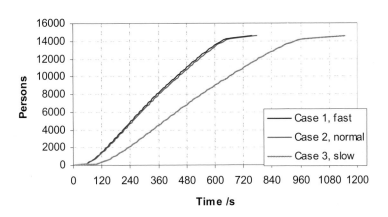

Fig. 5. Results for the different scenarios.

4 Summary and Conclusions

We have briefly described a two dimensional cellular automaton model for the simulation of crowd movement and evacuation processes. The model has been applied to simulate the egress from a sports stadium. The results obtained are within those provided by observations from the staff.

The approach provides two big advantages: 1) it is rather easy to handle and therefore efficient, 2) animations can be generated from the simulation which allow to analyse the evacuation in an intuitive way. Of course, the detailed analysis based on curves like the one shown in Fig. 4 also provides valuable information for improving procedural or geometrical aspects of the structure.

References

1. Roth, G., „Gleichtakt im Neuronennetz", Gehirn und Geist 1, 2002.
2. Weidmann, U., „Transporttechnik der Fußgänger", Schriftenreihe des IVT 90, ETH Zürich, 1992.
3. Yamori, K., "Going with the flow: Micro-macro dynamics in the macrobehavioral patterns of pedestrian crowds", Psychological Review 105(3), 530-557, 2001.
4. IMO, "Interim Guidelines for Evacuation Analyses for New and Existing Passenger Ships", IMO, MSC/Circ. 1033, 2002.
5. AlGadhi, S., H. Mahmassani, and R. Herman, "A speed concentration relation for bi-directional crowd movements with strong interaction", In: 6.
6. Schreckenberg, M. and S.D. Sharma, "Pedestrian and Evacuation Dynamics", Springer, Berlin, 2002.
7. Burstedde, C., K. Klauck, A. Schadschneider, and J. Zittartz, „Simulation of pedestrian dynamics using a 2-dimensional cellular automaton", Physica A 295, 507—525, 2001.
8. Fruin, J., "Pedestrian Planning and Design", New York, Metropolitan Association of Urban Designers and Environmental Planners, 1971.
9. Galea, E., "A general approach to validating evacuation models with an application to EXODUS", Journal of Fire Sciences 16(6), 414—436, 1998.
10. Hamacher, H. and S. Tjandra, "Mathematical modelling of evacuation problems – a state of the art", In: 6.
11. Helbing D., "Traffic and related self-driven many particle systems", Rev. Mod. Phys. 73(4), 1067—1141, 2001.
12. Henderson, L., "The statistics of crowd fluids", Nature 229, 381—383, 1971.
13. Kirchner, A., H. Klüpfel, K. Nishinari, A. Schadschneider, and M. Schreckenberg (2002). Simulation of competitive egress behaviour. Physica A 324, p. 689—697, 2003.
14. Klüpfel, H., T. Meyer-König, J. Wahle, and M. Schreckenberg, "Microscopic simulation of evacuation processes on passenger ships", In Proc. Fourth Int. Conf. On Cellular Automata for Research and Industry", pp. 63—71, London, Springer, 2000.
15. Pauls, J., "Movement of People", Chapter 3-13, pp. 3-263—3-285, In: DiNenno, P. (Ed.), "SFPE Handbook of Fire Protection Engineering", (2nd ed.), National Fire Protection Association, 1995.
16. TraffGo GmbH, "PedGo Users' Manual (v. 2.1.1)", Duisburg, 2003.
17. Transportation Research Board, "Highway Capacity Manual", Washington D.C., TRB, 1994.
18. Weckman, H., S. Lehtimäki, and S. Männikö, „Evacuation of a theatre: Exercise vs. calculations", Fire and Materials 23(6), 357—361, 1999.
19. Galea, E. (Ed.): Pedestrian and Evacuation Dynamics 2003 – Proceedings of the second international conference. CMS Press, London, 2003.
20. Klüpfel, H., A Cellular Automaton Model for Crowd Movement and Egress Simulation. Dissertation, Universität Duisburg-Essen, Duisburg, 2003. http://www.ub.uni-duisburg.de/ETD-db/theses/available/duett-08012003-092540/

Granular

Statistical Properties of Dense Granular Matter

R.P. Behringer[1], E. Clément[2], J. Geng[1], R. Hartley[1], D. Howell[1], G. Reydellet[2], and B. Utter[1]

[1] Dept. of Physics – Duke University
Box 90305, Durham, NC – USA
bob@phy.duke.edu
[2] Université Pierre et Marie Currie
Paris 75231 – France
erc@ccr.jussieu.fr

Summary. We review recent work characterizing force fluctuations and transmission in dense granular materials. These forces are carried preferentially on filimentary structures known as force chains. When a system is deformed, these chains tend to resist further deformation; with continued deformation, chains break and rearrange, leading to large spatio-temporal fluctuations. We first consider experiments on force fluctuations, diffusion and mobility under steady-state shear. We then turn to force transmission in static systems as determined by the response to a small point force. These experiments show that the packing structure and friction play important roles in determining the force transmission. Disordered highly frictional packings have responses that are similar to that of an elastic solid. Ordered packings show responses that may be described either by anisotropic elasticity or by a wave-like description.
Keywords: Fluctuations, Granular Materials, Force Transmission

It is convenient to think of granular materials as existing in phases that are roughly comparable to those of molecular matter: in gaseous states, grains are separated, and interact during collisions of short duration. In the solid state, granular matter is jammed, i.e. it requires a finite stress to displace particles relative to each other. When stress is applied, grains move irreversibly, i.e. by plastic deformation. If there is a bit more separation between grains, the material behaves somewhat as a fluid/dense gas. Necessarily, these analogies are limited; granular systems do not absorb energy from a surrounding heat bath–i.e. they are a-thermal.

Although granular systems are a-thermal, they nevertheless show strong fluctuations. For granular gases, fluctuations occur in the grain velocities. If ordinary statistical mechanics were applicable, it would be simple to characterize fluctuations in the denser phases. However, there is no Canonical

Ensemble, for these materials, and characterization of fluctuations in dense granular materials is an open subject.

In the first part of this work, we focus on the nature of force fluctuations, as determined in several different flow/shear experiments both in 2D and in 3D. Since forces are carried along relatively long filamentary structures known as force chains in dense granular materials, a small displacement of a grain can have a relatively long range effect. The interplay between grain displacements and force fluctuations is therefore also interesting. In the second part of this work, we consider combined diffusion and mobility studies. We find for a system undergoing strong shear deformation that diffusion is normal Brownian motion, but that the mobility of a particle being pushed at constant velocity through the medium behaves in a very non-fluid-like way that is reminiscent of Coulomb friction. In the third and final part of this work, we present a brief discussion of the nature of force transmission through static granular assemblies in 2D. Recently, there have been a number of new models that, when added to classical soil mechanics models, offers a bewildering range of possible descriptions.

Force Fluctuations Although in the past, force fluctuations in slowly flowing granular media have been assumed small, experiments have shown that this need not be the case. One of the first demonstrations of this, to our knowledge, was in the experiments of Baxter et al. for flow out of a conical hopper[1, 2]. A force sensor embedded in the interior of the hopper wall detected signals whose spectrum was typically power-law like with an exponent of about -2. In steady-state shearing experiments, Miller et al. found force fluctuations such as those seen in Fig. 1. The small bar towards the left side shows the mean force on the force sensor; fluctuations can be over an order of magnitude greater than that value. Distributions of detected pressure, given in Fig. 2, do not change appreciably as the ratio of the particle size to detector size is varied over a range for which only a few particles touch at a given time, to a value for which ~ 100 particles touch at one time.

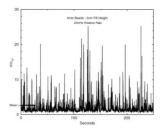

Fig. 1. Stress fluctuations, normalized by the static stress, vs. time in seconds. The small bar in the lower left corner indicates the mean stress.

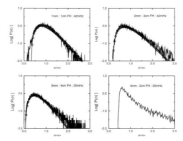

Fig. 2. Log of the distribution of stress vs. $\sigma/<\sigma>$ for particles of size 1 to 4 mm. The diameter of the detector is about 1 cm.

The root cause of strong force fluctuations is evident by turning to studies using photoelastic particles in 2D. Experiments by Howell et al. and by Hartley et al. have probed force fluctuations in 2D Couette shear, as depicted in Fig. 4. Particles, either disks or flat particles with a pentagonal cross section lie flat on a transparent smooth surface (powder-lubricated Plexiglas). Shearing is sustained in steady state by rotating a rough inner wheel. The whole is placed between crossed polarizers, and, because the particles are made of a photoelastic polymer, it is possible to visualize forces and kinematic properties at the particle scale. In Fig. 3, we show a false-color image that demonstrates the nature of the force chains in this system. Here, red corresponds to large forces and blue to small forces. As shearing continues, the strong force chains form and disappear in various portions of the experiment. It is the variation of these structures that is at the heart of the force fluctuations seen in experiments where force visualization is not directly possible.

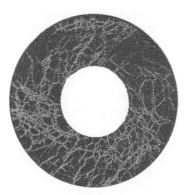

Fig. 3. Image from a 2D Couette experiment by Howell et al. showing force chains. Here, bright (red) and dark (blue) correspond to particles that are subject to respectively large and small forces.

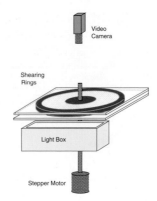

Fig. 4. Sketch of the apparatus for carrying out 2D shear experiments using photoelastic particles. Particles lie on a smooth Plexiglas sheet that is sandwiched between crossed polarizers. A CCD camera then images the details of the photoelastic response and the particle motion.

Two aspects of these experiments are of additional interest. First, we can vary the mean density, characterized by the packing fraction, γ, of the material in the Couette cell. We find that there is a transition as a function of γ where forces are just percolate from one boundary of the cell to the other. This transition at γ_c, which is presumably related to jamming, shows a number of interesting features, including critical slowing down, a divergent compressibility, and intermittency[8]. In addition, we find that the mean force chain length grows as we approach γ_c from above. The second feature that is of note here is that there is logarithmic rate-dependence in the mean stress as a function of the shearing rate, characterized by Ω, the rotation rate of the inner shearing wheel, as shown by Hartley et al.[4].

Diffusion and Mobility It is also interesting to ask about the particle kinematics, as characterized by velocity profiles and also by diffusion. First, it is important to note that the particles next to the inner shearing wheel dilate substantially to form a shear band. It is in this region that particle motion predominantly occurs, and the velocity profile decays roughly exponentially with distance from the shearing wheel. In Utter et al.[7] we have probed the nature of particle diffusion by tracking individual particles and then characterizing the variance of these displacements. A typical example of these variances is given in Fig. 5. At first sight, it appears that radial motion is subdiffusive because the variance appears to grow slower than linearly in time for relatively long times, and the azimuthal variances appear to demonstrate super-diffusion. However, in a simple random walk model, we show that this impression is probably not correct. Rather, the observed variances can be accounted for by the fact that the mean azimuthal velocity profile is nearly exponential, and because there is an impenetrable boundary at the surface of the shearing wheel. When both these effects are included in an otherwise unbiased random-walk model, we obtain model variances that are in agreement with Fig. 5. It is also interesting to note that the diffusion coefficients that we obtain scale as $D \propto d^2 \dot{\gamma}$, where d is the particle diameter, and $\dot{\gamma}$ is the local shear rate.

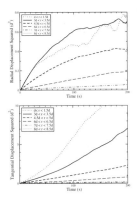

Fig. 5. Variance of displacement vs. time for different distances (in particle diameters) from the shearing wheel in a 2D Couette experiment.

It is also interesting to ask whether one can obtain diffusion information in a system that is not characterized by the rather non-uniform velocity profiles characteristic of Couette shear. To that end, we constructed a different sort of apparatus in which the material rested within the confines of a rotating annulus. Two pins inserted from above created a rather uniform deforming region in the interior of the annulus. The results of this work are described in Geng et al.[6] Briefly, we again find Brownian diffusion. An additional feature of this experiment is that we measure the mobility, B, of a tracer particle. The tracer is suspended in the moving sample from an electronic force gauge. As we rotate/stir the sample, the tracer moves at constant velocity, v through the sample. We then recorded the force felt by the tracer, and a typical example is given in Fig. 6a. This force is strongly fluctuating as force chains form to resist the tracer's motion but then break. For the present purposes, perhaps the most interesting feature is that the mean force, $\bar{f}$, is nearly independent of the speed, v. We show the mean force vs. v in Fig. 6b. Note that there is once more a logarithmic variation with speed that is quite similar to what was seen in the Couette shear experiments of Hartley et al. Thus, under constant-v conditions, this experiment yields a force that is reminiscent of Coulomb friction in its (near) rate-independence, and that leads to a mobility that is nearly a linear function of velocity: $B = |v/\bar{f}|$. This result differs rather substantially from recent model studies by Makse and Kurchan, and the likely difference is that in the present experiments, the tracer can break force chains and other cage-like structures that surround the tracer, whereas in the simulations of Makse and Kurchan, a very weak constant *force* was applied that could only push the tracer when cages were open. We note that Albert et al. found rate-independence for the force needed to push a rod through a container of particles in 3D.

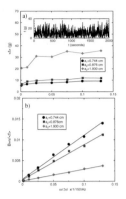

Fig. 6. (a) inset: time series of force vs. time. body: mean force vs. tracer speed showing roughly logarithmic rate-dependence. (b) Mobility, $v/ < f >$, vs. tracer speed.

Force Transmission As a final aspect of this work, we consider recent experiments that determine the nature of force transmission in static granular assemblies. Force in granular materials is a fundamental, but unresolved problem which has received much recent attention[3, 9–16]. This problem is particularly challenging due to the fact that friction and extra contacts beyond what is required for force balance overdetermine forces on a grain. Thus, even if the exact position of all particles and contacts were specified, in the general case, it would not be possible to calculate the forces on an assembly of grains. The exception is for the marginally rigid case, when there are just enough contacts per particle to provide precise force and torque balance. This means that even a spatially ordered system can contain disorder in the contact forces, due to small shape and size variation of disks and, more importantly, to the existence of friction[3]. If an experiment is repeated multiple times in a static granular arrangement, the expectation is that the results will differ from realization to realization. Hence, we find that an ensemble of measurements is required to adequately characterize the nature of force transmission for a given set of macroscopic conditions.

There is a striking range of models[11–14, 17–21] that predict how forces are transmitted in dense granular materials. These include lattice[12, 18, 20] and continuum[17] descriptions. The choices are really quite varied: in the continuum case, various models involve PDE's of totally different type, including elliptic PDE's for elasto-plastic models[17] (below the onset of yield) parabolic PDEs in the continuum limit of the Coppersmith et al. q-model[18] and a hyperbolic PDE for the oriented stress linearity (OSL) model[19] of Bouchaud et al. There are also a number of other models that have appeared very recently, inluding the double-Y model of Bouchaud et al.[11, 12] and calculations based on microscopic anisotropic elasticity by Goldenberg and Goldhirsch[13]. The difference in modeling form also impacts the numbers of required boundary conditions.

We use a determination of the Greens function at the local scale and average over an ensemble of realizations to determine the way in which forces are transmitted for a variety of different types of 2D granular systems. We work in the regime of small deformation, so that to a reasonable approximation, our results represent a true Green's function. In these experiments, we have probed the force response in: 1) ordered hexagonal packings of monodisperse disks, 2) bidisperse packings with different amount of disorder, 3) packings of pentagonal particles, 4) rectangular packings of monodisperse disks with variable inter-particle friction, 5) for various arbitrary angles to the surface, and 6) in textured/anisotropic system that was prepared by simple shear. In this work, we can focus on only a couple of these results. A longer description is available in Geng et al.[6].

Regarding experimental techniques, we note that the particles used here are the same as those used in the 2D shear experiments of Howell et al.[8] and Hartley et al.[4]. In particular, we obtain forces at the particle scale using the G^2 technique. The samples are typically prepared by placing flat particles of different cross sectional design against a nearly vertical piece of powder-lubricated plate glass (thus removing any effect from friction with the glass). We obtain photoelastic images before, with, and after the removal of a small point force applied to a single particle at the top surface. The response is given in terms of ΔG^2, which measures through the photoelastic effect, the change in force at the particle scale. We work in a regime which we experimentally determine corresponds to a locally linear response. After each set of three images, we gently stir the system so as to rearrange the contacts. We typically obtain an ensemble of 50 different trials for each macroscopic set of conditions and/or particle type. There are substantial differences from one realization of an ensemble to another, and it is necessary to work with the mean properties of each ensemble. This is demonstrated in Fig. 7. This figure is for regular hexagonal packings; the left side shows the color/greyscale representation of the mean response, and the right side shows the corresponding data for the standard deviation over the ensemble. In fact, the mean and standard deviations of the ensembles tend to resemble each other. When we compute the distribution of rms to mean values at each point, the peak of this distribution tends to occur around 2 to 3–underscoring the range of forces that can be observed at the particle scale, even for nominally identical systems.

Below, we focus on the role of changing the geometric disorder in the system by first considering bidisperse packings and then spatially disordered packings of pentagons. For bidisperse systems, we control the amount of disorder by the difference in radius of the two particle types. We do this by mixing nearly equal amounts of the two particle sizes. We prepare a packing

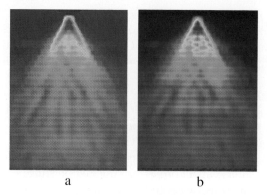

a   b

Fig. 7. The mean response, (a), and the standard deviation, (b), of G^2 for a hexagonal packing of disks. The standard deviation image has a similar shape to the mean image.

by drawing randomly from bins containing the two different size particles. We parameterize the disorder using the parameter $\mathcal{A} = \langle a \rangle^2 / \langle a^2 \rangle$, proposed by Luding et al.[3, 22], where a is the particle radius. The deviation from unity of this parameter is then a measure of geometric disorder. We have considered $\mathcal{A} = 1$, the case for ordered monodisperse packing, and packings with 3 different diameters (0.597 cm, 0.744 cm and 0.806 cm). The resulting values of $\mathcal{A}$ are $\mathcal{A} = 0.993$, 0.988 and 0.965.

In Fig. 8, we show using greyscale/color the average response, ΔG^2, for monodisperse disks, a mixture of disks, and pentagons, for a point load of 50 g. (We give more quantitative data in Geng et al.[3, 6]) These data are clearly different, depending on the packing geometry. The monodisperse packing has two peaks emerging from the central source, broadening with z, but remaining clearly identifiable, as one would expect for a wave-like (hyperbolic) model. The peaks are also present for the bidisperse construction, but much less sharply resolved. For the pentagons, there is a single broad region of response, with no evidence for wave-like stress propagation.

When we increase the amount of disorder in the bidisperse systems, the two-peak structure broadens to a broad single peak. Again, more quantitative data is given in Geng et al.[6]. Additional experiments[6] show that by decreasing the number of contacts and/or the friction at each contact, for instance by using square packings, the 2-peaked structure becomes sharper. This observation is consistent with either models such as the OSL model or anisotropic elastic models.

For strongly disordered packing of pentagonal particles, we find[3] that an applied normal force at a boundary produces a response that resembles that of an elastic solid. For instance, when a force is applied to an elastic plate of thickness t at angle θ with respect to the horizontal direction, the stress tensor components are[23]:

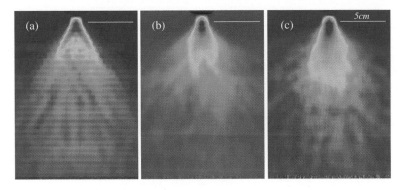

Fig. 8. The mean responses, ΔG^2, for (a) hexagonal, (b) bidisperse ($\mathcal{A} = 0.99$), and (c) pentagonal packings. Each set of data corresponds to an ensemble of 50 trials and shows the response to a $50g$ point load. The bar indicates a length of $5cm$.

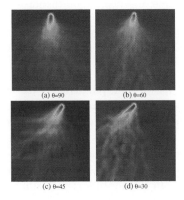

Fig. 9. Non-normal force responses for a system of pentagonal particles. A force of 0.5N was applied on the surface of the sample. The force directions are: (a). $90°$, (b) $60°$, (c) $45°$ and (d) $30°$, with respect to the horizontal.

$$\sigma_{rr} = \frac{2Ft}{\pi r}\cos\phi, \quad \sigma_{r\phi} = \sigma_{\phi\phi} = 0, \tag{1}$$

where the angle ϕ is measured from the direction of the applied force, r is the distance from the point under consideration to the point of contact. In Cartesian coordinates, with z aligned along the applied force, the stress components have a simple scaling form, such as

$$\sigma_{zz}(x, z) = \frac{2F}{z\pi}\frac{1}{[(x/z)^2 + 1]^2}. \tag{2}$$

Note that $z\sigma_{zz}(x, z)$ depends only on the ratio of x to the depth, z, a conclusion which applies to all components of the stress tensor.

We show in the greyscale/color the average responses for pentagonal particles in Fig. 9 for various angles of applied force. The responses follow the

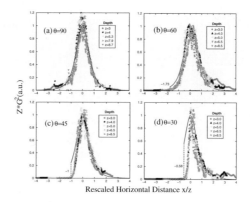

Fig. 10. Rescaled mean force responses in a system of pentagonal particles for non-normal forces applied at various angles: (a) $90°$, (b) $60°$, (c) $45°$ and (d) $30°$. We obtained this figure, by first rotating the coordinate axes so that the vertical axis is along the applied force; we then obtain the force responses along a series of horizontal lines at depths, z, measured from the source. We rescale these responses by normalizing the x coordinate by the depth, z in the rotated frame, and the stresses by z in the rotated frame. Solid lines are the semi-infinite elastic plate solution.

direction of the applied forces and, particularly for larger angles, resemble a rotated version of the response to a normal force. Fig. 10 shows this data in a rotated reference frame, with the z-axis along the direction of the applied force. We then rescaled these responses by multiplying the stresses by z in the rotated frame. And, we normalized the x coordinate by the depth, z in the rotated frame. The solid lines show the elastic plate solutions obtained from Eq. 1. Thus, the response in this system is very similar to that of a 2D elastic material. In particular, the widths of the responses vary linearly with depth[3]. However, there is some departure from elastic-like behavior: the response on the side towards which the force is directed clearly deviates from the elastic solution, which may be due to the fact that there are no tensile forces in a granular material.

Finally, we consider force transmission in a textured medium that has been subject to a modest amount of simple shear. The experimental method is sketched in Fig. 11(a). The particles, (pentagons) rested on a flat horizontal Plexiglas surface. Two parallel boundaries were hinged at their lower corners to allow shear. The other two boundaries remained parallel and at a fixed distance during the shearing process, so that the available area was constant. The system size was about $\sim 47\,\mathrm{cm} \times \sim 22\,\mathrm{cm}$. We slowly sheared this system by displacing the upper left corner. This experiment contained 1167 pentagonal particles that were $\sim 6.3\,\mathrm{mm}$ on an edge and had a packing fraction of 0.795. The experimental procedure[6] involved first preparing a sheared state, and then determining the response to a small applied force applied normally inward at one of the long boundaries.

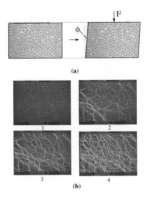

Fig. 11. (a) Sketch of 2D shear cell with images of the pentagonal particles over-layed. The small black dots on each particle denote their centers of mass. (b). A series of photoelastic images showing stress chain patterns for different amounts of shear deformation: $\phi=0$, 2.4°, 3.2° and 4.8° respectively for image 1, 2, 3 and 4.

Fig. 11 shows the evolution of the force chains beginning with $\phi = 0^\circ$ and ending with a total shear through an angle $\phi = 4.8^\circ$. The force chains tend to orient so as to oppose the deformation at an angle of close to 45°. The force chain orientation saturated following a small angular deformation, so that for $\phi \gtrsim 5^\circ$, the typical stress chain angle did not significantly change.

We characterize the force chain orientation and chain length through the spatial auto-correlation $c(\mathbf{r})$ for the stress, (G^2):

$$c(\mathbf{r}) = c(r, \theta) =< G^2(\mathbf{x})G^2(\mathbf{x} + \mathbf{r}) >, \tag{3}$$

where the brackets indicate an average over spatial coordinates $\mathbf{x}$. We keep angular information in $c(r, \theta)$ since this force chain structures are not isotropic.

In Fig. 12, we show data for $c(r, \theta)$ obtained by averaging 50 realizations. We consider two directions, one corresponding to the strong force chain direction, and the other normal to the predominant force chain direction. Clearly, and perhaps not surprisingly, these data show that correlation along the force chain directions is much longer range than along the perpendicular direction, even though the force chain directions span a finite range of angles. The strongest direction for $c(r, \theta)$ is 45° from the vertical. Along the perpendicular direction, the correlation is almost a δ function, dropping rapidly to a value close to zero over a distance of about 1 grain diameter; along the strong direction, the correlation function is consistent with a power-law with an exponent of -0.81, showing long range order over the size of the system. The angle 45° can be understood by the following simple argument. Since simple shear can be expressed as a solid-body rotation by $\phi/2$ plus compression along a line oriented at 45° and an expansion normal to that direction, force chains form along the compressional direction, and any pre-existing chains along the expansion direction tend to vanish.

When a small point force was applied at the boundary (in the inward normal direction) the force tended to propagate more strongly along the force chains, so that the response deviated from the normal direction, as shown in Fig. 13. These data show the response at various distances from the boundary for a 4.7° deformation. In Fig. 13b, we show this data with the x coordinate scaled by z. In Fig. 13b , all peaks of the responses at different depths are located around $x/z = 0.5$, corresponding to an angle of $25 \pm 4°$ from the vertical.

Summary and conclusions We have explored the nature of fluctuations in dense granular systems undergoing steady state shearing. Force fluctuations can be dramatically large, with distributions that are roughly exponential at large force, but have additional and yet to be explained structure at forces near or below the mean. These fluctuations are tied to variations in the force chain structure during the shearing process. It is interesting to note that the stresses in such a system grow logarithmically with the shear rate. Diffusion in these systems is consistent with ordinary Brownian motion, but large velocity gradients and the effect of a rigid boundary at the shearing surface lead to particle displacements that are more complex than in a simple fluid. Measurements of particle mobility in such a material, carried out at constant tracer particle velocity, v, show that the force needed to push a particle through the force network is nearly independent of v. This is reminiscent of Coulomb friction and the studies by Albert et al.[24]

Regarding force transmission, the structure of the granular packing plays a crucial role. Here, structure can refer to the geometric packing and also to

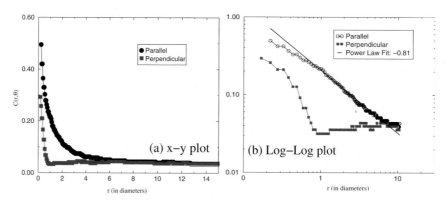

Fig. 12. (a) Spatial auto-correlation function $c(r, \theta = 135)$ (parallel to the stress chain direction) and $c(r, \theta = 45)$ (perpendicular to the stress chain direction). Here, θ is measured from the right horizontal direction. (b) same data shown on double logarithmic scales. The correlation function parallel to the stress chain direction can be fitted with a power-law: $c(r, \theta) \sim r^{-\gamma}$, where γ is 0.81, showing a persistent long range order. In the direction transverse to the chains, the correlation function falls to the background value over a length that is roughly one grain size. Note that distances are measured in particle sizes.

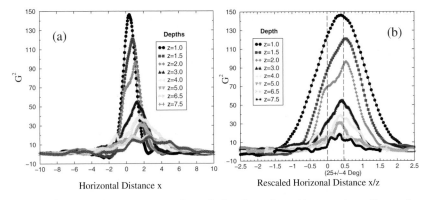

Fig. 13. Quantitative representation of the data from the previous figure for a pentagonal system with shear deformation of 4.7^o: (a) The averaged photoelastic response, G^2, v.s. horizontal distance, x, at various depths z. (b) The same data as (a), but with the x coordinate rescaled with the depth z. In both plots, x and z are measured in grain sizes, where a grain size is about 1 cm.

disorder in the contacts associated with static friction. We find, for disordered packings, that the response is elastic-like. For ordered packings, either an anisotropic elastic response or a wave-like response are possible descriptions. The resolution of which of these pictures is most appropriate remains as a future research question. A small amount of shear generates a highly textured system, with force chains aligned so as to resist further deformation. The force-force correlation function shows a power-law decay (exponent of ~ -0.8) for forces along the strong chain direction, and a delta function in the weak direction. In such a textured system, the response to an applied force is rotated toward the strong direction.

Acknowledgements

We appreciate helpful interactions with P. Claudin, S. Luding, C. Goldenberg, I. Goldhirsch, and J. Socolar. The work of GR and EC was supported by PICS-CNRS #563. The work of JG and RPB was supported by the US National Science Foundation under Grant DMR-0137119, and DMS-0204677, and by NASA under Grant NAG3-2372.

References

1. G.W. Baxter, R. Leone, G.A. Johnson, and R.P. Behringer. Time-dependence and pattern formation in flowing sand. *Eur. J. Mech. B: Fluids*, 10:181, 1991.
2. G.W. Baxter, R. Leone, and R.P. Behringer. Experimental test of time scales in flowing sand. *Europhys. Lett.*, 21(5):569–574, 1993.

3. J. Geng, D. Howell, E. Longhi, R.P. Behringer, G. Reydellet, L. Vanel, E. Clément, and S. Luding. Footprints in sand: The response of a granular material to local perturbations. *Phys. Rev. Lett.*, 87:035506, 2001.

4. R.R. Hartley and R.P. Behringer. Logarithmic rate dependence of force networks in sheared granular materials. *Nature*, 421(6926):928–931, 2003.

5. B. Miller, C. O'Hern, and R.P. Behringer. Stress fluctuations for continously sheared granular materials. *Phys. Rev. Lett.*, 77:3110–3113, 1996.

6. J. Geng, R.P. Behringer, G. Reydellet, and E. Clément. Green's function measurements in 2d granular materials. *Physica D*, 182:274–303, 2003.

7. B. Utter and R.P. Behringer. Self-diffusion in dense granular shear flows. *Phys. Rev. E*, 69:031308, 2004.

8. D. Howell, R.P. Behringer, and C. Veje. Stress fluctuations in a 2D granular Couette experiment: A continuous transition. *Phys. Rev. Lett.*, 82(26):5241–5244, 1999.

9. G. Reydellet and E. Clement. Green's function probe of a static granular piling. 2001.

10. M. Da Silva and J. Rajchenbach. Stress transmission through a model system of cohesionless elastic grains. *Nature*, 406:708–XXX, 2000.

11. J-P. Bouchaud, P. Claudin, D. Levine, and M. Otto. Force chain splitting in granular materials: A mechanism for large-scale pseudo-elastic behavior. *Euro. Phys. J.*, E4:451, 2001.

12. J.E.S. Socolar, D.G. Schaeffer, and P. Claudin. Directed force chain networks and stress response in static granular materials. *Euro. Phys. J.*, E7:353, 2002.

13. C. Goldenberg and I. Goldhirsch. Elasticity of microscopically inhomogeneous systems: The one-dimensional case. preprint, 2000.

14. A.V. Tkachenko and T.Q. Witten. Stress propagation through frictionless granular material. *Phys. Rev. E*, 60:687, 1999.

15. A.V. Tkachenko and T.Q. Witten. Stress in frictionless granular material: Adaptive network simulations. *Phys. Rev. E*, 62:2510, 2000.

16. N.W. Mueggenburg, H.M. Jaeger, and S.R. Nagel. Stress transmission through three-dimensional ordered granular arrays. *Phys. Rev. E*, 66:031304, 2002.

17. R.M. Nedderman and C. Laohakul. The thickness of the shear zone of flowing granular material. *Powder Technol.*, 25:91, 1980.

18. C.H. Liu, S.R. Nagel, D.A. Schecter, S.N. Coppersmith, S. Majumdar, O. Narayan, and T.A. Witten. Force fluctuations in bead packs. *Science*, 269:513, 1995.

19. J.-P. Bouchaud, M.E. Cates, and P. Claudin. Stress distribution in granular media and nonlinear wave equation. *J. Phys. I*, 5:639–656, 1995.

20. P. Claudin and J.-P. Bouchaud. Stick-slip transition in the scalar arching model. In H.J. Herrmann, J.-P. Hovi, and S. Luding, editors, *Physics of Dry Granular Media*, page 129, Dordrecht, 1998. Kluwer Academic Publishers.

21. M.E. Cates, J.P. Wittmer, J.-P. Bouchaud, and P. Claudin. Jamming, force chains, and fragile matter. *Phys. Rev. Lett.*, 81(9):1841–1844, 1998.

22. O. Straußand S. Luding. *Granular Gases*. Springer-Verlag, 2000.

23. L. Landau and E. Lifschitz. *Theory of elasticity*. MIR, Moscow, 1967.

24. R. Albert, M.A. Pfeifer, A.L. Barabási, and P. Schiffer. Slow drag in a granular medium. *Phys. Rev. Lett.*, 82:205, 1999.

Small Scale Modelling of Large Granular Systems in a Centrifuge

H.G.B. Allersma

University of Delft – Civil Engineering and Geosciences
Stevinweg 1, 2628CN Delft – The Netherlands
h.allersma@ct.tudelft.nl

Summery. The behaviour of granular material is strongly dependent on the stress level, so that small scale tests simulating large granular systems are not reliable in many cases. The stress dependent behaviour, however, can be simulated by placing the small scale test in a centrifuge. Due to the artificial gravity the stresses gradient in the small model can be made similar as in the prototype, resulting in a realistic behaviour of the granular material. Almost all types of granular systems can be investigated by this testing technique.
Keywords: geotechnical centrifuge modelling, soil mechanics, granular material

1 Introduction

In several systems the behaviour of a structure is strongly dependent on the properties and behaviour of granular material. The behaviour of granular material is complicated so that, in spite of powerfull computers, it is still difficult to make reliable predictions by calculations. In particular when a new area is under investigation physical tests are necessary to validate the calculation results. In some cases, however, systems are so complicated that they cannot be modelled mathematically at all, so that physical modelling is the only way to get insight in the problem. In the first instance the most reliable method of testing a system seems the performance of real scale tests. For some specific cases this is true. However, in most problems real scale test are not practical. In the first instance it would be very costly to test e.g. foundation elements with a diameter of 20 m. However, a good test is not soon too expensive, so that the cost will not be hampering the research.

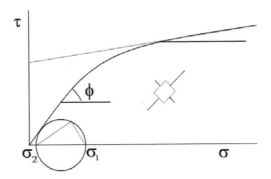

Fig. 1. Schematic visualisation of the stress dependent behaviour of the friction angle.

A more physical problem is that it would be rather difficult to subject huge elements to large well controlled loading programs. Furthermore the large size makes it difficult to find a suitable test location and it would be almost impossible to perform several test with exactly the same initial condition, necessary to perform a parametric study.

The disadvantages mentioned before are not longer valid if the test could be performed at small scale. In relation to the real scale tests it would be easy to prepare reproducible samples, and it would not be so difficult to load the structure. However, a problem is that in small scale models the stress level in the granular material, and so the shear stress, is much lower than in reality. This affects significantly the behaviour of granular materials in which cohesion has an important role. But also frictional materials, like dry sand, behave stress dependent. One of te reasons for that is the stress depedent value of the friction angle, caused by crushing of roughness at the particle surface. The stress dependent behaviour of the friction angle is illustrated in a Mohr diagram in Fig.1. Also many powders and agricultural granular matters show this property. This stress dependent behaviour causes that wrong predictions are made if the small scale tests are translated to prototype size. In order to take the stress dependent bahaviour into account the small scale model tests can be carried out in a centrifuge. The artificial acceleration increases the self weight of the granular material, so that a similar stress level can be simulated in the small scale model as is the case in the prototype. A higher stress level allows larger shear stresses, so that models with cohesive material behaves more realistic. An example is shown in Fig.2a, where a vertical cut in a soft clay is modelled. At earth gravity (1g) nothing will happen, because the shear stresses are far too low to cause failure. The model in Fig.2b was placed in a centrifuge and accelerated up to 150g. Actually the depth of the cut is simulated to increase, so that finally failure can be realized. The complicated failure mechanism demonstrates the need why model tests are necessary for beter understanding of the behaviour.

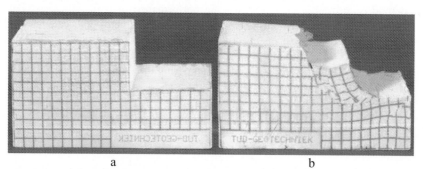

a   b

Fig. 2. Vertical cut modelled in a small clay sample (grid size is 5mm). a) at 1g; b) at 150g in a centrifuge.

The stress dependent behaviour is demonstrated in dry sand, where a circular footing is pushed into a sand layer. The test is performed at earth gravity and at different accelerations in a cenrifuge. The load displacement diagram of the different tests is shown in Fig. 3. In each tests the load is divided by the artificial gravity. As can be seen the bearing load is relatively much larger at low stress levels. It is also visible that the stress dependent behaviour is stronger in the low stress regions.

Furthermore there are some other reasons why small scale models at 1g are not so useful. This is for example the case when sand layers have to be modelled with a phreatic surface. At 1g the capillary rising of water in sand is approximately 300mm, so that samples with a height og 100mm are more or less saturated by capillary suction. Furthermore the capillary suction pressure is of the same magnitude as the self weight stresses in the granular, so that they influence the behaviour significantly. At 100g the capillary rising is only 3mm, so that the groundwater tabel is good visible.

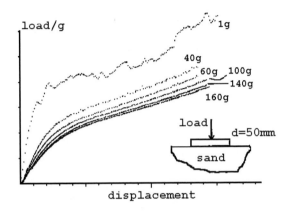

Fig. 3. Load displacement diagram at different gravities of circular footing pushed into a sand layer.

The late sixties can be considered as the beginning of a new area for centrifuge modelling. Several centrifuges were built for geotechnical work and a great variety of problems were studied by this technique (1). The tendency was to increase the size of the devices, so that the costs of tests became very high. For several geotechnical problems, however, the use of a small centrifuge is quite adequate. By making an optimal choice between size and facilities and using up-to-date electronics and computer control effective tests can also be performed in a small centrifuge.

In 1988 the development of a small geotechnical beam centrifuge with a diameter of two metres was started at the Geotechnical Laboratory of the University of Delft. The device was operational in 1990. Test devices with a dimension of 300x400x450 mm and a weight of 300 N can be accelerated up to 300 g. A small geotechnical centrifuge is relatively cheap to operate, and the development of the equipment did not take so long compared to a big centrifuge. To enable the performance of advanced tests in flight, the carriers of the centrifuge were made large enough to contain computer-controlled devices. Because the costs of operation are low, the device is suitable to perform trial and error tests. Modification of the centrifuge for different tests is simple, so that a flexible operation is obtained. The test containers and actuators are, in general, so small that they can be conducted by one person. This is convenient during the preparation of the tests and leads to good reproducibility of the soil samples. This is important if the results of similar tests have to be compared.

A disadvantage of a small centrifuge is the limitation in the use of sensors during a test. This restriction, however, can be compensated partly by using image processing techniques in video images taken with the on-board video camera.

Miniature devices have been developed for performing advanced tests in flight, such as: loading, displacement and controlling the supply of sand, water and air. The devices operate under software control, which runs in a single board PC compatible computer located in the spinning part of the centrifuge. The signals from load cells, pressure transducers and other sensors are received by the on-board computer without interference of slip rings. The computer is assessable in a normal way via slip rings and commercial available line drivers. The test devices are driven by small DC motors, which are manipulated by the on board computer.

Several devices have been developed to prepare sand and clay samples. To improve the reproducibility sample preparation is automated as much as possible. A special centrifuge has been built to consolidate clay slurry, in order to obtain a very soft normally consolidated clay. Several different research projects were conducted in the centrifuge, i.e.: sliding behaviour of spudcan foundations, stability of dikes during wave overtopping, gas blowouts and cratering, stability of embankments during widening, shear band analysis, buckling behaviour large diameter piles, simulation suction pile installation, pile driving, pollution transport, land subsidense, flow in hoppers, etc. Several project were carried out as a 6 month graduate study. In recent years the centrifuge was in operation almost every day, hence the flexibility is demonstrated by the fact that three quite different model tests were performed on some days.

2   The Small Geotechnical Centrifuge of the University of Delft

Mechanical part
The geotechnical centrifuge at the University of Delft was designed by the Geotechnical Laboratory of the Department of Civil Engineering and was built by the mechanical workshop of the University.

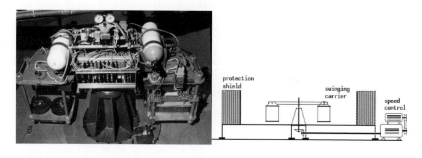

a   b

Fig. 4. a) Photograph of the small centrifuge of the University of Delft. b) Schematic drawing.

The electronic systems were designed and built by the Geotechnical Laboratory. The advantage of the in-house design is that the system can be expanded and modified under internal supervision, which guarantees a good interaction between the facilities of the device and the tests. The centrifuge frame is fixed to the floor and bears the vertical axis and the protection shield (Fig.4). A beam with a length of 1500 mm is connected to the axis, so that it can be rotated in the horizontal plane. Two swinging carriers are connected to the beam by means of brackets.

The carriers are formed by two plates at a distance of 450 mm apart, which are connected to each other by four cylindrical steel beams. The surface of the plates is 400 x 300 mm . Because the weight of the beam and carriers is large imbalance, which can occur during tests, has not a significant effect on the stability of the centrifuge. The potential danger from the spinning part of the centrifuge is minimized by a protection shield of steel (thickness= 5 mm) that forms a large cylindrical box. A second shield, 50 cm outside the first, is made of wooden plates. The gap between the two shields is filled with concrete blocks and granular material. This fill gives additional safety against flying projectiles and the weight stabilizes the device.

The centrifuge is driven by an electric motor of 18 kW via a hydraulic speed control unit. The hydraulic speed controller is manipulated by a step motor, which is interfaced to the speed control computer. A computer program has been developed to adjust the speed of the centrifuge using the signal of a tachometer. Several options are available to control the speed. It is, for example possible to make the acceleration dependent on time or on other test parameters, such as the pore water pressure in a clay sample.

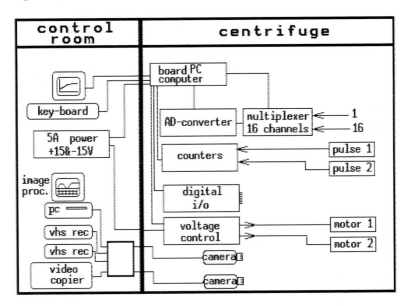

Fig. 5. Diagram of the electronic control and measuring system.

Measuring facilities
The system electronics enables the performance of computer-controlled tests in flight (Fig.5). To minimize the number of slip rings the control system is placed in the spinning part of the centrifuge. The control unit contains a small single board computer (180x120x25mm ; Pentium; 500 MHz; 128 Mbyte

RAM; 80 Mbyte solid state disk), a 12-bit analog to digital converter with a 16-channel multiplexer, two voltage controlled outputs of 8 Ampere each, two 16-bit counters and several digital input/output channels. For additional data storage a 1Gbyte hard disk unis is placed just in the center of the centrifuge. The signals from the sensors are conditioned by on-board amplifiers. Eight power slip rings are available to feed the electronics and the actuators. 24 high quality slip rings are used to transmit the more sensitive signals, such as, for example, two video lines and the connection between the on-board computer and the PC in the control room. By means of commercial available line driver units is was possible to realize a normal access to the board computer.

A special feature is that several phenomena can be measured using the video images. In this technique the video images of the in flight test are captured by the frame grabber in the PC and processed until the relevant parameters are isolated. Image processing can be used to visualize and digitize the surface deformation of clay and sand samples or to digitize the consolidation of a clay layer (2). This technique has proven to be very useful in several research projects.

3 Sample Preparation Devices

An important aspect of centrifuge research is sample preparation. To achieve good test results, the following is required:
- The ability to use samples of different soil types
- The ability to vary sand densities
- The ability to accurately reproduce samples, so that results from different tests can be compared.

Two different devices have been developed in order to prepare clay or sand layers.

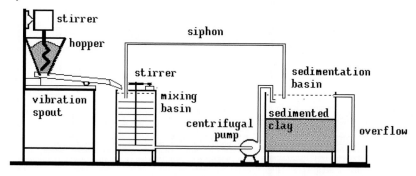

Fig. 6. Diagram of the clay mixing device.

Clay preparation
Up to now it was found that the best control over the samples is obtained by making artificial clay. In this technique clay powder (several types are commercial available) is mixed with water, where the air content is kept as low as possible. A technique has been developed in which an air free slurry is obtained under normal atmospheric pressure. The device operates more or less automatically and is self cleaning. The principle of the device is that the clay is added in a thin layer to a rotating water surface (Fig.6), so that no air is

included due to differences in capillarity. The water with a very low clay content is pumped to a basin where the clay is sedimented. The clay slurry, with a water content of approximately 100%, is homogenized in a mixer before it is put into the sample boxes. The best way to obtain a soft, normally consolidated soil with a smooth and realistic gradient of water content and strength over the height of the sample is to consolidate the slurry in the centrifuge at the same g level as will be used in the tests. The consolidation will take several hours or even days when a low permeable clay is used. Because the centrifuge will be occupied all that time no other tests can be performed. Therefore a special centrifuge has been built which is only used to consolidate the clay layers. This centrifuge has a diameter of 1 metre and can accelerate sample boxes with a weight of approximately 200N up to 200g.

The consolidation can be followed by pressure transducers via slip rings. Or by means of a small camera. The settlement of the clay surface can be digitized in real time by using image processing.

   To improve the reproducibility of the sample preparation a technique has been developed to copy a grid at the surface of a black or white clay without removing boundaries. A grid is plotted on a special sheet which is made waterproof by a thin cover. This sheet is placed in the sample box which is filled with slurry. After consolidation the protection cover is removed. Due to the water the grid is copied to the clay surface, were the special layer on the sheet became very smooth. A grid with a good contrast is required to derive the deformation of the clay by image processing.

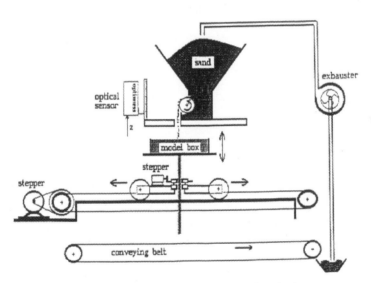

Fig. 7. Diagram of the sand preparation device.

Sand preparation machine
A computer controlled device has been developed to prepare well defined sand layers in the test containers (Fig.7). Since this device is completely automated, very good reproducible samples can be made. The sand, which is stored in a hopper can be sprinkled in a curtain by means of a rotating axis, following the same technique as the in flight sand sprinkler. The falling height of the sand can be adjusted. The distance to the sand surface is meas-

ured by means of an optical sensor and the height of the test box is adjusted by a step motor in order to keep the falling height constant during raining. The sample box is moved back and forth by means of a second step motor, while a smooth acceleration is realized in the turning points to prevent shocks. The sand supply system can be controlled in the computer program, so that only sand is sprinkled when the sample box is located under the outlet of the hopper. The wasted sand is transported by a belt to a container. The sand level in the container is detected by a photo cell. Depending on the sand level, a vacuum cleaner is started, so that the wasted sand is transported back to the hopper. Special precautions are taken to prevent the fine material from being extracted from the sand used, because it was found that small changes in the composition has a large influence on the mechanical properties of the sand.

Sand samples with a surface of 300x300 mm and a maximum thickness of approximately 150 mm can be made. The porosity, depending on the sand type, can be varied between 35% and 39%. The porosity of the sand layers can be reproduced with an accuracy of less than one percent. The preparation of the sample with a thickness of 100mm takes about 20 minutes.

4 The Test Equipment

Several devices have been developed by the Geotechnical Laboratory of the University of Delft to perform tests in flight. The mechanical equipment, electronic system and control software are designed in all details by the laboratory. The following devices are available:
- Gas supply system
- Water supply system
- Excavation
- Pollution transport simulator
- Two dimensional loading system
- Sand sprinkler

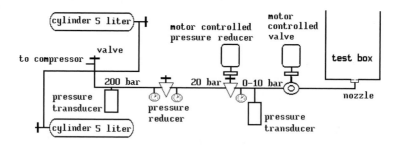

Fig. 8. Diagram of the air supply system.

Gas supply system
In some tests gas supply to a soil sample is required. Since the small centrifuge is not equipped with fluid slip rings the gas has to be stored in the spinning section of the centrifuge. To make the storage as compact as possible, two high pressure (200 bar) cylinders of 5 litres each are mounted on the beam of the centrifuge (Fig.4). Before a test is started the cylinders are filled

with air by means of a high pressure compressor. A computer controlled air supply system has been developed in order to regulate the gas flow from a distance. The system is shown schematically in Fig.8. The pressure of the supplied air is controlled by a conventional pressure regulator, which is modified in such a way that it can be driven by a small DC motor. The output pressure of the regulator is detected by a pressure transducer and used in the computer program to control the DC motor. A modified valve, which is also driven by a small DC motor is used to start or stop the gas flow quickly. The gas flow per unit of time is measured by measuring the pressure drop of the gas cylinders during the test. A computer program has been developed to interactively control the gas supply. During a test the cylinder pressure, the test pressure and the gas flow are plotted on the screen. Flow rates of 10 l/s can be reached. The gas in the high pressure cylinders represents a lot of power, which can be used, in principle, for tests in which large loads or energy are needed.

The gas supply system is used to simulate gas blowouts and cratering (Fig.9) in a sand layer with an equivalent thickness of approximately 20 - 30 metres.

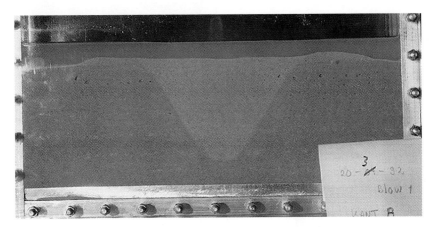

Fig. 9. Photograph of the simulation of cratering in a sand layer with a height of 20m.

The air system can be used in several other applications to manipulate a test in flight. Examples are: trapdoor system, flow in hoppers, pile driving, wave generation, etc.

Water supply system
In several geotechnical problems it is required to control water flow in the spinning centrifuge. It is not so easy to control the water supply because rather high pressures (and energy) are required to overcome the acceleration. At this moment two systems are available. In the most simple system the water is circulated by means of an air jet (Fig.10a). The air supply system is used to control the jet. The advantage of this system is that the water supply can be controlled smoothly from zero to maximum flow. The flow rate can be measured by means of a small turbine. A maximum flow of about 10 l/min can be obtained. The second system uses a small pump, which is commer-

cially available as an accessory for electric drilling machines. In the centrifuge there is not enough electric power available to drive such a pump. Therefore an air motor has been applied. An air motor delivers a lot of power per unit of weight and uses the air more efficiently than an air jet. The air motor has proven to be reliable up to at least 130g. An additional advantage of an air motor is that no electric power drop can hang up the control computer. The speed of the air motor cannot be controlled smoothly from zero, so that a large water flow is generated at start up. The water flow rate is measured by means of a small turbine.

The water circulation system is used to investigate the stability of dikes during water infiltration, which can occur by wave overtopping. In Fig.10b it is visualized how a sand slope, which was covered with a thin clay layer, collapses. Typically is the crack in the clay layer, caused by friction reduction due to uplift pressure.

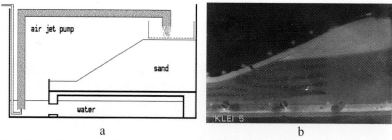

a   b

Fig. 10. a) Diagram of the air jet pump. b) Fialure of a sand slope, covered with clay, due to water infiltration at the crest.

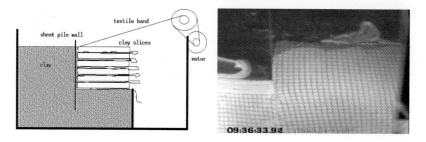

Fig. 11. a) Diagram of the in flight excavation device. b) Photograph of a stage in a test at 100g.

In flight excavation
For several applications, in flight excavation of soil is necessary to simulate a realistic prototype situation. In flight excavation is important if, for example, the behaviour of sheet pile walls or the construction of a clay slope is being examined. It is desirable for the cut to be made gradually, where the time interval between the extracted layers can be varied. In some centrifuge labs an excavation is simulated by lowering the level of a heavy liquid. Another technique, which can be used to realize the whole cut at once, is extraction of a soil body supported by e.g. textile net. Robots have also been developed to extract the soil in a more controlled manner.

It was found that a well controlled excavation could also be realized by a very simple technique. A diagram of a set-up to simulate the extraction of soil in a test with a sheet pile wall is shown in Fig.11a. At 1g, a cut is made in a clay model and the sheet pile wall is installed. Next the cut is placed back in the form of slices of clay, where the slices are separated by a textile band. The band is folded in such a way that the clay slices can be removed one by one simply by winding up the textile band. This process is shown in Fig.11b. After removing a layer, the process can be stopped to allow examination of the depth of the excavation at which the wall starts to deform. This excavation method can also be used in sand.

Simulation pollution transport
LNAPL transport in partly saturated soils can be physically modelled in a geotechnical centrifuge (3). Two dimensional tests at 30 g have been performed. In order to provide detailed information, the phenomena were observed and digitized in flight by using video cameras and digital image processing. The diagram of the test set-up is shown in Fig.12a. A two dimensional container with transparent walls at a distance of 40mm apart contains partly saturated sand (D50=0.2mm). At 30g the height of the capillary zone was approximately 10mm. The LNAPL is stored in a perspex container located at the surface of the sand layer. The container is devided into two compartments, of which one of them is equipped with a chink. In flight the LNAPL can be move to the perforated compartment by air pressure, in order to start the infiltration process. The infiltration of the LNAPL can be visualized thanks to the transparent walls and the on board camera. The plume at some stage and the digitized contours of the plume during the test are shown in Fig.12b. The influence of the capillary zone and the saturated zone is good visible. Digital image processing provided information about the speed of the front, the area of the contaminated zone, the amount of supplied NAPL and the distribution of NAPL concentration over the plume area.

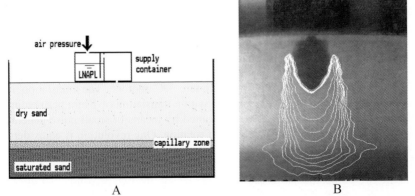

A   B
Fig. 12. a) Diagram of two dimensional test container to simulate pollution infiltration in sand. b) Behaviour of oil spill in partly saturated sand (course visualized by image processing).

Two dimensional loading system

The two dimensional loading system (Fig.13) can be considered as a universal tool, which can be used for several tests. Two guiding systems based on linear ball bearings and axes of tempered steel guarantee a low friction translation in two perpendicular directions.

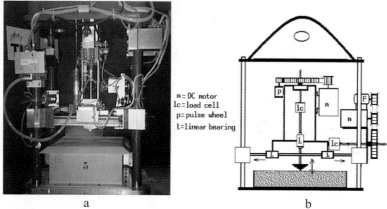

Fig. 13. a) Photograph of the two dimensional loading device, mounted in the centrifuge. b) Schematic drawing of the loading device, used for combined loading of footings.

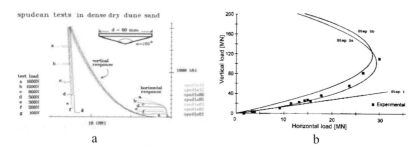

Fig. 14. a) Typical output of a sliding test of a conical footing on sand. b) Relation between horizontal and vertical loading; tests and calculations.

The system is driven by two miniature DC motors. The translation is achieved by means of a screw spindle with a translation of 1 mm per revolution. The number of revolutions is counted by means of small pulse generators, which also detect the direction of rotation. One revolution is equivalent to 200 pulses. A special interface has been built to make it possible for the pulses to be used in the control program in order to determine displacements in the two perpendicular directions. The loads in the two perpendicular directions are measured by means of load cells. The outputs of the load cells are multiplied and can be used in a computer program via the multiplexer and the analog-to-digital-converter.   Sufficient information is available to perform load or displacement controlled tests.

Furthermore the device can be used as a simple robot to manipulate a test during flight. Loads of more than 5 kN can be applied by the system. The accuracy of the measurement of the displacement (determined at 100g by image processing) is better that 0.1 mm, where the maximum displacement is about 50 mm. Up to now the device has been used at gravitation levels of more than 150 g. The weight of the two dimensional loading device is approximately 10 kg. It takes about ten minutes to install the loading system in the centrifuge. As an example, the typical output of a test with a conical footing (spudcan) foundation on sand is shown in Fig.14a. In this test series the sliding resistance of the foundation element at different vertical loads is investigated, in order to make predictions about the behaviour of offshore platforms during heavy storms. The relation between horizontal and vertical load is compared with theoretical calculations in Fig.14b. Spudcans with a diameter of more than 14 metres can be simulated. Thanks to the small size of the models it is very easy to examine the influence of modifications of the design, such as skirts, surface roughness, pins, hinge etc. Other tests that can be performed with the loading system are, for example: anchors, bulldozer, footing, sheet pile walls, expanding tubes, etc.

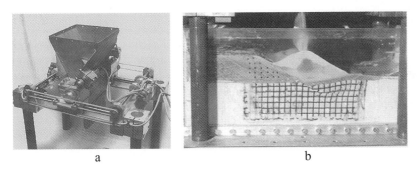

a   b

Fig. 15. a) Photograph of the sand sprinkler system. b) Deformation under an embankment on clay during widening of an road construction.

In flight sand sprinkler
A computer controlled sand sprinkler has been developed to make embankments in flight. The device consists of a hopper, which can be translated easily by means of linear ball bearings and axes of tempered steel (Fig. 15a). The weight of the device is approximately 10 kg. The translation (range is 150 mm) is realized by means of a small DC motor. The position of the hopper is detected by means of a pulse wheel and a 16 bit counter, which can be accessed in the program of the control computer. The sprinkler system is designed in such a way that no close seals are required. An axis is located in the outlet of the hopper in such a way that the granular material flows only when the axis is rotated. This mechanism has proven to be reliable up to 120g. The axis of the sprinkler system is also driven by a small DC motor and the amount of deposited sand is detected by counting the number of revolutions by means of a pulse wheel. Several options can be assessed in the control program. It is possible to sprinkle sand layer by layer or at one particular location. The disturbing effect of the Coriolis forces is minimized by means of hinged sheets, which guide the sand grains. On the other hand the Coriolis effect can be used to build an embankment with a gradient in height over the width of the sample box. This can be used to investigate time effects in the failure of clay under embankments. Optionally, the control program of the sand sprinkler also reads the output of pressure transducers, which can be

placed in the clay layer. The pore water pressure is plotted on the screen. An automatic link can be made between the pore water pressure and the sand supply scheme. The creation of a dike during sand supply and the deformation of the clay can be monitored by a video camera. The deformation of the clay is made visible by means of a grid. Software has been developed by the laboratory (2) to digitize the coordinates of the nodes of the grid automatically by means of image processing. In principle it is possible to make an automatic link between the sand supply program and the image processing system, so that an embankment can be built in flight where the images of the video camera are used to control the sand supply. The sand sprinkler system is used to investigate the stability of dikes and different methods of widening of embankments, which are founded on soft soils (Fig.15b). Since the centrifuge has two swinging platforms, the loading system and the sand sprinkler can both be mounted in the centrifuge. In this way the centrifuge can be used efficiently.

5 Conclusions

The centrifuge testing technique is a powerful method to examine a large variety of problems in which the properties of a granular material has a significant influence on the behaviour of the system. The behaviour of large structures can be modelled with small scale models. In this paper only a few examples are shwn. More applications and details can be found in the website: http://geo.citg.tudelft.nl/allersma/hgball.htm.

At the moment the applications are mainly devoted to civil engineering applications. It is believed, however, that the testing technique is applicable in several other branches, such as e.g. powder technology and geological phenomena.

References

1. H.Y. Ko. Centrifuge91, Proceedings of the International Conference Centrifuge91, Boulder, Colorado, Balkema, Rotterdam, 1991.
2. H.G.B. Allersma. On line measurement of soil deformation in centrifuge tests by image processing. Proc. Int. Conf. on Experimental Mechanics, Copenhagen, 1990, pp. 1739-1748.
3. M.A. Knight and R.J. Mitchell. A similitude and dimensional design guide for centrifuge modelling of multiphase contaminant transport. Envir. Geotechn., Kamon (ed.) Balkema, 1996, pp. 245-250.

Weakly Frictional Granular Gases

I. Goldhirsch[1], S.H. Noskowicz[1], and O. Bar-Lev[2]

[1] Dept. Fluid Mechanics and Heat Transfer, Faculty of Engineering –
  Tel-Aviv University
  Ramat-Aviv, Tel-Aviv 69978 – Israel
  isaac@eng.tau.ac.il, henri@eng.tau.ac.il
[2] School of Mathematics – Tel-Aviv University
  Ramat-Aviv, Tel-Aviv 69978 – Israel
  fnbarlev@post.tau.ac.il

Summary. Hydrodynamic equations of motion for a monodisperse collection of weakly frictional spheres have been derived from the corresponding Boltzmann equation, using a Chapman-Enskog expansion around the elastic smooth spheres limit. The hydrodynamic fields required in this case are: the velocity field, $\mathbf{V}(\mathbf{r}, t)$, the translational granular temperature, $T(\mathbf{r}, t)$, and the (infinite) set of number densities, $n(\mathbf{s}, \mathbf{r}, t)$, corresponding the continuum of values of the spin, $\mathbf{s}$. An immediate consequence of these equations is that the asymptotic spin distribution in the homogeneous cooling state for nearly smooth, nearly elastic spheres, is highly non-Maxwellian.

Keywords: Granular Gases, Kinetic Theory, Boltzmann Equation, Friction, Hydrodynamics.

1 Introduction

All macroscopic grains are frictional. Frictional interactions, including frictional restitution in granular collisions, are important and consequential (as examples consider the recent study of the effects of friction on granular patterns, cf., [1], note also the hysteretic effects induced by friction, [2]). Furthermore, it is known that friction induces non-equipartition, cf. e.g., [3, 4]. It is therefore curious that a rather small proportion of the granular literature, cf. e.g., [3–24] is concerned with the modeling of frictional granular gases. Even fewer articles are devoted to direct kinetic theoretical based studies of the full (i.e., not only e.g., the homogeneous cooling state) frictional granular hydrodynamics, cf. e.g., [5, 6, 9, 24] and references therein. Perhaps the fact that frictionless models have been rather successful in explaining many of the observed phenomena in granular gases [25], or the difficulties one encounters in the theoretical study of frictional granular gases are responsible for this state of affairs, and perhaps there are other reasons. In any case, we believe it is important to study the full kinetics and hydrodynamics of frictional granular gases. Such a study is presented below.

The study of gases whose constituents experience frictional interactions started in the realm of molecular gases. Indeed, the classic book by Chapman and Cowling [26] presents references on the kinetics of gases composed of "rough molecules" that date back to the year 1894. A Chapman-Enskog approach to the derivation of the hydrodynamics of such gases is presented in the same book. Later works are described in a review article [27]; see also studies of celestial granular systems [28, 29].

In previous kinetic theoretical based studies of the of granular hydrodynamics [6, 9, 24] it is assumed that the basic distribution function is Maxwellian in both the velocity and angular velocity (and usually different rotational and translational temperatures are allowed for), and corrections due to gradients are identified (on the basis of symmetry). The assumed distribution function is substituted in the Enskog equations, resulting in a closure for the constitutive relations. The above Maxwellian distribution corresponds (when both the rotational and translation temperature are taken to be equal) to the limit of rough molecules, in which there is equipartition between the rotational and translational kinetic energies. Our goal here is to study granular gases near a different limit, i.e., the smooth limit (as frictionless grains comprise the model most studied in the literature), and develop a systematic approach to the problem, i.e., a perturbative scheme.

2 Definition of the Model and Kinematics

Consider a monodisperse system of spherical grains of mass $m = 1$, diameter d, and moment of inertia I (for homogeneous spheres, $I = \frac{2}{5} \left(\frac{d}{2}\right)^2$), each. Denote the radius of gyration a grain by κ (with $\kappa \equiv \frac{4I}{d^2}$). The velocity of particle 'i' is denoted by $\mathbf{v}_i$, and its angular momentum by ω_i. It is convenient to define a 'spin variable', $\mathbf{s}_i$, whose dimension is that of velocity, by $\mathbf{s}_i \equiv \frac{d}{2}\omega_i$.

Consider a binary collision between sphere '1' and sphere '2'. Let $\mathbf{k}$ be a unit vector pointing from the center of sphere '2' to the center of sphere '1' (notice that an opposite convention is used in some papers). The relative velocity of sphere '1' with respect to sphere '2', at the point of contact (when they are in contact), is given by: $\mathbf{g}_{12} = \mathbf{v}_{12} + \mathbf{k} \times \mathbf{s}_{12}$, where $\mathbf{v}_{12} \equiv \mathbf{v}_1 - \mathbf{v}_2$, and $\mathbf{s}_{12} \equiv \mathbf{s}_1 + \mathbf{s}_2$ is the total spin of the two particles (notice: it is not the difference!). *In the following, precollisional entities are primed.* During a collision the normal component of the relative velocity changes according to: $\mathbf{k} \cdot \mathbf{g}_{12} = -e \, \mathbf{k} \cdot \mathbf{g}'_{12}$, where e is the coefficient of normal restitution. The effect of the tangential impulse at a collision is modeled in the same way as in [19, 24], with slight notational differences. Let $\mathbf{J} = \mathbf{v}_1 - \mathbf{v}'_1$ be the change in the momentum of particle "1" (recall that $m = 1$) in a collision. Linear and angular momentum conservation imply that: $\mathbf{v}_1 = \mathbf{v}'_1 + \mathbf{J}$, $\mathbf{v}_2 = \mathbf{v}'_2 - \mathbf{J}$, $\mathbf{s}_1 = \mathbf{s}'_1 - \frac{1}{\kappa}\mathbf{k} \times \mathbf{J}$, and $\mathbf{s}_2 = \mathbf{s}'_2 - \frac{1}{\kappa}\mathbf{k} \times \mathbf{J}$. Next, consider the decomposition of

$\mathbf{J}$ into its normal and tangential components: $\mathbf{J} = A\mathbf{k} + B\dfrac{(\mathbf{k}\times(\mathbf{k}\times\mathbf{g}'_{12}))}{|\mathbf{k}\times\mathbf{g}'_{12}|}$. Let γ be the angle between the direction, $-\mathbf{k}$, and the relative velocity of the grains at the point of contact (when they are in contact): $\cos\gamma \equiv -\dfrac{\mathbf{k}\cdot\mathbf{g}'_{12}}{g'_{12}} = -\dfrac{\mathbf{k}\cdot\mathbf{v}'_{12}}{g'_{12}}$, where $g'_{12} \equiv \|\mathbf{g}'_{12}\|$; it follows that $0 \le \gamma \le \frac{\pi}{2}$. Define γ_0 such that if $\gamma > \gamma_0$ there is sliding (Coulomb friction) during the collision and $B = \mu A$, where μ is the friction coefficient, while if $\gamma \le \gamma_0$ there is sticking (or the grain is 'rough') and $\mathbf{k}\times(\mathbf{k}\times\mathbf{g}_{12}) = -\beta_0\mathbf{k}\times(\mathbf{k}\times\mathbf{g}'_{12})$, with $1 \ge \beta_0 \ge -1$. (else energy is 'produced' in the collision). It is easy to show that in both cases: $A = -\frac{1+e}{2}\mathbf{k}\cdot\mathbf{v}'_{12}$. In the case of sliding one obtains:

$$\mathbf{k}\times(\mathbf{k}\times\mathbf{g}_{12}) = \left(1 - \frac{1+\kappa}{\kappa}(1+e)\mu\cot\gamma\right)\mathbf{k}\times(\mathbf{k}\times\mathbf{g}'_{12}) \qquad (1)$$

It follows that in both cases: $\mathbf{k}\times(\mathbf{k}\times\mathbf{g}_{12}) = -\beta(\gamma)\,\mathbf{k}\times(\mathbf{k}\times\mathbf{g}'_{12})$ where (re-quiring $\beta(\gamma)$ to be a continuous function of γ):

$$\beta(\gamma) = \min\left\{\beta_0, -1 + \frac{1+\kappa}{\kappa}(1+e)\mu\cot\gamma\right\} \qquad (2)$$

The transition angle, γ_0, between the two ranges of γ is determined by the requirement of continuity of $\beta(\gamma)$ to be: $\cot\gamma_0 = \frac{\kappa}{1+\kappa}\frac{1+\beta_0}{1+e}\frac{1}{\mu}$. Using the con-servation laws for the linear and angular momenta one obtains the the trans-formation between the precollisional and postcollisional velocities and spins of a colliding pair of grains:

$$\mathbf{v}_1 = \mathbf{v}'_1 - \frac{1+e}{2}(\mathbf{k}\cdot\mathbf{g}'_{12})\mathbf{k} + \frac{\kappa}{1+\kappa}\frac{1+\beta(\gamma)}{2}\mathbf{k}\times(\mathbf{k}\times\mathbf{g}'_{12})$$

$$\mathbf{v}_2 = \mathbf{v}'_2 + \frac{1+e}{2}(\mathbf{k}\cdot\mathbf{g}'_{12})\mathbf{k} - \frac{\kappa}{1+\kappa}\frac{1+\beta(\gamma)}{2}\mathbf{k}\times(\mathbf{k}\times\mathbf{g}'_{12})$$

$$\mathbf{s}_1 = \mathbf{s}'_1 + \frac{1}{1+\kappa}\frac{1+\beta(\gamma)}{2}\mathbf{k}\times\mathbf{g}'_{12}$$

$$\mathbf{s}_2 = \mathbf{s}'_2 + \frac{1}{1+\kappa}\frac{1+\beta(\gamma)}{2}\mathbf{k}\times\mathbf{g}'_{12} \qquad (3)$$

The Jacobian of this transformation is given by:

$$J(\gamma) \equiv \frac{\partial(\mathbf{v}_1,\mathbf{v}_2,\mathbf{s}_1,\mathbf{s}_2)}{\partial(\mathbf{v}'_1,\mathbf{v}'_2,\mathbf{s}'_1,\mathbf{s}'_2)} = \left\{\begin{array}{ll} e\beta_0^2 & \gamma < \gamma_0 \\ e\,|\beta(\gamma)| & \gamma > \gamma_0 \end{array}\right\} \qquad (4)$$

Notice that in the singular case $\beta_0 = 0$ or $\beta(\gamma) = 0$, when the Jacobian vanishes, the relation between the precollisional and postcollisional variables is not invertible (same as when $e = 0$ in the absence of friction), and the Boltzmann equation has to be revised.

3 The Boltzmann Equation and Hydrodynamics

Let $f(\mathbf{v},\mathbf{s},\mathbf{r},t)$ denote the single particle distribution (of the velocity and spin) function at point $\mathbf{r}$ and time t. The Boltzmann equation satisfied by f can be obtained either phenomenologically (as e.g., in [26]) or from the corresponding BBGKY hierarchy, the result being:

$$\frac{\partial f(\mathbf{v}_1, \mathbf{s}_1, \mathbf{r}, t)}{\partial t} + \mathbf{v_1} \cdot \nabla f(\mathbf{v}_1, \mathbf{s}_1, \mathbf{r}, t) = d^2 \int_{\mathbf{v} \cdot \mathbf{k} > 0} d\mathbf{v}_2 d\mathbf{s}_2 d\mathbf{k} \, (\mathbf{k} \cdot \mathbf{v}_{12}) \times$$

$$\left(\frac{1}{eJ(\gamma)} f(\mathbf{v}_1', \mathbf{s}_1', \mathbf{r}, t) f(\mathbf{v}_2', \mathbf{s}_2', \mathbf{r}, t) - f(\mathbf{v}_1, \mathbf{s}_1, \mathbf{r}, t) f(\mathbf{v}_2, \mathbf{s}_2, \mathbf{r}, t) \right) \quad (5)$$

where the Jacobian, $J(\gamma)$, is given in Eq. (4), and the integration is restricted to $\mathbf{v} \cdot \mathbf{k} > 0$ by the kinematics. Below we shall denote the right hand side of Eq. (5), which represents the collisions, by $B(f, f, \mathbf{v}, \mathbf{s})$, where the space and time dependence have been suppressed.

3.1 The Small Parameters and the Hydrodynamic Equations

The basic premise of the Chapman-Enskog expansion is that the dependence of the single particle distribution function on space (i.e., $\mathbf{r}$) and time, t, can be replaced by a dependence on the 'slow' fields. In other words, when the temporal resolution exceeds a few mean free times, and the spatial resolution exceeds a few mean free paths, the (local) dynamics of the particles is only constrained by the slowly changing fields. Clearly, the Chapman-Enskog expansion is justified only when scale separation exists (this does not necessarily imply that the results of applying this method cannot be valid beyond the nominal range of validity of the assumptions, cf. e.g., [30]). Therefore, the slow (or hydrodynamic) fields should be those that correspond to conserved (or nearly conserved) entities. In the case at hand, the momentum and the number of particles (or the total mass) are conserved, hence the momentum density and the particle number (or mass) density are hydrodynamic fields. The granular temperature corresponding to the translational degrees of freedom is not conserved, but it is nearly conserved in the near-elastic, near-frictionless case, and is therefore a hydrodynamic field. In the same limit the spin degrees of freedom are decoupled from the translational degrees of freedom and the number density corresponding to each value of the spin, $\mathbf{s}$, is conserved as well, hence the infinite set of spin dependent number densities should be taken as hydrodynamic fields. In the rough limit rotation is strongly coupled to translation and the total (translational plus rotational) energy is conserved in each collision. Therefore, in the latter case one only needs the total energy, momentum and mass densities as hydrodynamic fields. However, no harm is done by exaggerating the number of hydrodynamic fields, hence even in this case one can use the above set of spin dependent densities (which can be used to calculate the rotational temperature as the average of the square of the of the fluctuating spin), and in this case one can retain the translational granular temperature as well. All in all, the general set of hydrodynamic fields that is appropriate for all cases is that corresponding to the smooth elastic limit. They are given by:

$$\mathbf{V}\left(\mathbf{r},t\right) \equiv \frac{1}{n} \int \mathbf{v} f\left(\mathbf{v},\mathbf{s},\mathbf{r},t\right) d\mathbf{v} d\mathbf{s}$$

$$T\left(\mathbf{r},t\right) \equiv \frac{1}{n} \int u^2 f\left(\mathbf{v},\mathbf{s},\mathbf{r},t\right) d\mathbf{v} d\mathbf{s}$$

$$n\left(\mathbf{s},\mathbf{r},t\right) \equiv \int f\left(\mathbf{v},\mathbf{s},\mathbf{r},t\right) d\mathbf{v} \tag{6}$$

where $\mathbf{u}$ is the peculiar velocity $\mathbf{u} = \mathbf{v} - \mathbf{V}\left(\mathbf{r},t\right)$. Upon integrating $n(\mathbf{s},\mathbf{r},t)$ over the spin one obtains the number density:

$$n\left(\mathbf{r},t\right) = \int n\left(\mathbf{s},\mathbf{r},t\right) d\mathbf{s} = \int f\left(\mathbf{v},\mathbf{s},\mathbf{r},t\right) d\mathbf{v} d\mathbf{s} \tag{7}$$

Another field of interest is the velocity field corresponding to particles of spin $\mathbf{s}$. It is defined by:

$$\mathbf{V}\left(\mathbf{s},\mathbf{r},t\right) = \frac{1}{n\left(\mathbf{s},\mathbf{r},t\right)} \int \mathbf{v} f\left(\mathbf{v},\mathbf{s},\mathbf{r},t\right) d\mathbf{v} \tag{8}$$

The latter field is not a hydrodynamic field, as it does not correspond to a conserved entity in the smooth elastic limit, and indeed, like any such field it should be enslaved to the slow fields (one of the above mentioned assumptions of the Chapman-Enskog expansion) i.e., expressible in terms of these fields (see below). The resulting expression qualifies as an additional constitutive relation.

The hydrodynamic (or continuum equations) are (standardly) obtained by multiplying both sides of the Boltzmann equation by 1, $\mathbf{v}$, and $\mathbf{v}^2$, respectively, and integrating over the speed and the spin variables, except for the case of '1', where only an integration over the speed is performed (to obtain an equation for $n(\mathbf{s},\mathbf{r},t)$). The result is:

$$n\frac{DV_\alpha}{Dt} + \frac{\partial}{\partial r_\beta} P_{\alpha\beta} = 0$$

$$n\frac{DT\left(\mathbf{r},t\right)}{Dt} + 2\frac{\partial V_\alpha}{\partial r_\beta} P_{\alpha\beta} + 2\frac{\partial Q_\alpha}{\partial r_\alpha} = -2\Gamma$$

$$\frac{Dn\left(\mathbf{s}\right)}{Dt} + n\left(\mathbf{s}\right)\nabla \cdot \mathbf{V} = \int B\left(f,f,\mathbf{v},\mathbf{s}\right) d\mathbf{v} - \nabla \cdot \left(n\left(\mathbf{s}\right) \delta\mathbf{V}\left(\mathbf{s}\right)\right) \tag{9}$$

where $\frac{D}{Dt} \equiv \frac{\partial}{\partial t} + \mathbf{V}(\mathbf{r},t) \cdot \nabla$ is the material derivative, and the summation convention is assumed. Also, $\delta\mathbf{V}\left(\mathbf{s}\right) = \mathbf{V}\left(\mathbf{s}\right) - \mathbf{V}$ is the relative velocity between the particles of spin $\mathbf{s}$ and the overall velocity field (notice that the explicit dependence on $\mathbf{r}$ and t has been suppressed for notational simplicity). The stress tensor is given by: $P_{\alpha\beta} = \int u_\alpha u_\beta f d\mathbf{u} d\mathbf{s}$, the heat flux vector is given by: $Q_\alpha = \frac{1}{2} \int u_\alpha u^2 f d\mathbf{u} d\mathbf{s}$, and the energy sink term is:

$$\Gamma = \frac{d^2}{2} \int_{\mathbf{v} \cdot \mathbf{k} > 0} d\mathbf{v}_1 \, d\mathbf{v}_2 \, d\mathbf{s}_1 \, d\mathbf{s}_2 \, d\mathbf{k} \, v_1^2 \, (\mathbf{k} \cdot \mathbf{v}_{12}) \times$$

$$\left(\frac{1}{eJ(\gamma)} f(\mathbf{v}_1', \mathbf{s}_1', \mathbf{r}, t) \, f(\mathbf{v}_2', \mathbf{s}_2', \mathbf{r}, t) - f(\mathbf{v}_1, \mathbf{s}_1, \mathbf{r}, t) \, f(\mathbf{v}_2, \mathbf{s}_2, \mathbf{r}, t) \right) \quad (10)$$

Since clearly $\int B(f, f, \mathbf{v}, \mathbf{s}) \, d\mathbf{v} d\mathbf{s} = 0$ (number conservation) and also $\int \nabla \cdot (n(\mathbf{s}) \delta \mathbf{V}(\mathbf{s})) \, d\mathbf{s} = 0$ (this reduces to a surface integral and for $s \to \infty$, $n(\mathbf{s}) \mathbf{V}(\mathbf{s})$ are assumed to vanish), it follows from the third line in Eq. (9) that $\frac{Dn}{Dt} + n \nabla \cdot \mathbf{V} = 0$, i.e., one regains the total number density conservation.

4 The Perturbative Expansion

There are several ways to design a perturbative expansion for the distribution function in the present case; each of the expansions uses a different zeroth order limit, around which it is defined. It is preferred that this limit is an exact solution of the Boltzmann equation. In the realm of smooth friction-less granular systems there have been two such expansions. One was [31] the equilibrium state corresponding to the limit of zero inelasticity ($e = 1$) and no-gradients (i.e., the Knudsen number, K, was taken to be zero) and, the other [32] employed the (local) homogeneous cooling state (with $K = 0$) as a zeroth order. These two approaches agree with each other in the common domain of validity (near elastic collisions) and both can be used as zeroth order solutions in the presence of weak friction (i.e., $\beta \approx -1$). Another limit at which a solution of the Boltzmann equation can be identified is that of rough molecules ($\beta = 1$), in which case there is a zeroth order solution in which the rotational and translational temperature are equal to each other. The latter has been employed to describe molecules, cf. [26]. One of the important differences between the smooth particle limit and the rough limit is that in the former the spin degrees of freedom are decoupled from the translational degrees of freedom to zeroth order in the interaction, and serve as independent variables, whereas in the latter case the spin degrees of freedom are strongly coupled to the translational degrees of freedom (the conserved energy in collisions is the sum of the translational and rotational energies) and thus the spin degrees of freedom are not independent. For this reason, and since the smooth limit is the most studied case, we chose to focus on the smooth equilibrium limit.

Consider Eq. (2). When $\beta_0 = -1$, it follows that $\beta(\gamma) = -1$ for all allowed values of γ_0, and the collision is always "smooth". Energy is conserved if, in addition, $e = 1$. Next, consider the near-smooth case, $\beta_0 \approx -1$. Clearly, following Eq. (2), $\beta(\gamma) = \beta_0$ unless $\mu \cot(\gamma)$ is sufficiently small, i.e., either μ or $\cot(\gamma)$ is small. When μ is taken to be $O(1)$, $\beta(\gamma) = -1 + \frac{1+\kappa}{\kappa}(1+e)\mu \cot(\gamma)$ for γ close to $\frac{\pi}{2}$, i.e., for near grazing collisions. On the other hand, when μ is very small, this equality can hold for almost all values of γ (except for γ near zero. This implies that the small parameters corresponding to friction can be taken to be either $\epsilon_3 \equiv 1 - \beta_0^2$ and μ, or just ϵ_3, assuming μ to be $O(1)$. With the first choice, both the Coulomb and the non-Coulombic friction are small and the transition angle between them is finite, whereas the second choice assumes an $O(1)$ Coulomb friction (and the transition angle is near grazing). These two choices of expansion near the smooth limit give different expansions. Below, we choose to study the case of finite Coulomb friction. In this case, the small parameters are: the degree of normal inelasticity, $\epsilon \equiv 1 - e^2$, the small parameter corresponding to the friction, ϵ_3, and the Knudsen number, $K \equiv \frac{\ell}{L}$, where $\ell = \frac{1}{\pi n d^2}$ is the mean free path (n being the number density) and L is a macroscopic scale. The hydrodynamic fields are given in Eq. (6). Notice that in the considered limit, the system can be regarded as a mixture of particles characterized by their respective spins.

The zeroth order distribution function, f_0, (which solves the Boltzmann equation for $\epsilon = \epsilon_3 = K = 0$), is given by: $f_0 (\mathbf{u}, \mathbf{s}, \mathbf{r}, t) = f_M (\mathbf{u}, \mathbf{r}, t) \frac{n(\mathbf{s}, \mathbf{r}, t)}{n(\mathbf{r}, t)}$, where $f_M(\mathbf{u}, \mathbf{r}, t) = n(\mathbf{r}, t) \left(\frac{3}{2\pi T(\mathbf{r}, t)} \right)^{\frac{3}{2}} e^{-\frac{3u^2}{2T(\mathbf{r}, t)}}$, $\mathbf{u} \equiv \mathbf{v} - \mathbf{V}(\mathbf{r}, t)$ is the peculiar velocity, and the temperature field is denoted by $T(\mathbf{r}, t)$. The full single particle distribution function $f(\mathbf{v}, \mathbf{s}, \mathbf{r}, t)$ can be written as

$$f (\mathbf{v}, \mathbf{s}, \mathbf{r}, t) = f_0 (\mathbf{u}, \mathbf{s}, \mathbf{r}, t) (1 + \phi (\mathbf{u}, \mathbf{s}, \mathbf{r}, t)) \tag{11}$$

and ϕ is assumed to be expansible as follows:

$$\phi (\mathbf{u}, \mathbf{s}, \mathbf{r}, t) =$$
$$K\phi_K (\mathbf{u}, \mathbf{s}, \mathbf{r}, t) + \epsilon\phi_\epsilon (\mathbf{u}, \mathbf{s}, \mathbf{r}, t) + \epsilon_3\phi_3 (\mathbf{u}, \mathbf{s}, \mathbf{r}, t) + K\epsilon\phi_{K\epsilon} (\mathbf{u}, \mathbf{s}, \mathbf{r}, t)$$
$$+ K\epsilon_3\phi_{K3} (\mathbf{u}, \mathbf{s}, \mathbf{r}, t) + \epsilon\epsilon_3\phi_{\epsilon 3} (\mathbf{u}, \mathbf{s}, \mathbf{r}, t) + \epsilon_3^2\phi_{33} (\mathbf{u}, \mathbf{s}, \mathbf{r}, t)$$
$$+ \epsilon^2\phi_{\epsilon\epsilon} (\mathbf{u}, \mathbf{s}, \mathbf{r}, t) + K^2\phi_{KK} (\mathbf{u}, \mathbf{s}, \mathbf{r}, t) + \dots \tag{12}$$

Since $f_0 (\mathbf{u}, \mathbf{s}, \mathbf{r}, t)$ satisfies $\int d\mathbf{v} f_0 (\mathbf{u}, \mathbf{s}, \mathbf{r}, t) = n(\mathbf{s}, \mathbf{r}, t)$, and has (by construction) the correct temperature, (spin dependent) number density and average velocity dependence, it follows that $\int f_0 (\mathbf{u}, \mathbf{s}, \mathbf{r}, t) \phi (\mathbf{u}, \mathbf{s}, \mathbf{r}, t) d\mathbf{v} = 0$, for any value of $\mathbf{s}$, and $\int f_0 (\mathbf{u}, \mathbf{s}, \mathbf{r}, t) u^2 \phi (\mathbf{u}, \mathbf{s}, \mathbf{r}, t) d\mathbf{v} d\mathbf{s} = 0$. Also, notice that, for each value of the spin: $\mathbf{V}(\mathbf{r}, t) = \frac{1}{n(\mathbf{s}, \mathbf{r}, t)} \int f_0 (\mathbf{u}, \mathbf{s}, \mathbf{r}, t) \mathbf{v} d\mathbf{v}$, by construction. Next, using Eq. (8), define the spin dependent velocity field by: $\mathbf{V}(\mathbf{s}, \mathbf{r}, t) \equiv \frac{1}{n(\mathbf{s}, \mathbf{r}, t)} \int \mathbf{v} f (\mathbf{u}, \mathbf{s}, \mathbf{r}, t) d\mathbf{v}$. It follows that the velocity difference between the particles of spin $\mathbf{s}$ and the average velocity at the same point is given by:

$$\delta\mathbf{V}(\mathbf{s}) = \frac{1}{n(\mathbf{s}, \mathbf{r}, t)} \int f_0 (\mathbf{u}, \mathbf{s}, \mathbf{r}, t) \phi (\mathbf{u}, \mathbf{s}, \mathbf{r}, t) \mathbf{v} \, d\mathbf{v} \tag{13}$$

It is convenient to rescale the variables as follows: lengths are rescaled by a macroscopic length scale, L (understood to exceed the mean free path), spatial gradients are rescaled as $\nabla = \frac{1}{L}\tilde{\nabla}$, and the peculiar velocity is rescaled as $\tilde{\mathbf{u}} = \left(\frac{3}{2T} \right)^{\frac{1}{2}} \mathbf{u}$. The spin $\mathbf{s}$ is not rescaled, and the contact relative velocity is rescaled as $\widetilde{\mathbf{g}}_{12} = \widetilde{\mathbf{v}}_{12} + \left(\frac{3}{2T}\right)^{\frac{1}{2}} \mathbf{k} \times \mathbf{s}_{12}$. The rescaled distribution function, $\widetilde{f}$, is given by: $f = n \left(\frac{3}{2T}\right)^{\frac{3}{2}} \widetilde{f}\left(\widetilde{\mathbf{u}}, \mathbf{s}\right)$. It follows that $\widetilde{f}_0\left(\widetilde{\mathbf{u}}, \mathbf{s}\right) = \frac{1}{\pi^{\frac{3}{2}}} e^{-\mathbf{u}^2} \frac{n(\mathbf{s})}{n}$ and $f d\mathbf{u} d\mathbf{s} = n \widetilde{f} d\widetilde{\mathbf{u}} d\mathbf{s}$.

The perturbative solution of the Boltzmann equation, in powers of K, ϵ, and ϵ_3, involves the repeated solution of the equation $\widetilde{L}\phi_{\text{given order}} = R$, where, the value of ϕ, to a given order in the small parameters, is determined by a right hand side term that depends on previous orders, and the linearized Boltzmann operator, $\widetilde{L}$, is given by:

$$\widetilde{L}\phi = \frac{1}{\pi^{\frac{5}{2}}} \int d\tau_2 d\mathbf{k} \left(\mathbf{k}\cdot\widetilde{\mathbf{u}}_{12}\right) \frac{n(\mathbf{s}_2)}{n} e^{-\mathbf{u}_2^2} \left(\phi\left(\tau_1'\right) + \phi\left(\tau_2'\right) - \phi\left(\tau_1\right) - \phi\left(\tau_2\right)\right)$$

where τ' and τ are the incoming (outgoing) velocity-spin (6 - component) vectors in a smooth elastic collision. Since individual spins are conserved in a smooth elastic collision one has $\mathbf{s}_i' = \mathbf{s}_i$ for $i = 1, 2$. The algebra involved in these inversions (which requires the use of rather high orders in expansions in Sonine polynomials; convergence is obtained at 6th order in the polynomials) is rather heavy. We have devised a computationally-aided method, using a symbolic program, to overcome this difficulty. The details are too lengthy to be reported here, and will be presented elsewhere.

Clearly the zero eigenvalues of $\widetilde{L}$ are: 1 (or any constant), $\mathbf{u}$ and u^2. However, in the present case the constants may depend on $\mathbf{s}_1$ so that if $\phi\left(\tau_1\right)$ is a solution so is: $\phi\left(\tau_1\right) + \mathbf{a}(\mathbf{s}_1)\cdot\mathbf{s}_1 + \mathbf{b}(\mathbf{s}_1)\cdot\widetilde{\mathbf{u}}_1 + c(\mathbf{s}_1)\widetilde{u}_1^2$, where $\mathbf{a}\left(\mathbf{s}_1\right)$, $\mathbf{b}(\mathbf{s}_1)$, and $c(\mathbf{s}_1)$ are arbitrary functions of $\mathbf{s}_1$, to be determined by the normalization and solubility conditions (of the above mentioned inversion). The normalization conditions require that $\phi\left(\tau_1\right)$ does not change momentum and temperature nor the spin number density: $\int \frac{n(\mathbf{s})}{n} e^{-\mathbf{u}^2} \phi\left(\tau\right) \widetilde{\mathbf{u}} d\tau = 0$, $\int \frac{n(\mathbf{s})}{n} e^{-\mathbf{u}^2} \phi\left(\tau\right) \widetilde{u}^2 d\tau = 0$, and $\int e^{-\mathbf{u}^2} \phi\left(\tau\right) d\widetilde{\mathbf{u}} = 0$. Once ϕ is known to the desired order, one computes the appropriate averages of f to obtain the constitutive relations.

5 Constitutive Relations

The results of the application of the Chapman-Enskog expansion yield the following constitutive relations, presented below to second order in the small parameters (but only to first order in the Knudsen number, K, as the $O(K^2)$ terms are identical to those corresponding to frictionless particles, cf. [33]). The various undefined prefactors below are numbers, whose values will be presented elsewhere (with some specific exceptions; see below). All results presented below are dimensional.

To the above order in the small parameters, the stress tensor is given by:

$$P_{\alpha\beta} = -nT\delta_{\alpha\beta} - 2\overline{\mu}_{K}\, n\ell\sqrt{T}\overline{\frac{\partial V_\alpha}{\partial r_\beta}} - 2\overline{\mu}_{K\epsilon}\, n\ell\epsilon\sqrt{T}\overline{\frac{\partial V_\alpha}{\partial r_\beta}} - 2\mu_{K3}\, n\ell\epsilon_3\sqrt{T}\overline{\frac{\partial V_\alpha}{\partial r_\beta}}$$

$$+\epsilon_3^2 n\left((2S_\alpha S_\beta + Y_{\alpha\beta})\,\overline{\mu}^{(a)}_{K\epsilon_3\epsilon_3} + T\overline{B_{\alpha\beta}}\,\overline{\mu}^{(b)}_{K\epsilon_3\epsilon_3}\right)$$

where:

$$B_{\alpha\beta} = \int \left(\frac{\partial}{\partial s_\alpha}\left[\frac{n\,(\mathbf{s})}{n}\,(\mathbf{S}+\mathbf{s})_\beta\right]\right)\frac{n}{n\,(\mathbf{s})}\frac{\partial}{\partial s_\gamma}\left[\frac{n\,(\mathbf{s})}{n}\,(\mathbf{S}+\mathbf{s})_\gamma\right]d\mathbf{s},$$

$\mathbf{S}(\mathbf{r},t) \equiv \frac{1}{n(\mathbf{r},t)}\int \mathbf{s}\, n(\mathbf{s},\mathbf{r},t)\,d\mathbf{s}$ is the spin density, and an overlined tensor, $\overline{\mathbf{A}}$, denotes the traceless symmetric part of the tensor $\mathbf{A}$: $\overline{A}_{\alpha\beta} \equiv \frac{1}{2}(A_{\alpha\beta} + A_{\beta\alpha}) - \frac{1}{3}\delta_{\alpha\beta}Tr\mathbf{A}$, where $Tr\mathbf{A}$ is the trace of the tensor $\mathbf{A}$. The Einstein summation convention is assumed.

The heat flux is given by:

$$Q_\alpha = -2\left(\overline{\kappa}_K + \epsilon\overline{\kappa}_{K\epsilon} + \epsilon_3\kappa_{K3}\right)n\ell\sqrt{T}\frac{\partial T}{\partial r_\alpha} - 2\left(\epsilon\overline{\lambda}_{K\epsilon} + \epsilon_3\lambda_{K3}\right)n\ell T^{\frac{3}{2}}\frac{\partial n}{\partial r_\alpha}$$

$$-\epsilon_3 nr_{K3}\ell T^{\frac{3}{2}}\int\left(\frac{\partial}{\partial s_\gamma}\left[\frac{n\,(\mathbf{s})}{n}\,(\mathbf{S}+\mathbf{s})_\gamma\right]\right)\frac{\partial}{\partial r_\alpha}\log n\,(\mathbf{s})\,d\mathbf{s}$$

$$-\epsilon_3 ns_{K3}\ell T^{\frac{3}{2}}\int\overline{\frac{\partial}{\partial s_\alpha}\left[\frac{n\,(\mathbf{s})}{n}\,(\mathbf{S}+\mathbf{s})_\beta\right]}\frac{\partial}{\partial r_\beta}\log n\,(\mathbf{s})\,d\mathbf{s}$$

The energy sink term is given by:

$$\Gamma = \frac{nT^{\frac{3}{2}}}{\sqrt{6}\ell}\left[\epsilon\frac{2}{3}\sqrt{\frac{2}{\pi}} - \epsilon^2 1.3435 \times 10^{-2} + \epsilon_3\frac{2}{3}\sqrt{\frac{2}{\pi}}\frac{\kappa}{1+\kappa}\right.$$

$$-\epsilon\epsilon_3\frac{\kappa}{1+\kappa}\frac{4}{3\pi^2}\times 0.2492 + \epsilon_3^2\left(\frac{\kappa}{1+\kappa}\right)^2\left(\frac{1}{3}\sqrt{\frac{1}{2\pi}}\frac{1}{\kappa} - 0.01344\right)$$

$$\left.-\epsilon_3^2\frac{1}{3}\sqrt{\frac{1}{2\pi}}\left(\frac{\kappa}{1+\kappa}\right)^2\frac{1}{T}\left(T_{\text{rot}} + 2S^2\right)\right] \tag{14}$$

where T_{rot} is the rotational temperature, defined as:

$$T_{\text{rot}}(\mathbf{r},t) \equiv \frac{1}{n(\mathbf{r},t)}\int (\mathbf{s} - \mathbf{S}(\mathbf{r},t))^2\, n(\mathbf{s},\mathbf{r},t)\,d\mathbf{s}.$$

The relative velocity of particles of spin $\mathbf{s}$ with respect to the (local) velocity field, which appears in the equation of motion for $n(\mathbf{s})$, is given by:

$$n\left(\mathbf{s}\right)\delta V_{\alpha}\left(\mathbf{s}\right)=n\ell\sqrt{T}w_{\kappa}\frac{\partial}{\partial r_{\alpha}}\left(\frac{n\left(\mathbf{s}\right)}{n}\right)+\epsilon n\ell\sqrt{T}w_{\kappa_{\epsilon}}\frac{\partial}{\partial r_{\alpha}}\left(\frac{n\left(\mathbf{s}\right)}{n}\right)$$

$$+\epsilon_{3}n\ell\sqrt{T}\times\left\{w_{\kappa3}^{(1)}\frac{\partial}{\partial r_{\alpha}}\left(\frac{n\left(\mathbf{s}\right)}{n}\right)+w_{\kappa3}^{(2)}\frac{\partial\log T}{\partial r_{\alpha}}\frac{\partial}{\partial s_{\beta}}\left[\left(\mathbf{S}+\mathbf{s}\right)_{\beta}\frac{n\left(\mathbf{s}\right)}{n}\right]\right.$$

$$+w_{\kappa3}^{(3)}\frac{\partial\log T}{\partial r_{\beta}}\frac{\partial}{\partial s_{\alpha}}\left[\frac{n\left(\mathbf{s}_{1}\right)}{n}\left(\mathbf{S}+\mathbf{s}\right)_{\beta}\right]+w_{\kappa3}^{(4)}\frac{\partial}{\partial s_{\beta}}\left[\left(\frac{\partial}{\partial r_{\alpha}}\left(\frac{n\left(\mathbf{s}\right)}{n}\right)\right)\left(\mathbf{S}+\mathbf{s}\right)_{\beta}\right]$$

$$+w_{\kappa3}^{(5)}\left[\frac{\partial}{\partial s_{\beta}}\left(\frac{n\left(\mathbf{s}\right)}{n}\right)\right]\frac{\partial\mathbf{S}}{\partial r_{\alpha}}+w_{\kappa3}^{(6)}\frac{\partial\log n}{\partial r_{\alpha}}\frac{\partial}{\partial s_{\beta}}\left[\left(\mathbf{S}+\mathbf{s}\right)_{\beta}\frac{n\left(\mathbf{s}\right)}{n}\right]$$

$$+w_{\kappa3}^{(7)}\left[\frac{\partial}{\partial r_{\alpha}}\left(\log\frac{n}{n\left(\mathbf{s}\right)}\right)\right]\frac{\partial}{\partial s_{\beta}}\left[\left(\mathbf{S}+\mathbf{s}\right)_{\beta}\frac{n\left(\mathbf{s}\right)}{n}\right]+w_{\kappa3}^{(8)}\frac{\partial^{2}}{\partial r_{\alpha}\partial s_{\beta}}\left[\left(\mathbf{S}+\mathbf{s}\right)_{\beta}\frac{n\left(\mathbf{s}\right)}{n}\right]$$

$$+w_{\kappa3}^{(9)}\frac{\partial}{\partial r_{\beta}}\frac{\partial}{\partial s_{\alpha}}\left[\frac{n\left(\mathbf{s}\right)}{n}\left(\mathbf{S}+\mathbf{s}\right)_{\beta}\right]+w_{\kappa3}^{(10)}\frac{\partial\log n}{\partial r_{\beta}}\frac{\partial}{\partial s_{\alpha}}\left[\frac{n\left(\mathbf{s}\right)}{n}\left(\mathbf{S}+\mathbf{s}\right)_{\beta}\right]\right\}$$

Finally, the term $\int B\left(f,f,\mathbf{v},\mathbf{s}\right)d\mathbf{v}$, which appears in the equation of motion for $n(\mathbf{s})$, is given by:

$$\int B\left(f,f,\mathbf{v},\mathbf{s}\right)d\mathbf{v}=\frac{\epsilon_{3}}{\ell}\left(\frac{2T}{3}\right)^{\frac{1}{2}}\frac{1}{3}\sqrt{\frac{2}{\pi}}\frac{1}{1+\kappa}\frac{\partial}{\partial s_{\alpha}}\left[n\left(\mathbf{s}\right)\left(\mathbf{S}+\mathbf{s}\right)_{\alpha}\right]$$

$$+\frac{\epsilon\epsilon_{3}}{\ell}d_{\epsilon\epsilon3}\left(\frac{2T}{3}\right)^{\frac{1}{2}}\frac{1}{1+\kappa}\frac{\partial}{\partial s_{\alpha}}\left[\left(\mathbf{S}+\mathbf{s}\right)_{\alpha}n\left(\mathbf{s}\right)_{\alpha}\right]$$

$$+\frac{\epsilon_{3}^{2}}{\ell}\left(\frac{2T}{3}\right)^{\frac{1}{2}}\frac{1}{24}\sqrt{\frac{2}{\pi}}\frac{1}{\left(1+\kappa\right)^{2}}\times$$

$$\frac{\partial}{\partial s_{\alpha}}\left[n(\mathbf{s})(3+2\kappa)\left(\mathbf{S}+\mathbf{s}\right)_{\alpha}+\left(2T+\left(\mathbf{S}+\mathbf{s}\right)^{2}+T_{\rm rot}\right)\frac{\partial}{\partial s_{\alpha}}\left(n(\mathbf{s})\right)\right]$$

$$+\frac{\epsilon_{3}^{2}}{\ell}\left(\frac{2T}{3}\right)^{\frac{1}{2}}\frac{1}{6}\frac{1}{1+\kappa}\left(\frac{1}{\pi}\right)^{2}\times$$

$$\frac{\partial}{\partial s_{\alpha}}\left[\left(a_{00}^{(2)}\kappa-a_{01}^{(2)}\right)n(\mathbf{s})(S_{\alpha}+s_{\alpha})+b_{0}^{(2)}\left[\frac{\partial}{\partial s_{\beta}}\left(n(\mathbf{s})(S_{\beta}+s_{\beta})\right)\right]\left(S_{\alpha}+s_{\alpha}\right)\right.$$

$$\left.+c_{0}^{(2)}\frac{\partial}{\partial s_{\alpha}}\left(n(\mathbf{s})\left(S_{\beta}+s_{\beta}\right)\right)(S_{\beta}+s_{\beta})\right]$$

where $d_{\epsilon\epsilon3}=4.6522\times10^{-3}$, $a_{00}^{(2)}=0.9065$, $a_{01}^{(2)}=4.522\times10^{-2}$, $b_{0}^{(2)}=3.833\times10^{-2}$, and $c_{0}^{(2)}=4.253\times10^{-3}$. Notice that the spin number densities satisfy a (generalized) diffusion equation, as expected for a "mixture". The nature of this equation is made clearer in the analysis of the HCS presented next.

6 The Near Frictionless Near Smooth HCS

In the present section we apply the above equations of motion and constitutive relations to the study of the near frictionless near smooth (i.e. $\beta_0 \approx -1$ and $\mu = O(1)$) homogeneous cooling state (HCS). In this state all spatial derivatives vanish, as does the momentum density. One therefore needs to study the equations of motion for the translational temperature and the spin dependent densities alone. The equations (9) read:

$$n\frac{dT(t)}{dt} = -2\Gamma$$

$$\frac{\partial n(\mathbf{s})}{\partial t} = \int B(f, f, \mathbf{v}, \mathbf{s})\, d\mathbf{v}$$

(as a consequence $\frac{dn}{dt} = 0$). Upon multiplying the second equation by $\mathbf{s}$ or by $(\mathbf{s} - \mathbf{S})^2$, integrating over $\mathbf{s}$, performing some trivial integrations by parts, and changing time to $\tau \equiv \int_0^t \frac{T^{\frac{1}{2}}}{\ell} dt'$ (by now, a standard transformation), one obtains the following exact equations of motion:

$$\frac{dT}{d\tau} = -\alpha T + \beta T_{\text{rot}} + 2\beta S^2$$

$$\frac{dT_{\text{rot}}}{d\tau} = \frac{3}{\kappa^2}\beta T - \left(\delta - \frac{1}{4}\mu - \frac{3}{2\kappa^2}\beta\right) T_{\text{rot}} + \mu S^2$$

$$\frac{dS^2}{d\tau} = -\delta S^2 \qquad\qquad (15)$$

where α and β are read off Eq. (14) and

$$\mu = \frac{6}{\kappa^2}\beta + \epsilon_3^2\sqrt{\frac{2}{3}\frac{1}{\pi^2}\frac{1}{1+\kappa}}\left(\frac{4}{3}b_0^{(2)} + \frac{20}{9}c_0^{(2)}\right)$$

$$\delta = -\frac{1}{\kappa^2}\beta + 2\epsilon_3\sqrt{\frac{2}{3}\frac{1}{\pi^2}\frac{1}{1+\kappa}}\left[X - \epsilon_3\left(\frac{1}{6}b_0^{(2)} + \frac{5}{18}c_0^{(2)}\right)\right]$$

where

$$X = \frac{1}{2}\sqrt{2\pi^3} + \epsilon\pi^2 d_{\epsilon\epsilon_3} + \frac{1}{6}\epsilon_3\left(a_{00}^{(2)}\kappa - a_{01}^{(2)} + \frac{1}{4}\sqrt{2\pi^3}\frac{3+2\kappa}{1+\kappa}\right)$$

Eq. (15) is linear in T, T_{rot} and S^2. The three eigenvalues of this system are negative. The long-time decay rate of T and T_{rot} follows Haff's law, and:

$$\lim_{t\to\infty}\frac{T_{\text{rot}}}{T} \equiv r = \frac{\left(\left(\alpha - \delta + \frac{\mu}{4} + \frac{3\beta}{4\kappa^2}\right) + \sqrt{\left(\alpha - \delta + \frac{\mu}{4} + \frac{3\beta}{4\kappa^2}\right)^2 + \frac{12\beta^2}{\kappa^2}}\right)}{2\beta}$$

Also:

$$\lim_{t\to\infty}\frac{S^2}{T} = 0$$

Since S^2 decays to zero faster than T and $T_{\rm rot}$ it is justified to consider the asymptotic time dynamics for the case $\mathbf{S} = 0$. To simplify matters even further, we shall assume that the spin distribution is isotropic, i.e., $n(\mathbf{s}, t)$ depends on the scalar s (and time) alone. With these simplifications the equation of motion for $n(s, t)$ reads:

$$\frac{\partial n}{\partial \tau} = A_1 \left(s \frac{\partial n}{\partial s} + 3n \right) + A_2 (2T + T_{\rm rot}) \left(\frac{2}{s} \frac{\partial n}{\partial s} + \frac{\partial^2 n}{\partial s^2} \right) + A_3 \left(4s \frac{\partial n}{\partial s} + s^2 \frac{\partial^2 n}{\partial s^2} \right),$$

where

$$A_1 = \epsilon_3 \sqrt{\frac{2}{3} \frac{1}{\pi^2} \frac{1}{1 + \kappa}} \left(X + \frac{1}{2} \epsilon_3 b_0^{(2)} \right)$$

$$A_2 = \frac{1}{4\kappa^2} \beta$$

$$A_3 = \frac{1}{4\kappa^2} \beta + \epsilon_3^2 \sqrt{\frac{2}{3} \frac{1}{\pi^2} \frac{1}{1 + \kappa}} \left(\frac{1}{6} b_0^{(2)} + \frac{1}{9} c_0^{(2)} \right)$$

It is convenient to define a scalar spin density, $N(s, t)$ by $N(s, t) \equiv 4\pi s^2 n(s, t)$. Clearly: $\int N(s, t) ds = n$, and $\frac{1}{n} \int s^2 N(s, t) ds = T_{\rm rot}(t)$. It follows from the above equation that:

$$\frac{\partial N}{\partial \tau} = \frac{\partial}{\partial s} \left[A_1 s N + (2T + T_{\rm rot}) A_2 \left(\frac{\partial N}{\partial s} - \frac{2N}{s} \right) + A_3 \left(s^2 \frac{\partial N}{\partial s} - 2sN \right) \right] \tag{16}$$

In the large time limit, this equation possesses a scaling solution of the form $N(s, \tau) = \frac{n}{\sqrt{T}} F(\zeta)$ where $\zeta = \frac{s}{\sqrt{T}}$, with $T_{\rm rot}/T$ replaced by its asymptotic value r. $F(\zeta)$ satisfies:

$$F_{\zeta\zeta} + \frac{F_\zeta}{\zeta} \frac{\left(A_1 - \frac{\alpha}{2} + \frac{\beta}{2} r \right) \zeta^2 - 2(2 + r) A_2}{(2 + r) A_2 + A_3 \zeta^2}$$

$$+ \frac{F}{\zeta^2} \frac{\left(A_1 - 2A_3 - \frac{\alpha}{2} + \frac{\beta}{2} r \right) \zeta^2 + 2(2 + r) A_2}{(2 + r) A_2 + A_3 \zeta^2} = 0 \tag{17}$$

In the large ζ limit the above equation reduces to

$$F_{\zeta\zeta} + \frac{F_\zeta}{\zeta} C + \frac{F}{\zeta^2} (C - 2) = 0 \tag{18}$$

with $C \equiv \frac{A_1 - \frac{\alpha}{2} + \frac{\beta}{2} r}{A_3}$. Following Eq. (18) the normalizable solution of Eq. (17) satisfies $F(\zeta) \sim \zeta^{2-C}$ i.e. the distribution decays as a power law: $n(s) \sim \frac{n}{4\pi T^{3/2}} \left(\frac{s}{\sqrt{T}} \right)^{-C}$. For instance for $\epsilon = \epsilon_3 = 0.1$ $F(\zeta) \sim \zeta^{-5.935}$. In figure 1 the solution of Eq. (17) is compared to a Gaussian distribution.

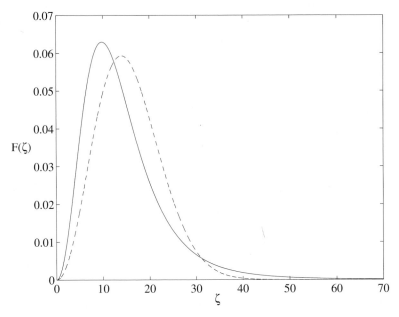

Fig. 1. The rescaled spin distribution, $F(\zeta)$, for the homogeneous cooling state (full line) and a Gaussian distribution (dashed) with the same value of T_{rot}/T, for the following values of the small parameters: $\epsilon = \epsilon_3 = 0.1$. Recall that the rescaled spin ζ equals $s/\sqrt{T}$.

7 Concluding Remarks

It has been shown that the dynamics of frictional spheres near the smooth limit requires the consideration of an infinite number of fields, the spin-dependent densities, in addition to the standard hydrodynamic fields. An appropriate Chapman-Enskog expansion yields the hydrodynamic equations, which explicitly couple the spin and translational degrees of freedom. These equations are rather cumbersome, but they simplify in particular cases. The Homogeneous Cooling State (HCS) of frictional granular spheres was studied. It turns out that (expectedly) the average spin decays with time, as do the two temperatures (we did not study the parameter range in which the spin temperature converges to a constant [19]). Contrary to assumptions in most theories of frictional granular matter, the spin distribution function in the considered case is very different from Maxwellian. As the hydrodynamics was developed on the basis of the Boltzmann equation, without the Enskog correction, the stress tensor (being the average of a symmetric entity) is symmetric, i.e., one does not obtain a micropolar or Cosserat theory. Finite density corrections will be considered elsewhere. Many open questions still remain, among them the effects of friction on shear and vibrated flows, the matching of the expansion presented above with results obtained from an expansion around the rough limit, and, in general, the effects of friction in many types of granular flows.

Acknowledgments

One of us (I.G.) would like to gratefully acknowledge support from the Israel Science Foundation, grant no. 53-01, the US-Israel Binational Science Foundation, grant no. 1999-417, and INTAS, grant no. 577.

References

1. S. J. Moon, J. B. Swift, and H. L. Swinney. Role of friction in pattern formation in oscillated granular layers. *cond-mat/0308541*, 2003.
2. S. Nasuno, A. Kudrolli, and J. P. Gollub. Friction in granular layers: Hysteresis and precursors. *Phys. Rev. Lett.*, 79:949–952, 1997.
3. S. Luding, M. Huthmann, S. McNamara, and A. Zippelius. Homogeneous cooling of rough dissipative particles: Theory and simulations. *Phys. Rev. E*, 58:3416–3425, 1998.
4. M. Huthmann and A. Zippelius. Dynamics of inelastically colliding rough spheres: Relaxation of translational and rotational energy. *Phys. Rev. E*, 56(6):6275–6278, 1998.
5. J. T. Jenkins and M. W. Richman. Kinetic theory for plane shear flows of a dense gas of identical, rough, inelastic, circular disks. *Phys. of Fluids*, 28:3485–3494, 1985.
6. C. K. K. Lun and S. B. Savage. A simple kinetic theory for granular flow of rough, inelastic, spherical particles. *J. Appl. Mech.*, 54(1):47–53, 1987.
7. P. C. Johnson, P. Nott, and R. Jackson. Frictional-collisional equations of motion for particulate flows and their application to chutes. *J. Fluid Mech.*, 210:501–535, 1990.
8. H. M. Jaeger, C.-H. Liu, S. R. Nagel, and T. A. Witten. Friction in granular flows. *Europhys. Lett.*, 11(7):619–624, 1990.
9. C. K. K. Lun. Kinetic theory for granular flow of dense, slightly inelastic, slightly rough spheres. *J. Fluid Mech.*, 233:539–559, 1991.
10. S. Abu-Zaid and G. Ahmadi. Analysis of rapid shear flows of granular materials by a kinetic model including frictional losses. *Powder Technol.*, 77(1):7–17, 1993.
11. S. Luding. Granular materials under vibration: Simulations of rotating spheres. *Phys. Rev. E*, 52(4):4442–4457, 1995.
12. J. T. Jenkins and M. Y. Louge. On the flux of fluctuating energy in a collisional grain flow at a flat, frictional wall. *Physics of Fluids*, 9(10):2835–2840, 1997.
13. M. Alam and P.R. Nott. The influence of friction on the stability of unbounded granular shear flow. *J. Fluid Mech.*, 343:267–301, 1997.
14. M. Müller, S. Luding, and H. J. Herrmann. Simulations of vibrated granular media in 2d and 3d. In D. E. Wolf and P. Grassberger, editors, *Friction, Arching and Contact Dynamics*, pages 335–340, Singapore, 1997. World Scientific.
15. T. Elperin and E. Golshtein. Effects of convection and friction on size segregation in vibrated granular beds. *Physica A*, 247:67–78, 1997.

16. L. S. Mohan, P. R. Nott, and K. K. Rao. A frictional Cosserat model for the flow of granular materials through a vertical channel. *Acta Mech.*, 138(1-2):75–96, 1999.

17. A. Goldshtein and M. Shapiro. Mechanics of collisional motion of granular materials. Part 1. General hydrodynamic equations. *J. Fluid Mech.*, 282:75–114, 1995.

18. S. McNamara and S. Luding. Energy nonequipartition in systems of inelastic, rough spheres. *Phys. Rev. E*, 58:2247–2250, 1998.

19. O. Herbst, M. Huthmann, and A. Zippelius. Dynamics of inelastically colliding spheres with Coulomb friction: relaxation of translational and rotational energy. *Granular Matter*, 2(4):211–219, 2000.

20. S. G. Bardenhagenj, J. U. Brackbill, and D. Sulsky. Numerical study of stress distribution in sheared granular material in two dimensions. *Phys. Rev. E*, 62(3):3882–3890, 2000.

21. R. Cafiero and S. Luding. A mean field theory for a driven granular gas of frictional particles. *Physica A*, 280(1-2):142–147, 2000.

22. R. Cafiero, S. Luding, and H. J. Herrmann. Rotationally driven gas of inelastic rough spheres. *Europhys. Lett.*, 60(6):854–860, 2002.

23. H. Hayakawa N. Mitarai and H. Nakanishi. Collisional granular flow as a micropolar fluid. *Phys. Rev. Lett.*, 88(17):174301–1–4, 2002.

24. J. T. Jenkins and C. Zhang. Kinetic theory for identical, frictional, nearly elastic spheres. *Physics of Fluids*, 14(3):1228–1235, 2002.

25. I. Goldhirsch. Rapid granular flows. *Annual Reviews of Fluid Mechanics*, 35:267–293, 2003.

26. S. Chapman and T. G. Cowling. *The mathematical theory of nonuniform gases.* Cambridge University Press, London, 1960.

27. J. S. Dahler and M. Theodosopulu. The kinetic theory of dense polyatomic fluids. *Advances in Chem. Phys.*, 31:155–229, 1975.

28. I. G. Shukhman. The collisional dynamics of particles in the Saturn rings. *Astronomicheskii Journal (Russian).*, 61(5):985–1004, 1984.

29. S. Araki. The dynamics of particles disks ii: effects of spin degress of freedom. *Icarus*, 76:182–198, 1988.

30. M.-L. Tan and I. Goldhirsch. Rapid granular flows as mesoscopic systems. *Phys. Rev. Lett.*, 81(14):3022–3025, 1998.

31. N. Sela and I. Goldhirsch. Hydrodynamic equations for rapid flows of smooth inelastic spheres to Burnett order. *J. Fluid Mech.*, 361:41–74, 1998.

32. J. J. Brey, J. W. Dufty, C. S. Kim, and A. Santos. Hydrodynamics for granular flow at low density. *Phys. Rev. E*, 58(4):4638–4653, 1998.

33. N. Sela, I. Goldhirsch, and S. H. Noskowicz. Kinetic theoretical study of a simply sheared two-dimensional granular gas to Burnett order. *Phys. Fluids*, 8(9):2337, 1996.

The Physics of Overcharging

C. Holm

Max-Planck-Institut für Polymerforschung
Ackermannweg 10, 55128 Mainz – Germany
holm@mpip-mainz.mpg.de

Summary. In this article we review the two basic mechanisms responsible for overcharging a single spherical colloid in the presence of aqueous salts. The first mechanism rests on energetic arguments, and deals only with the ground state configuration of the counterions. This mechanism can be qualitatively explained in very simple terms using freshmen electrostatics, and a very good quantitative description can be obtained using Wigner crystal theory. The second mechanism is driven by a combination of excluded volume correlations and the preceeding energetic arguments, and in the following called entropic overcharging. We demonstrate the validity of the proposed mechanisms with results of molecular dynamics simulations within the primitive model of electrolytes.

1 Introduction

There has been a recent interest in the study of systems which are coupled by Coulomb interactions[1, 2], since these systems can show, under certain circumstances, a variety of surprising behaviors. For example, there are attractions between like charged objects [3], and even a charge reversal of macroions can occur when they are looked at from some distance. This means that there are more ions of the opposite charge within a certain radius around the macroion present than necessary to neutralize it. This overcompensation is called "overcharging" and will be the topic of the present article. This effect was already discovered in the beginning of 80's, both by computer simulations [4, 5] and analytical studies [6, 7]. Based on *reversed* electrophoretic mobility, some experimental [8] and numerical (MD) [9] studies provide some hints for the manifestation of overcharging and its possible experimental relevance. More recently, it has regained a considerable attention on the theoretical side [10–19].

We first present a mechanism which rests on the ground state picture of a single colloid with excess counterions in a regime where excluded volume effects play no role. The underlying physics at zero temperature of such non-neutral systems can be quantitatively explained with Wigner crystal (WC) theory [10–12, 15]. The basic concept is that the counterions form a two-dimensional lattice on the macroion surface, and when overcharging counterions are present on this layer the energy of the system is lowered compared to the neutral case. This feature can be directly and *exactly* computed for a small number of counterions at zero temperature, and was illustrated in Refs. [15]. This WC approach remains qualitatively correct for finite temperature as long as the Coulomb coupling is very high. The crystal then melts into a strongly correlated liquid, where the local order is still strong enough to lower the free energy for the overcharged state.

The situation becomes much more complicated for aqueous systems, where the coupling is weak and in addition salt is present. One-particle inhomogeneous integral equation theories can describe some of these situations fairly well [6, 7, 18, 20] but the computed correlation functions do not necessarily give direct insight into the physical mechanisms behind these effects. The presence of excluded volume interactions can lead to layering effects near the macroion, which are known from simple fluid theories. Here, due to the presence of charge carriers of both signs, this can even lead to layers of oscillating charge inversions [7, 18, 21]. For this situation we present a mechanism that gives at least qualitative inside into the physics of entropic overcharging.

2 The Energetic Ground State Approach to Overcharging

In this section we want to demonstrate that there are situations for charged colloidal objects in which one can understand the phenomenon of overcharging by very simple energetic arguments. By overcharging, in general, we mean that the bare charge of the macroion is overcompensated at some distance by oppositely charged "microions". To achieve this in nature we have to add salt to the system. For the sake of simplicity, however, we will consider non-neutral systems, because they can on a very simple basis explain why colloids prefer to be overcharged.

2.1 The Model

Our model is solely based on electrostatic energy considerations, meaning that we only look at the ground state of a system of charges. We consider a colloid of radius a with a central charge Z. In the ground state the counterions of this colloid are located on the surface, because there they are closest to the central charge. On the other hand they want to be in such a configuration that they minimize their mutual repulsion. For two, three, and four counterions these configurations correspond to a line, an equilateral triangle, and a tetrahedron, respectively, regardless of the central charge magnitude. The problem of the minimal energy configuration of electrons disposed on the surface of a sphere dates back to Thomson [22], and is actually unsolved for large N. The reason is, that there are many metastable states which differ only minimally in energy, and their number seems to grow exponentially with N. Also chemists developed the valence-shell electron-pair repulsion (VSEPR) theory [23] which uses similar arguments to predict the molecular geometry in covalent compounds, also known as the Gillespie rule.

A simple illustration of energetically driven overcharging is depicted in Fig. 1. The central charge is $+2e$, and the neutral system has two counterions of valence 1. If we add successively more counterions of the same valence, and put them on the surface such that their mutual repulsion is minimized, we can compute the total electrostatic energy according to

$$E(n) = k_B T(l_B/a)\left[-nZ_m + f(\theta_i)\right],\qquad(1)$$

where $l_B = e^2/(4\pi\epsilon_0\epsilon_r k_B T_0)$ is the Bjerrum length that is the length at which two unit charges have an interaction energy of $k_B T$, and $f(\theta_i)$ is the repulsive energy part which is only a function of the ground state configuration. We surprisingly find, that actually the minimal energy is obtained when *four* counterions are present, hence we overcharged the colloid by two counterions, or by 100 %! That is, the excess counterions gain more energy by assuming a energetically favorable configuration around the macroion than by escaping to infinity, the simple reason behind overcharging. In our example, the min-

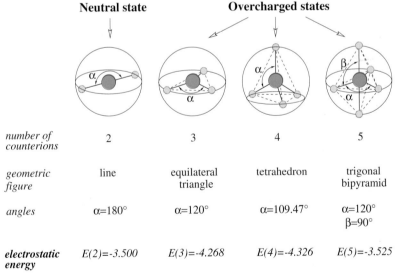

	Neutral state	Overcharged states		
number of counterions	2	3	4	5
geometric figure	line	equilateral triangle	tetrahedron	trigonal bipyramid
angles	$\alpha=180°$	$\alpha=120°$	$\alpha=109.47°$	$\alpha=120°$ $\beta=90°$
electrostatic energy	E(2)=-3.500	E(3)=-4.268	E(4)=-4.326	E(5)=-3.525

Fig. 1. Ground state configurations for two, three, four and five electrons. The corresponding geometrical figure repulsion and their typical angles are given. The electrostatic energy (in units of $k_B T l_B/a$) is given for a central charge of $+2e$.

imum is reached when four counterions are present. The colloid radius and the Bjerrum length enter as prefactors and change only the energy difference between neighboring states.

The spatial correlations of the counterions are fundamental to obtain overcharging. Indeed, if we apply the same procedure and smear Z counterions onto the surface of the colloid of radius a, we obtain for the energy

$$E = l_B \left[\frac{1}{2} \frac{Z^2}{a} - \frac{Z_m Z}{a} \right]. \tag{2}$$

The minimum is reached for $Z = Z_m$, hence no overcharging can occur.

The important message to be learned is that, from an energetic point of view, a colloid *always* tends to be overcharged by discrete charges. Other important geometries like infinite rods or infinitely extended plates cannot be treated in such a simple fashion because they are not finite in all directions. One needs therefore enough screening charges in the environment to limit the range of the interactions in the infinite directions, there is a need for a *minimal* amount of salt present to allow for overcharging[19], which is not the case for a colloid.

Obviously, for a large number of counterions the direct computation of the electrostatic energy by using the exact equation (1) becomes unfeasible. Therefore we resort to simulations for highly charged spheres.

2.2 One Colloid

The electrostatic energy as a function of the number of overcharging counterions n is displayed in Fig. 2. We note that the maximal (critical) acceptance of n (4, 6 and 8) increases with the macroionic charge Z_m (50, 90 and 180 respectively). Furthermore for fixed n, the gain in energy is always increasing with Z_m. Also, for a given macroionic charge, the gain in energy between two successive overcharged states is decreasing with n.

In the ground state the counterions are highly ordered. Rouzina and Bloomfield [24] first stressed the special importance of these crystalline arrays for interactions of multivalent ions with DNA strands, and later Shklovskii [25] showed that the Wigner crystal (WC) theory can be applied to determine the interactions in strongly correlated systems. In two recent short contributions [15, 26] we showed that the overcharging curves obtained by simulations of the ground state, like Fig. 2, can be simply explained by assuming that the energy ε per counterion on the surface of a macroion scales as $\sqrt{c}$, where c denotes the counterion concentration $c = N/A$, N is the *total* number of counterions on the surface and A the total macroion area. This can be justified by a simple argument, where each ion interacts in first approximation only with the oppositely charged background of its Wigner-Seitz (WS) cell, which can be approximated by a disk of radius h, yielding the same WS cell area.

For fixed macroion area we can write the energy per counterion as

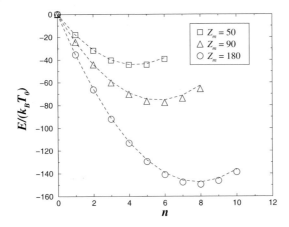

Fig. 2. Electrostatic energy (in units of $k_B T_0$) for *ground state* configurations of a single charged macroion as a function of the number of *overcharging* counterions n for three different bare charges Z_m. The neutral case was chosen as the potential energy origin, and the curves were produced using the theory of Eq. (6), compare text.

$$\varepsilon^{(h)}(N) = -\frac{\bar{\alpha}^{(h)}\ell}{\sqrt{A}}\sqrt{N} = -\bar{\alpha}^{(h)}\ell\sqrt{c}, \tag{3}$$

where $\ell = l_B Z_c^2$ and the simple hole theory gives $\bar{\alpha}^{(h)} = 2\sqrt{\pi} \approx 3.54$[16].

For an infinite plane, where the counterions form an exact triangular lattice, one obtains the same *functional* form as in Eq. (3), but the prefactor $\bar{\alpha}^{(h)}$ gets replaced by the numerical value $\bar{\alpha}^{WC} = 1.96$[27].

Not knowing the precise value of $\bar{\alpha}$ we can still use the simple scaling behavior with c to set up an equation to quantify the energy gain ΔE_1 by adding the first overcharging counterion to the colloid. To keep the OCP neutral we imagine adding a homogeneous surface charge density of opposite charge $\left(\frac{-Z_c e}{A}\right)$ to the colloid[25]. This ensures that the background still neutralizes the incoming overcharging counterion and we can apply Eq. (3). To cancel our surface charge addition we add another homogeneous surface charge density of opposite sign $\frac{Z_c e}{A}$. This surface charge does not interact with the now neutral OCP, but adds a self-energy term of magnitude $\frac{1}{2}\frac{\ell}{a}$, so that the total energy difference for the first overcharging counterion reads as

$$\Delta E_1 = (N_c + 1)\varepsilon(N_c + 1) - N_c\varepsilon(N_c) + \frac{\ell}{2a}. \tag{4}$$

By using Eq. (3) this can be rewritten as[28]

$$\Delta E_1 = -\frac{\bar{\alpha}\ell}{\sqrt{A}}\left[(N_c + 1)^{3/2} - N_c^{3/2}\right] + \frac{\ell}{2a}. \tag{5}$$

Completely analogously one derives for the energy gain ΔE_n for n overcharging counterions

$$\Delta E_n = -\frac{\bar{\alpha}\ell}{\sqrt{A}}\left[(N_c + n)^{3/2} - N_c^{3/2}\right] + \frac{\ell}{a}\frac{n^2}{2}. \qquad (6)$$

Using Eq. (6), where we determined the unknown $\bar{\alpha}$ from the simulation data for ΔE_1, we obtain a curve that matches the simulation data almost perfectly (Fig. 2). The second term in Equation (6) also shows why the overcharging curves of Fig. 2 are shaped parabolically upwards for larger values of n. If one successively removes each of n counterions from a neutral colloid, one can derive in a similar fashion the ionization energy cost

$$\Delta E_n^{ion} = -\frac{\bar{\alpha}\ell}{\sqrt{A}}\left[(N_c - n)^{3/2} - N_c^{3/2}\right] + \frac{\ell}{a}\frac{n^2}{2}. \qquad (7)$$

Using the measured value of $\bar{\alpha}$ we can simply determine the maximally obtainable number n_{max} of overcharging counterions by finding the stationary point of Eq. (6) with respect to n:

$$n_{max} = \frac{9\bar{\alpha}^2}{32\pi} + \frac{3\bar{\alpha}}{4\sqrt{\pi}}\sqrt{N_c}\left[1 + \frac{9\bar{\alpha}^2}{64\pi N_c}\right]^{1/2}. \qquad (8)$$

The value of n_{max} depends only on the number of counterions N_c and $\bar{\alpha}$. For large N_c Eq. (8) reduces to $n_{max} \approx \frac{3\bar{\alpha}}{4\sqrt{\pi}}\sqrt{N_c}$ which was derived in Ref. [11] as the low temperature limit of a a neutral system in the presence of salt. What we have shown is that the overcharging in this limit has a pure electrostatic origin, namely it originates from the energetically favorable arrangement of the ions around a central charge. We also showed in Ref. [28] that $\bar{\alpha}$ reaches the perfect WC value of 1.96 if the colloid radius a gets very large at fixed c, or when c becomes large at fixed a.

If instead of a central charge scheme one uses discrete charge centers distributed randomly over the colloidal surface we find counterion structures which are quite far away from the WC array, especially when the counterions are pinned to their counter charges. This depends on the interaction energy at contact, which depends of course on l and distance of closest approach. However, we still find overcharging, although reduced in value, of the form given by Eq. 6 [16, 29]

Asymmetrically Charged Colloids

The most interesting phenomenon, however, appears when the two colloids have different counterion concentrations, here $c_A > c_B$, since then **stable ionized states** can appear. The physical reason is that a counterion can gain more energy by overcharging the colloid with c_A then it loses by ionizing colloid B. A straight forward application of the procedure outlined for the barrier calculation [26, 28] yields a simple criterion (more specifically a sufficient condition), valid for large macroionic separations, for the charge asymmetry $\sqrt{N_A} - \sqrt{N_B}$ to produce an ionized ground state of two unlike charged colloids with the same size:

$$\left(\sqrt{N_A} - \sqrt{N_B}\right) > \frac{4\sqrt{\pi}}{3\bar{\alpha}^A} \approx 1.2. \tag{9}$$

We have also demonstrated that the ground state phenomena survive for finite temperatures, i.e. an ionized state can also exist at room temperature T_0. The left part of Figure 3 shows the time evolution of the electrostatic energy of a system $Z_A = 180$ with $Z_B = 30$, $R/a = 2.4$ and a colloidal volume fraction of $7 \cdot 10^{-3}$, where the starting configuration is the neutral state $(DI = 0)$. One clearly observes two jumps in energy, $\Delta E_1 = -19.5$ and $\Delta E_2 = -17.4$, which corresponds each to a counterion transfer from colloid B to colloid A. These values are consistent with the ones obtained for the ground state, which are -20.1 and -16.3 respectively. Note that this ionized state $(DI = 2)$ is more stable than the neutral but is expected to be metastable, since it was shown previously that the most stable ground state corresponds to $DI = 5$. The other stable ionized states for higher DI are not accessible with reasonable computer time because of the high energy barrier made up of the correlational term and the monopole term which increase with DI. In the right part of Fig. 3 we display a typical snapshot of the ionized state $(DI = 2)$ of this system at room temperature.

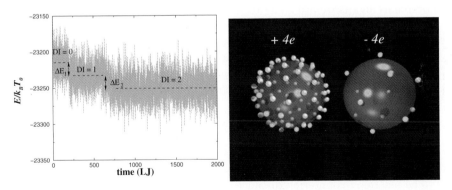

Fig. 3. Relaxation of an initial unstable neutral state towards ionized state at room temperature. Plotted is the total electrostatic energy versus time (LJ units), for $Z_B = 30$ and $R/a = 2.4$. Dashed lines lines represent the mean energy for each DI state. Each jump in energy corresponds to a counterion transfer from the macroion B to macroion A leading to an ionized state that is lower in energy than the neutral one. The right figure is a snapshot of the final ionized state, with net charges $+4e$ and $-4e$ as indicated.

3 Entropic Overcharging

The goal of this part is (i) to motivate in detail the role of the excluded volume contribution for the overcharging of a colloidal macroion in the presence of salts and (ii) to provide a qualitative insight into the mechanism behind these effects. In the presence of salt the contribution of excluded volume can be so important that the size of the small ions dominates the occurrence of overcharging and the overcharging strength increases with increasing ionic size when the electrolyte concentration is fixed. Even for monovalent ions with high enough ionic size, overcharging is observed, which can not be explained with a salt-free WC picture due to the low Coulomb coupling strength. In order to have the simplest system we consider only the cases where the coion and the counterions have the same size. This will reduce the effects of depletion forces which lead to nontrivial features already in neutral hard sphere fluids.

Our proposed mechanism will rest on the following arguments: For a fixed salt concentration, the available volume in the fluid is function of the electrolyte particle size. More precisely, the entropy of the solution is decreased by enlarging the size of the salt-ions [1], which enhances inter-particle correlations. On the other hand the interface provided by the macroion leads to an increase of microion density close to the macroion, and promotes there also *lateral* ordering, even in the absence of strong electrostatic coupling, similar to a prefreezing phenomenon. However entropy alone can never lead to overcharging, since in this limiting case coions and counterions have the same radial distribution. But ordering and weak electrostatic correlations can lead to overcharging, as we are going to show.

In Ref. [30] we carried out MD simulations and HNC/MSA (hypernetted chain/mean spherical approximation) integral-equation to study the overcharging in spherical colloidal systems within the primitive model. In particular, the solvent enters the model only by its dielectric constant and its discrete structure [31] is ignored. The system is made up of (i) a large macroion with a bare central charge $Q = -Z_m e$ (with $Z_m > 0$) and (ii) symmetric salt ions of diameter σ and valence Z_c. The system is globally electrically neutral.

For the simulation procedure, all these ions are confined in an impermeable cell of radius R and the macroion is held fixed at the cell center. For more details please consult Ref. [30].

The macroion-counterion distance of closest approach is defined as a. Energy units in our simulations are fixed by $\epsilon = k_B T$ (with $T = 298$ K). We have chosen to work at *fixed* distance of closest approach (between the centers of macroion and the salt-ion) $a = 2l_B = 14.28$ Å in order to give prominence to the effect of ionic size. In this way, the electrostatic correlation induced

---

[1] This argument holds far enough from the solid-liquid transition as in the present study (i. e. the highest fluid volume fraction f we consider is 0.23). Indeed entropy alone can drive crystallization as in hard-sphere systems

by the colloid remains the same (i.e. fixed macroion electric field at contact) no matter what the ionic size is. Thus by changing the ionic size σ one also changes that colloidal radius accordingly so that a remains constant. The salt concentration ρ is given by N_-/V where $V = \frac{4}{3}\pi R^3$ is the cell volume and N_- is the number of coions. We will restrict the present study to $\rho = 1$M salt concentration. The fluid volume fraction f is defined as $(N_+ + N_-)\left(\frac{\sigma}{2R}\right)^3$ where N_+ is the number of counterions. To avoid size effect induced by the simulation cell we choose a sufficiently large radius $R = 8.2 l_B$ yielding to more than 1000 mobile charges. Simulation run parameters are gathered in Table 1. The HNC/MSA calculations were performed using the technique presented in [20] and references therein. Here it is assumed that the system size is infinite, and the bulk salt concentration is fixed. In practice there should be no observable difference in the correlation functions (between HNC/MSA and MD) close to the macroions, because the wall effects die off sufficiently fast.

Table 1. Simulation run parameters for the charged fluid $(A - F)$ and the neutral fluid $(G - H)$.

parameters	salt valence	Z_m	σ/l_B	f
run A	2:2	48	1	2.3×10^{-1}
run B	2:2	48	0.5	2.9×10^{-2}
run C	2:2	48	0.25	3.6×10^{-3}
run D	1:1	10	1	2.3×10^{-1}
run E	1:1	48	1	2.3×10^{-1}
run F	1:1	48	0.5	2.9×10^{-2}
run G	-	-	1	2.3×10^{-1}
run H	-	-	0.5	2.9×10^{-2}

We first illustrate the excluded volume correlations present for a *neutral* hard sphere fluid (runs $G - H$) [identical to systems $A - B$ and $E - F$, but uncharged (see Table 1)]. Although these kind of systems are "simple" fluids [32], it is fruitful to elucidate what exactly happens at this "low" fluid density upon varying the fluid particle size σ in the presence of a single large spherical particle. To characterize the fluid structure we consider the pair distribution function $g(r)$ between the colloid and the fluid particle, which is just proportional to the local density $n(r)$: $n(r) = \rho g(r)$. Results are depicted in Fig. (4). For the large value $\sigma = l_B$, one clearly observes a relative high local concentration as well as a *short range ordering* nearby the colloidal surface. When the particle size is reduced by a factor 2 (holding a fixed), the behavior is qualitatively different and the system is basically uncorrelated. By increasing σ at fixed fluid density ρ the mean surface-surface distance between particles is reduced which in turn leads to higher collision probability and thus to higher correlations. In other words, by reducing the *available* volume one promotes *ordering* [20]. This seems to be trivial in the bulk, but the presence of the colloidal surface induces an even enhanced ordering and the system can

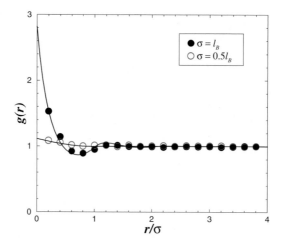

Fig. 4. Pair distribution function $g(r)$ between the colloid and the fluid particle in the neutral state (runs $G - H$) for two particle sizes σ. The origin of the abscissa is taken at the distance of closest approach $a = 2l_B$. Lines and symbols correspond to HNC theory and simulation respectively.

prefreeze (order) close to the colloid. Note that the same effects are naturally present for a fluid close to a planar wall [32].

To characterize the overcharging we introduce the fluid integrated charge $Q(r)$ which corresponds to the total *net* charge in the fluid (omitting the macroion bare charge Z_m) within a distance r from the distance of closest approach a. Results are sketched in Fig. 5(a) and Fig. 5(b) for divalent and monovalent salt ions respectively. Theoretical and numerical analysis are in very good qualitative agreement.

For the divalent electrolyte solutions (runs A-C) we observe that overcharging is strongly dictated by the ionic size σ [see Fig. 5(a)]. For small ions (run C) *no* overcharging occurs (i. e. $Q(r)/Z_m < 1$), which is a non trivial effect probably related to the forming of ionic pairs (counterion-coion pairing) due to the strong pair interaction of $8k_BT$. This delicate point has also been observed in [19]. Upon increasing σ, the degree of overcharging increases. We carefully checked that the distance $r = r^*$, where $Q(r^*)$ assumes its maximal value, corresponds to a zone of coion depletion (in average there are less than 2 coions within r^*). This implies that also the absolute number of counterions at the vicinity of the macroion surface increases with increasing σ. This is qualitatively in agreement with what we observed above for neutral systems. However we are going to show later that electrostatic correlations are also *concomitantly* responsible of this extra counterion population (increasing with σ) in the vicinity of the colloidal surface.

For the monovalent electrolyte solution [see Fig. 5(b)] overcharging occurs for $\sigma = l_B$ (runs D and E). In respect to the salt-free WC picture this is

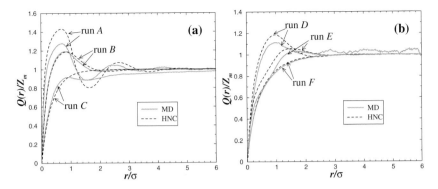

Fig. 5. Reduced fluid integrated charge $Q(r)/Z_m$ as a function of distance r for for three different particle sizes σ. (a) divalent salt ions (runs A-C), (b) monovalent salt ions (runs D-F). The origin of the abscissa is taken at the distance of closest approach $a = 2l_B$

rather unexpected since here the "plasma" parameter $\Gamma_{cc} = l_B Z_c^2/a_{cc}$, where $a_{cc} = (\pi c)^{-1/2}$ (c standing for the two-dimensional surface counterion concentration) is the mean distance between counterions on the surface, is small. More precisely for $Z_m = 10$ (run D) we find $\Gamma_{cc} \approx 0.8$, and for $Z_m = 48$ (run E) we have $\Gamma_{cc} < 1.0$, which is due to a packing effect which imposes that a_{cc} cannot be smaller than σ. But following this salt-free approach, it is necessary to have at least $\Gamma_{cc} > 2$ to get overcharge [24]. Note that from run D ($Z_m = 10$) to run E ($Z_m = 48$) one increases the macroion surface-charge density leading (for $Z_m = 48$) to a higher *absolute* overcharging $Q(r^*)$ but a weaker ratio $Q(r^*)/Z_m$, which is qualitatively in accord with the WC picture, since the maximal overcharging is proportional to $\sqrt{a_{cc}}$. A closer look on Fig. 5(b) reveals that for $Z_m = 48$ (run E) r^*/σ is shifted to the right compared to the divalent case. This is merely a packing effect and it is due to the fact that for monovalent counterions the macroion charge (over)compensation involves twice more number of counterions particles than in the divalent case. Therefore for the macroion charge density under consideration ($Z_m = 48$), more than one counterion-layer is needed to compensate the macroion charge. Again for a smaller ionic salt size $\sigma = 0.5l_B$ (run F) the overcharging effect is canceled.

For the WC picture to be effective we need strong lateral correlations which cannot come from pure electrostatic effects. To see if such correlations are present we consider the local two-dimensional surface counterion structure. We analyzed in our simulations the *two-dimensional* counterion pair distribution $g_{cc}(s)$, where s is the arc length on the macroion sphere of radius a. All counterions lying at a distance $r < a + 0.5\sigma$ from the macroion center are radially projected to the contact sphere of radius a. Predominantly counterions are present in the first layer. For the neutral system (run G) we analyzed the small neutral species. Results are given in Fig. 6. We observe

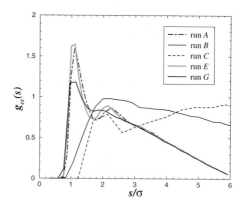

Fig. 6. Two-dimensional surface counterion correlation functions $g_{cc}(s)$ for divalent salt (run A - C), monovalent salt (run E), and a neutral system (run G).

that all systems with an ion diameter $\sigma = l_B$ show their first peak at roughly 1σ, and a weaker second peak at about 2σ, exhibiting long range surface correlations. The second peak is very weak in the neutral system G, but clearly visible in system A and E. Due to the stronger electrostatic repulsion the second peak for the divalent system A is somewhat further apart than for the monovalent system E. The systems B and C with smaller ion diameters show a correlation hole of size $\approx 2\sigma$, which is of purely electrostatic origin. Therefore lateral correlations can be produced either by pure entropy effects (run G) or pure electrostatic effects (run C), or we can have an enhanced lateral ordering due to the interplay of both (run A, E). Qualitatively, WC arguments are still applicable to monovalent systems (such as run E) if one considers an effective low local surface temperature stemming from the strong surface ordering.

In summary, the observed overcharging for low Coulomb coupling can be traced back to the complicated interplay of entropy and energy by the following two effects. First, by enlarging the excluded volume of the salt ions at fixed concentration, one lowers the accessible volume to the fluid particles. Second, the presence of the large macroion provides an interface near which the density of the fluid is increased compared to the bulk, and the solution can prefreeze due to entropic effects, provided the available volume gets low enough. The closest layer to the interface possesses already strong lateral correlations, even for a neutral system. If the system is additionally charged, then even weak Coulomb correlations can lead to the formation of a strongly correlated liquid, where the overcharged state is energetically favorable, as shown in [11, 15, 16, 20]. The order of this counterion layer is however not created by electrostatic interactions as in the normal WC picture, but it is largely due to entropic effects. The observed overcharging effect might have implications for the stability of colloidal suspensions. Additions of monovalent salt will eventually make colloidal suspensions unstable due to the onset of the van-der-Waals attractions. Upon addition of even more monovalent salt there is the possibility of salting the precipitate again in, as has been seen for poly-electrolyte systems[33]. The observation of such a reentrant transition could be an important hint towards the existence of overcharging with monovalent salt ions.

Acknowledgements

Collaborations with E. Gonzáles-Tovar, R. Messina, K. Kremer, M. Lozada-Cassou, and financial support by the DFG with in the SFB 625, TR6 and the "Zentrum für Multifunktionelle Werkstoffe und Miniaturisierte Funktionsein-heiten", grant BMBF 03N 6500, are gratefully acknowledged.

References

1. C. Holm, P. Kékicheff, and R. Podgornik, editors. *Electrostatic Effects in Soft Matter and Biophysics*, volume 46 of *NATO Science Series II - Mathematics, Physics and Chemistry*. Kluwer Academic Publishers, Dordrecht, NL, December 2001.
2. Y. Levin. Electrostatic correlations: from plasma to biology. *Rep. Prog. Phys.*, 65:1577–1632, 2002.
3. W.M. Gelbart, R.F. Bruinsma, P.A. Pincus, and V.A. Parsegian. Dna-inspired electrostatics. *Physics Today*, 53(9):38–44, 2000.
4. W. van Megen and I. Snook. *J. Chem. Phys.*, 73:4656, 1980.
5. G.M. Torrie and J.P. Valleau. *J. Chem. Phys.*, 73:5807, 1980.
6. M. Lozada-Cassou, R. Saavedra-Barrera, and D. Henderson. *J. Chem. Phys.*, 77:5150, 1982.
7. E. Gonzales-Tovar, M. Lozada-Cassou, and D. Henderson. Hypernetted chain approximation for the distribution of ions around a cylindrical electrode. ii. numerical solution for a model cylindrical polyelectrolyte. *J. Chem. Phys.*, 83(1):361–72, 1985.
8. R. Hidalgo-Álvarez, A. Martin, A. Fernandez, D. Bastos, F. Martinez, and F.J. de las Nieves. Electrokinetic properties, colloidal stability and aggregation ki-netics of polymer colloids. *Adv. Colloid Interface Sci*, 67:1, 1996.
9. M. Tanaka and A.Y. Grosberg. Electrophoresis of charge inverted macroion complex: Molecular dynamics study. *Euro. Phys. J. E*, 7(4):371, 2002.
10. B.I. Shklovskii. Wigner crystal model of counterion induced bundle formation of rodlike polyelectrolytes. *Phys. Rev. Lett.*, 82(16):3268–3271, 1999.
11. B.I. Shklovskii. Screening of a macroion by multivalent ions: Correlation-induced inversion of charge. *Phys. Rev. E*, 60(5):5802–5811, 1999.
12. T.T. Nguyen, A.Y. Grosberg, and B.I. Shklovskii. Macroions in salty water with multivalent ions: Giant inversion of charge. *Phys. Rev. Lett.*, 85:1568–1571, 2000.
13. M. Lozada-Cassou, E. González-Tovar, and W. Olivares. Nonlinear effects in the electrophoresis of a spherical colloidal particle. *Phys. Rev. E*, 60:R17–R20, 1999.

14. T.T. Nguyen and B.I. Shklovskii. Adsorption of multivalent ions on a charged surface: Oscillating inversion of charge. *Phys. Rev. E*, 64:041407, 2001.
15. R. Messina, C. Holm, and K. Kremer. Strong attraction between chared spheres due to metastable ionized states. *Phys. Rev. Lett.*, 85:872–875, 2000.
16. R. Messina, C. Holm, and K. Kremer. Effect of colloidal charge discretization in the primitive model. *Euro. Phys. J. E.*, 4:363–370, 2001.
17. M. Tanaka and A.Y. Grosberg. Giant charge inversion of a macroion due to multivalent counterions and monovalent coions: Molecular dynamics study. *Journal of Chemical Physics*, 115(1):567–574, 2001.
18. H. Greberg and R. Kjellander. Charge inversion in electrical double layers and effects of different sizes for counterions and coions. *J. Chem. Phys*, 108:2940–2953, 1998.
19. M. Deserno, F. Jiménez-Ángeles, C. Holm, and M. Lozada-Cassou. Overcharging of dna in the presence of salt: Theory and simulation. *Journal Phys. Chem. B*, 105(44):10983 –10991, October 2001.
20. M. Lozada-Cassou and F. Jiménez-Ángeles. Overcharging by macroions: above all, an entropy effect. *eprint physics/0105043*.
21. M. Deserno, C. Holm, and S. May. The fraction of condensed counterions around a charged rod: Comparison of Poisson-Boltzmann theory and computer simulations. *Macromolecules*, 33:199–206, 2000.
22. J.J. Thomson. *Philos. Mag.*, 7:237, 1904.
23. D.W. Oxtoby, H.P. Gillis, and N.H. Nachtrieb. *Principles of Modern Chemistry*, chapter 3, page 80. Saunders College Publishing, Philadelphia, 1999.
24. I. Rouzina and V.A. Bloomfield. Macroion attraction due to electrostatic correlation between screening counterions. 1. mobile surface-adsorbed ions and diffuse ion cloud. *Journal of Phys. Chem.*, 100(23):9977–9989, 1996.
25. A.Y. Grosberg, T.T. Nguyen, and B.I. Shklovskii. *Rev. Mod. Phys*, 74:329, 2002.
26. R. Messina, C. Holm, and K. Kremer. Ground state of two unlike charged colloids: An anology with ionic bonding. *Europhys. Lett.*, 51:461–467, 2000.
27. L. Bonsall and A.A. Maradudin. Some static and dynamical properties of a two-dimensional wigner crystal. *Phys. Rev. B*, 15:1959–1973, 1977.
28. R. Messina, C. Holm, and K. Kremer. Strong electrostatic interactions in spherical collidal systems. *Phys. Rev. E*, 64:021405, 2001.
29. R. Messina. Spherical colloids: Effect of discrete macroion charge distribution and counterion valence. *Physica A*, 308:59–79, 2002.
30. R. Messina, E.G. Tovar, M. Lozada-Cassou, and C. Holm. Overcharging: The crucial role of excluded volume. *Euro. Phys. Lett.*, 60:383–389, 2002.
31. E. Allahyarov and H. Löwen. Discrete solvent effects on the effective interaction between charged colloids. *Journal of Physics, Cond. Mat.*, 13(13):L277–L284, Apr. 2001.
32. J. P. Hansen and I. McDonald. *Theory of Simple Liquids*. Academic, London, 1990.
33. H. Eisenberg and G.R. Mohan. *J. Phys. Chem.*, 63:671, 1959.

Computer Simulations of Magnetic Grains

S. Fazekas[1,2], J. Kertész[1], and D.E. Wolf[3]

[1] Department of Theoretical Physics –
Budapest University of Technology and Economics
H-1111 Budapest – Hungary
[2] Theoretical Solid State Research Group of the Hungarian Academy of Sciences –
Budapest University of Technology and Economics
H-1111 Budapest – Hungary
[3] Institute of Physics – University Duisburg-Essen
47048 Duisburg – Germany

Summary. In dense arrangements of magnetic grains, cutting off the interaction potential gives a possibility to accelerate simulation algorithms. We argue that $R \approx 5$ particle diameters is a reasonable choice for dipole-dipole interaction cutoff in two-dimensional dipolar hard sphere systems, if one is interested in local ordering. As an application, we performed computer simulations based on a two-dimensional Distinct Element Method to study granular systems of magnetized spherical particles. The effect of the magnetization on the angle of repose, on the surface roughness of piles, and on particle avalanches were studied. We found a smooth transition in the avalanche formation from a *granular regime* to a *correlated regime* controlled by the magnetic interparticle force. This observation underlines the analogies between systems with magnetic and adhesive forces.

Keywords: magnetic grains, granular piles, avalanches

1 Introduction

The computer simulation of an ensemble of magnetic grains is hampered by the long range nature of the interaction. The size of the tractable systems (N particles) is limited through the fact that the order of N^2 calculations are to be carried out at each step, though for many purposes large systems need to be studied. Periodic boundary conditions, which are often helpful, can be implemented only by using sophisticated summation algorithms (if possible due to screening).

In this study we investigate the possibility of cutting off the magnetic interaction potential. For this, we consider two-dimensional ensembles of magnetic particles interacting with dipole-dipole interaction. It is crucial from the point of view of efficient programming to know if a reasonable cutoff can be introduced in these systems. We investigate the problem by comparing the stability of static configurations.

The dipole interaction between magnetic grains can be viewed as an anisotropic adhesion force. Because of the strong anisotropy and the longer interaction range one can expect differences between the results on magnetic grains and wet granular systems, however, the basic effects are expected to be the same.

Some time ago Hornbaker et al. [1] addressed the question how sand castles stand. They stated that already small quantities of wetting liquid can dramatically change the properties of granular media, leading to large increase in the angle of repose and correlation in grain motion. Theoretical studies on the angle of repose based on stability criteria have been done by Albert et al. [2]. They theoretically determined the dependence of the angle of repose on cohesive forces, and applied the results to wet granular material.

Experimental studies of Tegzes et al. on angle of repose using the draining crater method [3] and on avalanches using a rotating drum apparatus [4] identified three distinct regimes as the liquid content is increased: a *granular regime* in which the grains move individually, a *correlated regime* in which the grains move in correlated clusters, and a *plastic regime* in which the grains flow coherently.

Experiments of Quintanilla et al. [5] using the rotating drum apparatus addressed the question of self-organized critical behavior in avalanches of slightly cohesive powders. Their results show that avalanche sizes do not follow a power-law distribution, however, they scale with powder cohesiveness.

To study the transition from noncohesive to cohesive behavior, Forsyth et al. [6] suggested a method based on magnetized particles. The particles placed in an external magnetic field become magnetized, all having the same magnetic orientation parallel to the field. Varying the strength of the field allows to continuously vary the resulting interparticle magnetic force. Using nonmagnetic perspex walls the particle-wall interaction remains unchanged relative to the noncohesive state. Using particles under same packing conditions it is ensured that the initial conditions are as uniform as possible.

The magnetic interaction of magnetized grains is highly anisotropic, and the fixed external field introduces even more anisotropy as the grains are aligned to the field. A similar experimental setup [7], but with particles carrying a remanent magnetization in the absence of an external magnetic field, would partly diminish the mentioned anisotropy, however in this case the magnetizations and the interparticle forces are not as well defined as in the experiments of Forsyth et al. [6].

We carried out computer simulations on a system corresponding to the experiments of Forsyth et al. [6] and studied the angle of repose, the surface roughness, and the effect of magnetization on avalanching in two-dimensional particle piles.

2 The Role of an Interaction Cutoff

Brankov and Danchev [8] used the Luttinger-Tisza method [9] to study the crystalline state of a system of two-dimensional dipole moments with identical scalar strength located at the sites of an infinite rhombic lattice with an arbitrary rhombicity angle. They considered that the ground state of this system has a translational symmetry corresponding to discrete translations along the lattice lines. They found that the ground state depends on the rhombicity angle and the interaction range.

We repeated their calculations [10] with the consideration that the dipoles are carried by identical hard spherical particles of diameter equal to the lattice constant, and according to this geometrical constraint we limited the rhombicity angle to $60° \leq \alpha \leq 90°$. We confirmed their results and we repeated the calculations taking into consideration the interaction of only two neighboring lines. This corresponds to the interaction of two lines of dipolar hard spheres shifted according to the α rhombicity angle. The Luttinger-Tisza method can be applied in a straightforward way also in this case. At infinite interaction range, the system has a ferromagnetic ground state for $60° \leq \alpha \lesssim 75.67°$, and an anti-ferromagnetic ground state for $75.67° \lesssim \alpha \leq 90°$. As there is no continuously degenerate ground state, the ground state of the two-line system is always defined by the Luttinger-Tisza *basic arrangement* with the lowest eigenvalue. In particular there is a ferromagnetic and an anti-ferromagnetic state, which – depending on the α rhombicity angle – defines the ground state.

Due to the dipole-dipole interaction, dipolar spheres tend to aggregate into chain like structures (see for example [11–13] and references therein) in which the energies of intrachain interactions are much greater than those of interchain interactions. Based on our calculations, the interaction of two neighboring lines almost saturates the long-range dipole-dipole interaction, confirming that the study of a two-line system gives valuable results for the dipole-dipole interaction in general.

Brankov and Danchev [8] observed that the ground state of a system of dipoles is sensitive to the dipole-dipole interaction range. We introduced an R interaction cutoff measured in particle diameters. Our numerical results for the two-line system at large $(R \geq 8)$ and at low $(R < 8)$ interaction cutoff distances are shown in Fig. 1. The figure shows the lowest energy per dipole of the ferromagnetic and the anti-ferromagnetic states as function of the α rhombicity angle. It can be observed that below a certain α the ground state of the system corresponds to the ferromagnetic state, and above this α to the anti-ferromagnetic state. The results corresponding to $R = 10^6$ are close to the infinite-range interaction limit within the numerical errors of 64 bit floating point arithmetic.

On the right panel of Fig. 1 the ferromagnetic lines are shifted upward as R decreases, while the anti-ferromagnetic lines remain almost unchanged. This is a consequence of the strong coupling of neighboring dipoles of opposite orientation, which makes the interaction cutoff irrelevant in the anti-ferromagnetic

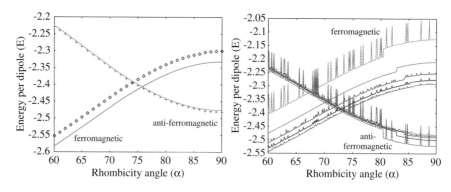

Fig. 1. Numerical results for a two-line system at different R interaction cutoff values: 10^6 (lines) and 8 (points) particle diameters on the left panel, and 7, 6, 5, 4, and 3 particle diameters on the right panel. The diagrams show the lowest energy per dipole of the ferromagnetic and the anti-ferromagnetic state as function of the rhombicity angle. The energy is measured in S^2/D^3 units, where S is the scalar strength of the dipoles and D is the particle diameter. On the right panel the ferromagnetic lines are shifted upward as R decreases, while the anti-ferromagnetic lines remain almost unchanged. (Observe the different scale on the two diagrams.)

state. According to this as R decreases the anti-ferromagnetic state becomes more and more dominant.

At low interaction cutoff distances ($R \lesssim 8$) the discrete nature of the system becomes more and more relevant and both the ferromagnetic and anti-ferromagnetic energy per dipole begin to exhibit sudden jumps in function of the rhombicity angle (see right panel of Fig. 1). As the interaction cutoff decreases the energy jumps become larger and larger. This behavior can introduce numerical instabilities in simulations using a badly chosen cutoff distance.

For large R the ferromagnetic state at $\alpha = 60°$ has lower energy per dipole than the anti-ferromagnetic state at $\alpha = 90°$. Our numerical results show that at $R \approx 4$ the situation is reversed, and at $R \approx 2$ the ferromagnetic ground state disappears. Brankov and Danchev [8] found that in case of an infinite rhombic lattice with rhombicity angle $\alpha = 60°$ the ferromagnetic ground state disappears at $R \approx 3$.

For characterizing the finite size behavior of the two-line system we introduced a finite size coefficient [10], and we observed that it is sensitive to the interaction cutoff for both ferromagnetic and anti-ferromagnetic states. The introduced finite size coefficient gives a good description of finite dipole systems at $N > 10$ (where N is the number of dipoles per line), but it is inaccurate at small N. For a better understanding of the system, we studied numerically finite systems (at small N) investigating the effect of bending two lines of dipolar hard spheres in ferromagnetic and anti-ferromagnetic states. We introduce the γ bending parameter and we define the bent system as com-

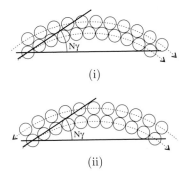

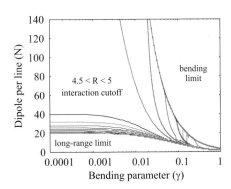

Fig. 2. Numerical results on bending a two-line system at different interaction cutoff values. The left panel shows finite systems of two lines of dipolar hard spheres in ferromagnetic and anti-ferromagnetic states. The right panel shows (γ, N) state diagrams (see text for description) for R interaction cutoff values ranging from infinity to 2 particle diameters. The lines move upward as R decreases.

posed of particles placed on an arc of angle $2N\gamma$ with dipole vectors tangential to the arc (see (i) and (ii) on the left panel of Fig. 2). The definition of the γ angle involves a so called 'bending limit' as the arc's angle is limited to 2π, and thus γ must satisfy the $\gamma \le \pi/N$ condition.

Our numerical results show that for bending either a ferromagnetic or anti-ferromagnetic two-line system some physical effort is needed. We observed that the two-line system in ferromagnetic state can be bent easier than in the corresponding anti-ferromagnetic state. This is a consequence of the strong coupling of neighboring dipoles oriented anti-parallel. We also observed that as the anti-ferromagnetic state is bent it becomes less and less stable. Fig. 2 shows numerical results on bending a two-line system at different interaction cutoff values. As function of the bending parameter γ and system size N we compared the energy per dipole of the ferromagnetic and anti-ferromagnetic states and we identified the points (γ, N) at which these two states are energetically equivalent. We repeated this procedure at different R interaction cutoff values. The right panel of Fig. 2 shows corresponding (γ, N) state diagrams.

In the long-range limit (i.e. at $R = \infty$) the anti-ferromagnetic state has lower energy per dipole only for small system sizes. As the interaction cutoff decreases, at R between 5 and 4.5 the (γ, N) state diagrams change rapidly, and for $R \le 4.5$ the anti-ferromagnetic state remains more stable even at large N. This means that at this point the general characteristics of the dipole system is substantially changed. Based on Fig. 2 and on our previous results we argue that $R \approx 5$ is a reasonable choice for dipole-dipole interaction cutoff in two dimensional systems of dipolar hard spheres, if one is interested in local ordering.

3 Piling and Avalanches of Magnetized Particles

Using the previous results on a reasonable interaction cutoff, we performed computer simulations based on a two-dimensional Distinct Element Method (DEM) [14, 15] to study granular systems of magnetized spherical particles. The particles are magnetized by a constant external field. The magnetization is modeled with dipoles. We neglect any coupling between the magnetic orientation and particle rotation. The external field allows only for ferromagnetic dipole arrangements (i.e. the dipoles are parallel to the field) at any time.

For characterizing the strength of the interparticle force, we introduce a dimensionless quantity f defined by the ratio of the maximum magnetic interparticle force at contact and the gravitational force. According to this

$$f = \frac{F_m}{F_g} = \frac{\mu_0}{4\pi} \frac{6S^2}{mgD^4}, \tag{1}$$

where S is the magnetic dipole strength, D is the particle diameter, m denotes the particle mass, and g is the gravitational acceleration.

The diameter of the spherical particles was taken from the $0.7-0.9$ mm interval, with $D = 0.8$ mm average particle diameter and a slight polydispersity, resembling real experimental setups. The long range magnetic interaction was taken into consideration within $6.25\ D$ interaction cutoff distance. The angle of repose, the surface roughness, and the particle avalanches depend crucially on local orderings inside the pile, as noted for example by Altshuler et al. in [16]. The used cutoff keeps the character of local orderings and changes the magnetic energy per particle by less than 5% [10].

We calculated the collision interaction of particles using the Hertz contact model with appropriate damping [17, 18]. We implemented Coulomb sliding friction for large relative translational velocities and for numerical stability viscous friction for small velocities, with continuous transition between the two. We did not use any static or rolling friction model in this simulation. The parameters of the Hertz contact model were chosen such that they approximately correspond to Young modulus of 0.015 GPa and restitution coefficient of 0.86. The particle-particle and the particle-wall sliding friction coefficients were 0.5 and 0.7.

The simulation setup can be seen in Fig. 3. The external magnetic field is vertical. The particles are added one by one with constant rate along vertical trajectories at small (maximum one particle diameter) random distance from the left wall. They either reach the pile with a given velocity (*they are fired into the pile*), or their impact velocity is set to zero (*they are placed gently on the pile*). Any particle touching the bottom wall sticks to the wall. This builds up a *random base* (see the experimental setup of Altshuler et al. [16]). The particles can leave the system on the right side.

As part of our investigations, with a special *side wall model*, we also simulated the effect of front and back walls in a Hele-Shaw cell geometry encountered in experimental studies. We took into consideration the frictional

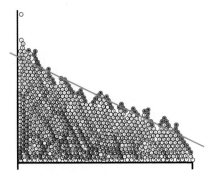

Fig. 3. Simulation setup. The particles are introduced with constant rate one by one at small random distances from the left wall. The particles can leave the system on the right side. The surface angle (i.e. the angle of repose) is measured by fitting a straight line over the positions of the surface particles (marked with black). The figure also shows the normal contact forces. The thickness of the lines connecting the centers of the particles in contact is proportional to the normal contact force. The sample corresponds to $f = 6$ interparticle force ratio.

interaction with side walls by summing the magnitude of normal forces acting on one particle, directing a certain percentage of this *pressure* on the walls, and deriving a frictional force using the already mentioned friction model. The percentage of the total force directed on the side walls was a parameter of our simulations.

We performed three sets of simulations: (a) the particles were fired into the pile, (b) the particles were placed gently on the pile, and (c) the particles were fired into the pile, while 4% of the internal *pressure* was directed on the *front and back walls* (see the side wall model). In both (a) and (c) the particles reached the pile with 0.5 m/s impact velocity. In all three simulation sets we did runs for different interparticle force ratios. In each run we started with an empty system, and introducing 12000 particles, one particle every 3000 integration steps, we numerically integrated the system for 3 minutes (simulated time).

In the first part of the process the number of particles in the system increased monotonically. After a pile was built, avalanches started, which in a pulsating manner moved particles out of the system. In this way the number of particles began to oscillate around some well defined value. In this latter part we identified the surface particles (marked with black in Fig. 3) every 500 integration steps, and we measured the slope of the fitted surface line and the standard deviation of surface points from this line. The average of this quantities over the simulated time gave the measured angle of repose and surface roughness.

We also measured the avalanche sizes and avalanche durations. We defined an avalanche by a number of individual events on the time scale of the integration steps (i.e. the smallest simulation time step) in which one particle leaves the system, and the time between two consecutive events is smaller than a well defined value. We take this value equal to the time corresponding to 3000 integration time steps. Our choice is based on the system's observed dynamic time scale, and the fact that one new particle is introduced (i.e. the system is perturbed) every 3000 integration time steps, and thus on a larger time scale there are surely uncorrelated events.

3.1 Angle of Repose and Surface Roughness

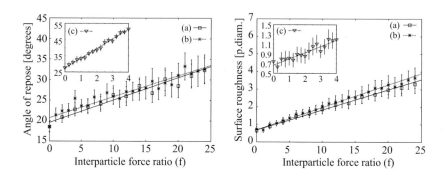

Fig. 4. Angle of repose (left panel) and surface roughness (right panel) at different magnetic interparticle force ratios. The angle of repose is measured in degrees. The surface roughness is measured in (average) particle diameters. We did three sets of simulations (see text for details).

In all cases both the angle of repose and the surface roughness (in the examined domain) exhibit a linear dependence on f (see Fig. 4). The angle of repose in case (a) and (b) increases by approximately 0.5 degree per unit change of interparticle force ratio (see left panel of Fig. 4). This is in good accordance with the experimental results of Forsyth et al. [6], however the angle of repose at zero magnetization in our case is about 10 degrees smaller.

As a consequence of our side wall model, the angle of repose in case (c) is in agreement with the experimentally observed value. However, the way we model the side walls leads to a stronger increase of the angle of repose with f than in the experiments of Forsyth et al. [6] (see inset of left panel of Fig. 4). The side wall effect does not influence the surface roughness (see inset of right panel of Fig. 4).

3.2 Effect of Magnetization on Particle Avalanching

We analyzed the effect of magnetization on particle avalanching, and we found that there is a difference in avalanche formation at small and at large interparticle force ratios. We identified a *granular* and a *correlated regime*. The transition between the two regimes is not sharp. Similar regimes were identified experimentally by Tegzes et al. [3,4] in case of wet granular materials. At high liquid content they could also identify a third *plastic regime*.

Studying the recordings from our simulations [19], it can be observed that for small magnetization the avalanches are formed by small vertical chains following each other at short times, giving the impression of a quasi-continuous flow (*granular regime*). As the particle magnetization increases, at $f \approx 7$ the previously quasi-continuous flow is replaced by individual narrow and long particle clusters falling at the system's boundary (*correlated regime*).

The difference in avalanche formation in different regimes can be clearly observed in avalanche duration to avalanche size relation. In all three simulation sets, at given interparticle force ratios, for each avalanche size we collected the measured avalanche durations and calculated the corresponding average avalanche durations. We also examined the avalanche size and duration distributions. Our results are summarized in the next two subsections.

3.3 The Granular Regime

In the granular regime the avalanche sizes are proportional to the corresponding average avalanche durations. The proportionality factor (i.e. the ratio of avalanche sizes and average avalanche durations) defines an average avalanche flow, which increases linearly with f (see left panel of Fig. 5). This linear dependence is explained by the fact that the avalanches in granular regime are formed by quasi-continuous flows of small particle chains, and the height of these chains increases linearly with f.

The avalanche size distribution at interparticle force ratios $1 < f < 7$ can be scaled together reasonably well (see right panel of Fig. 5) using the ansatz

$$P(s, f) = f^{-1}Q(s/f), \tag{2}$$

where s denotes avalanche sizes, $P(s, f)$ is the probability associated with an avalanche size, and $Q(\cdot)$ is a function with integral 1 on the $[0, +\infty)$ interval. Based on the avalanche size distributions, we argue that the magnetic cohesion introduces a well-defined characteristic size in particle avalanches. From the scaling property we conclude that the characteristic avalanche size increases linearly with the interparticle force ratio. Qualitatively similar results were found in the experiments by Szalmás et al. [7].

As both the characteristic avalanche size and the average avalanche flow increase linearly with f, the characteristic average avalanche duration (equal to the ratio of the former two) in leading order is independent of f. The dependence of the avalanche duration distribution on f is contained in higher order corrections which could not be captured by our simulations.

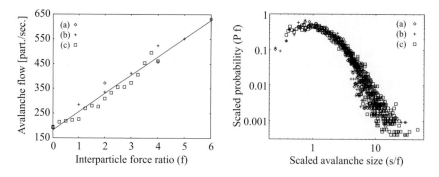

Fig. 5. Dependence of average avalanche flow on interparticle force ratio (left panel) and scaled avalanche size distribution (right panel) in granular regime. The avalanche flow is measured in particles per second. We examined three different simulation setups (see text for details). A well-defined characteristic size can be observed on the right panel.

3.4 The Correlated Regime

In the correlated regime the avalanche durations are given by the free fall of long particle clusters, and accordingly the square of the measured avalanche durations are proportional to the length of the clusters, and thus proportional to f. At the same time, the width of the falling particle clusters is small, and thus the avalanche sizes in leading order are proportional to the cluster length. In consequence the avalanche sizes are proportional to the square of the corresponding average avalanche durations, contrary to the linear dependence found in granular regime.

The width of the falling particle clusters can take arbitrary values up to some maximum. In other words, the particle clusters can have different number of *layers* while having the same length. This introduces large fluctuations in avalanche sizes. These fluctuations are proportional to the cluster length, and thus proportional to f.

The avalanche duration distributions at interparticle force ratios $7 < f < 25$ can be scaled together reasonably well (see Fig. 6) using the ansatz

$$P(\tau, f) = f^{-1/2}Q(\tau/f^{1/2}), \tag{3}$$

where τ denotes avalanche durations, $P(\tau, f)$ is the probability associated with an avalanche duration, and $Q(\cdot)$ is a function with integral 1 on the $[0, +\infty)$ interval. A well-defined characteristic duration can be observed in Fig. 6. Based on the scaling property, the square of the characteristic avalanche duration increases linearly with the interparticle force ratio. As the square of the avalanche duration is proportional to a corresponding cluster length, and a cluster length in leading order defines an avalanche size, we can conclude that there is also a mean characteristic avalanche size, which consequently

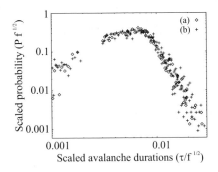

Fig. 6. Scaled avalanche duration distribution in correlated regime. A well-defined characteristic duration can be observed. We examined two simulation setups (see text for details).

is proportional to f. We could not find this avalanche size explicitly in the avalanche size distribution because of the large fluctuations of the same order of magnitude as the mean value.

4 Conclusions

Numerical results on a dipolar two-line system show that the ferromagnetic state is much more sensitive to the interaction cutoff than the corresponding anti-ferromagnetic state. For $R \gtrsim 8$ there is no substantial change in the energetical balance of the ferromagnetic and anti-ferromagnetic state, and the ferromagnetic state slightly dominates. The situation is changed rapidly for lower interaction cutoff values, leading to the disappearance of the ferromagnetic ground state at $R \approx 2$. Brankov and Danchev [8] found that in case of an infinite triangular lattice the ferromagnetic ground state disappears at $R \approx 3$.

We studied the effect of bending ferromagnetic and anti-ferromagnetic two-line systems. We created (γ, N) state diagrams separating energetically favorable ferromagnetic and anti-ferromagnetic states. We observed that there is a substantial change on these state diagrams for $R \lesssim 4$, and – in accordance with our previous results – we argue that $R \approx 5$ is a reasonable choice for dipole-dipole interaction cutoff.

As an application, we studied basic effects in two-dimensional granular piles formed by magnetized particles in simulations similar to the experiments of Forsyth et al. [6]. We measured the angle of repose, the surface roughness, and the effect of magnetization on particle avalanching. Dropping the particles from a given height or placing them gently, and switching on or off the effect of the side walls gave no qualitative difference. In accordance with the results of Forsyth et al. [6], we found that (in the examined domain) both the angle of repose and the surface roughness exhibits linear dependence on the ratio f of the maximum magnetic force at contact and the gravitational force.

As reported by Tegzes et al. [4] in case of wet granular media there is a difference in avalanche formation at small and at large interparticle force ratios. We could also identify a *granular* and a *correlated regime* in case of magnetized particles. The granular regime is characterized by quasi-continuous granular flows, while the correlated regime is characterized by long particle clusters falling at the system's boundary. The transition between the two regimes is not sharp. In simulations we found that the transition is at $f \approx 7$. Calculations based on stability criteria [20] indicate a transition at $f \approx 6$.

In the granular regime the avalanche sizes are proportional to the corresponding average avalanche durations. According to this, there is a well defined average granular flow characterizing the avalanches. We found that this increases linearly with f. Analyzing the avalanche size distributions, we also found that there is a well-defined characteristic size in particle avalanches. Based on scaling properties of the avalanche size distributions, we argue that the characteristic avalanche size increases linearly with f. The characteristic average avalanche duration in leading order seems to be independent of f.

In the correlated regime the avalanche sizes are in leading order proportional to the square of the corresponding average avalanche durations. This is explained by the free fall of long particle clusters. The avalanche durations are defined by the length of the falling particle clusters. The width of the falling clusters is small, and can take arbitrary values up to a maximum. According to this the avalanche sizes are defined in leading order by the cluster lengths, but there are large fluctuations which are also proportional to the cluster lengths. The cluster length increases linearly with f. The avalanche duration distributions show evidence of a characteristic avalanche duration, indicating also a characteristic avalanche size. We could not identify this avalanche size explicitly because of the large fluctuations in the avalanche size distribution.

Our results regarding the avalanche size distributions in both the granular and the correlated regime are very close to the experimental results of Tegzes et al. [4] on wet granular materials. They could identify characteristic avalanche sizes in both regimes, and they state that the avalanche size distributions in the correlated regime are *broad* and both small and large avalanches may occur. This seems to confirm our finding that in the correlated regime there are large avalanche size fluctuations.

Acknowledgements

This research was carried out within the framework of the "Center for Applied Mathematics and Computational Physics" of the BUTE, and it was supported by BMBF, grant HUN 02/011, and Hungarian Grant OTKA T035028.

References

1. D. J. Hornbaker, R. Albert, I. Albert, A.-L. Barabasi, and P. Schiffer. What keeps sandcastles standing? *Nature*, 387:765, 1997.
2. R. Albert, I. Albert, D. Hornbaker, P. Schiffer, and A.-L. Barabási. Maximum angle of stability in wet and dry spherical granular media. *Phys. Rev. E*, 56(6):R6271–R6274, 1997.
3. P. Tegzes, R. Albert, M. Paskvan, A. L. Barabási, T. Vicsek, and P. Schiffer. Liquid-induced transitions in granular media. *Phys. Rev. E*, 60:5823–5826, 1999.
4. P. Tegzes, T. Vicsek, and P. Schiffer. Development of correlations in the dynamics of wet granular avalanches. *Phys. Rev. E*, 67:051303, 2003.
5. M. A. S. Quintanilla, J. M. Valverde, A. Castellanos, and R. E. Viturro. Looking for self-organized critical behavior in avalanches of slightly cohesive powders. *Phys. Rev. Lett.*, 87:194301, 2001.
6. A. J. Forsyth, S. R. Hutton, M. J. Rhodes, and C. F. Osborne. Effect of applied interparticle force on the static and dynamic angles of repose of spherical granular material. *Phys. Rev. E*, 63:031302, 2001.
7. L. Szalmás, J. Kertész, and M. Zrínyi. Study of systems of magnetized particles. (unpublished), L. Szalmás, Diploma Work (in Hungarian), BUTE (2000).
8. J. G. Brankov and D. M. Danchev. Ground state of an infinite two-dimensional system of dipoles on a lattice with arbitrary rhombicity angle. *Physica A*, 144:128–139, 1987.
9. J. M. Luttinger and L. Tisza. Theory of dipole interaction in crystals. *Physical Review*, 70:954–964, 1946.
10. S. Fazekas, J. Kertész, and D. E. Wolf. Two-dimensional array of magnetic particles: The role of an interaction cutoff. *Phys. Rev. E*, 68:041102, 2003.
11. J. J. Weis. Orientational structure in a monolayer of dipolar hard spheres. *Molecular Physics*, 100:579–594, 2002.
12. Weijia Wen, F. Kun, K. F. Pl, D. W. Zheng, and K. N. Tu. Aggregation kinetics and stability of structures formed by magnetic microspheres. *Physical Review E*, 59:R4758–R4761, 1999.
13. D. L. Blair and A. Kurolli. Magnetized granular materials. In H. Hinrichsen and D. E. Wolf, editors, *The Physics of Granular Media*. Wiley-VCH, 2004.
14. P. A. Cundall and O. D. L. Strack. A discrete numerical model for granular assemblies. *Géotechnique*, 29(1):47–65, 1979.
15. S. Luding. Molecular dynamics simulations of granular materials. In H. Hinrichsen and D. E. Wolf, editors, *The Physics of Granular Media*. Wiley-VCH, 2004.
16. E. Altshuler, C. Martínez O. Ramos, L. E. Flores, and C. Noda. Avalanches in one-dimensional piles with different types of bases. *Phys. Rev. Lett.*, 86:5490–5493, 2001.
17. G. Kuwabara and K. Kono. Restitution coefficient in a collision between two spheres. *Japanese Journal of Applied Physics*, 26(8):1230–1233, 1987.
18. Th. Schwager and Th. Pöschel. Coefficient of normal restitution of viscous particles and cooling rate of granular gases. *Phys. Rev. E*, 57:650–654, 1998.
19. http://maxwell.phy.bme.hu/~fazekas/magaval. Avalanche movies of magnetized particles, 2004.
20. S. Fazekas, J. Kertész, and D. E. Wolf. Piling and avalanches of magnetized particles. (submitted to Phys. Rev. E).

Simulation of Particle Transport and Structure Formation by Deposition

D. Hänel and U. Lantermann

Institute of Combustion and Gasdynamics – University Duisburg-Essen
47048 Duisburg – Germany
haenel@ivg.uni-duisburg.de

Summary. This paper concerns with a solution concept for three-dimensional simulations of the transport of discrete particles and their deposition to irregular surface structures. The gas phase, governed by the Navier-Stokes or Stokes equations, and potential fields of electrical forces are solved by lattice-Boltzmann methods. The motion of particles is described by a particle Monte-Carlo method including convection, stochastic diffusion (Brownian diffusion), van der Waals forces or electrical forces. A number of simulation results show the influence of different effects on the particle motion and deposition.

Keywords: particle Monte-Carlo method, Lattice-Boltzmann methods, particle transport, deposition, structure formation

1 Introduction

Particles moving towards rigid surfaces form irregular, granular structures by depositing on the surface or on former deposited particles. The formation of deposits is essential governed by the transport of particles to surfaces, which is influenced by convection, diffusion or external forces. The structures and the containing amount of particles are of interest in filter techniques, e.g. for determining the filter efficiency or pressure losses due to obstruction, but play also a role in experimental and industrial processes for controlled manufacturing of surface structures or for extraction and analysis of the particle phase.

This paper concerns with the modeling and three-dimensional computation of transport and deposition of particles in a gas phase environment. The mathematical modeling of the physical problem is based on the assumption of dilute particle concentrations which enables decoupling of the phases. The transport of particles is governed by the flow of a gas phase, i.e. by convection and stochastic Brownian diffusion, and by external forces, as gravitation, electrical or magnetic forces. Interaction between the phases is considered near surfaces where particles are depositing and changing the boundary conditions of flow and potential fields and thus influencing the succeeding deposition. This influence is taken into account by recomputing the fluid and potential fields after a fixed amount of deposited particles.

The flow field and the potential fields of electric or magnetic forces are calculated here with lattice-Boltzmann (LB) methods, which enable not only very efficient algorithms but also the treatment of the very complex boundary conditions, e.g. resulting from the deposited particle structures. The algorithmic concept and further improvement of LB methods are published by the authors e.g. in [1, 2]. First applications of these methods to filtration problems of submicron particles were published by the authors in [3–5]. The discrete particle phase is computed using a particle Monte-Carlo method, including convection, stochastic diffusion (Brownian diffusion), van der Waals forces or electrical and magnetic forces.(The latter influence is discussed elsewhere.)

The mathematical models and numerical method are described briefly in the next section. A few typical results are presented finally.

2 Mathematical Models

2.1 Lattice-Boltzmann Methods

Lattice-Boltzmann approaches are employed for computing the flow field and the potential fields of electric or magnetic forces. Lattice-Boltzmann (LB) methods, used in an efficient variant, the so called lattice-BGK (LBGK) method [6, 7], is employed and modified here using grid refinement, [1], for local high resolution and higher order boundary conditions on molecular level for treating curvilinear contours, [2]. The LBGK method in detail and its analysis is described elsewhere, e.g. in [3, 8], especially for the case of viscous flows at low Mach number, governed by the Navier-Stokes equations. With a few modifications this LBGK method is able to compute the solution of a Poisson equation, e.g. for calculating the electric or magnetic field around a solid body.

$$\frac{\partial^2 \phi_\alpha}{\partial x_\beta^2} = -RS_\alpha \qquad (1)$$

where $RS_\alpha(t, \mathbf{x})$ is a given right hand side of a Poisson equation and ϕ_α are the potentials of electrical or magnetic fields.

Further extension of the LBGK method to the low Mach number approximation of the Navier-Stokes equations are presented by the authors in [9] for flows with external or internal (reactive) heat sources, where the density is a function of temperature and concentration.

Summarizing, the numerical advantages of the LB methods are the easy and granular, parallel algorithms, the capability to deal with different physical applications and in particular to deal with complex geometries.

2.2 Particle Monte Carlo Method

The transport of particles, their deposition and the formation of structures on surfaces under the influence of gaseous flow and electrical or magnetic fields is computed by tracking individual solid particles of spherical shape in dilute concentration. The particle motion is described in Lagrangian form by the Langevin equations for a particle of mass m_p:

$$m_p \frac{d\mathbf{v}(t, \mathbf{x})}{dt} = \mathbf{F}_d(t, \mathbf{x}) + \mathbf{F}_{ext}(t, \mathbf{x}) + m_p \mathbf{A}(t) \tag{2}$$

where the particle path is given by

$$\frac{d\mathbf{x}(t)}{dt} = \mathbf{v}(t, \mathbf{x}) \tag{3}$$

The motion of particles is influenced by the flow drag $\mathbf{F}_d(t, \mathbf{x})$, by external forces $\mathbf{F}_{ext}(t, \mathbf{x})$ and by random diffusion acceleration $m_p \mathbf{A}(t)$ representing Brownian motion.

The drag force $\mathbf{F}_d$ is proportional to the difference of fluid velocity $\mathbf{u}$ and particle velocity $\mathbf{v}$

$$\mathbf{F}_d = m_p \beta (\mathbf{u} - \mathbf{v}) \quad \text{with} \quad \beta = \frac{6\pi\eta r_p}{C m_p} \tag{4}$$

The friction coefficient β, related to the particle mass, is determined from the Stokes law with the dynamic viscosity of the gas η, the particle radius r_p and the particle mass m_p. The slip correction factor C has to be taken into account for particles whose diameters are of the same order or smaller than the mean free path λ of the gas molecules. It is calculated depending on the Knudsen number $\text{Kn} = \lambda/2r_p$ by $\quad C = 1 + 2\text{Kn}\left[A_1 + A_2 \exp\left(-\frac{A_3}{\text{Kn}}\right)\right] \quad$ with constants A_1, A_2 and A_3 from [10].

The equations Eq. (2) and (3) are rearranged in a coupled system of differential equations. These equations have to be integrated for determining the velocity $\mathbf{v}(t, \mathbf{x})$ and the position $\mathbf{x}(t)$ of each particle.

$$\begin{pmatrix} \frac{d\mathbf{v}}{dt} \\ \frac{d\mathbf{x}}{dt} \end{pmatrix} \underbrace{\begin{pmatrix} -\beta\mathbf{I} & 0 \\ \mathbf{I} & 0 \end{pmatrix}}_{\mathbf{T}_1} \begin{pmatrix} \mathbf{v} \\ \mathbf{x} \end{pmatrix} + \underbrace{\begin{pmatrix} \beta\mathbf{u} + \mathbf{F}_{ext}/m_p \\ 0 \end{pmatrix}}_{\mathbf{T}_2} + \underbrace{\begin{pmatrix} \mathbf{A}(t) \\ 0 \end{pmatrix}}_{\mathbf{T}_3} \tag{5}$$

Assuming the time step Δt is in the order of β^{-1} and is sufficient small then fluid velocity and forces can be considered as constant during one Δt. The system (5), neglecting first the stochastic term $\sim \mathbf{A}(t)$, can be solved analytically similar as in [11] and yields:

$$\begin{pmatrix} \mathbf{v}(t) \\ \mathbf{x}(t) \end{pmatrix} = \mathbf{B}(t) \begin{pmatrix} \mathbf{v}(t_0) \\ \mathbf{u}(t_0) + \frac{1}{m_p\beta}\mathbf{F}_{ext}(t_0) \end{pmatrix} + \begin{pmatrix} 0 \\ \mathbf{x}(t_0) \end{pmatrix} \tag{6}$$

with

$$\mathbf{B}(t) = \begin{pmatrix} e^{-\beta(t-t_0)}\mathbf{I} & (1 - e^{-\beta(t-t_0)})\mathbf{I} \\ \frac{1}{\beta}\left(1 - e^{-\beta(t-t_0)}\right)\mathbf{I} & \left[(t - t_0) - \frac{1}{\beta}\left(1 - e^{-\beta(t-t_0)}\right)\right]\mathbf{I} \end{pmatrix} \tag{7}$$

The stochastic term $\mathbf{T}_3$ can be determined in analogous manner due to Itô's rule [11]. The resulting integrals cannot be solved directly, since the random function $\mathbf{A}(t)$ has a mean equal to 0 and is not continuous. The statistical properties of the integrals are discussed in detail in [12]. Application to this particular case as in [13], yields a formulation of the random movement, validated and used in the succeeding computations:

$$\int_{t_0}^{t} \mathbf{T}_3(\tau)\,d\tau = \begin{pmatrix} z_1\sigma_v \\ z_1\frac{\sigma_{vx}}{\sigma_v} + z_2\sqrt{\sigma_x^2 - \frac{\sigma_{vx}^2}{\sigma_v^2}} \end{pmatrix} \tag{8}$$

where

$$\sigma_v = \sqrt{\frac{kT}{m_p}\left(1 - e^{-2\beta(t-t_0)}\right)} \qquad \sigma_{vx} = \frac{kT}{m_p\beta}\left(1 - e^{-2\beta(t-t_0)}\right)^2 \tag{9}$$

$$\sigma_x = \frac{1}{\beta}\sqrt{\frac{kT}{m_p}\left(2\beta(t - t_0) - 3 + 4e^{-\beta(t-t_0)} - e^{-2\beta(t-t_0)}\right)} \tag{10}$$

with the Boltzmann constant k and the temperature T. The components of z_1 and z_2 are normally distributed random numbers with mean equal 0 and variance equal 1.

The complete solution of equation Eq. (5) results from the superposition of Eq. (6) and Eq. (8). With index n for temporal and index i for spatial discretization level the solution is given by

$$\begin{pmatrix} \mathbf{v}_i^{n+1} \\ \mathbf{x}_i^{n+1} \end{pmatrix} = \mathbf{B}_i^n \begin{pmatrix} \mathbf{v}_i^n \\ \mathbf{u}_i^n + \frac{1}{m_p\beta}\mathbf{F}_{ext,i}^n \end{pmatrix} + \begin{pmatrix} 0 \\ \mathbf{x}_i^n \end{pmatrix} + \begin{pmatrix} z_1\sigma_v \\ z_1\frac{\sigma_{vx}}{\sigma_v} + z_2\sqrt{\sigma_x^2 - \frac{\sigma_{vx}^2}{\sigma_v^2}} \end{pmatrix} \tag{11}$$

where

$$\mathbf{B}_i^n = \begin{pmatrix} e^{-\beta\Delta t}\mathbf{I} & \left(1 - e^{-\beta\Delta t}\right)\mathbf{I} \\ \frac{1}{\beta}\left(1 - e^{-\beta\Delta t}\right)\mathbf{I} & \left[\Delta t - \frac{1}{\beta}\left(1 - e^{-\beta\Delta t}\right)\right]\mathbf{I} \end{pmatrix} \tag{12}$$

The stochastic approach of the Brownian motion, Eq. (8), was confirmed by comparing the diffusion coefficients obtained from the theoretical Stokes-Einstein equation for mean square of the particle displacement $\overline{\mathbf{x}^2}$.

External Forces

The gravitational force and the buoyancy force due to the displaced fluid volume are combined in one force

$$\mathbf{F}_g = -(m_p - m_g)\,\mathbf{g}. \tag{13}$$

The displaced fluid mass of spherical particles of volume $V_p = \frac{4}{3}\pi r_p^3$ is $m_g = \rho_g V_p$ with the gas density ρ_g, the particle radius r_p and the gravitational constant $\mathbf{g}$. This force is proportional to r_p^3 and is negligible in general for very small particles as considered here. Additional forces, as lift forces or added mass force, are neglected with the same argument.

Electrical forces are relevant for charged particles moving in an electric field. Two forces, the Coulomb and the image force, are included. The Coulomb force

$$\mathbf{F}_{\mathrm{cou}} = -qe\,\nabla\phi \tag{14}$$

describes the migration of a particle in an external electric field. Here e is the electronic charge, q is the number of electronic charges accumulated by the particle and ϕ is the electric potential which can be determined by a Laplace equation. It is well known that a charged particle moving in a fluid of dielectric constant ϵ_3 induces an image charge in the vicinity of a surface with the dielectric constant ϵ_2. The image charge is located in a distance D on the other side of the boundary, where D is the distance between the center of the particle and the surface. The force of this interaction is given by

$$\mathbf{F}_{\mathrm{im\_ps}} = \frac{-(qe)^2}{(4\pi\epsilon_0\epsilon_3)(2D)^2}\left(\frac{\epsilon_2 - \epsilon_3}{\epsilon_2 + \epsilon_3}\right)\mathbf{n} \tag{15}$$

where ϵ_0 is the static dielectric constant and $\mathbf{n}$ is the normal vector directed from the particle center to the surface.

More details for modeling of external forces are found in the literature, as e.g. in [10, 14].

Deposition Model, van der Waals Forces and Interactions

Van der Waals forces act over distances short compared to the particle diameter and play therefore a role in the deposition process. These forces cannot be neglected for very small particles and if the distance between the particles or rigid surface is sufficiently small.

Two different numerical approaches for modeling van der Waals forces were tested. The first, simpler approach assumes an effective collision cross-section to model the effect of van der Waals forces. The particle radius or the wall surface is increased then by a distance d of order of magnitude of particle radius. Deposition is assumed if a particle contacts a rigid wall or a previously deposited particle in this distance. This model is computationally easier but suffers on the realistic determination of the effective cross-section. The additional collision cross-section can be neglected for larger particles, then a "touch and stop" model is used for modeling deposition, as described in [3].

The second approach substitutes van der Waals forces by a continuous force model described in the following. For particle-surface interactions the van der Waals force is described by

$$\mathbf{F}_{\mathrm{vdw\_ps}} = \frac{2A}{3} \frac{r_p^3}{d^2(2r_p - d)^2} \mathbf{n} \tag{16}$$

and for particle-particle interaction by

$$\mathbf{F}_{\mathrm{vdw\_pp}} = \frac{32A}{3} \frac{r_p^6}{d^3(4r_p^2 - d^2)^2} \mathbf{n}, \tag{17}$$

with the Hamaker constant A, the particle diameter r_p, the minimal distance d between the surface of the moving particle and the one of the macroscopic body or of the deposited particle. In general the distance d approaches to zero if the particle is deposited. The vector $\mathbf{n}$ is the normal unit vector directed from the moving particle to the surface of the deposited particle.

The interaction of particle deposits with fluid flow and potential forces is taken into account via changed boundary conditions. Particles deposited on the rigid surface or on previously deposited particles form irregular and complex structures, depending on the fluid flow and the fields of the external forces. Thus the changed surface structures alter the boundary conditions and influence interactively the next particle deposition. These interactions are taken into account by recomputing the flow and electrical fields with the new boundary conditions each time after a certain number of particles has been deposited during the computations. The relatively easy, numerical realization of such complex boundary geometries, as well for fluid flow as for potential fields, is one of the advantages of the lattice-Boltzmann methods.

3 Numerical Results

Fiber filters are considered here as test geometries. A three-dimensional model filter consists of crossing fibers, periodically extended in their plane, such that an infinite filter mat is represented, see e.g. Fig. 1. Other more complex configurations are possible.

The gas-particle flow through fibrous filters and deposition of particles in the submicron range $d_p > 0.5\,\mu$m was considered by the authors in [3–5]. The motion of these particles is influenced essentially by inertia forces, the corresponding particle trajectories are deterministic. Since these particles are relatively large, the obstruction of the filter and the change of boundary conditions by deposited particles plays a role. The interaction between already deposited particles and the changing of hydrodynamics and thus change of particle paths is taken into account by recomputing the flow field over the resulting complex surfaces. Fig. 1 shows the deposits with and without interaction for particles of diameter $d_p = 1\,\mu m$ and correspondingly in Fig. 1,

  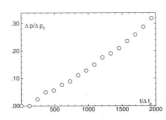

Fig. 1. Deposited layers of particles, $d_p = 1\,\mu m$, on fibrous filter with $D_f = 50\,\mu m$ and $Re = 1$, $St = 0.14$. Structures without interaction between deposit and flow (left) and with interaction (middle). Resulting pressure loss versus time due to increasing filter obstruction by deposition (right).

right, the growing pressure loss due the obstruction of the filter, which is a result if interaction is taken into account.

Recent considerations deal with nano particles with diameters smaller than 100 nm. The effect of inertia forces becomes smaller in this range, but random diffusion and thus the Brownian random motion of particles plays an increasing role. In the following examples the fluid field and the electric field, if required, is calculated by the lattice-Boltzmann method. The particle motion is determined by the integration of the Langevin equation Eq. (5) including stochastic Brownian motion. The deposition of the particles is influenced by van der Waals forces or image forces in the case of charged particles. The interaction between the deposited particles and the fluid field or the electric field respectively is taken into account iteratively. Numerically, higher resolution near the surface is realized by local grid refinement. Averaging of multiple solutions with different initial configurations of particles is required to achieve reliable statistical results.

Figure 2 shows geometry and deposited particles for two different particle diameters of 10 nm (left) and 30 nm (right). Packages of particles are started at the inflow boundary, due to different influences some of them deposit, the other are flying through. From these the penetration values is calculated with 0.764 for10 nm and 0.519 for 30 nm particles. The penetration value is defined as ratio of non-deposited to oncoming particles and is equal one minus filter efficiency. Irregular dendrite-like formations arise in both cases but in different strength. The dependence of penetration from the diameter is governed by the balance between Brownian fluctuations and inertia forces. Fig. 3, left, shows calculated values of penetration values versus particle diameter. The results in Fig. 2 for 10 nm correspond to values on the right branch of curve Fig. 3, i.e. the penetration decreases with increasing particle diameter. The penetration versus diameter for smaller particles (left branch) shows the opposite behavior, which is explained by the fact that for smaller particles the Brownian diffusion dominates, while the inertia forces have an increasing influence for larger particles. This behavior is confirmed by other investigations, too.

Comparison is made with experimental data of particle deposition in fibrous

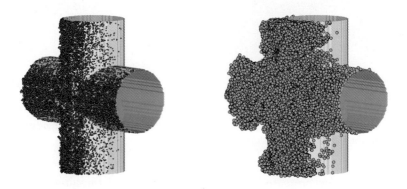

Fig. 2. Deposited layers of nano particles on fibrous filter with $D_f = 650\,nm$, flow with Re $= 0.89$. Left: particle diameter $d_p = 10\,nm$ and right $d_p = 30\,nm$.

filters, [15]. The experiments are performed with particles of 10, 15, 20, 25 and 30 nm diameter and a fiber diameter of the filter of $65\,\mu$m in nitrogen with a fluid velocity of $0.2126\,$m/s. Fig. 3, right, shows sufficient agreement for the penetration as function of the particle diameter between experimental values (crosses) and calculated values (circles). Comparing with diagram Fig. 3, left, this behavior belongs qualitatively to left branch near maximum.

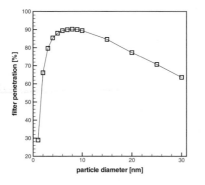 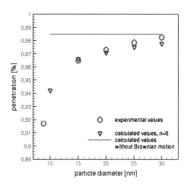

Fig. 3. Calculated penetration values for different particle diameters with a fiber diameter of 650 nm at Re $= 0.89$ (left) and comparison between calculated penetration values and experimental values, [15] (right).

One example with electric forces shows the deposition of charged particles on a flat plate. The main part of the plate has a different potential as

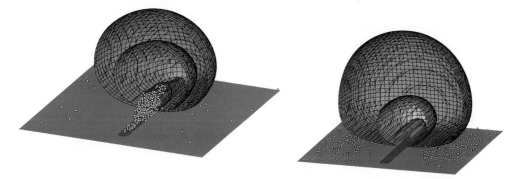

Fig. 4. Iso-surfaces of the electric potential and deposited layer after 1000 deposited 80 nm particles influenced by an attracting electric force (left) and repulsive forces (right).

the narrow stripe in the middle. According to the charge of the particles either attracting or repulsive effects are acting on the particles. Image forces and van der Waals forces are considered in the vicinity of the surface. Three-dimensional stagnation point flow from the top to the plate is assumed. The Fig. 4, left, shows the deposition behavior for opposite charged middle stripe and particles. The particles are attracted essentially by the stripe and deposited there. The electric potential of the deposited particle is taken from the place where it touches the surface, i.e. the plate or an earlier deposited particle. Therefore it changes the boundary conditions for the potential as seen by the iso-surfaces of the electric potential. Fig. 4, right, illustrates the deposited layer for the same parameters but with rejecting electric forces in the middle range. Most of the particles are deposited outside of the stripe.

4 Conclusion

The paper presents a solution concept for simulating the three-dimensional processes of transport and deposition of particles in complex geometries. The use of lattice-Boltzmann methods for the carrying gas phase and potential fields coupled with a particle Monte-Carlo method enables calculations with high numerical efficiency and high flexibility with respect to geometries and different physical aspects. Work is in progress for magnetic particles, which require extension of the method to the rotational motion of particles.

Acknowledgements

This work is supported by the "Deutsche Forschungsgemeinschaft" in the frame of the SFB 445

References

1. Filippova, O. & Hänel, D. (1998). Grid refinement for Lattice-BGK models. *J. Comput. Phys.*, *147*, 219 - 228.
2. Filippova, O. & Hänel, D. (1998a). Boundary fitting and local grid refinement for lattice-BGK models. *Int. J. Mod. Phys. C*, *9*, 1271 - 1279.
3. Filippova, O. & Hänel, D. (1997). Lattice-Boltzmann simulations of gas-particle flow in filters. *Comput. Fluids*, *7*, 697 - 712.
4. Filippova, O. & Hänel, D. (1998c). *Numerical Simulation of Gas-Particle Flow in Filters by Lattice Bhatnagar-Gross-Krook Model*. In: K.R. Spurny (Ed.): Advances in Aerosol Filtration. Lewis Publishers, Boston London New York.
5. Krafczyk, M., Lehmann, P, Filippova, O., Hänel, D. & Lantermann, U. (2000). *Lattice Boltzmann simulations of complex multiphase flows*. In: A.M. Sändig, et al. (eds.), Multifield problems - state of the art. Springer-Verlag.
6. Qian, Y.H., d'Humiéres, D. & Lallemand, P. (1992). Lattice BGK models for Navier-Stokes equation. *Europhys. Lett.*, *17*, 479 - 484.
7. Chen, H., Chen, D. & Matthaeus, W. (1992). Recovery of the Navier-Stokes equations through a lattice gas Boltzmann equation method. *Phys. Rev. A*, *45*, R5339 - R5342.
8. Filippova, O. & Hänel, D. (2000b). Acceleration of Lattice-BGK schemes with grid refinement. *J. Comput. Phys.*, *165*, 407 - 427.
9. Filippova, O. & Hänel, D. (2000a). A novel lattice BGK approach for low Mach number combustion. *J. Comput. Phys.*, *158*, 139 - 160.
10. Friedlander, S.K. (2000). *Smoke, dust, and haze - fundamentals of aerosol dynamics* Oxford University Press.
11. Karatzas, I & Shreve, S.E. (1988). *Brownian motion and stochastic calculus.* Springer-Verlag.
12. Chandrasekhar, S. (1943). Stochastic problems in physics and astronomy. *Rev. Mod. Phys.*, *15*, 1 - 89.
13. Zarutskaya, T. & Shapiro, M. (2000). Capture of nanoparticles by magnetic filters. *J. Aerosol. Sci.*, *31*, 907 - 921.
14. Brown, R.C. (1993). *Air filtration, an integrated approach to the theory and applications of fibrous filters.* Pergamon Press.
15. Kauffeldt, T. (2003). Personal communication.

Experiment and Simulation of Electrospray Particle-Flows for Controlled Release of Drugs

J.C.M. Marijnissen[1], T. Ciach[2], M.-K. Müller[1], T. Winkels[1], K.B. Geerse[1,3], A. Schmidt-Ott[1], and S. Luding[1]

[1] Particle Technology, DelftChemTech – TU Delft
Julianalaan 136, 2628 BL Delft – The Netherlands
j.c.m.marijnissen@tnw.tudelft.nl, s.luding@tnw.tudelft.nl
[2] Faculty of Chemical and Process Engineering –
Warsaw University of Technology
Warynskiego str. 1, 00-645 Warsaw – Poland
ciach@ichip.pw.edu.pl
[3] PURAC division
Arkelsedijk 46, P.O.Box 21, 4200 AA Gorinchem – The Netherlands

Summary. Electro HydroDynamic Atomisation (EHDA) disperses a liquid into small, highly charged droplets. We show that this method can be used to produce particles that release a drug at a desired rate. This is done by spraying a solution of bio-degradable polymers and an enzyme, which represents the effective drug. The release rate can be varied by modification of the polymer matrix. It is further demonstrated that the enzyme fully retains its functionality in the EHDA process. Practical use of this technique for medicine production requires a scaled-up design, which must be based on an adequate model of the particle flow in the charged droplet spray plume. As a step in this direction, the most important result is a scale-up relation that allows simulations of an experimental spray with millions of particles, using only a few thousand model particles. The experimental spray is examined with a Phase Doppler Particle Analyser (PDPA) set-up, and the resulting density and velocity profiles are compared to the numerical results. There is a qualitative agreement between experiment and model.

Keywords: aerosol particle flows, electrospray, drug release, discrete particle flow simulation, verification with experiments

1   Introduction

A spray of uniformly charged particles is an example of a granular system, the flow behavior of which is governed by the repulsive particle-particle interaction. This many-body problem is not easy to solve but the investigation of charged sprays is of special interest, because a spray process can be used for the production of particles for a wide variety of materials. In particular, the phenomenon of electro hydrodynamic atomization (EHDA) or electrospray allows dispersion of a liquid into equally sized droplets. Since they are highly charged, their mutual repulsion avoids aggregation, and they retain their uniform size. EHDA has the potential of producing uniformly sized particulate products from a large variety of materials [1,2]. One of the most interesting options is the production of new drugs with designed, well-defined release characteristics [3-6]. A condition for this field of applications is that the dispersion process is "soft" enough to avoid chemical modification of fragile bio-molecules [7]. The present study will prove this for a special but probably representative case. It will also be shown that EHDA enables design of particles that release the effective agent slowly and with an adjustable rate. In developing EHDA into a production technique satisfying the quantitative needs for drug production, we meet the challenge of scaling up the present laboratory set-ups. This requires modeling of a system containing millions of particles mutually interacting via a second-order force law. The present study represents an important step in this direction, by applying a simulation technique that assumes a smaller, feasible number of particles and finding a scale-up relation that translates the result to the real world system. The numerical simulation results are compared to experimental results obtained through phase Doppler particle analysis.

2   Electro Hydro Dynamic Atomisation (EHDA)

If a fluid forms a droplet at the end of a capillary tube, the droplet deforms into a cone (Taylor cone) under the action of an electric field. Under suitable conditions, a jet is formed at the cone tip (cone jet mode). The jet is unstable due to its high charge density and breaks into small, charged droplets. These droplets have a charge that is close to the so-called Rayleigh limit charge, i.e. the highest charge a droplet can have without exploding. In the cone jet mode the flow and field conditions can be adjusted in a way that equally sized particles are formed [8,9]. Particle sizes from nanometers till tens of micrometers can be achieved. Figure 1 shows a magnified photo of the spray formation and a high-speed camera picture of the the break-up of the jet into droplets

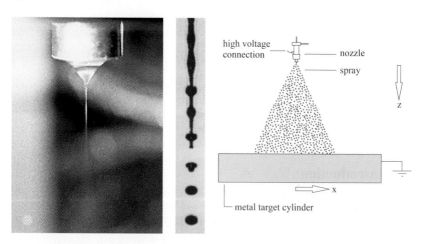

Fig. 1. Electrospraying of the liquid in the cone-jet (left) mode and the jet break-up process (center). The outer nozzle diameter is 5 mm, the inner nozzle is 1 mm. Scheme of the experimental set-up (right) with a spray below the cone. The distance from the nozzle to the target cylinder is about 20 cm.

An experimental set-up used for the present study is schematically shown in Fig. 1 (right). Droplet velocity and size measurements are done with a Phase Doppler Particle Analyser (PDPA). The PDPA equipment measures the number, the size and the velocity of the droplets at a single spot [8,9,10].

A 9:1 ethanol/triethylene glycol mixture is sprayed onto a grounded cylindrical metal target with a flow rate of 7.0 ml/h. The nozzle is set to an electric potential of 18 kV, the target is grounded. The PDPA transceiver and receiver are in front of and behind the plane shown in Fig. 1, respectively. By moving the equipment and thus the point of focus, velocity profiles can be measured. This set-up uses simple geometries and primarily serves the purpose of delivering experimental data to be compared with the model described in Sect. 5.

The occurrence of droplet evaporation adds considerable complexity to the process, because the droplets exceed the Rayleigh limit by shrinking and thus explode to form smaller droplets. Monodispersity is then lost. In the experiment, we are reasonably sure that evaporation does not occur upstream of the measuring zone. If evaporation is too strong corona discharge can be used to reduce the charge of the droplets.

3   Proof of Enzyme Survival Under EHDA

Since in EHDA high voltages are applied and high shear forces occur, we examined if easily decomposed substances, like enzymes, are not affected by electrospraying. For this purpose, a solution of the enzyme α-amylase in water was atomised by EHDA. The composition of the solution by weight was 5% of glucose, 10% of polyethylene glycol (PEG, 4 600 mw. Aldrich) and 0.03% of the enzyme protein (bacterial α-Amylase (Sigma)). The enzyme content in the dry particles was thus 0.2%. To check the enzyme activity after atomization, a standard colorimetric test with the dye Amylopectin Azure (Sigma) was performed. About 80 mg of particles were collected on a glass fiber filter. Then the filter was immersed in 100 ml of a pH7 buffer solution. After 15 minutes of gentle agitation, dissolving the particles, 10 ml of this solution was added to 100 ml of a pH7 buffer containing 0.5 g of Amylopectin Azure. The concentration of Amylopectin Azure was checked every 5 minutes by colorimetry. The whole test was performed at 22°C. The same test was done with 0.5 ml of the primary, not electrosprayed solution. The comparison showed that the enzyme activity was the same within the experimental error before and after electrospraying. Thus the EHDA process proves to be "soft" enough to retain the protein structure. We presume that this experiment is representative for all proteins used in drug formulation.

4   Production/Test of Particles for Controlled Drug Release

A solution of a drug and a biodegradable polymer is electrosprayed in a set-up with a corona discharge unit in order to neutralize the droplets. The spray is then passed through a diffusion dryer, and the resulting solid particles are deposited in a glass fibre filter. As precursors for a polymer matrix, PLGA (poly-DL-lactide-co-glycolide 50:50, Aldrich) and PEG (polyethylene glycol, 4 600 mw, Aldrich) were selected. Both polymers are biodegradable and have been approved by the American Food and Drug Administration authorities as drug components. PLGA slowly decomposes in the human body. PEG is a waxy solid. This polymer was added to the precursor solution to modify the hydrolysis rate of the PLGA particles. Hydrolysis is an important mechanism in the drug release process. As an example of a drug, we used paclitaxel (taxol), a medicine against certain types of cancer. It is extracted from yew trees. The drug content in the dry particles was one percent by weight. A mixture of dichloromethane and acetone was used as a solvent. The filters with collected particles were left in a dryer at 40°C for thorough drying overnight.

Figure 2 (left) shows a micrograph of the particles. The size distribution was fairly narrow, but there was also a fraction of much smaller particles. It is not clear if these very small particles originate from the droplet break-up process or from Rayleigh disintegration. The contribution of these particles to the volume of the system is very small. The average particle size as measured with an optical particle counter (Topas, volumetric display) was 13 μm.

The drug release characteristics of both PLGA/Paclitaxel particles and PLGA/PEG (10:1 weight ratio)/Paclitaxel particles were investigated. To simulate the release in the human body liquid environment, the particle-loaded collection filter was immersed in 200 ml of a pH7 buffer solution with a small addition of sodium azide to prevent bacteria growth. 1 ml samples were periodically taken from the solution to measure the paclitaxel release into the liquid as a function of time. They were analyzed with respect to the paclitaxel content by means of liquid chromatography. The cumulative paclitaxel release with time, measured with liquid chromatography, is shown in Fig. 2 (right).

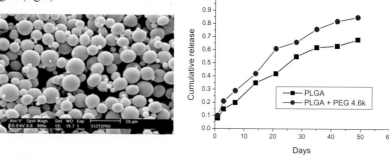

Fig. 2. (Left) Paclitaxel containing polymer (PLGA) particles produced by EHDA. (Right) Cumulative release of paclitaxel from polymer (PLGA, PEG) microparticles.

The release rate is not far from constant over a time span of 30 days. The initial burst of the active substance might be associated with the very small particles and from drug on the particle surfaces which can be avoided by coating particles with a inert layer. PEG has the effect of increasing drug release. Thus the release characteristics could be tailored to the needs by adjusting the polymer mixture.

5   Particle Flow Model

While the potential of the EHDA principle for the production of slow release medicine is evident from the results above, the problem of scaling up the process to deliver sufficient quantities has not been solved yet. Since we will deal with a completely different electric field and space charge situation in any multi-nozzle system, this requires a model that adequately reflects the spray evolution between the nozzle and the target. Our approach consists in modeling the individual trajectories of the charged droplets in the spray, taking into account gravity, drag, external fields, and self-interaction due to the uni-polar charge on the droplets [8,10,11]. The experimental data that have been obtained with the PDPA measurement described in Sect. 2 is used to test the model.

For each droplet produced, the equation of motion,

$$m_i \frac{d\vec{v}_i}{dt} = q_i \vec{E} + \vec{f}_D + \sum_{j \neq i} \frac{f_q^2 q_i q_j \vec{r}_{ji}}{4\pi\varepsilon_0 r_{ji}^3} + m_i \vec{g} \ , \tag{1}$$

with the drag force

$$\vec{f}_D = C_D \frac{\pi}{8} \rho_{air} d_i^2 (\vec{v}_{air} - \vec{v}_i)|\vec{v}_{air} - \vec{v}_i| \ , \tag{2}$$

is solved for each time-step dt, for each particle i. The particles are inserted with a certain rate and initial velocity at the tip of the nozzle [10]. The droplet production time (or the inverse production rate) was decreased in comparison with reality, while the Coulomb interaction charge q was systematically increased by the factor of f_q. For the numerical solution of the equations of motion for each particle, several assumptions lead to the terms in Eq. (1):

(i)   The first term corresponds to the force on the particles due to the external electric field between the charged nozzle and the target cylinder. For the el.-field computation MATLAB/FEMLAB was used [10,11].

(ii) The second term is the drag force, where the atmosphere is here assumed to be at rest. The drag coefficient leads to Stokes drag in the laminar regime and to turbulent drag for large relative velocities, with a transient regime in between [9,10].

(iii) The third term is the particle-particle self-interaction, where the sum extends over all charged particles with charge q and charge correction factor f_q. Image charges are not taken into account here.

(iv) The fourth term is the gravitational force.

(v)   Neglect other forces and changes of droplet properties (evaporation).

Starting the simulation with single particles being produced one by one, it takes some time until a steady state situation evolves. The spray shape is shown (right).

Besides the other forces, which are not affected by other particles, the strong, long-range Coulomb forces cause every droplet to interact with all other droplets. This leads to a many-body problem with immense computational effort for large particle numbers. The limits of our computing power (single processor) were reached with about 1000 droplets, for a steady state simulation. Our approach is therefore to reduce the droplet  number (in experiment, typically 10^5 droplets are in the steady-state spray) by reducing the droplet production frequency. To compensate for the resulting reduction in space charge and Coulomb interaction, the particle charge (in the particle-particle interaction term) is increased by a factor f_q. The problem is then reduced to finding a scaling relation between charge and concentration/production-rate and verifying its validity.

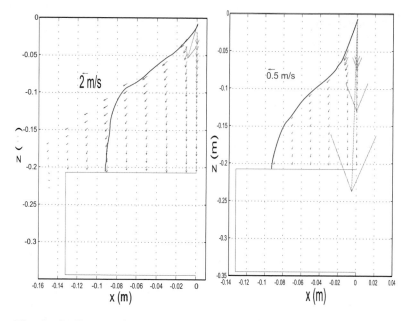

Fig. 3. (Left) Experimental velocity profile in the left-half of the system. (Right) Corresponding velocity profile from the simulations.

In order to find the correct charge correction factor f_q, a set of simulations with various production-rates and charge correction factors was performed. The results were examined with respect to the size of the spray cloud/plume, as measured at the target, where the droplets are deposited. Two simulations were found to lead to identical deposits, if Eq. (3) describes the relation between the varied parameters:

$$\frac{f_q(1)}{f_q(2)} = \left(\frac{t_{prod}(1)}{t_{prod}(2)}\right)^{\left(\frac{0.57}{1.06}\right)}.$$

(3)

Furthermore, it is possible to extrapolate from the range of simulated parameters, which charge factor has to be used for the scaled simulation (with about 1000 droplets) of an experiment (with assumed 10^5 droplets). Assuming that for the experiment, $f_q(2)=1$ is inserted, so that one obtains $f_q(1) \approx 12$ for the ratio of production times of about 100.

The simulated and experimental velocity fields in horizontal and vertical direction are shown for a half-plane in Fig. 3, left and right, respectively. The unit of velocity is displayed as an arrow in both figures. Note that the velocity in the experiment is systematically larger than the simulation velocity. The quantitative disagreement is due to the simplifications of the model [10], the most severe of which are: (i) no image charge is used – with an image charge, particles would be accelerated more strongly towards the target, and (ii) the gas is assumed at rest – in reality the gas is accelerated by the particles and moves with them, thus reducing the drag force on the particles and allowing for larger velocities. These points are to be addressed in more detail in a future publication [12].

6   Conclusions

The present study has shown that Electro Hydrodynamic Atomization (EHDA) is a suitable tool of producing drug particles in the size range of several microns. The process is "soft" enough to retain complicated organic molecules such as enzymes. Bio-degradable polymers added to the sprayed solution form a matrix enclosing the drug and guaranteeing slow release. In contrast to wet chemical methods, the EHDA spray drying technique exhibits great flexibility in mixing different components. It has been shown that the rate of drug release can be varied by the choice of the polymer mixture. Practical use of EHDA for medicine production depends on successful scaling up of the process. A prerequisite for designing a multinozzle system is a simulation model that adequately describes spray evolution. We have presented a simple modelling approach for the particle flow in the charged sprays. It takes into account the electric field, gravity, the drag between droplets and the gas (which is at rest), and the Coulomb interaction of the particles with each other. The model was tested, and a scale-up relation was proposed that enables simulation of realistic sprays with a much smaller number of particles as present in the experiment. Comparison with experiment showed a qualitative discrepancy between model and reality, which can be attributed to missing image charges, the static background gas or the effect of other simplifications.

References

1.   J.M. Grace and J.C.M. Marijnissen, A review of liquid atomization by electrical means, J. Aerosol Sci., 25, 1005-1019 (1994).
2.   R.P.A. Hartman, D.J. Brunner, D.M.A. Camelot, J.C.M. Marijnissen, and B. Scarlett, Electrohydrodynamic atomization in the cone-jet mode physical modelling of the liquid cone and jet, J. Aerosol Sci., 30, 823-849 (1999).
3.   V.P. Torchilin, Structures and design of polymeric surfactant-based delivery systems, Journal of Controlled Release, 73, 137-172 (2001).
4.   G.W. Bos, R. Verrijk, O. Franssen, J. Bezemer, W.E. Hennink, and D.J.A. Crommelin, Hydrogels for controlled release of pharmaceutical proteins, Biopharm. Europe 13, 64-74 (2001).
5.   W.E. Hennink, H. Talsma, J.C.H. Borchet, S.C. De Smet, and J. Demeester, Controlled release of protein from dextran hydrogels, J. of Controlled Release 39, 47-55 (1996).
6.   T. Ciach, L. Diaz, E. Ijsel, and J.C.M. Marijnissen, Application of electro hydro dynamic atomization in the production of engineered drug particles, in: Optimization of Aerosol Drug Delivery, Kluwer Academic Publishers (2003).
7.   T. Ciach, J. Wang, and J.C.M. Marijnissen, Production of protein loaded microparticles by EHDA, J. Aerosol Sci., 32, s1, 1001 (2001).
8.   K.B. Geerse, From people to plants, PhD Thesis, TU Delft, Faculty of Applied Sciences, Particle Technology, DCT (2002).
9.   R.P.A. Hartman, J.-P. Borra, D.J. Brunner, J.C.M. Marijnissen, and B. Scarlett, The evolution of electrohydrodynamic sprays produced in the cone-jet mode, a physical model, J. of Electrostatics 47, 143-170 (1999).
10. T. Winkels, Modeling and Measuring Droplet Trajectories in an EHDA Spray, Master Thesis, TU Delft, Faculty of Applied Sciences, Particle Technology, DCT (2002).
11. K.B. Geerse, T. Winkels, S. Luding, and J.C.M. Marijnissen, Modeling droplet trajectories of a single electrospray, J. Aerosol Science 2, 1267-1268, 2004; M.-K. Müller, T. Winkels, K.B. Geerse, J.C.M. Marijnissen, A.

Schmidt-Ott, and S. Luding, Experiment and Simulation of Charged Particle Sprays, PARTEC proceedings, Nuremberg, Germany, 16-18 March, 2004.

12. M.-K. Müller and S. Luding, Long-range interactions in ring-shaped particle aggregates, Powders and Grains 2005, Stuttgart, July 2005, Balkema, Dordrecht, Netherlands, in press.

The Branly Effect and Contacting Grains in a Packing

S. Dorbolo and N. Vandewalle

GRASP, Department of Physics – University of Liège
4000 Liège – Belgique
S.Dorbolo@ulg.ac.be

Summary. The electrical conductivity of a granular matter has been studied. Electromagnetic perturbations have been experimentally produced at the vicinity of the packing. The Branly experiment has been performed and quantified. It appears that the soldering of grains is induced by the electromagnetic waves. This explains the drop of electrical resistance, i.e., the Branly effect. The contacts between the grains is enhanced because the electromagnetic waves induce soldering between grains.
Keywords: Electrical conductivity, Branly effect, granular materials

1 Introduction

In 1890, Branly discovered that electromagnetic perturbations are able to change the electrical properties of a granular material located in the vicinity of the perturbations [1]. A glass tube containing metallic filings is called a Branly's coheror. This system has been used by Marconi to transmit the first transatlantic message using electromagnetic waves. From discoveries to discoveries, emitter and receptor enhancements were such that 100 years later everybody is able to transmit sound and video through the air. On the other hand, very few papers consider the Branly's coheror problem [2].

Granular packing are characterized by arches and force chains through the system. The electrical current is a good tool to probe those paths since the electrical resistance is lowered by the pressure induced by the force chains [2]. The current has to find the lower electrical resistance through the system and therefore the current preferably uses the chain force path. Any new contact decreases the total electrical resistance because this new connection is in parallel with the lowest electrical resistance path. Obviously, any enhancement of this latter path decreases the global resistance of the packing.

Moreover, the nature of the electrical contact also play an important role. Oxide layers or weak contacts induce particular behaviors to the global resistance. Indeed, some non-linear behavior and memory effects are observed in electrical properties of metallic granular materials.

2 Electrical Properties of a Granular Matter

Contrary to a piece of metal, the electrical resistance of a granular material is not a fixed value. It depends on the electrical history of the system, on the injected current (or on the imposed voltage) and on the state of compaction. In this paper, we consider a packing of 14000 lead beads of 2 mm diameter contained in a insulating vessel ($40 \times 50 \times 50$ mm^3) [3]. The electrodes are rectangular and placed at opposite vertical faces of the parallelepiped.

Figure 1 shows the voltage V versus the injected current i. This $i - V$ diagram is very different from that of a bulk material since the voltage V is not univocally given by the current i. To emphasize this fact, ramps of current have been injected and the voltage has been recorded during those ramps. Starting for $i =0$ A, the current was increased until a maximum current value. The current value was then decreased until the same maximum current but with the opposite polarity. Finally the current was set back to zero. The amplitude of the ramps has been increased after each loop. On the right of the Fig. 1, a sketch of the i-V diagram has been depicted. The different areas of the sketch are described here below. When we wish to set the current to

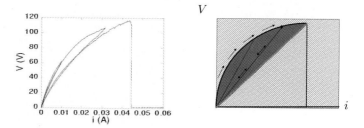

Fig. 1. (left) Current-voltage diagram of a granular material composed of 14 000 lead beads. Ramps of current (with increasing amplitude) have been applied in order to point out the memory effect behavior. (right) Schematic i-V diagram. The grey zone and the dashed zone materialize the reachable and the unreachable points respectively.

a given value for the first time, the voltage follows an exponential law (black curve). When the current is then decreased, the voltage linearly decreases as for a bulk material, i.e. Ohms law is of application (black line that links the black point to the origin).

This particular behavior is due to the structure of the grain contacts. Figure 2 shows a sketch of a contact between two beads. Beads are covered by a thin layer of oxide. A semiconducting junction is formed. This layer provides a non-linear behavior as far as electrical properties are concerned. This behavior can be modeled by two opposite diodes having exponential law behavior in the i-V plane.

When the voltage between two adjacent beads is increased, the electric field can be such that microscopic dielectric breakdown can occur. In this case, the oxide layer is destroyed and some well-conducting electrical paths are created. Those conducting channels follow Ohm's law and are modeled by the resistance in Fig. 2. This resistance part of the contact explains the reversibility and the Ohm like behavior of the packing under certain conditions.

In summary, a region of the i-V plane is accessible (grey area in Fig. 2) by tuning the injected current.

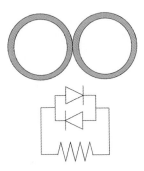

Fig. 2. Sketch of the electrical contact between two adjacent beads. The grey area represents an oxide layer which is semi-conducting.

The vertical line that occurs at high current is the brutal Calzecchi-Onesti transition [4]. At a given current, the resistance sharpely drops. Beads solder to each other. The resistance becomes then low. The system is similar to a bulk material with a resistance of about 20 Ω. The dashed area represents the unreachable points.

3 Branly Experiment

First of all, a fixed current is injected through the sample, 10 mA for instance. The system is then located in the i-V plane at a certain representative point in the i-V plane (black point in Fig.3 (right)). The system is characterized by a subsequent resistance $R_0 = V/i$. The resistance is recorded versus time and normalized by R_0. Firstly, R_0 slightly decreases. Then sparks are produced at 1 m from the parallelepiped. The resistance consequently drops as shown between the two vertical lines in Fig.3 (left). As soon as the spark production ends, the resistance stops decreasing.

4 Path in the i-V Diagram

The electrical resistance variations can be interpreted in view of the i-V plane. As we work with a constant current, the system follows a vertical line in the i-V plane (arrow 1 in Fig. 3 (left)). When sparks are produced, the resistance

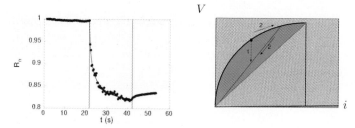

Fig. 3. (left) Normalized resistance of the packing as a function of time. The two vertical lines delimit the spark production period. (right) Motion of the representative point of the system in the i-V diagram.

decreases and the system reaches a point located in the accessible region (grey area). Such point can be also reached by tuning the current (arrows 2 in Fig.3 (left)): by increasing the current following the black curve (exponential) and by decreasing the current following the Ohm law line . Such trajectory is characterized by the microsoldering of the beads since the resistance decreases. That means that irreversible processes (like a microsoldering) are also induced by the electromagnetic waves since the same situation can be obtained by tuning the injected current. Note that the change of resistance is continuous contrary to the Calzecchi-Onesti transition.

5 Conclusion

The decrease of the resistance of a metallic granular packing is due to microsolderings induced by external electromagnetic waves. This transition is continuous and saturates towards a minimum of the resistance.

Acknowledgements

SD thanks FNRS for a financial support. This work is also supported by the contract ARC 02/07-293.

References

1. E. Branly, C. R. Acad. Sci. Paris **111**, 785 (1890).
2. E. Guyon, Pour la Science **300**, 130 (2002); S. Dorbolo, M. Ausloos, and N. Vandewalle, Physical Review E 67 (2003), 040302(R).
3. S. Dorbolo, M. Ausloos, and N. Vandewalle, Appl. Phys. Lett. **81** 936 (2002); *ibid.* Europhys. J. B **34** (2003), 201.
4. T. Calzecchi-Onesti, Il Nuovo Cimento **16**, 58 (1884).

Universal Wide Shear Zones in Granular Bulk Flow

D. Fenistein and M. van Hecke

Kamerlingh Onnes Laboratory – Leiden University
2300 RA Leiden – The Netherlands
fenistein@phys.leidenuniv.nl

Summary. Granular matter exhibits an intricate mix of solid and liquid-like phenomena, some familiar, others remarkable, but almost always poorly understood [1-5]. In particular, when submitted to external stress, granular matter does not flow homogeneously like an ordinary fluid would. Instead, it forms rigid regions separated by narrow shear bands where the material yields and flows [1-7]. This shear localization may also be relevant for dense colloids, emulsions and foams [8-10], but for granular media it is ubiquitous-think of geological faults [11,12], avalanches [13,14] and silo discharges [2,15-17]. Empirically, shear bands are observed to be narrow, particle-shape dependent and often localize near a wall [1,2,6,7,11-21]. Here in contrast to this behaviour, we present experiments in which much wider and universal shear zones can be created in the bulk of the material. These shear zones exhibit Gaussian strain rate profiles, with position and width tunable by the experimental geometry and particle properties.
Keywords: shear zone experiment, ring shear tester, strain rate

1 Introduction

Inhomogeneity is a crucial thread connecting many of the unusual properties of granular matter [1-5]. For slowly flowing grains, the localization of deformations into narrow shear bands is a prime manifestation of this heterogeneity. Despite its omnipresence and crucial industrial importance, there is no deep understanding of this phenomenon. The narrow width of shear bands [1-7,11-21] (5-10 grain diameters) forms a major obstacle for continuum theories of granular flows, hinders mixing and makes grain flows hard to predict.

Probing the mechanisms responsible for shear localization is difficult because there is little room for experimental manipulation – shear bands are very robust. For example, in Couette cells [6,7,19-21], (two coaxial cylinders that rotate with respect to each other), the same narrow shear band is observed to form near the inner cylinder, irrespective of the driving rate [7,20] and the details of the geometry [21]. Its width is simply equal to a few grain diameters [6,7,19-21]. The tail of the velocity profile across the shear band is subtly influenced by the particle shape via boundary induced layering effects [7]. Wall localized shear bands may not be representative of granular bulk behaviour.

2 Experimental Setup and Results

Here we present experiments in which wide and tunable shear zones are created away from the side walls. By avoiding wall localization and narrow shear bands, we probe a completely different regime of granular flow which we find to exhibit surprisingly universal features. A Couette cell is modified by splitting the bottom at radius R_s and attaching the two resulting concentric rings to the inner and outer cylinder (Fig.1a). Grains, similar to those used in the bulk, are glued to the side walls and bottom rings to obtain rough boundaries. The cell is then filled with grains up to a given height H, the outer cylinder and its co-moving ring are rotated and the resulting flow at the top surface is observed from above by a fast CCD camera. This surface flow rapidly relaxes to a steady state where it is purely in the azimuthal direction. It is therefore a function of the radial coordinate only, in agreement with previous studies [6,7,19-21]. In addition, the surface velocities are proportional to the driving rate Ω [7,20] (Ω was varied from 0.16 to 1.5 rad/s). We subsequently fix Ω at 0.16 rad/s and focus on the velocity profile $\omega(r)$, the dimensionless ratio of the average angular velocity and Ω.

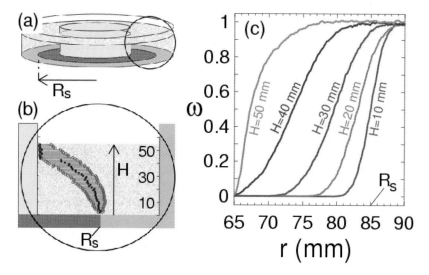

Fig. 1. Split-bottom Couette cell and main features of flow. (a) Schematic side-view of the experimental setup, showing the inner cylinder (radius 65 mm), the outer cylinder (radius 105 mm) and the split at R_s in the bottom plate. R_s can be varied by mounting various sets of bottom rings. By removing the inner cylinder we obtain the "disk" geometry in which R_s can vary from 45 to 95 mm. Stationary parts are drawn in green. After the cell is filled, an adjustable blade flattens the surface at the desired height H. As granular media we used four different mixtures (I - IV) of spherical glass beads of size distributions 0.3±0.1 mm (I), 0.65±0.1 mm (II), 1.1±0.1 mm (III) and 2.2±0.2 mm (IV) and irregularly shaped bronze beads of 1.1±0.2 mm. Movies of a rectangular area of the surface flow consisting of typically 2000 subsequent frames are recorded by a Pulnix TM-6710 CCD camera at a rate of 120 frame/s with pixel resolution ≈100μm. The average azimuthal velocity as function of the radius r is then obtained by applying a time-lag image correlation method. (b) Extension of the shear zone (going from 10\% to 90\% of the angular velocity) as function of H for small glass beads (mixture I) and R_s =85 mm. (c) Examples of normalized angular surface velocities $\omega(r)$ for the same beads and same R_s value.

Figure 1 illustrates the main phenomenology of these velocity profiles. For shallow layers, a narrow shear zone develops above the split at R_s. When H is increased, this shear zone broadens continuously and without any apparent bound - the widest observed zones are 50 grain diameters wide. Additionally, the shear zone shifts away from R_s when H is increased, presumably because the curved geometry leads to larger shear stresses for decreasing radius [22]. For sufficiently large height, the shear zone then reaches the inner wall. As shown in Fig.1, the maximum strain rate rapidly localizes there and the asymptotic shear band regime observed in earlier work is approached [7,20,21]. Before this wall localization occurs, however, there is a substantial range of layer heights where wide and symmetric bulk shear zones are observed.

The main result of our work is illustrated in Fig.2: After appropriate rescaling, all bulk profiles collapse on a universal curve, which is extremely well fitted by an error function.

$$\omega(r) = 1/2 + 1/2 \; erf((r - R_c)/W). \qquad (1)$$

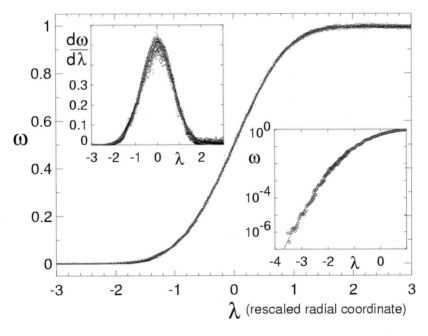

Fig. 2. Collapse of all 35 bulk profiles obtained for H=4,5,6,. .38 mm by appropriately rescaling the width and shifting the center. The rescaled dimensionless radial coordinate $(r - R_c)/W$ is denoted by α. The red curve is an error function. The left inset illustrates that the appropriately rescaled radial derivatives of the bulk velocity profiles, the strain rates, are Gaussian. In the right inset we compare the error function to the left tail of the velocity profile, which is obtained from a much longer run (30 min) for H= 20 mm. This provides strong evidence for agreement to Eq.1 down to $\omega \approx 10^{-6}$.

Accurate measurement of the left tail of the velocity profile (right inset in Fig.2) confirms this form. The strain rate is therefore Gaussian, and the shear zones are completely determined by their centres R_c and widths W. The fit to Eq.1 is similarly good for particles of different size and shape, and in contrast to wall-localized shear bands [7], bulk shear zones are not qualitatively influenced by the granular microstructure. Note the remarkably simple and universal form of the velocity profiles – for granular systems, universality is rare.

At this stage, our observation range of bulk profiles is mainly limited by the presence of the inner cylinder (shown in light green in Fig.1). Removing this cylinder but keeping the stationary bottom disk (dark green in Fig.1) we find that the bulk velocity profiles remain unaffected - the wide shear zones are insensitive to the presence of the side walls. In the following, we present the functional dependencies of R_c and W on the parameters R_s, H and particle type for bulk profiles measured in both "cylinder" and "disk" geometries.

The location of the centre of the shear zone, R_c, is found to be independent of the material used, so the only relevant lengthscales appear to be H and R_s. We obtain that the dimensionless displacement of the shear zone, $(R_s-R_c)/R_s$, is a function of the height (H/R_s) only. As illustrated in Fig.3a, the simple relation

$$(R_s-R_c)/R_s = (H/R_s)^{(5/2)} \tag{2}$$

fits the data well.

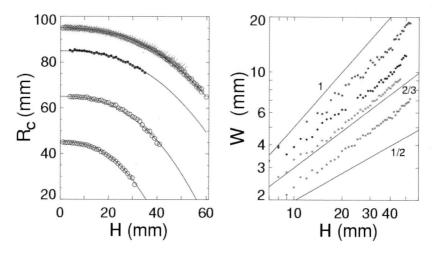

Fig. 3. Shear zone center R_c and width W as function of height H. (a) $R_c(H)$ for a variety of particle sizes and geometries. Mixture I-IV are distinguished by color (green, orange, blue and red respectively), and disk and cylinder geometry are represented by closed and open symbols. The values of R_s (45, 65, 85, 95 mm) are given by $R_c(H=0)$, and the curves represent Eq. 2. (b) $W(H)$ for mixtures I to IV (same colour coding), $R_s =95$ mm in the disk geometry. As guide to the eye we have included powerlaws with exponents 1/2, 2/3 and 1.

The width of the shear zones W depends on the particle size and type, but not on R_s - quite the contrary to the parameter dependence of R_c. In Fig.3b, $W(H)$ is shown for glass beads of different sizes. Clearly, the width increases for larger particles and grows with H. We found that irregular bronze particles display smaller shear zones than spherical glass ones of similar diameter.

The velocity profile is a step function at the bottom and an error function at the surface, which is reminiscent of a diffusive process along the vertical axis. We have found, however, that W(H) grows faster than $\sqrt{H}$ as diffusion would suggest, but slower than H - intriguingly, we obtain the best fit for W $\propto H^{2/3}$ over the limited range where we have reliable data.

3   Discussion and Conclusions

What limits the universal bulk regime? Apart from wall-localization (see Fig.1), we find that in the disk geometry, $\omega(r)$ starts to deviate from Eq. (1) when H exceeds $R_s /2$. Here the symmetry of the velocity profile is weakly broken and the tail towards larger r becomes steeper than the tail towards smaller r. Such deviations can be detected by a simple χ^2 test, which sets a clear criterion to properly identify the universal parameter range shown in Figs. 2 and 3. We thus expect that for large R_s, shear zones of arbitrary width can be created by increasing the depth of the granular layer.

In this work we have probed a regime of granular flow that exhibits surprisingly broad shear zones. We have uncovered their universal features (velocity profile, location) and found some particle dependent properties as well (width). This addresses a key issue, namely to identify those features which are amenable to a continuum description of granular matter [1-5]. We believe that continuum theories should be able to capture the universality of the velocity profile, in particular for very wide shear zones. Crucial input for such a description is the shear zone width, which depends on particle size and shape. This characteristic lengthscale forms therefore a bridge between microscopic and coarse grained descriptions. The development of theories of granular flows can further be guided by the universal relation for the shear zone position, and should incorporate the strong influence of the boundary. As we have shown here, avoiding the proximity of the side wall can turn the shear zones from narrow to wide, and from particle dependent to universal.

The simple experimental protocol that we provide for generating tunable shear zones of "unbounded" width may find applications in, e.g., mixing, and suggests that novel regimes of granular flows can easily be probed, thus addressing the basic question: "How does sand flow?".

References

1. Duran J., Sand, powders and grains Eyrolles, Paris (1997).
2. Nedderman R., Statics and Kinematics of Granular Materials (Cambridge University Press 1992).
3. Jaeger H.M., Nagel S.R., and Behringer R.P. Granular solids, liquids and gases. Rev. Mod. Phys. 68, 1259-1272 (1996).
4. De Gennes P.G., Granular matter: a tentative view . Rev. Mod. Phys. 71, 374-382 (1999).
5. Jaeger H.M. and Nagel S.R., Physics of the granular state. Science 255, 1523-1531 (1992).
6. Hartley R.R. and Behringer R.P., Nature 421, 928-930 (2003).
7. Mueth D.M., Debrégeas G.F., Karczmar G.S., Eng P.J, Nagel S.R., and Jaeger H.M, Nature 406, 385-389 (2000).
8. Chen L.B., Zukoski C.F., Ackerson B.J., Hanley H.J.M., Straty G.C., Barker J., and Glinka C.J., Structural Changes and Orientational Order in a Sheared Colloidal Suspension. Phys. Rev. Lett. 69, 688-691 (1992).
9. Mason T.G., Bibette J., and Weitz D.A., Yielding and flow of monodisperse emulsions. J. of Coll. and Int. Sci. 19, 439-448 (1996).

10. Debrégeas G., Tabuteau H., and di Meglio J-M., Deformation and Flow of a Two-Dimensional Foam under Continuous Shear. Phys. Rev. Lett. 87, 178305 (2001).
11. Scott D.R., Seismicity and stress rotation in a granular model of the brittle crust. Nature 381, 592-595 (1996).
12. Oda M. and Kazama H., Microstructure of shear bands and its relation to the mechanisms of dilatancy and failure of dense granular soils. Géothechnique 48, 465-481 (1998).
13. Komatsu T.S., Inagaki S., Nakagawa N. and Nasuno S., Creep motion in a granular pile exhibiting steady surface flow. Phys. Rev. Lett. 86, 1757-1760 (2001).
14. Daerr A. and Douady S., Two types of avalanches in granular media. Nature 399, 241-243 (1999).
15. Nedderman R.M. and Laohakul C., The thickness of the shear zone of flowing granular materials. Powder Technol. 25, 91-100 (1980).
16. Bridgewater J., On the width of failure zones Géothechnique 30, 533-536 (1980).
17. Muhlhaus H.B. and Vardoulakis I., The thickness of shear bands in granular materials. Géotechnique 37, 271-283 (1987).
18. Pouliquen O. and Gutfraind R., Stress fluctuations and shear zones in quasistatic granular chute flows. Phys. Rev. E 53, 552-560 (1996).
19. Howell D.W., Behringer R.P., and Veje C.T., Stress Fluctuations in a 2D Granular Couette Cell experiment: A Continuous Transition. Phys. Rev. Lett. 82, 5241-5244 (1999).
20. Losert W., Bocquet L., Lubensky T.C., and Gollub J.P., Particle dynamics in sheared granular matter. Phys. Rev. Lett. 85,1428-1431 (2000).
21. Losert W. and Kwon G., Transient and steady-state dynamics of granular shear flow Adv. in Comp. Syst. 4, 369 (2001).
22. Lätzel M., Luding S., Herrmann H.J., Howell D.W., and Behringer R.P., Comparing Simulation and Experiment of a 2D Granular Couette Shear Device Eur. Phys. J. E. 11, 325-333 (2003).

Numerical Stress Response Functions of Static Granular Layers

A.P.F. Atman and P. Claudin

Laboratoire des Milieux Désordonnés et Hétérogènes – UMR 7603
4, Place Jussieu, case 86, 75252 Paris Cedex 05 – France
atman@ccr.jussieu.fr, claudin@ccr.jussieu.fr

Summary. We investigate the stress response function of a layer of grains, i.e. the stress profile in response to a localized overload. The shape of the profile is very sensitive to the packing arrangement, and is thus a good signature of the preparation procedure of the layer. This study has been done by the use of molecular dynamics numerical simulations. Here, for a given rain-like preparation, we present the scaling properties of the response function, and in particular the influence of the thickness of the layer, and the importance of the location of the overload and measurement points (at the boundaries, in the bulk).
PACS: 81.05.Rm, 83.70.Fn, 45.70.Cc, 45.70.-n
Keywords: Granular Materials, Simulation, Response Functions

The statics of granular materials has been a rich field of research over the last few years. One of the reasons of this interest is that a system of grains reaches its mechanical equilibrium after the particles are 'jammed' in configurations in relation to the previous dynamics of the grains. As a consequence, the distribution of the stresses in a static piling of grains depends, in a subtle manner, on the preparation procedure of the system. A now famous example is that of the sand pile for which the pressure profile at the bottom is different when built from a hopper, or by successive horizontal layers (a 'rain' of grains). In the former situation, this profile exhibits a clear 'dip' below the apex of the pile [1], whereas it shows a flat 'hump' in the latter [2].

In fact, more interesting than the case of a pile, is the study of the stress response function of a layer of grains, i.e. the stress profiles in response to a localized overload, see Fig.1. This geometry allows much more variety in the preparation of the layer. Experiments have been performed for instance with compacted, loose or sheared layers [3–5], whose response pressure profiles are different enough from each other to be a kind of 'signature' of their texture. Besides, such a response test is the elementary 'brick' with which other situations can be deduced, e.g. that of the pile [6]. It is then also well adapted for the comparison of the different models of stress distribution in granular media.

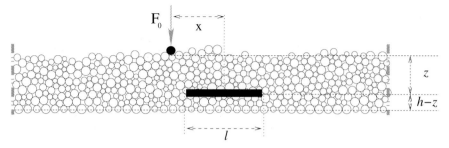

Fig. 1. System geometry and notations. The stress measure is done at a position (x, z) from the applied force F_0 and at a distance $h - z$ from the bottom of the layer. To compute the stress, contact forces are integrated over a linear extension ℓ. Note that we use horizontal periodic boundary conditions.

Our aim is to perform extensive simulations of assemblies of grains, in order to provide precise two-dimensional numerical data of stress response functions. The control of all the parameters of the simulations, as well as the ability of measuring both micro (grain size) and macro (system size) quantities, ensure a useful and interesting feed back to the experiments and the models.

Although our ambition is to be as general and systematic as possible, we shall in this paper, present only few results concerning some scaling and size effects of this response function. In particular we shall restrict to a single (rain-like) preparation history and study the influence of the thickness of the layer, as well as that of the location of the overload and measure points. Besides, the overloading procedure requires careful and important validity tests (e.g. linearity, additivity, reversibility) that will be also described below.

1 The Simulations

In Fig. 2, we show a part of a typical simulated layer, where we can see the initial force chain network and the response produced by a localized overload applied to a single grain after the initial forces have been subtracted. The simulations are performed using a classical molecular dynamics (MD) algorithm in three successive stages: preparation, deposition and overloading that are described in the following.

We start with the preparation of N polydisperse grains, with radii homogeneously distributed between R_{min} and $R_{max} = 2R_{min}$. These N particles are put on the nodes of a grid with aspect ratio $1 : 4$ – this aspect ratio is needed for the appropriate study of the response functions, see below. This lattice ends at its bottom with a fixed horizontal line of similar particles which will be used as the support for deposition – the distances between these bottom grains are small enough to avoid grain evasion.

The deposition stage consists of untying the particles from the grid and, under gravity and horizontal periodic boundary conditions, letting them evolve

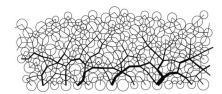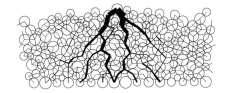

Fig. 2. Example of the granular layers produced by our simulations. Left: Force chain network after the deposition under gravity of $N = 1600$ grains. This picture corresponds to the equilibrium configuration after 628100 MD steps. Right: Response due to an overload localized over a single grain after subtraction of the initial forces. The amplitude of this overload corresponds to the weight of five average grains. This picture has been obtained after 29500 additional MD steps to reach a new equilibrium configuration. In black, contact forces have been increased in response to the overload, whereas in gray they have been decreased.

with random initial velocities. In this 'rain-like' deposition the grains interact with elastic and friction forces, and the system evolves under classical Newton equations and molecular dynamics rules (e.g. predictor-corrector, Verlet algorithm). The time to reach the equilibrium depends on the characteristics of the particles. We chose $\mu = 0.5$ for the contact friction coefficient. The rheology of the contacts is that of Kelvin-Voigt with $k_n = 1000$ N/m and $k_t = 750$ N/m for the normal and horizontal contact stiffness. The viscosity g_n is chosen in order to get a critical damping. At last, the gravity is set to unity. The equilibrium criteria consist of the five following tests which are applied after each period of 100 MD time steps: (1) the number of gained/lost contacts during this period has to be zero; (2) the number of sliding contacts between particles also has to vanish; (3) the integrated force measured at the bottom of the layer must be equal to the sum of the weight of all the grains; (4) all the particles have to have, at least, two contacts; and (5) the total kinetic energy has to be lower than some low threshold. Once these criteria are all satisfied, the deposition is stopped, and the overloading phase can begin.

The overload is applied to a single grain with a force F_0 that we wish to be small enough to not cause any rearrangement of the layer structure. The determination of the amplitude of the loading force is then a crucial part of the simulation. For that purpose we made several tests which check the reversibility, the additivity and the linearity of the stress response. These tests (see below for more details) let to conclude that the optimal magnitude of F_0 is of the order of few times the weight of an average grain. In the overloading procedure, F_0 is split in 10000 MD time steps, i.e. in each time step the force is increased in $1/10000$ of F_0, and it remains constant and equal to F_0 after that. Again, the equilibrium criteria are those explicited above, except that the integrated force at the bottom of the layer has to be equal to the weight of all the grains *plus* F_0.

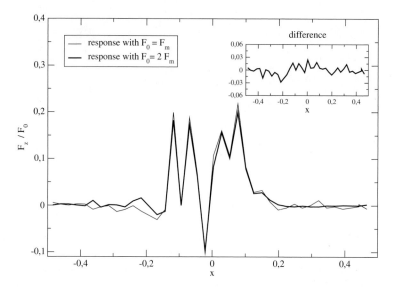

Fig. 3. In this picture we show one of the tests applied to the layers (linearity, additivity and reversibility) to settle the optimal value for F_0. We present here the linearity test which consists in comparing the force response profiles with different values of F_0. These curves have been obtained on a layer with $N = 400$ grains ($h \sim 7D$), and $F_m = 2.5$ average grain weight. Rescaled by the overload force, the difference of the two responses is reasonnably small. This difference is due to small slips in the contacts between the grains.

Only the linearity test is illustrated here, in Fig. 3: it consists in comparing the force response after loading with different values of F_0. We find that the response is satisfactorily linear in F_0 as long as its magnitude does not exceed ~ 40 times the average grain weight. With the other tests we have checked that loading and unloading with the very same F_0 gives the original contact force distribution with a good precision, and that two simultaneous overloads located at two different positions give a response which is perfectly equivalent to the sum of the two responses computed from the two corresponding single overloads.

2 Results and Discussion

In order to measure the vertical stress σ_{zz}, we integrate the contact forces over a set of grains. In fact, we do not make use of the usual formula $\sigma_{\alpha\beta} = 1/S \sum_i f_\alpha^i r_\beta^i$, where the sum is computed over the contacts between the grains in the 'volume' S (here we are two-dimensional), f_α being α^{th} component of the considered contact force, and r_β the βth component of the corresponding distance vector between the grains in contact. Rather, we take a horizontal

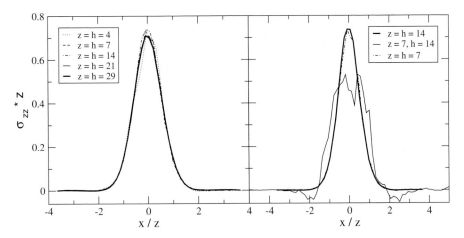

Fig. 4. Response functions. Several curves are shown for different values of z and h. On the left, the response profiles collapse when measured at the bottom of the layers and rescaled by h. On the right, the response in the bulk. Although more averaging is needed, this graph suggests interesting features for $z < h$. All these curves have been computed with $\ell = 0.77z$.

'line' of grains of length ℓ. The corresponding stress σ_{zz} is then equal to the sum of the vertical components of the forces carried by the contacts of one of the sides of the line – say the upper one – divided by ℓ. This stress measure can be done at any depth z ($z = h$ means that the measure is done on the bottom) and centered at any horizontal distance x from the overload point. As we are particulary interested in σ_{zz} profiles along x at a given z, our choice is here better adapted than the usual stress formula which would mix together grains of (slightly) different depths in S. This is particulary important at small z. Besides, with this stress definition, the integral of σ_{zz} over x at a fixed z is exactly equal to F_0 at any scale. This property is crucial to normalize and compare data from layers of different thickness.

The response function at depth z is obtained by making the difference of the stress profile $\sigma_{zz}(x, z)$ measured on the layer with the additional force F_0, with a copy of the very same layer without overload. This difference is ensemble averaged over several overload positions and layer samples. We present here the results for five different layer sizes, with $N = 100, 400, 1600, 3600$ and 6400 grains. The $1 : 4$ aspect ratio gives layers of average thickness of $4, 7, 14, 21$ and 29 mean particle diameters D, respectively. The $10, 5, 5, 4$ and 4 different realizations give in the end $62, 60, 65, 84$ and 86 different overload points – on average, there are six grains between two consecutive overload points.

The results are shown on Fig. 4. One can see that the normalized profiles measured at the bottom of layers of different thickness h collapse when they are linearly rescaled by h, and show a single peak whose width is of the order of h. This is precisely the reason why it is important to have layers four times wider than thick – $2h$ on both sides of the overload point. When the response is computed in the bulk, the structure of the response profile looks more complicated. The present curve needs however more ensemble averaging and should be taken as preliminary. These results were all obtained using a measure scale $\ell = 0.77z$ for all layer sizes. A systematic study of the importance of this integration measure length ℓ is under way. We are also working on the calculation of the other stress components σ_{xx} and σ_{xz}.

These scaling studies are currently extended to other preparation histories, and in particular to the cases of sheared or more anisotropic layers of grains. At last, we plan to test these data against the predictions of anisotropic elasticity which, depending on the values of the different parameters, can give various stress response profiles [7].

Acknowledgements

We are indebted to G. Combe for a decise help on the MD codes. We also thank the granular group of the LMDH for stimulating discussions.

References

1. J. Šmid and J. Novosad. Pressure distribution under heaped bulk solids. *I. Chem. E. Symposium Series*, 63:D3/V/1–12, 1981.
2. Loic Vanel, Daniel Howell, D. Clark, R. P. Behringer, and E. Clément. Memories in sand: Experimental tests of construction history on stress distributions under sandpiles. *Phys. Rev. E*, 60(5):5040–5043, 1999.
3. D. Serero, G. Reydellet, P. Claudin, É. Clément, and D. Levine. Stress response function of a granular layer: Quantitative comparison between experiments and isotropic elasticity. *Eur. Phys. J. E*, 6:169–179, 2001.
4. J. Geng, D. Howell, E. Longhi, R. P. Behringer, G. Reydellet, L. Vanel, E. Clément, and S. Luding. Footprints in sand: The response of a granular material to local perturbations. *Phys. Rev. Lett.*, 87:035506, 2001.
5. J. Geng, G. Reydellet, E. Clément, and R.P. Behringer. Green's function measurements of force transmission in 2d granular materials. *Physica D*, 182:274, 2003.
6. G. Reydellet, P. Brunet, J. Geng, A. P. F. Atman, P. Claudin, R. P. Behringer, and É. Clément. From the stress response function (back) to the sandpile pressure 'dip'. in preparation, 2003.
7. M. Otto, J.-P. Bouchaud, P. Claudin, and J. E. S. Socolar. Anisotropy in granular media: classical elasticity and directed force chain network. *Phys. Rev. E*, 67:031302, 2003.

Granular Segregation in Hydraulic Ripples

H. Caps[1], G. Rousseaux[2,3], and J.-E. Wesfreid[2]

[1] GRASP – University of Liège
    Department of Physics B5, 4000 Liège – Belgium
    `herve.caps@ulg.ac.be`
[2] LPMMH – Ecole Supérieure de Physique et de Chimie Industrielles
    Rue Vauquelin 10, 75231 Paris – France
[3] Institut Non-Linéaire de Nice – Sophia-Antipolis
    Route des Lucioles 1361, 06560 Valbonne – France

Summary. We report an experimental study of a binary sand bed under : (i) an oscillating flow and (ii) under a circular flow. In both cases, the appearance of a granular segregation is shown to strongly depend on the sand bed preparation. A reproducible protocol is proposed. In the oscillating case, a segregation in volume is observed in the final steady state. The correlation between this phenomenon and the fluid flow is emphasised. In the unidirectional case, a segregation in surface in observed instead of a segregation in volume. The relative difference in size between both sand species is highlighted as the physical parameter leading to the phase segregation. The difference between the characteristic times of appearance of both types of ripples is shown to be the major reason for the different behaviours leading to either a bulk segregation or a surface segregation.
Keywords: Patterns formation, Phase segregation, Instability

1 Introduction

When sand beds are eroded by a fluid (e.g. air or water), pattern formation is generally observed [1]. Coastal areas as well as desert landscapes are so covered by ripples and dunes. Despite their familiar aspect, the involved physical mechanisms are related to complex granular transport processes such as saltation, reptation, suspension and avalanches [2].

Recently, sand ripples created by water shear flows have received much attention [3, 4]. Many experiments have been performed in rectangular [2] or circular [3–6] setups. If the fluid motion is oscillatory, the created ripples are symmetric (both ripple faces have the same length). This happens on beaches when the water depth is large enough. On the contrary, the ripples created under a non-oscillating fluid motion are generally asymmetric [2]. Indeed, the upstream face (stoss slope) is generally larger than the downstream face (lee slope) as observed on beaches for small water depths.

Natural sand beds are generally composed of different granular species. Broad granulometric distributions are indeed observed [7]. As a consequence, phase segregation and stratigraphy may occur [7, 8].

In this paper, we report an experimental study of a binary sand bed under : (i) an oscillating flow and (ii) under a circular flow. In section II, the experimental setups are presented. The necessity of a protocol for the interface preparation is pointed out in Section III. The experimental results are presented in Section IV. Eventually, a summary of our findings is given in Section V.

2 Experimental Setups

Ripples are created in different experimental setups depending on their type: oscillatory (Setup 1) or not (Setup 2) [see Fig. 1].

The Setup 1 is composed of two concentric cylinders joined by two circular plates on the top and on the bottom (for further details, see [4]). The so-created annular channels is fully filled with a layer of sand and water. The mean radii of the setups is 7.31 cm and the width of channel is 0.8 cm. The sand bed is put into oscillations at fixed amplitude A and frequency f. After several oscillations (typically $A = 2$ cm, $f = 1$ Hz) small ripples appear all around the perimeter.

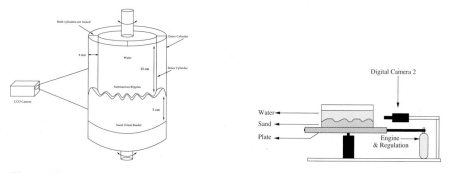

Fig. 1. Experimental setups for the ripple formation and analysis: (left) Setup 1 for the oscillatory case, and (right) Setup 2 for the unidirectional case.

Setup 2 consists in a cylindrical vessel connected by a belt to an engine. The rotation speed can be adjusted from 8 rpm to 100 rpm. The container (7 cm radius) is filled with water and sand and is put into rotation. When the rotation is brutally stopped, ripples are observed after a short time (typically 2s). In both cases, a CCD camera is horizontally placed at the same height as the sand/water interface and records images of the interface. The main advantage of such circular geometries, compared to linear channels [8], is a strict mass conservation and periodic boundary conditions.

3 Interface Preparation

For reproducibility reasons, special care has to be taken to the initial sand bed preparation. Indeed, the homogeneity as well the compaction are parameters conditioning the evolution of the ripples [5]. A systematic and reproducible procedure is thus needed to flatten the sand bed prior to each experiment.

The sand bed is flattened by the way of strong oscillations which destroy the former ripples. Some problems may come from this method : (i) if the sand bed is violently shaken, the grains are put into suspensions and fall down with a velocity depending on their size [10]. A larger amount of small grains will thus be found on the top of the sand layer [see Fig. 2 c]. (ii) If the oscillations are strong enough to fluidize the sand bed but not sufficient to put them into suspension, a Brazil nut effect [9] occurs and causes the rising of larger grains to the top of the sand/water interface [see Fig. 2 d]. From a segregation point of view, the ideal case is thus in between these two ones. However, we have observed that the compaction is smaller when the oscillations are abruptly stopped. The reproducibility of measurements is very bad in such configurations [5]. We have thus decided to start from a compacted bed [see Fig. 2 (a, b)].

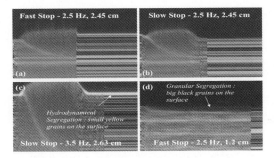

Fig. 2. Different configurations for the interface preparation: (a) A rapid stop of the fluidization leads to a loose packed sand bed. (b) The 'ideal' case corresponds to a well compacted bed with a quasi homogeneous grain distribution. (c) Phase segregation coming from a very strong acceleration followed by a slow stop. (d) Phase segregation resulting from a 'small' acceleration. Images are 1.1 cm height and 50 s wide.

The procedure we used is the following [see Fig. 2 b]. First, the setup is accelerated at a high frequency in order to destroy remaining patterns. Then, we decrease the amplitude of oscillations keeping the frequency constant and we let the setup oscillate for a while. During this step, the height of the sand/water interface is observed to decrease, this is the compaction step. After typically 10 s of oscillations at low amplitude the compaction is nearly constant and the sand bed is quasi homogeneous. This method is largely qualitative, but the quantitative determination of the bed compaction and homogeneity are difficult tasks and need special apparatus.

4 Results

Herebelow, we describe the experimental results concerning both kinds of ripples. As a general observation we may state that starting from an homogeneous mixture the sand bed always evolves through a segregated state. however, this segregation depends on the way the ripples are created.

4.1 Oscillatory Case

After many hours ($\approx 5\,\text{h}$) of oscillations, stable "vortex ripples" [2] are found to cover the entire sand/water interface. The wavelength of these ripples only depends on the amplitude A of oscillations and their morphology does not depend on the grain size. Moreover, the mean ripple amplitude h decreases as the frequency f is increased. This is clearly visible in Fig. 3, where we report images of a final configuration for different frequencies of oscillations.

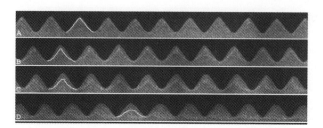

Fig. 3. Side views of the final steady state. Different frequencies are illustrated ($A = 3\text{cm}$): (a) $f = 0.8\,\text{Hz}$, (b) $f = 0.85\,\text{Hz}$, (c) $f = 0.9\,\text{Hz}$, (d) $f = 1\,\text{Hz}$. Solid white lines emphasise the limit of the segregated zone (the vertical scale is stretched 2.5 times with respect to the horizontal).

Our experimental setup allows us to directly look inside the sand bed, by contrast to [8] where only top observations were performed. We have so noticed the formation a thin layer of small (white) grains under the sand/water interface. This layer corresponds to the limit of a *segregated zone*. As the frequency f is increased (from top image to bottom one), the distance between the small grain layer and the sand/water interface at the crests increases. In the bottom parts, this layer remains nearly unchanged.

The formation of this kind of segregation may be understood as follows. As the fluid flow shears the granular bed, the compaction of the bed decreases. The spaces created between the larger grains allow the smaller ones to fall down by percolation. Thus, the thin layer of small grains corresponds to the thickness of the layer where percolation may occur. This hypothesis may be comforted by averaged close-up movies of vortex ripples created in monodisperse ($d = 500\,\mu$ m) sand beds [see Fig. 4]. A layer of grains in motion is observed over a layer of static grains. The depth of the layer of grains in motion is maximum near the ripple crests and decreases near the bottom parts of the ripples.

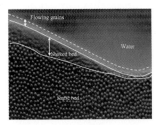

Fig. 4. Average of different images of a ripple during the fluid oscillations.

4.2 Unidirectional Case

Figure 5 presents a typical pattern of *phase segregation* in non-oscillatory ripples. The larger grains (black) segregate on the lee slopes of the ripples, while the small grains (white) are found on the stoss slopes. The segregation is superficial here, by contrast to vortex-ripples. We have noticed that any finite

Fig. 5. Transverse view of one ripple. A phase segregation is observed: the larger grains (black) are found on the lee slope, while the smaller grains (white) mainly segregate on the stoss slope. The water flows from right to left.

size difference between both granular species leads to a phase segregation. On the contrary, the angle of repose seems to be irrelevant. Moreover, the larger grains are always found on the lee slopes of the ripples for all experimental conditions.

Actually, the mechanism of phase segregation should be understood as follows. Because of their weight, the smaller grains are more easily carried by the fluid than the larger ones. Those small grains create a saltation fog over the sand bed. Although the fluid speed is not large enough to carry the large grains in a saltation motion, this fog captures and transports them. As the vessel motion is stopped the larger grains are observed to move further by reptation during a short time (less than 1 s) because of their inertia. Arriving on the crest of a ripple, they roll on the lee slope. Because of frictional forces between those grains and the static ones, they are soon stopped. As a consequence, larger grains are found near the ripple crests, while the smaller ones are mainly found near the stoss slope bottoms.

5 Summary

We have studied experimentally the erosion of a binary sand bed submitted to water flows. The importance of the interface preparation with a well-defined protocol has been demonstrated. Two kinds of phase segregation (bulk and surface) have been shown as a function of the time of ripple formation.

Acknowledgements

This work is financially supported by the ACI "Jeunes chercheurs" contract n°2314 and by the ARC contract n°02/07 − 293. HC benefits a FRIA (Brussels, Belgium) grant. The authors would like to thank H. Herrmann, and N. Vandewalle for fruitful discussions.

References

1. P. Ball, *The Self-Made Tapestry - Pattern Formation in Nature*, (Oxford Univ. Press, Oxford, 2001).
2. R.A. Bagnold, *Proc. Roy. Soc. A* **187**, 1 (1946).
3. A. Betat, V. Frete, and I. Rheberg, *Phys. Rev. Lett.* **83**, 88 (1999).
4. A. Stegner and J.E. Wesfreid, *Phys. Rev. E* **60**, R3487 (1999).
5. G. Rousseaux, H. Caps, and J.-E. Wesfreid, submitted for publication (2003).
6. H. Caps and N. Vandewalle, Physica A **314**, 320 (2002).
7. R.E. Hunter, *Sedimentology* **24**, 261 (1977).
8. E. Foti and P. Blondeaux, *Coastal Engineering* **25**, 237 (1995).
9. L. Trujillo and H. Herrmann, cond-mat/0202484 (2002).
10. E. Guyon, L. Petit, J.P. Hulin, and C. Mitescu, *Physical Hydrodynamics*, (Oxford University Press, Oxford, 2001).

Grain Motion Under Air Flow

C. Becco, H. Caps, S. Dorbolo, C. Bodson, and N. Vandewalle

GRASP, Dept. of Physics, Sart-Tilman B5 – University of Liège
4000 Liège – Belgium
c.becco@ulg.ac.be

Summary. The common experiment of a granular flow in a vertical tube is modified. The grains are submitted to the joint action of both gravity and upward air flux. While searching their equilibrium, the grains form clogs. A space-time analysis of the phenomenon is conducted. One-dimensional simulations of the experiment are presented. The influence of several parameters is discussed.

1 Introduction

Granular chute flows are of great interest in many practical applications, from pharmaceutical industries to their similarity with traffic flow. Numerous experimental and numerical studies [1–4] were conducted concerning granular flow in a vertical pipe. These are based on the following experiment : a reserve of particles is poured into a vertical tube. The beads fall trough the tube and get out at the bottom. Typically, three different situations may occur. As a function of the flow rate, the granular flow can be either in a stationary free-fall regime or a density-wave regime or a compact regime. Clogs are zones of higher density. In the density-wave case, clogs form and move at different speeds. The aim of our study is to emphasize the role played by the air in the density-wave regime.

2 Experimental Setup

The experimental setup consists in a vertical glass tube partially filled with polystyrene beads. The particle diameters are between 2 mm and 5 mm with a peaked maximum in the distribution of diameters around 3 mm. The glass tube is 1.5 m high and the inner diameter is 2.2 cm. Air is injected at the bottom of the tube. By adjusting the incoming air flux, a stationary state can be reached, in which clogs are formed and are moving upwards (Fig. 1).

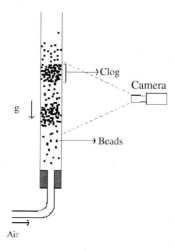

Fig. 1. Sketch of the experimental setup. Air is blown by a flexible tube connected to the bottom of the vertical glass tube. The combined action of gravity and air flow produce the motion of the particles along the tube. Density waves are observed. A CCD camera records the middle part of the tube.

For each experiment, a space-time diagram is created using the collected images. A typical diagram is shown in Fig. 2. This type of diagram emphasizes the evolution of the density of particles over time and space. In experiments, we put a white backscreen so the particles appear in black. When the density of particles is high, a shadow region appears. This corresponds to the formation of a clog in the stream of particles. Clogs appear as dark diagonal lines on these diagrams, as shown on Fig. 2.

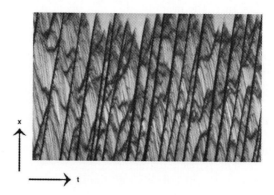

Fig. 2. Typical space-time diagram obtained from our experiment.

3 Results

The clogs are moving at a nearby constant speed. By looking at the slopes of their trajectories in the space-time diagrams, one can see that clogs are moving upward and that all clogs have the same velocity. The remainder of the space-time diagram is filled with two patterns : (i) fine traits with a NNW-SSE direction that are the mark of a flux of falling individual particles. Those traits are mostly parallel and correspond to particles reaching their terminal speed. (ii) The second feature is constituted of black lines which have variable concavity occuring between the straight lines of the main clogs. Those represent smaller clogs having characterictic speeds smaller than the main ones, sometimes negative (small clogs go down). Those clogs seem to be mostly the result of small perturbations at the bottom of the main clogs. They detach from them and are fed by the falling particles. If such a clog ever encounter one of the main clogs, it is generally absorbed. One could also observe a few merging events between clogs.

Within the experimental setup, we can easily tune the air flux. In Fig. 3, two space-time diagrams are shown : for low and large air fluxes. The modification of the incoming air flux has a deep impact on the observed pattern. Globally, a low air flow is less perturbative ; clogs are more stable and less small clogs appear. On the contrary, at higher air flux [see Fig. 3 (right)], most of the clogs are fine and elusive. If a clog appears, it is unstable.

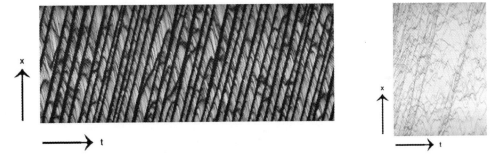

Fig. 3. (left) Low air flux, leading too well marked clogs - (right) : Higher air flux, most of the clogs are destroyed, the density is notably lower.

3.1 Clog Speeds

Clog speed is the most straightforward physical parameter to measure. On the space-time diagrams, the speed of the main clogs appears to be mostly a constant. The first step of our analysis is to transform the diagram into a black and white image by establishing a threshold. Then, the front position of each clog is recorded. A clog is defined and found by the algorithm when x

black pixels are in a row, the number x being fixed by the algorithm. From the position of the clogs at different times, we calculate their velocity. This *global* method is a simple version of the one used in [5, 6] for similar space-time diagrams obtained in another experiment. The result from a typical space-time diagram is shown in Fig. 4. The speed of the main clogs is determined despite the presence of noise in the signal. On the displayed plot in Fig. 4, the mean velocity for the main clogs is close to $v = 25\,\mathrm{cm/s}$.

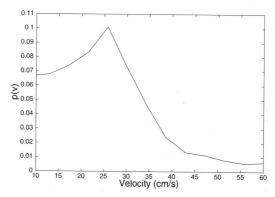

Fig. 4. Distribution $p(v)$ of the clog speeds computed by the global method. The peak marks the mean velocity of the main clogs. The remain of the distribution comes from noise and/or small clogs.

3.2 Clog Occurence

A fundamental question concerns the possible periodicity of density waves. From the images of Fig. 2 and 3, no simple time frame in the passage of clogs seems to appear. A possible correlation has been searched between the successive times Δt separating the passage of two clogs. More exactly, we plot Δt_i versus Δt_{i+1}. Fig. 5 tends to stress the lack of direct correlation between the behavior of the different clogs. No straightforward conclusion comes from the treated data.

3.3 Fundamental Diagram

Using the velocities computed in Section 3.1, a graph of the flow versus the density was plotted. The density ρ is calculated from the space-time diagram. The situation $\rho = 0$ corresponds to the total emptiness of the tube, while $\rho = 1$ corresponds to a state of the tube completely filled with beads. The flow (ρv) is computed with the speed of the clogs. Fig. 6 displays the flow of the clogs versus the density of beads in the tube. One should also remark that

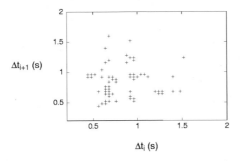

Fig. 5. The ellapsed time Δt_i between the passage of two successive clogs versus the time Δt_{i+1} between the passage of the second of them and the third one.

the global flow (not shown), computed from all the beads, has a mean value equal to zero due to conservative law.

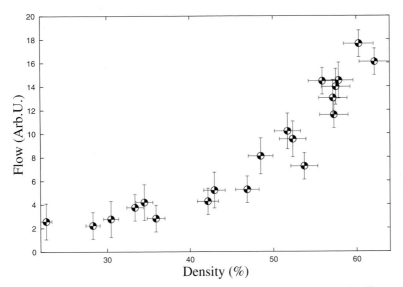

Fig. 6. The clog flow-density graph. Results from 20 experiments are illustrated.

Figure 6 shows an increase of the flow of the clogs when the air flow is lower. With the decrease of the air flow, the stabilization of the clogs makes them bigger and more rapids (as shown by a non-displayed graph of the velocity of the clog as a function of the air flow).

4 One Dimensional Model

Numerical simulations have been also performed. They are based on the Nagel and Schreckenberg's cellular-automaton [7] with some modifications. It includes non-integer velocities and a particular semi-parallel update of the array of particles. To simulate the glass tube, an array of N sites (N is even) is used. Each site is either empty (marked as 0) or occupied by a particle (marked 1). The length of each cell is supposed to be the length of a particle. We associate a velocity to each occupied site (each particle). This velocity may be positive or negative i.e. the particle may go up or down. A maximum velocity must be fixed, so we may choose to limit it to the interval $[-1; 1]$. At each timestep, a particle may change from a site to one of its nearest neighbors with a probability equal to its velocity. The relevant parameters of our simulation are g, θ and e_n, which represent respectively the gravitation, the air flow and the coefficient of restitution for the beads collisions. We choose an arbitrary value ($g = 0.5$) for the gravitation which allows us to scale the time of the simulation. The parameter θ plays mostly the role of a drag coefficient. It will decide of the upwards modification of the velocity of the particles (in probability and intensity).

As shown in [8], if we use a simple parallel updating of the array, problems of conservation of momentum appear. So we use the same principle of semi-parallel update as [8], each timestep updating the sites by pair :

1. Modification of the velocity of the particles, depending of the two parameters g and θ. The parameter g is an automatic modification to the velocity $(\mathbf{v(t+1)} = \mathbf{v(t)} + \mathbf{g})$ and θ fixes the probability that the particle is driven upwards by the air.
2. In the treated pair, if the two sites are occupied, check the probability of a collision linked to their relative velocity.
3. If only one is occupied, the particle jumps in the next site with a probability equal to its velocity, if it is of the good sign.
4. If there is a collision, the new velocities are calculated via the parameter e_n.

The particles velocities after collision v_1^f, v_2^f are calulated using $\varepsilon = (1 - e_n)/2$ and the following relations.

$$v_1^f = \varepsilon v_1^i + (1 - \varepsilon) v_2^i \tag{1}$$

$$v_2^f = (1 - \varepsilon) v_1^i + \varepsilon v_2^i$$

Also, if the first site (near the incoming air) is occupied, its particle receives a velocity 1. Every 10 timesteps, the location of each particles is recorded. A space-time diagram of the simulated tube can be drawn, as shown in Fig. 7.

With a definite set of parameters, we can easily obtain a satisfying result from our numerical simulations, as shown on Fig. 7. Taking into account the

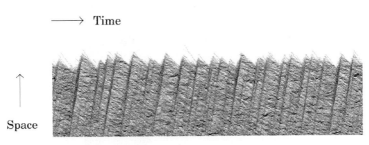

Fig. 7. Typical space-time diagram from a simulation. The main clogs are clearly visible. Small perturbations are also displayed

arbitrary chosen value of $g = 0.5$ in our simulations and scaling the size of a cell to the actual typical size of a grain ($d = 3\,\text{mm}$), we calulate the velocity of the clogs in the Fig. 7 to be $v \approx 20\,\text{cm/s}$, in good agreement with the value ($25\,\text{cm/s}$) experimentally obtained in Section 3.1.

4.1 Phase Diagram for the 1D Simulation

To determine the range of the parameters values leading to the apparition of clogs in our simulation, we created a phase diagram (see Fig. 8) in the (e_n, θ) plane. Both dimensionless parameters are ranging from 0 to 1. From this phase diagram, one sees that the behavior of the particles depends mostly on the value of θ, that is to say on the action of the air. In the gray zone of the phase diagram, low air flux leads to zero motion, [see Fig. 9 (left)]. It could be called the *frozen* state of the system. Since nothing happens, results corresponding to this particular state were not represented in Fig. 6. They would correspond to a 100% density-zero flow on this latter one. The black zone of Fig. 8 corresponds to high air flux, values leading to the expulsion of particles [see Fig. 9 (right)]. The remaining particles are in a *gaz* state. The third white zone is the *clogs* zone. The diagram presented in Fig. 7 corresponds to parameters values taken in the white zone of the phase diagram.

5 Conlusion and Prospects

We have experimentally shown that for low air flux values, the speed and flow of upgoing clogs are high. Moreover, in our simulations, the main parameter in the process of clogs formation is θ, which emphasizes the role played by the air. A possible explanation is the one proposed by [4]. First, the stability of the clogs is higher for lower air flow values. So, the permeability of the clogs (the easiness the fluid has to pass through the clog) is lower. A pressure drop between the top and bottom of the high density region of a clog appears.

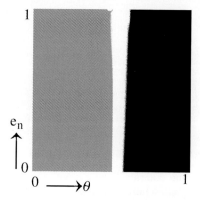

Fig. 8. State of the stack of beads in the tube, as a function of the parameters θ and e_n. In the gray zone, no significative motion is observed. The black region marks the value of the parameters leading to the ejection of most of the particles out of the tube. The white zone is the region of interest.

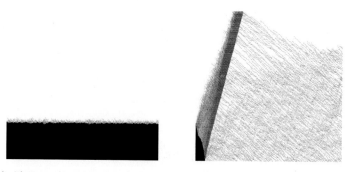

Fig. 9. (left) Typical space-time diagram of the gray zone of Fig. 8 - (right) Typical space-time diagram of the black zone of Fig. 8

This pressure gradient is more important for more stable, dense clogs. Thus the velocity and the flow of upwards clogs increase with lower air flow. This explanation has to be checked in our experiment by more accurate and precise measurements, but seem quite appealing.

In the future, our experimental study will focus on the correlation of the clogs in a tree-like shaped tube. It will also investigate other parameters such as the size of the polystyrene beads compared to the tube diameter.

Acknowledgements

HC thanks FRIA (Brussels, Belgium) for financial support. SD thanks FNRS (Brussels, Belgium) for a financial support. CB is financially supported by the contract ARC number $02/07 - 293$. The authors thank L.Delattre for fruitful discussions.

References

1. J. Lee. Density waves in the flows of granular media. *Phys. Rev. E*, 49(1):281, 1994.
2. T. Raafat, J. P. Hulin, and H. J. Herrmann. Density waves in dry granular media falling through a vertical pipe. *Phys. Rev. E*, 53:4345–4355, 1996.
3. J.-L. Aider, N. Sommier, T. Raafat, and J.-P. Hulin. Experimental study of a granular flow in a vertical pipe: A spatiotemporal analysis. *Phys. Rev. E*, 59(1):778–786, 1999.
4. Y. Bertho, F. Giorgiutti-Dauphiné, T. Raafat, E. J. Hinch, H. J. Herrmann, and J. P. Hulin. Powder flow down a vertical pipe: the effect of air flow. *J. Fluid. Mech.*, 459:317–345, 2002.
5. S. Hørlück and P. Dimon. Statistics of shock waves in a two-dimensional granular flow. *Phys. Rev. E*, 60:671–686, 1999.
6. S. Hørlück, M. van Hecke, and P. Dimon. Shock waves in two-dimensional granular flow: Effects of rough walls and polydispersity. *Phys. Rev. E*, 67:021304, 2003.
7. K. Nagel and M. Schreckenberg. A cellular automaton model for freeway traffic. *J. Phys. I France*, 2:2221, 1992.
8. D. Volk, G. Baumann, D. E. Wolf, G. Törner, and M. Schreckenberg. Simulating granular media unsing cellular automata. In M. Schreckenberg and D. E. Wolf, editors, *Traffic and Granular Flow '97*, page 145. Springer, 1997.

Bubble and Granular Flows :
Differences and Similarities

H. Caps, S. Trabelsi, S. Dorbolo, and N. Vandewalle

GRASP, Dept. of Physics, Sart-Tilman B5 – University of Liège
B-4000 Liège – Belgium
`nvandewalle@ulg.ac.be`

Summary. We have experimentally studied the dense flow of identical bubbles below inclined planes. The flow is driven by the fast motion of dislocations. As a function of the density of dislocations (controlled by the parameter θ), a transition occurs between a simple collective motion and a granular flow regime.
Keywords: dense granular flow, bubbles, defects

1 Introduction

Our main motivation was to study the granular flow of identical particules exhibiting weak friction. In order to reach this particular situation, we imagined to work with air bubbles in a liquid. This work led us to consider a dense bubble flow.

2 Experimental Setup

The experimental setup consists in a transparent inclined plane which is immerged into water. The tilt angle θ can be adjusted from 0° up to 5°. Spherical air bubbles are injected from below at the bottom of the tilted plane. The bubble diameter $2R$ is controlled by the air pump and is typically $2R = 2.3$ mm. In order to avoid the coalescence of the bubbles, a small quantity of surfactant is added into water. Due to buoyancy, the bubbles rise along the inclined plane. A rough obstacle has been glued at the top of the inclined plane such that bubbles aggregate there. Bubbles tend to pack in an ordered hexagonal structure [1] because the bubble size is nearly monodisperse. A sketch of the setup is drawn in Fig. 1.

Since the packing is continuously fed by new bubbles, a flow of bubbles is generated. When the bubble production is stopped, the pile slowly collapse. The equilibrium situation is a flat surface. Since the angle of stability of a granular pile depends on the friction between contacting grains [3], we could

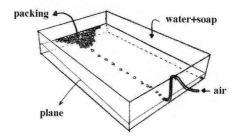

Fig. 1. Sketch of the experimental setup.

consider our system as a granular system with nearly zero friction. Of course, solid/solid friction does not really apply between those fluid objects but the analogy is of interest. Moreover, the bubble motion is quite slow such that the typical Reynolds number is close to $Re = 1$. The liquid flow around bubbles could thus be considered as a laminar flow. Viscous effects are thus negligeable in our experiment.

3 Results

Each bubble of the packing has been tracked through image analysis. Velocity fields have been measured. Typical velocity fields are illustrated in Fig. 2 for different values of the tilt angle θ. The motion of blocks of bubbles can be seen. Those blocks look like crystal domains and are sliced by fast moving defects (dislocations) along bubble lines. The main activity seems to occur near the surface of the bubble packing. Rare events are however observed deep inside the packing. Contrary to granular systems for which only surface grains are moving, all bubbles of the packing are in motion. When the angle θ increases, the mean velocity $\langle v \rangle$ increases. The density of dislocations seems also to rise with θ. Both defects and bubble motions are of course related.

The presence of dislocations is clearly emphasized when looking at the rotational velocity field (see Fig. 3). We have observed that a majority of defects are moving horizontally just below the surface. Moreover, dislocations begin to interact and to cooperate when the angle θ is large enough.

The bubble velocity fluctuations defined by

$$\sigma = \sqrt{\langle v^2 \rangle - \langle v \rangle^2} \tag{1}$$

have been also measured. Figure 4 presents this quantity as a function of θ. For low angle values ($\theta < 1.4°$), the velocity fluctuations are quite small. Only a few and independent dislocations propagate through the packing. The velocity of the bubbles could be considered as unique. Bubbles are moving within a single block motion, i.e. we observe a fully cooperative motion. When the angle becomes larger than a critical value θ_c, the velocity fluctuations increase

Fig. 2. Typical pictures of the bubble packing. The velocities associated to the bubbles are emphasized. (top row) Three successive pictures for $\theta = 0.7°$. (middle row) Three successive pictures for $\theta = 1.3°$. (bottom row) Three successive pictures for $\theta = 2.3°$.

suddenly and a broad distribution of velocities exists. The moving defects are interacting since we observe numerous events such as collision, merging, and annilihation of dislocation pairs. The cooperative motion of these dislocations lead to the motion of smaller clusters of bubbles. Thus, we observe a transition occuring at this critical angle $\theta_c = 2°$. The onset of the transition takes place at $\theta_{onset} = 1.4°$.

We have checked whether this transition around θ_c is observed or not in the velocity profiles. The velocity of every bubble has been averaged over

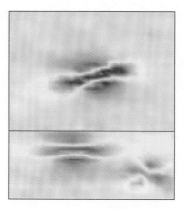

Fig. 3. Two typical density representations of $\vec{\nabla} \times \vec{v}$. Only the rotational component of the velocity field appears. Defects and their displacement are clearly revealed. (top) A dislocation moving along a diagonal line towards the surface of the bubble pile. (bottom) Two dislocations are interacting.

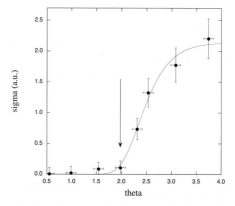

Fig. 4. The fluctuations σ of the bubble velocities as a function of θ. For low angle values ($\theta < 1.4°$), the velocity of the bubbles is unique. They are moving in a single block motion. For high angle values ($\theta > 2°$), the velocity fluctuations increases with θ meaning that the velocity profile along the flow should be curved. Error bars are indicated. The continuous curve is only a guide for the eye.

several pictures (about 600 images) and has been plotted as a function of the dimensionless depth $h/2R$ in the packing (see Fig.5). For small values of θ, the velocity profile is a constant. This is a signature of the simple block motion. However, above θ_c, the velocity profiles are curved like for a granular flow.

The occurence of the transition is unclear and needs further studies. Our work suggests also that a key ingredient for the study of granular flows is the presence of fast moving defects as recently discussed by Pouliquen and coworkers [2]. Additional information about our work can be found in [4].

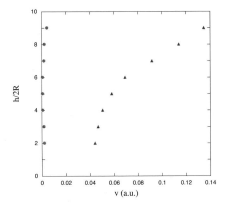

Fig. 5. Two typical velocity profiles. (dots) Vertical velocity profile for $\theta = 0.7°$. (triangles) Curved velocity profile $\theta = 2.3°$.

4 Summary

We have experimentally studied an ideal granular system since the friction is nearly zero. The flow is mainly due to the fast motion of defects (dislocations). The major part of them is moving just below the surface of the packing. When the density of defects is low, a block motion takes place. When the density of defects increases, a transition occurs leading to a granular flow regime.

Acknowledgements

HC thanks FRIA (Brussels, Belgium) for financial support. SD thans FNRS (Brussels, Belgium) for a financial support. This work is also financially supported by the contract ARC 02/07-293. NV thanks A.Pekalski and O.Pouliquen for fruitful discussions.

References

1. W.L. Bragg and J.F. Nye, *Proc. Roy. Soc. Lon.* **190**, 474 (1947).
2. O. Pouliquen and R. Gutfraind, *Phys. Rev. E* **53**, 552 (1996).
3. J. Alonso and H.J. Herrmann, *Phys. Rev. Lett.* **76**, 4911 (1996).
4. N. Vandewalle, S. Trabelsi, and H. Caps, cond-mat/0212250 (2003).

The Global Property of the Dilute-to-Dense Transition of Granular Flows in a 2D Channel

M. Hou[1], W. Chen[1], Z. Peng[1], T. Zhang[1], K. Lu[1], and C.K. Chan[2]

[1] Institute of Physics – Chinese Academy of Sciences
   Beijing 100080 – China
   mayhou@aphy.iphy.ac.cn
[2] Institute of Physics – Academia Sinica
   NanKang, Taipei 11529 – Taiwan

Summary. The dilute-to-dense transition of granular flow of particle size d_0 is studied experimentally in a two dimensional channel (width D) with confined exit (width d). It is found that there exists a maximum inflow rate Q_c, above which the outflow changes from dilute to dense and the outflow rate $Q(t)$ drops abruptly from Q_c to a dense flow rate Q_d. The re-scaled critical rate $q_c(\equiv Q_c/(D/d_0))$ is found to be a function of a scaling variable λ only, i.e. $q_c \sim F(\lambda)$, and $\lambda \equiv \dfrac{d}{d_0}\dfrac{d}{D-d}$. This form of λ suggests that the dilute-to-dense transition is a global property of the flow; unlike the jamming transition, which depends only on $\dfrac{d}{d_0}$. The transition is found to occur when the area fraction of particles near the exit reaches a critical value 0.65 ± 0.03.

Keywords: granular flow, scaling law, phase transition, hopper flow

1 Introduction

1.1 Two Important Transitions in Granular Flows

The phenomena of crowding or jamming are common experiences in our daily lives. However, advances in the understanding of these processes have only begun recently after the physics of granular materials have been under intense investigations [1,2]. It is now believed that the diverse phenomena of traffic flow, pedestrian flow and floating ice [3-5] are related to the nonlinear behaviours of granular materials which can exhibit both solid-like and fluid-like behaviours [6,7]. These peculiar properties [8,9] give rise at least to three important "states" in granular flows; namely, the dilute flow, dense flow and the jammed state. The phenomenon of crowding can then be understood as a transition from dilute to dense flows and that of jamming is a transition from dense flows to a jammed phase.

1.2 Difference Between Jamming and Dilute-to-Dense Transition

Obviously, the nature and properties of these transitions are governed by the interactions among the granular particles in the flow. In principle, these transitions and "states" can be well characterized when these interactions are known such as in the cases of equilibrium systems [10,11]. Unfortunately, since interactions among the granular particles are highly nonlinear and characterized by dissipations which can be density and even history dependent, it is not surprising to find that these transitions and their states are still not well understood. Recently, there are some progresses in the understanding of the transition from dense to jammed states [12,13] and even an equilibrium-like jamming phase diagram have been proposed [14-16]. However, still very little is known about the dilute to density flow transition.

An important characteristic of the dense-to-jam transition is that there is usually only one dominant length scale in the problem; namely the size of the grains (d_0) [12]. In contrast, it seems that the global scale of the order of the size of the system is important for the dilute-to-dense transitions. For example, a small bottle-neck of the size of the system can induce the dilute-to-dense transition [17]. In this aspect, the dilute-to-dense transition is similar to transitions in hydrodynamic flows. In the case of hydrodynamic flows, it is well known that different Reynolds numbers are associated with different flow configurations to characterize the flow as laminar or turbulent. The Reynolds number is a global parameter which scales with the system sizes. Intuitively, in the case of dilute-to-dense transitions, a similar global scaling parameter might exist. If such a scaling parameter can be found, its scaling form will probably provide a better understanding of the nature of the dilute-to-dense transitions.

In this paper, we report our results on experiments carried out in a 2D channel to look for a relation between the dilute-to-dense transition and system parameters. We find that the critical flow rate Q_c at the transition can be well characterized by the scaling variable $\lambda \equiv \dfrac{d}{d_0} \dfrac{d}{D-d}$ where the system parameters are the channel width (D), the opening of the channel (d) and the diameter of the grains (d_0). Q_c is the maximum dilute flow rate above which the flow change from dilute to dense. The re-scaled critical flow rate $q_c (\equiv Q_c/(D/d_0))$ is found to be a function of a scaling variable λ only as: $q_c \sim F(\lambda)$. This form of λ suggests that the dilute-to-dense transition is a global property of the flow; unlike the jamming transition, which depends only on $\dfrac{d}{d_0}$. The transition occurs when the area fraction of particles near the exit reaches a critical value $\sim 0.65\pm0.03$.

2 Experimental Setup

Our experiments are performed in a two-dimensional (2D) channel with an inclination angle of 20°. The 2D channel is established on a metal plate between two glass plates separated by specially shaped metal spacers to form a test section and a hopper as shown in Fig. 1. The gap between the two glass surfaces is kept at 1.2 mm (2.2 mm) to ensure an almost single-layer flow of steel beads of diameter $d_0=1\pm0.01$ mm (2 ± 0.01 mm) which are stored in the hopper at the top of the channel. The hopper, with an open angle of 60° is connected to a test section of width $D=24$ mm and length $l=500$ mm. At the end of the test section, there

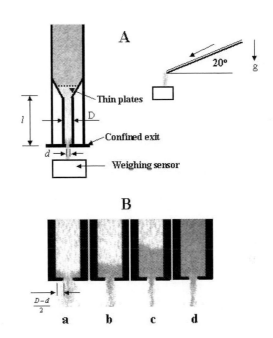

is an exit of width d which is controlled by micrometers to a precision of 0.01 mm. Granular flows in the test section is initiated by allowing the steel beads in the hopper to fall by gravity. A thin plate with a number of uniformly distributed holes is inserted close to the exit of the hopper to control the inflow rate and ensure the uniformity of particle distribution across the test section. The total mass M of the beads falling out of the exit is measured as a function of time t by an electronic balance with sensitivity of 0.02 g and a weighing period of 0.02s. The flow rate $Q(t)$ is obtained by the slope of the recorded $M(t)$ curve.

Fig. 1A. Top and side views of the inclined channel;
B. Photos (a-d) of dilute to dense flow transition in a time sequence.

3 Experimental Results

3.1   The Dilute-to-Dense Transition

The flow in the test section initiated from the hopper is dilute and accelerating. The typical velocity of the steel beads close to the exit is 1.0 ± 0.1 m/sec. Since d is smaller than D, two wedges (heaps) will be formed at both sides of the exit with a base length $(D-d)/2$. If the inflow rate Q_0 is small or d is not too small, there will be no net accumulation of beads in the test section other than the two wedges and $<Q(t)> = Q_0$. This is the regime of dilute flows as shown in (a) of Fig. 1B. However, for a given Q_0, there will be a critical d_c below which there will be net accumulation of beads in the test section. In other words, if d is decreased systematically, there will be a sudden drop of $Q(t)$ when d_c is reached. This accumulation of beads will proceed until the whole test section is filled with beads and this is the regime of dense flows. The process of this dilute-to-dense transition induced by reducing d is shown in Fig. 1B which is recorded by a video camera.

3.2   Flow Rates in Dilute and Dense Flow

Figure 2 shows the d dependence of $<Q(t)>$ when d is reduced or increased systematically at a given Q_0. When d is large (point A), $<Q(t)> = Q_0$, the flow is dilute. It can be seen that $<Q(t)>$ remains practically independent of d when d is larger than a critical size d_c (A to B in Fig.2). When d_c is reached, the dilute flow turns to dense flow and the flow rate can be reduced instanta-neously by several times to drop from Q_0 to Q_d (B to C). After the transition, Q_d decreases monotonically with reducing d (C to D). The flow jams when d is about the size of four-particle diameters, where permanent arching occurs to cause jamming of the flow [12] (D to E). When increasing d from the jammed phase, the flow starts as dense flow, and the rate $<Q(t)>$ increases gradually with increasing d until Q_d reaches Q_0 and turns back to dilute flow as shown by triangle points in Fig.2 (D through C to A). There is no sudden increase of the flow rate at the transition from dense to dilute flow. We have checked that the dense flow rate curve Q_d (d) (the CD part of curve ACD), follows the Beverloo empirical equation $(d-kd_0)^{3/2}$ with $k = 4$ [18,19] and is independent of Q_0.

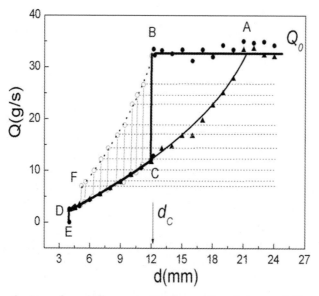

Fig. 2. At a given inflow rate Q_0, the outflow behaves differently when the flow is dilute or dense. For a dilute flow, a transition from dilute to dense occurs at a critical opening size d_c, and the outflow follows curve ABCDE. For a dense flow, as d increases the flow follows curve DC and extended to A

with no abrupt change. For different Q_0, the transition occurs at different d as shown by dash lines. The curve BF determines the optimal outflow rate at any given Q_0.

3.3   Scaling Relation

If the experiment of Fig. 2 is repeated with different Q_0, a family of paths similar to ABCD of Fig. 2 will be formed, shown as dotted lines in Fig.2. An important characteristic of this family of lines is that d_c decreases with Q_0 shown as broken line BF. That is: for a given d, there will be a Q_0 (denoted as $Q_c(d)$ at which a dilute-to-dense transition will occur. Fig. 2 is the result of experiments with fixed D but obviously $Q_c(d)$ will also depend on D and d_0. The $Q_c(d)$ curves for various D and d_0 have been measured and shown in Fig. 3. Four D's: $D = 30,25,20$ and 15 mm are tested for $d_0 = 1$ mm particles, and two D's: $D = 40$ and 30 mm are tested for 2 mm particles.

In Fig. 3 the upper six curves are $Q_c(d)$'s and lower six curves are the corresponding $Q_d(d)$ curves for particles $d_0 = 1$ mm and 2 mm. If the dilute-to-dense transition scales with the system size, one might expect that the exit width d should be scaled by channel width D as $Q_c(d)/D=F(d/D)$ for some scaling function F. Instead, a new scaling variable $\lambda \equiv \dfrac{d}{d_0}\dfrac{d}{D-d}$ is found to collapse the scaled critical flow rate q_c ($\equiv Q_c(d)/(D/d_0)$) into a single scaling curve as shown in Fig. 4. The result in Fig. 4 can therefore be expressed as: $q_c=q_0 F(\lambda)$, where q_0 is some constant.

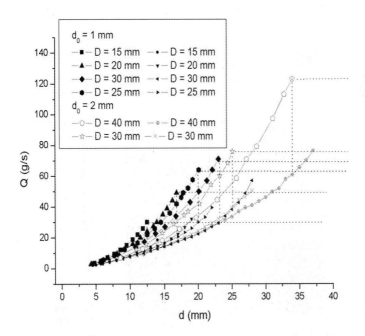

Fig. 3. The Q_c and Q_d versus d of particle size $d_0 = 1$ mm at channel widths $D = 30, 25, 20$ and 15 mm,and of particle size $d_0 = 2$ mm at channel widths $D = 40$ and 30 mm.

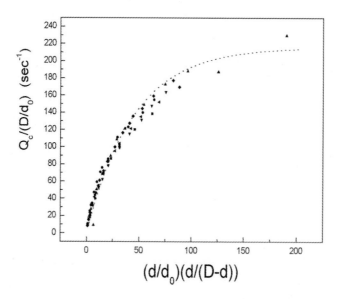

Fig.4. The scaled $q_c(\equiv Q_c/(D/d_0))$ vs. a new scaling variable $\lambda (= \dfrac{d}{d_0}\dfrac{d}{D-d})$. The dotted line is a fit of q_c by $q_m \left(1-e^{-(\lambda/\lambda_0)}\right)$, where $q_m =$ 216 and $\lambda_0 = 45$.

4 New Scaling Variable

4.1   Origin of Strong Density Fluctuations

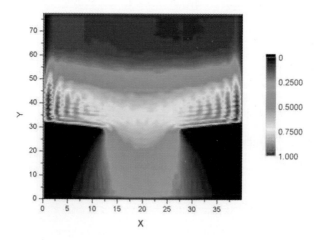

Fig. 5. The average occupancy probability of particles near the exit before transition occurs.

A remarkable feature of the dilute-to-dense transition is that there are strong fluctuations in $Q(t)$ when d is set to be close to d_c. Direct observations of the motions of the beads close to the exit reveal that there are avalanche-like events taking place at the two wedges on either side of the exit. After the wedges have been built up by the incoming flux, flows of surface layer in the form of avalanches will occur. The two wedges effectively act as collectors of incoming flux and direct them to the exit through the surface flow. Therefore, there are strong density fluctuations (in terms of area fraction p) near the exit. When the discharge from the two wedges meet at the exit, the outflow can sometimes be blocked leading to dense flow. This blockage is intermittent if the incoming flux is not large enough and therefore producing strong fluctuations in $Q(t)$. However, if the incoming flux exceeds Q_c, the blockage becomes permanent and there are accumulation of the beads in the test section until the whole section is filled. Particle density p of area 40 mm x 80 mm at the exit is shown by averaging video pictures in Fig. 5. It is found that a dilute-to-dense transition will occur if p reaches a critical value p_c. In our experiments, it seems that the value of p_c corresponds to that of an area fraction of 0.65±0.03 (see Fig. 5) and is independent of Q_c, d or D.

4.2 Simple Model for Scaling and Criteria for Dilute-to-Dense Transition

Following the observations discussed above, the flow across the test section can be divided into three regions; the central part with length d and the two wedges with base length $(D-d)/2$. If v_0 and p_0 are the velocity and density of the dilute flow initiated by the hopper just before particles reaching the wedges, the incoming flux on the wedge is $v_0 p_0 \dfrac{D-d}{2}$. The outgoing flux of the wedge will be carried away by the fluidized surface layer mentioned as $v_e p_c \delta$ where δ and v_e are the depth and characteristic velocity of the surface layer respectively. Conservation of flux gives: $\delta = \dfrac{v_0 p_0}{v_e p_c} \dfrac{D-d}{2}$. In this model, δ will increase with the incoming flux or p_0. When these two top layers meet at the exit, we have: $\delta \beta = d/2$ where β is some geometric factor which take care of the angle of repose. Therefore, at the transition, we have:

$$f_c \equiv (v_0 p_0)_c = v_e p_c \beta^{-1} \frac{d}{D-d} \quad \text{or} \quad \frac{f_c}{f^*} = \frac{v_e}{v^*} \frac{d}{D-d} \quad \text{where} \quad f^* \equiv v^* p_c \beta^{-1}.$$

One can consider v^* as some intrinsic velocity of the problem which is determined by physical properties of the system such as the angle of inclination or coefficient of sliding friction and etc. Presumably $\dfrac{v_e}{v^*}$ is a function of system parameters D and d. However, for the case of $D \gg d$, it is reasonable to assume that $\dfrac{v_e}{v^*}$ depends only on $\dfrac{d}{d_0}$. In such a case, one expect to see

$\dfrac{v_e}{v^*} \sim \dfrac{d}{d_0}$ because this is the first order expansion of v_e in terms of d when $d =$

0 gives $v_e = 0$. This later form of $\dfrac{v_e}{v^*}$ gives $\dfrac{f_c}{f^*} \sim \lambda \equiv \dfrac{d}{d_0} \dfrac{d}{D-d}$ which

agrees with our result in Fig. 4 for small d or λ. Physically $\dfrac{v_e}{v^*} \sim \dfrac{d}{d_0}$ means that the discharge velocity of the fluidized layer of the two wedges increases with d. Note that v_e for the dense flow, v_e with $\alpha = 1/2$ [19].

It can be seen from Fig. 4 that q_c increases with λ monotonically. Obviously, for a fixed D, Q_c must tend to a limit for large enough d/d_0 in the experiments because Q is given by $v_0 \rho_0 D$ and there must be an upper limit in ρ_0 to still have dilute flows in the test section. If q_m is the maximum dilute flow flux of the channel, we have $q_m = q_0 F(\infty)$. One can choose $F(0) = 0$ and $F(\infty) = 1$ to give $q_c = q_m F(\lambda)$. Note that there are two ways for λ to go to ∞. When $d = D$, the test section is just a straight pipe (2D), q_m is obviously just the maximum dilute flow capacity of the pipe. For fixed d/D, λ goes to ∞ when either d goes to ∞ or d_0 goes to zero. In both cases, we have a continuum limit in which the size of the particles can be neglected. Therefore, our model predicts that q_m is independent of d/D when d/d_0 is large enough. The functional form of $F(\lambda) = 1 - e^{-\lambda/\lambda_0}$, which gives the correct form of q_c for small λ, has been fitted to the data in Fig. 4 shown as dotted line. The scattering of the measured values of Q_c increases as λ increases, which makes it difficult to determine experimentally the limit value of q_c. Presumably, our model discussed above is valid only for small d.

4.3   Physical Picture of the Transition

The phenomenological model described above is based on the observation that the transition occurs when the flow density near the exit reaches $p = 0.65 \pm 0.03$ as is shown in Fig 5.

The abrupt change of the flow rate at the dilute-dense transition may be understood as the inelastic collisions of the particles with the two piles near exit. In some aspects, our system is similar to the wedge setup of Rericha's [20], where shocks identical to those in a supersonic gas are observed when a steady flow passes the wedge. Our system, however, with the piled heaps at the two sidewalls near the exit is equivalent to a system where particles pass two inward "soft wedges", a system more commonly seen in industrial transport of granules. The abrupt change in density, granular temperature, and velocity before and after the dense area near the exit may account for the flow rate drops at the dilute-dense transition, as the flow rate is a function of the product of flow density and the particle velocity.

5   Conclusion

In conclusion we have experimentally obtained a dilute-to-dense transition curve Q_c (Q_0, d_c), at which the flow rate Q_0 drops to Q_d at d_c. Instead of the intuitive scaling relationship d/D, experimental results show that a global scaling variable $\lambda \equiv \dfrac{d}{d_0} \dfrac{d}{D - d}$ determines if the flow is dilute or dense at a given Q_0. The scaled q_c can be fitted in an empirical form $q_m \left(1 - e^{-(\lambda/\lambda_0)}\right)$.

The ratio λ/λ_0 determines the critical flux of the system. Intuitively, λ_0 is determined by physical properties of the system, such as the elasticity of the particles, particle size and the inclination angle of the plate and etc. While the physics of jamming transition is determined by local scales close to d_0, our result suggests that the physics of dilute-to-dense transition is similar to hydrodynamic instability which is controlled by global scales such as $(d/(D-d))$. The discovery of this global scaling property may provide us with ideas of better designs for the transport of granules in industrial processing, and better understanding of similar flow systems such as systems in traffic flows.

This work was supported by the National Key Program for Basic Research and the Chinese National Science Foundation Project No. A0402-10274098.

References

1. L.P. Kadanoff, Rev. Mod. Phys. 71, 435 (1999).
2. P.G. de Gennes, Rev. Mod. Phys. 71, S374 (1999).
3. Traffic and granular flow '99. edited by D. Helbing, H.J. Herrmann, M. Schreckenberg, and D.E. Wolf (Springer, Singapore, 1999).
4. J. Rajchenbach, Advances in Physics 49, 229 (2000).
5. H.T. Shen and S. Lu, Proceeding of 8th Int'l Conference on Cold Regions Engineering (ASCE, Fairbanks, 1996), pp. 594-605.
6. G.H. Ristow, Pattern Formation in Granular Materials (Springer, New York, 2000).
7. X. Yan, Q. Shi, M. Hou, K. Lu, and C. K. Chan, Phy. Rev. Lett. 91, 014302 (2003).
8. J.H.M. Jaeger, S.R. Nagel, and R.P. Behringer, Rev. Mod. Phys. 68, 1259 (1996).
9. J. Duran, Sands, Powders, and Grains (Springer, New York, 2000).
10. D. Tabor, Gases, liquids and solids (Cambridge University Press, Cambridge, 1969).
11. Landau and Liftshitz, Statistical Physics (Pergamon Press, Oxford, 1959).
12. K. To, P.-Y. Lai, and H.K. Pak, Phys. Rev. Lett. 86, 71 (2001).
13. E. Longhi, N. Easwar, and N. Menon, Phys. Rev. Lett. 89, 45501 (2002).
14. A.J. Liu and S.R. Nagel, Nature 396, 21 (1998).
15. G. D'Anna and G. Gremaud, Nature 413, 407 - 409 (2000).
16. V. Trappe, V. Prasad, L. Cipelletti, P.N. Serge, and D.A. Weitz, Nature 411, 772 (2001).
17. W. Chen, M. Hou, Z. Jiang, K. Lu, and L. Lam., Europhys. Lett. 56 (4), 536 (2001).
18. W.A. Beverloo, H.A. Lengier, and J. Van de Velde, Chem. Engng. Sci. 15, 260 (1961).
19. R.M. Nedderman, Statics and Kinematics of Granular Materials (Cambridge University Press, Cambridge, 1992), Chapter 10.
20. E.C. Rericha, C. Bizon, M.D. Shattuck, and H.L. Swinney, Phys. Rev. Lett. 88, 014302 (2002).

From Granular Flux Model to Traffic Flow Description

K. van der Weele[1], W. Spit[2], T. Mekkes[1,2], and D. van der Meer[1]

[1] Physics of Fluids Group – University of Twente
   P.O. Box 217, 7500 AE Enschede – The Netherlands
   j.p.vanderweele@tn.utwente.nl
[2] Witteveen+Bos Consulting Engineers – Section Traffic Management
   P.O. Box 85948, 2508 CP Den Haag – The Netherlands
   w.spit@witbo.nl

Summary. A description of highway traffic flow is proposed, based on a flux model originally developed for granular gases in a compartmentalized setup. Results from a pilot study of the highway A58 in the Netherlands are presented, followed by a discussion of possible improvements and applications of the model.
Keywords: traffic flow, granular matter, flux model

1   Introduction

Traffic flow resembles granular flow nowhere more closely than on the highway. Here the individual behavior of the drivers forms a relatively small statistical perturbation on the deterministic part of the collective motion, and hence the cars can be treated as physical particles [1,2]. Both are many-particle systems far from equilibrium, in which the constant competition between driving forces and dissipative interactions leads to self-organized structures: Indeed, there is a strong analogy between the formation of traffic jams on the highway [3] and the formation of particle clusters in a granular gas [4].

  Here we propose a description of large-scale traffic flow based on a flux model originally developed for granular gases in a compartmentalized setup. Consider a row of compartments, divided by walls of a certain height h, in which granular matter is brought into a gaseous state through vertical shaking (see Fig. 1a). The particle flow from one compartment to the next is given by a *flux function* [5]:

$$F(n_k) \;=\; A\, n_k^2\, e^{-Bn_k^2} . \tag{1}$$

  Here n_k is the fraction of the total number of particles contained in compartment k, $\Sigma n_k = 1$, and the dimensionless parameter B determines whether the system will end up in the homogeneous situation (with equally filled compartments) or in a clustered state. It can be expressed as follows [5-7]:

$$B \;\propto\; \frac{gh}{(af)^2}\,(1-e^2)^2 \left(\frac{r^2 N_{tot}}{\Omega} \right)^2 , \tag{2}$$

with g the gravitational acceleration, h the height of the walls between the compartments, a and f the amplitude and frequency of the driving, e the coefficient of normal restitution of the particle collisions, r the radius of the particles, N_{tot} their total number, and Ω the ground area of each compartment.

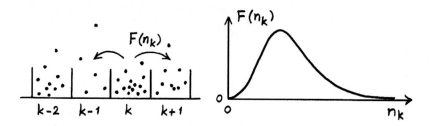

Fig. 1. (a) A granular gas, shaken vertically, in a compartmentalized setup. The arrows denote the particle flux from compartment k. (b) The flux function $F(n_k)$, i.e. the flux from compartment k (to any of its neighbors) as a function of the fraction n_k contained in it. The non-monotonic shape of $F(n_k)$ is due to the inelasticity of the particle collisions.

Note that, unless $B = 0$ (which is the limit of very strong shaking, $af \to \infty$, or completely elastic collisions, $e = 1$), the particle flux $F(n_k)$ is a one-humped function of n_k, see Fig. 1b. First it grows with increasing n_k, as in any ordinary molecular gas, but beyond a certain filling level it goes down again, since the increasingly frequent collisions make the particles so slow that they are hardly able to jump over the wall anymore. This non-monotonic behavior is essential for the clustering phenomenon: In the clustered situation the flux from a well-filled compartment (large n_k) equals the flux from its dilute neighboring compartments (small n_k).

The evolution of the system is given by the following balance equation:

$$\frac{dn_k}{dt} = F(n_{k-1}) - 2F(n_k) + F(n_{k+1}), \tag{3}$$

i.e., the rate of change of the fraction n_k is given by the flux coming into compartment k from left and right, minus the flux going out of it towards its two neighbors. Statistical fluctuations in the particle flux would add an extra term to Eq. (3), but we do not consider this here, so the present description is to be inter-preted as a mean-field theory. Equation (3) is written for a cyclic array of compartments; if the array is non-cyclic, the equations for the end compartments are modified accordingly [6].

The above model adequately describes the dynamics of the granular gas in the compartmentalized system. Its predictions have been shown to agree quantitatively with experiments, for the clustering phenomenon [6] as well as for the opposite process of declustering [7]. In the next sections we propose a similar (but of course unidirectional) model for traffic flow on the highway, focusing upon the A58 between the cities of Breda and Eindhoven, see Fig. 2. This two-lane highway, like many others in the Netherlands, is effectively *compartmentalized* by induction loops in the asphalt at distances of roughly 500 m, which constantly monitor the number and velocity of the cars passing over them [8, 9]. The distance of 500 m is very suitable, since this is the natural scale on which congestions form and develop. The corresponding temporal scale is in the order of several tens of seconds to a few minutes.

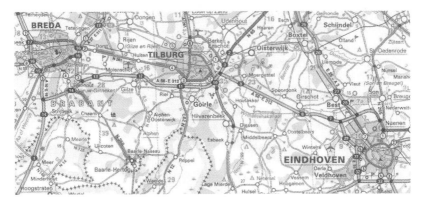

Fig. 2. Highway A58 (or E312) between Breda and Eindhoven, in the southern part of the Netherlands.

2   Traffic Phenomenology

A typical example of data collected by an induction loop is shown in Fig. 3. Here we see, for one specific location near Eindhoven, 5-minute averages of the measured velocity v (in km/h) and flux F (in veh/h/lane) plotted against the derived quantity ρ (the density at the loop, in veh/km/lane). The latter is determined via the relation $\rho = F/v$. The data were taken during the morning traffic (6 to 10 a.m.) on 15 standard working days in the autumn of 2001, without accidents or exceptional weather conditions. Two regimes can be discerned:

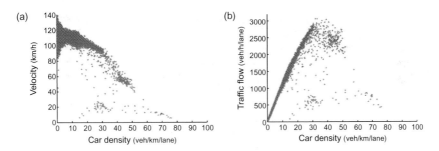

Fig. 3. Experimental data from the A58 at a detection loop between Oirschot and Batadorp, near Eindhoven, showing (a) the relation between the car velocity v and the density ρ, and (b) between the flow $\rho\, v(\rho)$ and ρ (the "fundamental diagram"). The points in these plots are 5-minute averages; the original 1-minute averages measured by the induction loop show more scatter.

    I. *Free flow*, at low densities, in which the cars drive at their desired velocity of roughly 110 km/h (with quite a large spread, partly due to the fact that the data include both passenger cars and trucks). This regime corresponds to the well-defined upward branch in the plot of $F(\rho)$, rising to nearly 3000 veh/h/lane at $\rho \approx 30$ veh/km/lane.

    II. *Congested traffic* at densities above 30 veh/km/lane. Here the distance between successive cars becomes such that the drivers can no longer maintain their desired velocity; they have to react, brake, and maneuver, and this causes a sudden drop in the velocity, and hence in the flux. The flux in Fig. 3b obviously depends in a non-monotonic way on the density: this is a crucial prerequisite for clustering, and is reminiscent of the granular flux function in Fig. 1b. The reason for the non-monotonic shape is the same in both cases, namely the dissipative nature of the particle interactions. However, the data in the F,ρ-plane do not follow a 1-dimensional function but are scattered over a 2-dimensional area, corresponding to various types of congested traffic (synchronized flow, jams, etc. [10,11]). In our traffic flux model we will therefore go beyond the 1-dimensional description.

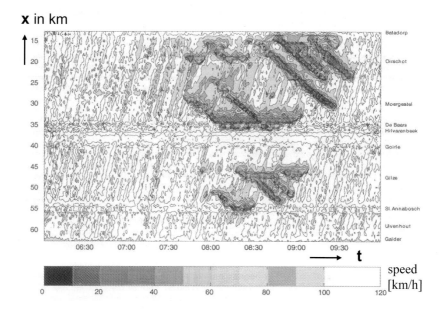

Fig. 4. An *x,t*-diagram of the morning traffic towards Eindhoven on Thursday, November 1, 2001. The colors represent the measured 1-minute averages of the velocity, and thereby the density, cf. Fig. 3a. The traffic jams (dark regions) are seen to move backward with an invariable group velocity of −17 km/h, and are occasionally pinned to the access and exit lanes near St Annabosch, Gilze, etc. The vehicles in free flow (light regions) go with an average speed of 110 km/h.

The data from all the induction loops together can be used to give an overall picture of the traffic as in Fig. 4. Here the measured average velocities are shown in an *x,t*-diagram, for the morning traffic towards Eindhoven on a typical working day. Many interesting congestion phenomena are reflected in this diagram, such as:
- All traffic jams move backward (against the flow) with the same *universal speed* of about −17 km/h [12].
- The *boomerang effect* [13], in which an emerging jam, originating at an on-ramp, first moves forward (with the flow) but changes direction as it grows larger and finally moves backward with the universal speed. See e.g. the second jam originating at Moergestel in Fig. 4.

- The *pinning phenomenon*, in which jams come to a standstill at an on-ramp. A good example is seen at De Baars, from 8:00 to 9:00.
- *Recurrent jam generation* at an on-ramp, giving rise to stop-and-go traffic (see in particular the pattern near Oirschot, between 8:30 and 9:30) [13,14], etc.

3 Traffic Flux Model

As stated above, we go beyond the 1-dimensional flux description, and model the traffic flow from cell k by a function $F(\rho_k, \rho_{k+1})$, which depends not only on the density in cell k itself, but also on the density in the target cell $k+1$ [15]. This means that we take into account the fact that car drivers *anticipate* to the situation ahead of them.

The time evolution of the system is then given by (cf. Eq. (3)):

$$\frac{d\rho_k}{dt} = \frac{1}{\Delta x}\left\{F(\rho_{k-1},\rho_k) - F(\rho_k,\rho_{k+1})\right\} + Q_k(t), \qquad (4)$$

where $\Delta x = 500$ m is the cell length. This is a continuity equation, with (in certain cells only) an additional term $Q_k(t)$ representing the in- and outflow at junctions and ramps. The model is related to the cell transmission model of Hilliges and Weidlich, in the limiting case that the velocity is adjusted instantaneously to the value $v(\rho)$ corresponding to the ambient density [16]. There is also a resemblance (if one ignores the dependence on the target cell density) with the continuum model of Lighthill and Whitham [17]; we refrain however from taking the continuum limit, since the discreteness of the cells is an essential feature of our approach.

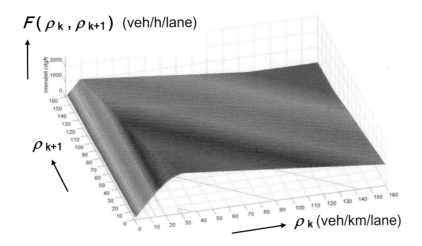

Fig. 5. The current version of the flux function $F(\rho_k, \rho_{k+1})$.

The current version of the flux function is depicted in Fig. 5: It is a one-humped function of ρ_k (of the form $A\rho_k\exp[-B\rho_k^\beta]$, with $A = 110$ km/h, and B and β two fit-parameters) and decreases monotonically for growing ρ_{k+1}. The downward slope, which has the physical meaning of a group velocity, can be calibrated to give the observed -17 km/h for jams [18]. The flux settles to an approximate value of 500 veh/h/lane when ρ_{k+1} exceeds 100 veh/km/lane, as indicated by our experimental data.

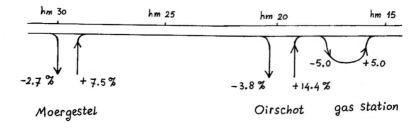

Fig. 6. The percentual in- and outflow at the various ramps between Moergestel and Oirschot during the morning hours (6 to 10 a.m.). Note that the two percentages at the gas station balance each other, as they should.

To model the term $Q_k(t)$, we make use of the following observation: While the absolute numbers of cars flowing in and out of the system at a given ramp vary widely from hour to hour, the *percentage* of ρ_k they represent is more or less constant during the morning hours - and again, but with different values, in the evening. As an example, in Fig. 6 the measured morning percentages at the on- and off-ramps between Moergestel and Oirschot (one of the most critical sections of the A58) are given.

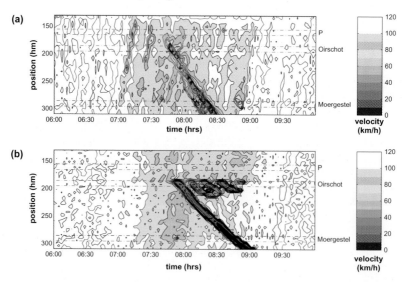

Fig. 7. (a) The morning traffic between Moergestel and Oirschot on Monday November 19, 2001. (b) Predicted traffic according to the current version of our model, given the actual inflow data at the bottom of the diagram. The good and weak points of the prediction are discussed in the text.

How well does the current model work in practice? To give an indication, in Fig. 7 the actual flow between Moergestel and Oirschot (on an ordinary working day) is compared with the flux model prediction, based only on the inflow data along the bottom of the diagram. The horizontal dashed lines in this figure stand for the ramps and "P" represents the gas station (cf. Fig. 6).

The model shows some very promising points: It correctly predicts the time *and* place where the first traffic jam occurs, and properly captures the shape of the blue region, which corresponds to comparatively heavy traffic. Also the internal structure of this region (e.g. the amplitude of the velocity variations) is quite realistic.

Figure 7 also shows some points on which we still have to work: For instance, the upstream movement of the jam needs fine-tuning. It starts out all right, but then slows down too much, and the density within the jam becomes considerably larger than in reality.

Fine-tuning is also required for the flux function near the ramps, as exemplified by the extra jams that are generated at the Oirschot ramp (even though such repetitive patterns are by no means unrealistic, see Fig. 4). The enhanced attentiveness, braking, and lane changing near the ramps can cause a significant reduction of the flux function. One may also anticipate that the local infrastructure (such as the length of the parallel access lane, traffic lights, etc.) will make the flux function near the ramps less universal than elsewhere.

4   Outlook

We plan to extend the model, and improve its performance further, by distinguishing between *different lanes*, and between *passenger cars and trucks*. In order to distinguish between the lanes we will study the traffic on roads with induction loops that measure the lanes individually. The highway is then treated as a two-dimensional array of cells, with forward flux functions that differ from lane to lane (because of the different mean velocities and velocity fluctuations) and cross-wise flux functions F_{cross} to connect the lanes.

With respect to the passenger cars and trucks, we intend to set up a bidisperse version of the flux model, in the same spirit as we did for the compartmentalized granular gas [19]. In this context, we will first concentrate on the two extreme cases of practically pure passenger-car traffic (on Sundays and holidays) and truck traffic (as e.g. observed on working days on the A15 between Europoort and Ridderkerk, near Rotterdam).

Once the flux model is perfected and properly tested, we want to go beyond the level of a single road and deal with *networks* of highways. Conveniently, the main part of the Dutch highway network is already equipped with induction loops, so there is an overwhelming abundance of experimental data. Parts of this network can be the playground for optimizing traffic on a large scale, or for traffic forecasts. Thanks to the computational speed of the present model, it will even be possible to make ensemble forecasts for a range of initial conditions, and hence to give an estimate for the reliability of the forecast.

Finally, the flux model may become a practical instrument for judging the effectiveness of strategies that are proposed to reduce traffic jams. For instance, what is the global effect of local no-pass rules for trucks? Do elongated ramps and "rush hour lanes" really improve the flow through the network as a whole? Or what are the large-scale consequences of locally imposing a certain speed limit? The flux model may help to provide quantitative answers to these and other questions.

Acknowledgements: We want to thank Detlef Lohse (University of Twente) for many fruitful discussions and Peter Veeke (Rijkswaterstaat Noord-Brabant) for providing us with the original traffic data of the A58. This work is part of the research program of the Stichting voor Fundamenteel Onderzoek der Materie (FOM), which is financially supported by the Nederlandse Organisatie voor Wetenschappelijk Onderzoek (NWO), and DvdM acknowledges financial support.

References

1. D. Helbing, *Traffic and related self-driven many-particle systems*, Rev. Mod. Phys. **73**, 1067 (2001).
2. D. Chowdhury, L. Santen, and A. Schadschneider, *Statistical physics of vehicular traffic and some related systems*, Phys. Rep. **329**, 199 (2000).
3. O. Biham, A.A. Middleton, and D. Levine, *Self-organization and a dynamical transition in traffic flow models*, Phys. Rev. A**46**, R6124 (1992); B.S. Kerner and P. Konhäuser, *Cluster effect in initially homogeneous traffic flow*, Phys. Rev. E**48**, R2335 (1993).
4. I. Goldhirsch and G. Zanetti, *Clustering instability in dissipative gases*, Phys. Rev. Lett. **70**, 1619 (1993).
5. J.Eggers, *Sand as Maxwell's demon*, Phys. Rev. Lett. **83**, 5322 (1999).
6. K. van der Weele, D. van der Meer, M. Versluis, and D. Lohse, *Hysteretic clustering in granular gas*, Europhys. Lett. **53**, 328 (2001); D. van der Meer, K. van der Weele, and D. Lohse, *Bifurcation diagram for compartmentalized granular gases*, Phys. Rev. E**63**, 061304 (2001).
7. D. van der Meer, K. van der Weele, and D. Lohse, *Sudden collapse of a granular cluster*, Phys. Rev. Lett. **88**, 174302 (2002).
8. The Dutch traffic is monitored by a special division of the Ministry of Transport, Public Works and Water Management, see http://avvisn0.rws-avv.nl/
9. J.W. Goemans, *Quickscan verkeersafwikkelingsonderzoek A58 Breda-Eindhoven*, Witteveen+Bos report, commissioned by Rijkswaterstaat Noord-Brabant (W+B-RW1163-1, 2002), in Dutch.
10. D. Helbing, A. Hennecke, and M. Treiber, *Phase diagram of traffic states in the presence of inhomogeneities*, Phys. Rev. Lett. **89**, 4360 (1999).
11. B.S. Kerner, S.L. Klenov, and D.E. Wolf, *Cellular automata approach to three-phase traffic flow*, J. Phys. A: Math. Gen. **35**, 9971 (2002).
12. B.S. Kerner and B.S. Kerner and P. Konhäuser, *Structure and parameters of clusters in traffic flow*, Phys. Rev. E**50**, 54 (1994); B.S. Kerner and H. Rehborn, *Experimental features and characteristics of traffic jams*, Phys. Rev. E**53**, R1297 (1996); *Experimental properties of complexity in traffic flow*, Phys. Rev. E**53**, R4275 (1996).
13. D. Helbing and M. Treiber, *Gas-kinetic-based traffic model explaining observed hysteretic phase transition*, Phys. Rev. Lett. **81**, 3042 (1998); *Jams, waves, and clusters*, Science **282**, 2001 (1998).
14. H.Y. Lee, H.W. Lee, and D. Kim, *Origin of synchronized traffic flow on highways and its dynamic phase transitions*, Phys. Rev. Lett. **81**, 1130 (1998); *Dynamics of a continuum traffic equation with on-ramp*, Phys. Rev. E**59**, 5101 (1999).
15. The density $\rho_k(t)$ in cell k is defined as the mean of the two values for F/v measured by the induction loops at both ends of the cell.
16. M. Hilliges and W. Weidlich, *A phenomenological model for dynamic traffic flows in networks*, Transportation Research B **29**, 407 (1995).

17. M.J. Lighthill and G.B. Whitham, *On kinematic waves: II. Theory of traffic flow on long crowded roads*, Proc. Roy. Soc. London A **229**, 317 1955); G.B. Whitham, *Linear and Nonlinear Waves* (Wiley and Sons, New York, 1974).
18. For dense jams this value can be understood as follows: On the average, the cars occupy 7.5 m each, and they leave the front of the jam at a rate of one per 1.5 s (the reaction time). So the front moves backward at a speed of 7.5 m per 1.5 s = 5 m/s, which is 18 km/h.
19. R. Mikkelsen, D. van der Meer, K. van der Weele, and D. Lohse, *Competitive clustering in a bidisperse granular gas*, Phys. Rev. Lett. **89**, 214301 (2002); R. Mikkelsen, K. van der Weele, D. van der Meer, M. Versluis, and D. Lohse, *Competitive clustering in a granular gas*, Phys. Fluids **15**(9), S8 (2003).

Simple Models for Compartmentalized Sand

U.M.B. Marconi[1], F. Cecconi[2], and A. Puglisi[3]

[1] INFM – University of Camerino
    Via Madonna delle Carceri, Camerino – Italy
    `umberto.marinibettolo@unicam.it`
[2] INFM – University of Rome
    Piazzale A.Moro 2, Rome – Italy
    `cecconif@roma1.infn.it`
[3] INFM Center for Statistical Mechanics and Complexity (SMC) –
    University of Rome
    Piazzale A.Moro 2, Rome – Italy
    `andrea.puglisi@roma1.infn.it`

Summary. We study the behavior of inelastically colliding particles moving in a bistable potential, and driven by a stochastic heat bath. The system has the tendency to cluster at low drivings, and to fill completely the available space when vigorously shaked. In the case of just two particles, we show that the hopping over the potential barrier occurs following the Arrhenius rate, if the temperature is replaced by the granular temperature. For systems containing many particles we observe a strong competition between the excluded volume effect, which favors a symmetric distribution between the two wells, and inelasticity which on the contrary induces clustering.
Keywords: Clustering, activated process, granular temperature.

1 Introduction

The study of the properties of nonuniform granular gases [1], i.e. rarefied systems of macroscopic particles experiencing inelastic collisions, is currently a subject of vivid interest for a variety of reasons which range from technological applications, including grain separation, jam formation, to fundamental issues for the statistical mechanics of far from equilibrium systems [2]. Uniformity and homogeneity of large assemblies of fluidized grains are the exception rather than the rule [3]. That for two main reasons: a) to balance the dissipation due to collisions some energy must be injected into the "gas" from moving boundaries; b) the homogeneous state is intrinsically unstable and velocity and density correlations lead to the appearance of vortices and clusters.

In the presence of external fields, a granular fluid will also develop inhomogeneities as an ordinary fluid. Moreover, the inelasticity determines new features. In a well known experiment, Nordmeier and Schlichtting [4] have shown that a collection of grains undergoes a separation process where the majority of the particles migrates to one side, when the container, partitioned into two sections by a dividing wall, is subjected to a mild shaking.

In the present paper we propose as a model [5] for such a system a collection of inelastic hard rods moving on a line and subjected to a potential $U(x)$, having two equivalent minima, as shown in Fig. 1.

2 The Driven Inelastic Hard Rod System

The inelastic hard sphere model is, perhaps, the simplest model able to capture the two salient features of granular fluids, namely the hard core repulsion between grains and the dissipation of kinetic energy due to the inelastic collisions. Since many of the equilibrium properties of the 1D elastic hard rods are known in closed analytical form, such a system represents an excellent reference model. Let us consider N identical impenetrable rods, of coordinates $x_i(t)$, mass m and size σ.

When two inelastic rods collide, their post-collisional velocities (primed symbols) are related to pre-collisional velocities (unprimed symbols) through the rule:

$$v_i' = v_i - \frac{1+r}{2}(v_i - v_j) \tag{1}$$

where r indicates the coefficient of restitution. Between collisions, the particles obey the following equations of motion:

$$m\frac{dv_i}{dt} = -m\gamma v_i + \xi_i(t) - \frac{dU(x_i)}{dx} \tag{2}$$

$$\frac{dx_i}{dt} = v_i \tag{3}$$

where γ is the viscous friction coefficient, $U(x)$ an external potential, $\xi_i(t)$ a Gaussian white noise with zero average and correlation

$$\langle \xi_i(t)\xi_j(s)\rangle = 2\gamma m T_b \delta_{ij}\delta(t-s) , \tag{4}$$

T_b is the "heat-bath temperature" and $\langle \cdot \rangle$ indicates the average over a statistical ensemble of realizations. The interaction of each particle with the heat-bath is the combination of a viscous force, proportional to the particle velocity, and a stochastic force [6]. A double well potential, of the form $U(X) = -\frac{a}{2}x^2 + \frac{b}{4}x^4$, illustrated in Fig. 1.a, mimics the two compartments. The distance between the bottoms of the wells is $L = 2\sqrt{a/b}$, the height of the barrier between the wells is $\Delta U = \frac{a^2}{4b}$.

In order to capture the basic mechanisms, we first consider the simplest granular gas consisting of $N = 2$ particles. The relative distance, $y = x_2 - x_1$, between the two particles fluctuates in time. During time intervals of average length τ_2 these are confined to the same well ($y \sim d$). Such intervals are alternated with periods of average length τ_1, where the particles sojourn in separate wells ($y \sim L$). The basic phenomenology of the average escape time τ from a well, in the case of a single particle, is given by the well known Kramers formula [7] that, apart from prefactors, reads

$$\tau \propto \exp\left(\frac{\Delta U}{T_b}\right). \tag{5}$$

We shall illustrate that, with two particles, the escape time can still be described by a similar formula, after an appropriate redefinition of parameters. As the driving intensity, T_b, and the inelasticity, r, vary, one observes a crossover from a regime where the particles are apart most of the time ($\tau_1 > \tau_2$) to a clustered regime ($\tau_1 < \tau_2$). Moreover, in the case of inelastic particles ($r < 1$), the kinetic temperature $T_2 = \langle v^2 \rangle$ measured taking averages only when the particles are together in the same well is different (lower) than the kinetic temperature T_1 measured when the particles are in different wells.

We find that the dependence of τ_2 and τ_1 on the model parameters can be captured by a simple extension of formula (5), with T_b replaced by one of the two different kinetic temperatures, T_2 and T_1.

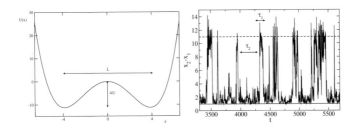

Fig. 1. a: The confining potential $U(x)$. **b**: Behavior of distance between the two particles as a function of time.

Two physical effects contribute: the hard core repulsion and the inelasticity of collisions. Let us begin with an elastic system ($r = 1$), where the most probable configuration, constituted by a single particle in each well, is the one which minimizes the free energy. This situation is reflected by $\tau_1 > \tau_2$ (see ig. 2, open symbols.) for all driving strengths, in spite of the fact that the measured T_2 and T_1 are equal and coincide with the heat bath temperature T_b. How do we understand two different escape times? As displayed in Fig. 2, τ_2 and τ_1 still follow the Arrhenius exponential behavior of Eq. (5), provided we write $\tau_i \propto \exp(W_i/T_b)$, where $W_1 = \Delta U$ and $W_2 = \Delta U - \delta U < \Delta U$. The correction δU takes into account the effect of the excluded volume repulsion. The smaller the ratio between the well width and the particle diameter, the stronger the reduction of the escape time.

The presence of inelasticity ($r < 1$) modifies such a scenario, because dissipation comes into play, temperature T_2 is lower than T_1 and the mean escape time, τ_2, may become larger than τ_1, a signal that the two particles tend to form a cluster [3]. As the temperature increases, a crossover occurs from a clustered state to a symmetric state (Fig. 2). For $r = 0.9$ one observes a crossover at $T_b \sim 4.0$ between τ_1 and τ_2. In other words, at sufficiently low temperatures the two particles spend more time in the same well. Indeed the overcoming of the barriers becomes less likely because of collisional dissipation. We notice that whereas in the elastic system the repulsion always renders the double occupancy less likely, for moderate inelasticity, instead, we observe a crossover from $\tau_2 < \tau_1$ at high T_b to $\tau_2 > \tau_1$ for low T_b. For small driving and small r, $\tau_2 > \tau_1$ means a small outflow from a doubly occupied well, i.e. clustering, whereas $\tau_2 < \tau_1$ corresponds to having the particles uniformly distributed.

As already anticipated, the collisional cooling makes T_2 lower than T_1. For moderate driving intensity, T_1 is nearly equal to T_b, while T_2 is lower than T_b by a factor which depends on the inelasticity. In Fig. 2b we show these temperatures as functions of T_b. It can be observed in Fig. 2b that T_2 varies almost linearly with T_b with a slope decreasing with inelasticity. We find

$$T_2 = \frac{T_b}{1 + \frac{\nu}{4\gamma}(1 - r^2)} \tag{6}$$

where the collision frequency ν is estimated as $\sqrt{U''(x_{min})}/\pi = \sqrt{2a}/\pi$.

Can we relate the two kinetic temperatures to the characteristic escape times defined above? Obviously, since T_1 is practically equal to T_b, τ_1 is described by Eq. (5). More interestingly, we observe that the same Kramers formula holds for the escape time τ_2 from a doubly occupied well, provided T_b is now replaced by T_2, and the excluded volume correction δU, is also taken into account. In both situations we write:

$$\tau_k \approx \exp\left[\frac{W_k}{T_k}\right] \tag{7}$$

where $k = (1, 2)$ indicates single or double occupation.

In the lower inset of Fig. 2a we plot the probability distributions of escape times τ_1 and τ_2 for several simulations. All the distributions are characterized by a peak at the origin and an exponential tail. When rescaled to have the same average, all the tails collapse to a single curve. Such exponential tails are typical of the original Kramers model for thermally activated barrier crossing.

2.1 Many Particles

Laboratory experiments, involve at least few hundred particles [8], thus we turn to consider a 1D system containing a larger number, N, of rods. The behavior of the system can be characterized by considering the evolution of the

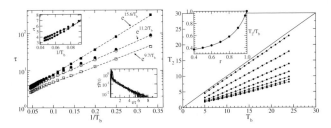

Fig. 2. a: Arrhenius plot of mean escape times τ. Open symbols refer to elastic case: the escape time is τ_1 (circles) when a well is singly occupied, τ_2 (squares) when a well is doubly occupied. Full symbols, instead, correspond to the inelastic system ($r = 0.9$). Linear behaviour indicates the validity of Kramers theory with renormalized parameters and the slopes agree with values obtained from Eq. (7). Upper inset: enlargement of the crossover region. Lower inset: Data collapse of distribution of escape times for different temperatures. The arguments of the exponentials (dashed lines) in the same figure have been obtained by applying formula (7). **b**: Granular temperature against T_b, for different choices of the coefficient of restitution.

absolute difference between the left and right populations, which is shown in Fig. 3a, with a pattern which recalls Fig. 1b. Again, below a certain temperature, T^*, one observes the appearence of an asymmetric "phase', related to such a population unbalance. In Fig. 3b we show the corresponding histogram which exhibit a peak around a non vanishing value. Remarkably, the majority of the particles in the inelastic system spend most of the time together in the same well. Again, the unbalance occurs because the more particles belong to the same well the higher is the energy dissipated, as the number of collisions increases dramatically. It is worth to mention that the measured average density profiles are expected to be symmetric only when the observation time window is much longer than the typical lifetime of the cluster, which can be estimated from Fig. 3a, as the distance between two successive zeroes. The situation is different in the elastic case where the histogram is peaked around the zero.

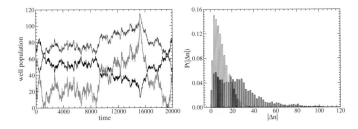

Fig. 3. a: Population (red) in the right side and population difference (green) with 128 particles.. **b**: Histogram for the population difference.

Conclusions

In simple dilute gases with hard-core interactions, symmetry breaking phenomena may appear as a consequence of inelasticity. We have found that two grains in a double well potential, follow approximately a Kramers law for the escape times, with a suitable renormalizations of the parameters due to excluded volume effects and inelasticity. We observed a crossover between a homogeneous phase and a clustered phase. Finally, we have considered the case where the system contains many particles. Again we found a crossover from a symmetric to an asymmetric phase as the driving is reduced.

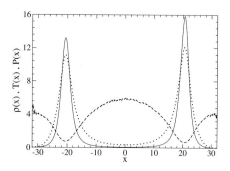

Fig. 4. Occupation, pressure and temperature profiles for and inelastic particles for $N = 128$

References

1. T. Pöschel and S. Luding, editors. *Granular Gases*, Berlin, 2001. Springer. Lecture Notes in Physics 564.
2. H.M. Jaeger, S.R. Nagel, and R.P. Behringer. Granular solids, liquids, and gases. *Reviews of Modern Physics*, 68(4):1259–1273, 1996.
3. I. Goldhirsch and G. Zanetti. Clustering instability in dissipative gases. *Phys. Rev. Lett.*, 70(11):1619–1622, 1993.
4. H.J. Schlichtting and V. Nordmeier. Strukturen im sand. kollektives verhalten und selbstorganisation bei granulaten. *MNU*, 49(6):323–332, 1996.
5. F. Cecconi, U.M.B. Marconi, A. Puglisi, and A. Vulpiani. Noise activated granular dynamics. *Physical Review Letter*, 90:064301, 2003.
6. A. Puglisi, V. Loreto, U.M.B. Marconi, and A. Vulpiani. Clustering and non-gaussian behavior in granular matter. *Phys. Rev. Lett.*, 81:3848, 1998.
7. H.A. Kramers. Brownian motion in a field of force and diffusion model of chemical reactions. *Physica (Utrecth)*, 7:284, 1940.
8. D. van der Meer, K. van der Weele, and D. Lohse. Sudden collapse of a granular cluster. *Phys. Rev. Lett.*, 88:174302, 2002.

Several Numerical Approaches of Granular Flows Applied to Inclined Plane Studies

L. Oger

Groupe Matière Condensée et Matériaux – Université de Rennes1
263 Avenue Général Leclerc, F35042 Rennes – France
`luc.oger@univ-rennes1.fr`

Summary. We compare different techniques to modelize granular flows on two-dimensional inclined planes. An example of each simulation is described. some characteristics and the limitations of the different models are mentionned.
Keywords: granular flows, DEM, PIC, SPH

1 Introduction

Understanding the mechanics of granular flows are a great subject of interest for numerous industrial domains and also natural events. Granular materials occur in a large range of applications like concrete, ceramic, pharmaceutical produtcs, agricultural grains, soils, powder metallurgy, and so on... Numerous approaches exist to simulate the behaviors of flowing granular materials. These models use mostly the so-called "Discrete Element Method" approach. In this technique, each individual grain exists and has well defined properties and interacts with its neighbors. On other hand the classical soil mechanic studies use Finite Element Method to analyze the global behavior of soil or dense granular materials. In this paper, we want also to focus our interest on innovative approaches such as Multiphase Flow with Interphase eXchanges (MFIX) which deals with dense or dilute fluid-solids flows or methods using either meshless technique or Eulerian-Lagrangian combination. These techniques are already knowned as SPH (Smoothed Particle Hydrodynamics) and PIC (Particle-In-Cells) and start to be used for granular purpose since ten years. After describing briefly the two methods, comparison of their results with DEM Molecular Dynamic simulations are performed.

2 Descriptions of the Different Models

The major focus of recent granular flow research has been on the formulation of the governing constitutive equations. During the past decades, a competition between the continuum treatment and the micro-mechanical description of a granular flow was the main concern of a large numbers of laboratories involved in this field. In an early stade, the continuum approach was often adopted because it was easy to use some averaged values of the properties. These values were also the only ones available experimentally, Most measurements were dealing with mean properties such as flow velocities or global results like height of the flows, mean packing fractions, and so on... Some other results were made at free surfaces or next to transparent walls but, because of the wall friction, these observations could give a misleading impression of what transpires in the interior of the flow. So the main problems of the granular studies are to compare the experimental results with the values obtained from the different available numerical simulations. The first ones are collectives while the other ones are mainly locals.

2.1 Molecular Dynamics Model or Discrete Element Method

In Molecular Dynamics model, each grain can be individually defined as a particle and can behave independently as the other (i.e. no permanent structure exists between the grains).

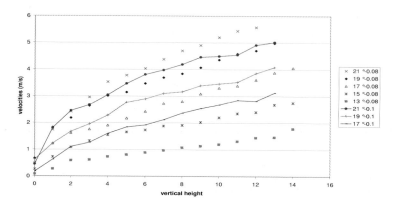

Fig. 1. Vertical velocities fluctuations for different angles of inclinaison of the slope for DM-DEM code.

The present model is similar to the two-dimensional formulation of Savage [1]: the particles are modeled as circular disks. A "soft-particle" approach is used, where each disk can have multiple contacts that can persist for extended durations. The size of the time step is chosen so that about 50 time steps elapse over a typical "rapid" collision. Both normal and tangential forces develop at the contact between two disks. When only compressive forces are allowed, the simulations represent dry, non-cohesive particle assemblies. No long range interactions are included in this model. In our test experiment, a layer of disks are glued on the basement of the inclined lane and the disks are initially closed packed on the upper part of the setup. At time $t = 0$ we released the particles and let them falling along the 2D structure without having periodic boundary conditions : So the disks can reach the lower part and exit. The basement layer is long enough that the disks reach a constant velocities as shown in Fig. 1.

2.2 MFIX

Multiphase Flow with Interphase eXchanges (MFIX) is a general-purpose hydrodynamic model developed at the National Energy Technology Laboratory (NETL) [2–4] that describes chemical reactions and heat transfer in dense or dilute fluid-solids flows. The MFIX code is defined on regular arrays and the calculations are performed on the node of these arrays (like Finite Element Method).

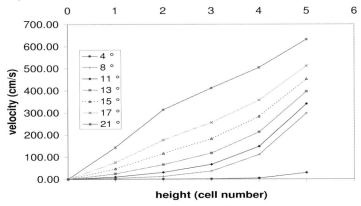

Fig. 2. Vertical velocities fluctuations for different angles of inclinaison of the slope for MFIX code.

The MFIX code has been developed over a number of years to simulate complex fluidization conditions. The code is based on hydrodynamic modeling and, thereby, is able to describe the time-dependent distribution of fluid and solids volume fractions, velocities, pressure, temperatures and species mass fractions. The last interest and not the smallest is that the code is available freeely from www.mfix.org. The code is mainly devoted to the transport of multi phase systems so it cannot handle a static or quasi-static behavior of disk packing. The runs are defined with friction wall effect and air and grains interactions and the local velocites are measured on the node as shown on the Fig. 2.

2.3 SPH

Smoothed Particle Hydrodynamics (SPH) is an N-body integration scheme introduced by Lucy [5] and Gingold and Monaghan [6] as an attempt to model continuum physics avoiding the limitation of grid based finite difference methods.

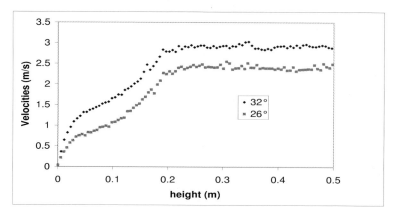

Fig. 3. Vertical velocities fluctuations for different angles of inclinaison of the slope for SPH code. The internal friction angle is 25 degrees and the wall friction angle is equal to 18 degrees

They developed this model to study problems in astrophysics such as the formation of planetoids, star formation and collisions between galaxies. These problems are usually characterized by masses that move in unbounded three-dimensional spaces. Most of the advantages of SPH arise directly from its Lagrangian nature. There are no constraint imposed either on the geometry of the system or how far it may evolve from the initial conditions. The first applications of SPH to problems involving other components of the stress tensor, besides the pressure, have appeared recently in the literature. Libersky and Petschek [7] have formulated an elastic, perfectly plastic constitutive model within the framework of SPH. We use the implementation of SPH to treat a viscous-plastic rheology and to implement the momentum equations in SPH for a Mohr-Coulomb rheology. In our case we have adopted a periodic boundary condition on the flowing direction in order to avoid the discontinuities of the kernel calculation close to the free surface, especially in the flow direction. For the vertical case the decrease of the density is more realistic. The Fig. 3 shows the different horizontal velocities for two inclined angle values.

2.4 PIC

The Particle-In-Cell (PIC) method was originally developed for interfacial flows and shock in fluid dynamics [8]. The PIC model is semi-Lagrangian approach in which the elements of material and/or fluid are presented by particles. So this PIC approach was extensively used in plasma simulations [9]. Recently the PIC method was extended to study inhomogeneous compressible flows [10], multiphase flows [11], sea ice dynamics [12, 13]. An other succesfull

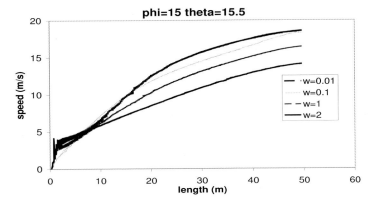

Fig. 4. Comparison of the horizontal velocities for different friction coefficients w with a Particle-In-Cell two-dimensional mesh. The internal friction coefficient is equl to 15 and the inclinaison angle is 15.5.

extension to the problem of solid mechanics was pushed in the last decade [14–19].

It uses discrete particles to model grain advection, while solving the momentum equations over an Eulerian grid. Information is exchanged between particles and the Eulerian grid by using an interpolating scheme. The use of discrete particles reduces numerical diffusion, and improves the accuracy of modelling boundary conditions. At the same time, the use of an Eulerian grid makes it possible to utilize an explicit numerical solution scheme for the momentum equations. The PIC method can deal with cases of larges deformations that are difficult to handle by classical Lagrangian approaches. In our example the particle is moved back to the top of the plane without thier velocities and the results are shown in the Fig. 4.

Conclusion and Acknowledgements

The different numerical appraoches show the same kind of behavior for the granular flow but none of them are perfectly relevant in order to simulate complex flows like with a stagnant zone at the bottom of it. This research was supported by the Centre National de la Recherche Scientifique through a *Mise à Disposition* in the group of M. Sayed at the CHC-CNRC.

References

1. S.B. Savage. Disorder, diffusion and structure formation in granular flows. In D. Bideau, editor, *Disorder and Granular Media*, page 1, Amsterdam, 1992.

North Holland.

2. M. Syamlal, W. Rogers, and T.J. O'Brien. Mfix documentation : Theory guide. Technical Report DOE/METC-94/1004, U.S. Department of Energy (DOE), Morgantown Energy Technology Center, Morgantown, West Virginia, 1993.

3. M. Syamlal. Mfix documentation : User's manual. Technical Report DOE/METC-95/1013, U.S. Department of Energy (DOE), Morgantown Energy Technology Center, Morgantown, West Virginia, 1994.

4. M. Syamlal. Mfix documentation : Numerical technique. Technical Report DOE/MC31346-582, U.S. Department of Energy (DOE), Morgantown Energy Technology Center, Morgantown, West Virginia, 1998.

5. L.B. Lucy. A numerical approach to the testing of the fission hypothesis. *The Astronomical Journal*, 82:1013–1024, 1977.

6. R.A. Gingold and J.J. Monaghan. Smoothed particle hydrodynamics: theory and application to non-spherical stars. *Monthly Notes of the Royal Astronomical Society*, 181:375–389, 1977.

7. L.D. Libersky and A.G. Petschek. Smooth particle hydrodynamics with strength of materials. In H.E. Trease, J. W. Fritts, and W. P. Crowley, editors, *The Next Free Lagrange Conference*, pages 248–257, Jackson Hole, WY, 1992. Springer-Verlag.

8. F.W. Harlow. The particle-in-cell computing method for fluid dynamics. *Methods Comput. Phys.*, 3:319–343, 1964.

9. C.K. Birdsall and A.B. Langdon. *Plasma Physics via Computer Simulation*. McGraw–Hill, New York, 1985.

10. A.B. Konstantinov and S.A. Orszag. Extended lagrangian particle-in-cell (elpic) code for inhomogeneous compressible flows. *J. Sci. Comput.*, 10(2):191–231, 1995.

11. M.J. Andrews and P.J. O'Rourke. The multiphase particle-in-cell (mp-pic) method for dense particulate flows. *Int. J. Multiphase Flows*, 2(2):379–402, 1996.

12. G.M. Flato. A particle-in-cell sea-ice model. *Atmos. Ocean*, 31(3):339–358, 1993.

13. S.B. Savage. Analyses of slow high-concentration flows of granular materials. *J. Fluid Mech.*, 377:1–26, 1998.

14. W.H. Lee and D. Kwak. Pic method for a two-dimensional elastic-plastic-hydro code. *Comput. Phys. Comm.*, 48:11–16, 1988.

15. D. Burgess, D. Sulsky, and J.U. Brackbill. Mass matrix formulation of the flip particle-in-cell method. *J. Comput. Phys.*, 103:1–15, 1992.

16. D. Sulsky, S.J. Shou, and H.L. Schreyer. Application of a particle-in-cell method to solid mechanics. *Comput. Phys. Comm.*, 87:236–252, 1995.

17. D. Sulsky and H.L. Schreyer. Axisymmetric form of the material point method with application to upsetting and taylor impact problems. *Comput. Methods in Appl. Mech. and Eng.*, 139:409–429, 1996.

18. D.M. Snider, P.J. O'Rourke, and M.J. Andrews. Sediment flow in inclined vessels calculated using a multiphase particle-in-cell model for dense particle flows. *Int. J. of Multiphase Flow*, 24:1359–1382, 1998.

19. Z. Wieckowski, S.K. Youn, and J.H. Yeon. A particle-in-cell solution to the silo discharging problem. *Int. J. Numer. meth. Engng.*, 45:1203–1225, 1999.

A Modified Tetris Model Including the Effect of Friction

F. Ludewig, S. Dorbolo, H. Caps, and N. Vandewalle

GRASP, Dept. of Physics, Sart-Tilman B5 – University of Liège
B-4000 Liège – Belgium
f.ludewig@student.ulg.ac.be

Summary. In the present work, we have introduced the presence of granular arches in the Tetris model. In this modified Tetris model, the friction between grains is the key parameter and a slipping threshold is defined. Two different regimes have been observed as a function of the slipping threshold.
Keywords: compaction, friction, tetris model

1 Introduction

In our industrial world, most of the products are processed, transported and stocked in their granular state. The volume of those granular systems appears therefore to be a crucial parameter for economical reasons. The study of compaction is thus revelant.

Experimental studies of granular compaction have underlined different behaviors for compaction: slow and fast dynamics. Slow dynamics have been reproduced by the Tetris model [1].

Our goal is to reproduce the large number of compaction behaviors by extending the Tetris model.

2 The Tetris Model

The Tetris model [1] considers a regular square lattice tilted at 45 degrees. The grains are rectangular blocks placed at the sits of the lattice. Grains can take two orientations on the lattice. The more important rule (depicted in Fig. 1) is the non-overlapping of the grains on the lattice. This rule is nothing else than a geometrical constraint: two adjacent blocks cannot have the same orientation along the principal axis of those blocks. This rule creates some frustation in the granular packing, and thus the density ρ_0 of a "loose" packing is low (typically $\rho_0 \approx 0.75$). The second consequence of this geometrical rule is the

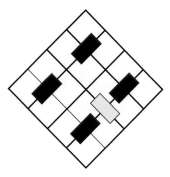

Fig. 1. The non-overlapping of adjacent grains is the first rule of the Tetris model.

following : the dense states are obtained for some anti-ferromagnetic order of grains.

Moreover, the grains are allowed to rotate. If grain has at least 3 free neighbors sites, it may change its orientation (Fig. 2). Note that the translation upward or downward is only subject to the first rule. The initialisation of the granular packing is done as follows : each grain is placed on the top of the lattice and falls by the effect of gravity ; until the grain reaches a stable position. This operation is re-executed until there is no place for a grain on the top of the lattice. The simulated tap consists in two stages. First, one grain is randomly chosen, it goes upward with a probability P_{up} and downward with a probability P_{down} ($P_{up} + P_{down} = 1$). It may rotate with a probability equal to 0.5. The grain is allowed to move if the geometric constraint is respected. In the second stage, a new grain is randomly chosen and a downward movement is tested. This represents some relaxation stage.

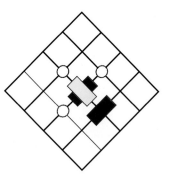

Fig. 2. When a grain has 3 neighboring empty sites, it may change its orientation.

The Tetris model succeed in reproducing a slow compaction behaviour observed in some experiment [2]. However, other experiments [3] have found that the density evolves following another law. The different dynamics wiil be

presented in the next section. This difference in behaviours is certainly related
to the friction between grains.

3 The Modified Tetris Model

A third rule has been added to the Tetris model in order to take the friction
into account. This rule is very important because it creates additional constraints for grain movements. This rule tries to estimate if the forces applied
on each grain are greater than the force of friction. An apparent weight is
given to each block. The weights are calculated starting from the top og the
packing and using the rules from the so-called "q-model" [4, 5]. Each block
has a real weight equal to the unity. The apparent weight distribution in the
pile is estimated with the "q-model". Finally, we introduce a slipping rule :
if the apparent weight W of a grain i is lower than a given threshold W_c the
block i can move but if W is larger than W_c the block i will be paralyzed.
The slipping threshold is linked to the friction coefficient. The friction force
between two blocks is indeed :

$$F_i = \frac{\sqrt{2}}{2}\mu_s W_i$$

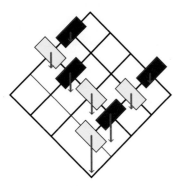

Fig. 3. Contacting blocks and apparent weights.

Figure 4 shows six simulations for 6 different slipping thresholds. The differents figures have been taken after the same number of times steps (t=10^4).
From left to right and from top to bottom : W_c = 2, 4, 8, 12, 16 and 32.
We observe that the surface position is higher when the slipping threshold is
large. In the two figures at the bottom of Fig. 4, the packing is ordered, but
defects are still there. In the middle of the figure, an ordered phase can be

observed near the top of the packing. On the other hand, in the top figures, we can see some holes due to arches.

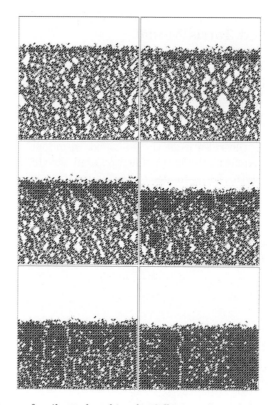

Fig. 4. Six pictures of a vibrated packing for different values of the slipping threshold W_c : 2, 4, 8, 16 and 32.

At high slipping thresholds, the evolution of the packing is fluid and the final state corresponds to a perfect order reached after infinite time. In opposition, at low slipping thresholds the packing is a rigid system having a low denisty. The Fig. 5 presents in a semi-log scale the evolution of the density ρ as a function of the number of taps t. The six curves correspond to differents slipping thresholds. Two trends can be pointed out : the first one at large slipping thresholds (16, 32 and 64) shows a fast and important compaction, the second one for the lower slipping thresholds (2, 4 and 8) shows a slow and small compaction. The solid curves correspond to a fit by an empirical law :

$$\rho = < \rho > + (\rho_\infty - \rho_0) \tanh(a \, log_{10}(t))$$

The fit with the empirical law allow to estimate the asymptotic density ρ_∞. The Fig. 6 shows the evolution of the maximum density with respect to the

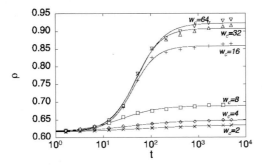

Fig. 5. Evolution of the density ρ as a function of the tap number t. Different slipping thresholds W_c are illustrated.

normalize slipping threshold W_c/W_{max}. W_{max} is the lower slipping threshold for which all grains are not paralysed. The different curves correspond to different sizes of simulated lattices (20×20, 50×50 and 100×100). The curves are normalized by the lattice size. Those curves present an inflection point corresponding to a critical value of slipping threshold W_c^*. This latter one separats two different regimes. When $W_c < W_c^*$, the compaction is not very important because many arches appear. When $W_c > W_c^*$, a fast phenomenon of compaction is obtained.

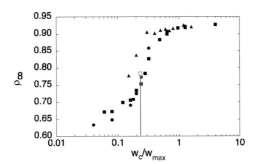

Fig. 6. Saturation density ρ_∞ as a function of normalized slipping threshold W_c/W_{max}.

4 Summary

A slipping threshold rule has been introduced in the model Tetris. The density has been studied with respect to the number of taps. Two regimes separated by a critical point have been found. For low values of the slipping threshold, the system is completely frustrated. Low compaction is obtained. The packing is characterized by large arches and large holes. For large values of the slipping threshold W_c, the system compacts easily. Arches are small. The weight distribution tends to an uniform distribution.

Acknowledgements

SD thanks FNRS for a financial support. HC benefits a FRIA grant. This work is also financially supported by the ARC contract 293-02/07.

References

1. E. Caglioti, V. Loreto, H.J. Herrmann, and M. Nicodemi, *A "Tetris-Like" Model for the compaction of Dry Granular Media*, Phys. Rev. Lett. <u>79</u>, 1575 (1997)
2. J.B. Knight, C.G. Frandrich, Chun Ning Lau, H.M. Jaeger, and S.R. Nagel, *Density relaxation in a vibrated granular material*, Phys. Rev. Lett. <u>51</u>, 3957 (1995).
3. S. Dorbolo and N. Vandewalle, *Electrical investigations of granular arches*, Physica A <u>311</u>, 307 (2002).
4. S.N. Coppersmith, C.-h. Liu, S. Majumdar, O. Narayan, and T.A. Witten, *Model for force fluctuations in bead packs*, Phys. Rev. E <u>53</u>, 4673 (1996).
5. S.N. Coppersmith, *Force fluctuations in granular media*, Wolf and Grossberger eds, pp 149-151, World - Scientific (1997).

An Aspect of Granulence in View of Multifractal Analysis

N. Arimitsu[1] and T. Arimitsu[2]

[1] Department of Computer Engineering – Yokohama National University
   Kanagawa 240-8501 – Japan
   arimitsu@ynu.ac.jp
[2] Institute of Physics – University of Tsukuba
   Ibaraki 305-8571 – Japan
   arimitsu@cm.ph.tsukuba.ac.jp

Summary. The probability density function of velocity fluctuations of *glanulence* observed by Radjai and Roux in their two-dimensional simulation of a slow granular flow under homogeneous quasistatic shearing is studied by the multifractal analysis for fluid turbulence proposed by the present authors. It is shown that the system of granulence and of turbulence have indeed common scaling characteristics.
Keywords: multifractal analysis, velocity fluctuation, turbulence, granulence

1 Introduction

In this paper, we apply the multifractal analysis (MFA) [1–4] of fluid turbulence to granular turbulence (*granulence* [5]) in order to see how far MFA works in the study of the data observed by Radjai and Roux [5] in their two-dimensional simulation of a slow granular flow subject to homogeneous quasistatic shearing. Radjai and Roux reported that there is an evident analogy between the scaling features of turbulence and of granulence in spite of the fundamentally different origins of fluctuations in these systems. MFA is a unified self-consistent approach for the systems with large deviations, which has been constructed based on the Tsallis-type distribution function [6] that provides an extremum of the *extensive* Rény [7] or the *non-extensive* Tsallis entropy [6, 8] under appropriate constraints.

2 Multifractal Analysis

MFA of turbulence rests on the scale invariance of the Navier-Stokes equation for high Reynolds number, and on the assumption that the singularities due to the invariance distribute themselves multifractally in physical space.

The velocity fluctuation $\delta u_n = |u(\bullet + \ell_n) - u(\bullet)|$ of the nth multifractal step satisfies the scaling law $|u_n| \equiv |\delta u_n/\delta u_0| = \delta_n^{\alpha/3}$ with $\delta_n = \ell_n/\ell_0 = \delta^{-n}$ ($n = 0, 1, 2, \cdots$). We call n the multifractal depth which can be real number in the analysis of experimental data. We will put $\delta = 2$ in the following in this paper that is consistent with the energy cascade model. At each step of the cascade, say at the nth step, eddies break up into two pieces producing the energy cascade with the energy-transfer rate ϵ_n that represents the rate of transfer of energy per unit mass from eddies with diameter ℓ_n to those with ℓ_{n+1}. Then, we see that the velocity derivative $|u'| = \lim_{n\to\infty} u'_n$ with the nth velocity difference $u'_n = \delta u_n/\ell_n$ for the characteristic length ℓ_n diverges for $\alpha < 3$. The real quantity α is introduced in the scale transformation [9, 10] $\vec{x} \to \vec{x}' = \lambda\vec{x}$, $\vec{u} \to \vec{u}' = \lambda^{\alpha/3}\vec{u}$, $t \to t' = \lambda^{1-\alpha/3}t$, $p \to p' = \lambda^{2\alpha/3}p$ that leaves the Navier-Stokes equation $\partial\vec{u}/\partial t + (\vec{u}\cdot\vec{\nabla})\vec{u} = -\vec{\nabla}p + \nu\nabla^2\vec{u}$ of incompressible fluid invariant for a large Reynolds number $\mathrm{Re} = \delta u_{\mathrm{in}}\ell_{\mathrm{in}}/\nu$. Here, ν is the kinematic viscosity, $p = \check{p}/\rho$ with the thermodynamical pressure $\check{p}$ and the mass density ρ, and δu_{in} and ℓ_{in} represent, respectively, the rotating velocity and the diameter of the largest eddies in turbulence. The largest size of eddies is, for example, about the order of mesh size of the grid, inserted in a laminar flow, which produces turbulence downstream.

Within MFA, it is assumed that the singularities due to the scale invariance distribute themselves, multifractally, in physical space with the Tsallis-type distribution function, i.e., the probability $P^{(n)}(\alpha)d\alpha$ to find in real space a singularity with the strength α within the range $\alpha \sim \alpha + d\alpha$ is given by [2, 11, 12] $P^{(n)}(\alpha) = (Z_\alpha^{(n)})^{-1}\{1 - [(\alpha - \alpha_0)/\Delta\alpha]^2\}^{n/(1-q)}$ with $(\Delta\alpha)^2 = 2X/[(1-q)\ln 2]$. Here, q is the entropy index introduced in the definitions of the Rényi and the Tsallis entropies. This distribution function provides us with the multifractal spectrum $f(\alpha) = 1 + (1-q)^{-1}\log_2[1 - (\alpha - \alpha_0)^2/(\Delta\alpha)^2]$ which, then, produces the mass exponent

$$\tau(\bar{q}) = 1 - \alpha_0\bar{q} + 2X\bar{q}^2(1 + \sqrt{C_{\bar{q}}})^{-1} + (1-q)^{-1}[1 - \log_2(1 + \sqrt{C_{\bar{q}}})] \quad (1)$$

with $C_{\bar{q}} = 1 + 2\bar{q}^2(1-q)X\ln 2$. The multifractal spectrum and the mass exponent are related with each other through the Legendre transformation [10]: $f(\alpha) = \alpha\bar{q} + \tau(\bar{q})$ with $\alpha = -d\tau(\bar{q})/d\bar{q}$ and $\bar{q} = df(\alpha)/d\alpha$.

The formula of the probability density function (PDF) $\Pi^{(n)}(u_n)$ of velocity fluctuations is assumed to consists of two parts, i.e., $\Pi^{(n)}(u_n) = \Pi_{\mathrm{S}}^{(n)}(u_n) + \Delta\Pi^{(n)}(u_n)$ where the first term is related to $P^{(n)}(\alpha)$ by $\Pi_{\mathrm{S}}^{(n)}(|u_n|)du_n \propto P^{(n)}(\alpha)d\alpha$ with the transformation of the variables $|u_n| = \delta_n^{\alpha/3}$, and the second term is responsible to the contributions coming from the dissipative term in the Navier-Stokes equation violating the invariance under the scale transformation given above. Then, we have the velocity structure function in the form $\langle\langle|u_n|^m\rangle\rangle \equiv \int du_n|u_n|^m\Pi^{(n)}(u_n) = 2\gamma_m^{(n)} + (1 - 2\gamma_0^{(n)})a_m\delta_n^{\zeta_m}$ with $2\gamma_m^{(n)} = \int du_n|u_n|^m\Delta\Pi^{(n)}(u_n)$, $a_m = \{2/[\sqrt{C_{m/3}}(1 + \sqrt{C_{m/3}})]\}^{1/2}$ and scaling exponent $\zeta_m = 1 - \tau(m/3)$ given with the mass exponent (1).

The PDF $\hat{\Pi}^{(n)}(\xi_n)$ both of velocity fluctuations and of velocity derivative to be compared with observed data is the one defined through $\hat{\Pi}^{(n)}(\xi_n)d\xi_n = \Pi^{(n)}(u_n)du_n$ with the variable $\xi_n = u_n/\langle\langle u_n^2\rangle\rangle^{1/2}$ scaled by the standard deviation of velocity fluctuations. For the velocity fluctuations larger than the order of its standard deviation, $\xi_n^* \leq |\xi_n|$ (equivalently, $|\alpha| \leq \alpha^*$), the PDF is given by [3, 4]

$$\hat{\Pi}^{(n)}(\xi_n)d\xi_n = \Pi_S^{(n)}(u_n)du_n$$

$$= \bar{\Pi}^{(n)}\frac{\bar{\xi}_n}{|\xi_n|}\left[1 - \frac{1-q}{n}\frac{(3\ln|\xi_n/\xi_{n,0}|)^2}{2X|\ln\delta_n|}\right]^{n/(1-q)}d\xi_n \qquad (2)$$

with $\xi_{n,0} = \bar{\xi}_n\delta_n^{\alpha_0/3-\zeta_2/2}$ and $\bar{\Pi}^{(n)} = 3(1 - 2\gamma_0^{(n)})/(2\bar{\xi}_n\sqrt{2\pi X|\ln\delta_n|})$. This *tail part* represents the large deviations, and manifests itself the multifractal distribution of the singularities due to the scale invariance of the Navier-Stokes equation when its dissipative term can be neglected. The entropy index q should be unique once a turbulent system with a certain Reynolds number is settled. For smaller velocity fluctuations, $|\xi_n| \leq \xi_n^*$ (equivalently, $\alpha^* \leq |\alpha|$), we assume the Tsallis-type PDF of the form [3, 4]

$$\hat{\Pi}^{(n)}(\xi_n)d\xi_n = \left[\hat{\Pi}_S^{(n)}(u_n) + \Delta\hat{\Pi}^{(n)}(u_n)\right]du_n$$

$$= \bar{\Pi}^{(n)}\left\{1 - \frac{1-q'}{2}(1+3f'(\alpha^*))\left[\left(\frac{\xi_n}{\xi_n^*}\right)^2 - 1\right]\right\}^{1/(1-q')}d\xi_n \qquad (3)$$

where a new entropy index q' is introduced as an adjustable parameter. This *center part* is responsible to smaller fluctuations, compared with its standard deviation, due to the dissipative term violating the scale invariance. The entropy index q' can be dependent on the distance of two measuring points.

The two parts of the PDF, (2) and (3), are connected at $\xi_n^* = \bar{\xi}_n\delta_n^{\alpha^*/3-\zeta_2/2}$ with the conditions that they have a common value and that their slopes coincide. The value α^* is the smaller solution of $\zeta_2/2 - \alpha/3 + 1 - f(\alpha) = 0$. The point ξ_n^* has the characteristics that the dependence of $\hat{\Pi}^{(n)}(\xi_n^*)$ on n is minimum for $n \gg 1$. With the help of the second equality in (3) and (2), we obtain $\Delta\Pi^{(n)}(x_n)$, and have the analytical formula to evaluate $\gamma_m^{(n)}$. Their explicit analytical formulae and the definition of $\bar{\xi}_n$ are found in [3, 4].

3 Turbulence

The dependence of the parameters α_0, X and q on the intermittency exponent μ is determined, self-consistently, with the help of the three independent equations, i.e., the energy conservation: $\langle\epsilon_n/\epsilon\rangle = 1$ (equivalently, $\tau(1) = 0$), the definition of the intermittency exponent μ: $\langle\epsilon_n^2/\epsilon^2\rangle = \delta_n^{-\mu}$ (equivalently,

$\mu = 1 + \tau(2))$, and the scaling relation: $1/(1 - q) = 1/\alpha_- - 1/\alpha_+$ with $\alpha_\pm$ satisfying $f(\alpha_\pm) = 0$. Here, ϵ is the energy input rate to the largest eddies. The average $\langle \cdots \rangle$ is taken with $P^{(n)}(\alpha)$.

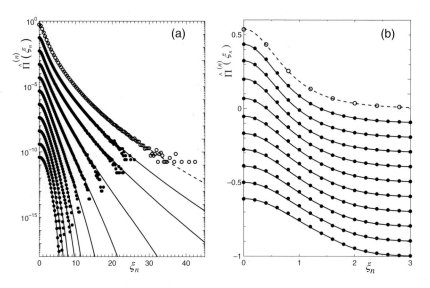

Fig. 1. Analyses of the PDF's of *velocity fluctuations* (closed circles) and of *velocity derivatives* (open circles) measured in the DNS by Gotoh et al. at $R_\lambda = 380$ with the help of the present theoretical PDF's $\hat{\Pi}^{(n)}(\xi_n)$ for *velocity fluctuations* (solid lines) and for *velocity derivatives* (dashed line) are plotted on (a) log and (b) linear scales. The DNS data points are symmetrized by taking averages of the left and the right hand sides data. The measuring distances, $r/\eta = \ell_n/\eta$, for the PDF of velocity fluctuations are, from the second top to bottom: 2.38, 4.76, 9.52, 19.0, 38.1, 76.2, 152, 305, 609, 1220. For the theoretical PDF's of velocity fluctuations, $\mu = 0.240$ ($q = 0.391$), from the second top to bottom: $(n, \bar{n}, q') = (20.7, 14.6, 1.60)$, $(19.2, 13.1, 1.60)$, $(16.2, 10.1, 1.58)$, $(13.6, 7.54, 1.50)$, $(11.5, 5.44, 1.45)$, $(9.80, 3.74, 1.40)$, $(9.00, 2.94, 1.35)$, $(7.90, 1.84, 1.30)$, $(7.00, 0.94, 1.25)$, $(6.10, 0.04, 1.20)$, and $\xi_n^* = 1.10 \sim 1.43$ ($\alpha^* = 1.07$). For the theoretical PDF of velocity derivatives, $(n, \bar{n}, q') = (22.4, 16.3, 1.55)$, and $\xi_n^* = 1.06$ ($\alpha^* = 1.07$). For better visibility, each PDF is shifted by -1 unit in (a) and by -0.1 in (b) along the vertical axis.

The PDF's extracted by Gotoh et al. from their DNS data [13] at $R_\lambda = 380$ are shown, on log and linear scales, in Fig. 3 both for *velocity fluctuations* and for *velocity derivatives*, and are analyzed by the theoretical formulae (2) and (3) for PDF's. We found the value $\mu = 0.240$ by analyzing the measured scaling exponents ζ_m of velocity structure function with the formula given above, which leads to the values $q = 0.391$, $\alpha_0 = 1.14$ and $X = 0.285$. Through the analyses of the PDF's for velocity fluctuations in Fig. 3, we extracted quite a few information of the system [3, 4, 14–16]. Among them, we only quote here the dependence of q' on r/η: $q' = -0.05 \log_2(r/\eta) + 1.71$ [4].

4 Granulence

Let us now analyze the velocity fluctuations in glanulence simulated by Radjai and Roux [5]. Since they observed that the fluctuations share the scaling characteristics of fluid turbulence, we try to investigate the system by means of MFA which extracted, successfully, the rich information out of turbulence as was seen in the previous section. The power spectrum of the fluctuating

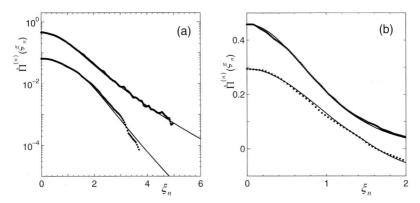

Fig. 2. Analysis of the experimental PDF of *fluctuating velocities*, measured in the quasistatic flow of granular media by Radjai and Roux, with the help of the present theoretical PDF $\hat{\Pi}^{(n)}(\xi_n)$ for *velocity fluctuations* (solid lines) are plotted on (a) log and (b) linear scales. The experimental data points are symmetrized by taking averages of the left and the right hand sides data. The integration time τ, normalized by a shear rate, for the experimental PDF are, from top to bottom, 10^{-3}, 10^{-1}. For the theoretical PDF, $\mu = 1.347$ ($q = 0.930$), from top to bottom: $(n, q') = (88.0, 1.28)$, $(40.0, 1.22)$, and $\xi_n^* = 1.14$, 1.14 ($\alpha^* = 0.364$). For better visibility, each PDF is shifted by -1 unit in (a) and by -0.1 in (b) along the vertical axis.

velocity field on one-dimensional cross sections exhibits a clear power-law shape with the slope $-\beta$ with $\beta \approx 1.24$ [5], which is quite similar to the power-law behavior with the slope $-5/3$ in the inertial range of the Kolmogorov spectrum [17]. However, the granular model is an assembly of frictional disks, the power-law observed in granulence does not mean the energy conservation in contrast with the case of the energy cascade model for fluid turbulence.

For the conditions to determine the parameters α_0, X and q, we adopt, instead of the energy conservation, the slope of the power spectrum, i.e., $\beta = 1 + \zeta_2 = 2 - \tau(2/3)$ in addition to the definition of the intermittency exponent and the scaling relation. The latter two are the same as those for turbulence. As there is no experimental data, for the present, to determine the intermittency exponent μ for granulence, we cannot have the values of the three parameters through the three conditions. Therefore, we determine the value of the intermittency exponent by adjusting the observed PDF with the theoretical formulae (2) and (3), since the accuracy of the formulae in the analysis of PDF's for turbulence is quite high as was shown in the previous section. The best fit of the observed PDF of fluctuating velocities by the formulae (2) and (3) is shown in Fig. 4. We found the value $\mu = 1.347$ giving $q = 0.930$, $\alpha_0 = 0.377$ and $X = 0.050$. By making use of the mass exponent with these values, we have $\langle \epsilon_n/\epsilon \rangle = \delta_n^{-\tau(1)}$ with $\tau(1) = 0.648$ representing a breakdown of energy conservation. It is attractive to see that the result is quite close to $\langle \epsilon_n/\epsilon \rangle = 3/2$ which may be consistent with the coefficient of friction 0.5 for the simulation [5]. We further extract the relation between τ and ℓ_n as $\tau = 1.3\, \delta_n^{0.131}$ by comparing the observed flatness and the one with the theoretical PDF's (2) and (3). This relation may be a manifestation of the fact that Taylor's frozen turbulence hypothesis does not work for granulence.

5 Prospects

We showed with the help of MFA that the system of turbulence and of granulence have, actually, common scaling feature in their velocity fluctuations as was pointed out by Radjai and Roux [5]. We expect that various observation of granulence will be reported at higher statistics, and that one can extract more information out of the data to determine the underlying dynamics for granulence in the near future.

References

1. T. Arimitsu and N. Arimitsu. *Phys. Rev. E*, 61:3237–3240, 2000.
2. T. Arimitsu and N. Arimitsu. *J. Phys. A: Math. Gen.*, 33:L235–L241, 2001.
3. T. Arimitsu and N. Arimitsu. cond-mat/0210274 2002.
4. T. Arimitsu and N. Arimitsu. In C. Noce A. Avella, R. Citro and M. Salerno, editors, *Highlights in Condensed Matter Physics (AIP Conference Proceedings 695)*. American Institute of Physics, 2003.
5. F. Radjiai and S. Roux. *Phys. Rev. Lett.*, 89:064302, 2002.
6. C. Tsallis. *J. Stat. Phys.*, 52:479–487, 1988.
7. A. Rényi. *Proc. 4th Berkeley Symp. Maths. Stat. Prob.*, 1:547, 1961.
8. J.H. Havrda and F. Charvat. *Kybernatica*, 3:30–35, 1967.
9. U. Frisch and G. Parisi. In R. Benzi M. Ghil and G. Parisi, editors, *Turbulence and Predictability in Geophysical Fluid Dynamics and Climate Dynamics*, page 84, New York, 1985. North-Holland.
10. C. Meneveau and K.R. Sreenivasan. *Nucl. Phys. B*, 2:49–76, 1987.
11. T. Arimitsu and N. Arimitsu. *Prog. Theor. Phys.*, 105:355–360, 2001.
12. T. Arimitsu and N. Arimitsu. *Physica A*, 295:177–194, 2001.
13. D. Fukayama T. Gotoh and T. Nakano. *Phys. Fluids*, 14:1065–1081, 2002.
14. T. Arimitsu and N. Arimitsu. *J. Phys.: Condens. Matter*, 14:2237–2246, 2002.
15. T. Arimitsu and N. Arimitsu. *Physica A*, 305:218–226, 2002.
16. N. Arimitsu and T. Arimitsu. *Europhys. Lett.*, 60:60–65, 2002.
17. A.N. Kolmogorov. *Dokl. Akad. Nauk SSSR*, 30:301–305, 1941.